Cultural Identity and New Communication Technologies:

Political, Ethnic and Ideological Implications

D. Ndirangu Wachanga
University of Wisconsin-Whitewater, USA

Information Science
REFERENCE

Senior Editorial Director: Kristin Klinger
Director of Book Publications: Julia Mosemann
Editorial Director: Lindsay Johnston
Acquisitions Editor: Erika Carter
Development Editor: Mike Killian
Production Editor: Sean Woznicki
Typesetters: Keith Glazewski and Natalie Pronio
Print Coordinator: Jamie Snavely
Cover Design: Nick Newcomer

Published in the United States of America by
Information Science Reference (an imprint of IGI Global)
701 E. Chocolate Avenue
Hershey PA 17033
Tel: 717-533-8845
Fax: 717-533-8661
E-mail: cust@igi-global.com
Web site: http://www.igi-global.com/reference

Library of Congress Cataloging-in-Publication Data
Cultural identity and new communication technologies : political, ethnic and ideological implications / D. Ndirangu Wachanga, editor.
p. cm.
Includes bibliographical references and index.
ISBN 978-1-60960-591-9 (hbk.) -- ISBN 978-1-60960-592-6 (ebook) 1. Information technology--Social aspects--Africa. 2. Information technology--Political aspects--Africa. 3. Telecommunication--Social aspects--Africa. 4. Telecommunication--Political aspects--Africa. I. Wachanga, D. Ndirangu, 1975-
HN780.Z9I5639 2011
303.48'3309--dc22
2011001308

British Cataloguing in Publication Data
A Cataloguing in Publication record for this book is available from the British Library.

All work contributed to this book is new, previously-unpublished material. The views expressed in this book are those of the authors, but not necessarily of the publisher.

Dedicated with affection to my parents – Simon Wachanga Muraya and Irene Njeri Wachanga.

To my late grandmother, Nyanya, for teaching me the value of hard work

Table of Contents

Section 3
Emerging Media, Global Politics, and Cultural Transformation

Detailed Table of Contents

The top-bottom model espoused by the traditional media structures is being problematized by the emerging technological changes. Kenya presents an example of bottom-up model, which is a challenge to the hegemonic potential of top-down model. We seek to establish how media in Kenya have been operating within a top-bottom model until recently. The emergence of the new communication technologies (NCTs) have allowed individuals to challenge dominant voices, mythologies, symbols, vocabulary, and politico-social structures around which the luminal rituals of the national have been interwoven for the sole purpose of fostering national cohesion. We submit that the proliferation of NCTs and communication strategies have dismembered the nation by stimulating exponential multiplication of discursive regimes that may have been impossible previously when media technologies were confined to the control of a central authority. Although there is a fundamental shift from 19th and 20th century media, there remains a strong top-down component in the media but bottom-up is now significant. It is from this background that we point to the vulnerability of NCTs to manipulation and co-option by the dominant, but hasten to mention that these technologies serve as a potential conduit to alter dominant hegemonic designs.

New technology is not always used to portray and perpetuate narrow ethnic identities; rather, instances exist when the users want to project professional and national identities. Supported with data that focus on the pre- and post-2007 elections, the chapter argues that inasmuch as the New Communication Technologies are good, great caution should be exercised when using them since unchecked use might bring untold suffering to society.

The way the media process, frame, and pass on information either to the government or to the people shall affect the function of the political system. The media may lead to dysfunction in some cases. This chapter seeks to outline and discuss the interaction between new media and ethnicity in Kenya. The chapter investigates ways in which the new media reinforced issues relating to ethnicity in Kenya prior to the 2007 presidential election.

The role NICTs are playing in the relations between Uganda's central government and four of the kingdoms in the country is analyzed and placed in the historical context. Within this contet, each kingdom has pursued a strategy for preservation and development of its people's unique culture and ethnic aspirations. A convergence of cell phone and FM radio designed by the Buganda kingdom to confront the central government with specific demands and the government's response are examined. The less combative use of NICTs by Bunyoro-Kitara, Busoga, and Toro kingdoms is described.

In this study we investigate the extent to which the cell phone technology has impacted the lives of young Kenyans, especially within the context of a university in Kenya. Kenya has had a tremendous technological leap from the 1990s where landlines were the main tools of communication, to mobile telephony where people can now communicate any time, any place.

Section 3
Emerging Media, Global Politics, and Cultural Transformation

Using a combined method of surveying and personal interviews, several African immigrants, their children, and friends living in the Dallas/Fort Worth area of Texas, USA were interviewed for their views on the role of the nascent Nigerian video industry in the way they sustain and straddle their multiple identities and culture in their society of settlement. Results indicate that most of the immigrants view the videos as affirmation of the values they grew up with and with which they still identify.

This chapter addresses mass media and cultural identity dynamics that impact U.S. military policy in the execution of the Iraq war and related cross-cultural communication issues that have evolved during the course of the war. The inability to consider, let alone plan for, cross-cultural ramifications has been a central communication failure that has proven tremendously problematic. The mass media, via traditional channels and new communication technologies, have been keen to report on these matters which, in turn, thave modified public understanding of the matters being reported.

The effects of the 'new media' may not be what we expect with respect to the democratization of society. The analysis in this chapter shows that ideological hegemony is not only alive and well, but might even thrive in these new media. To scrutinize these effects, the chapter focuses on the impact of Pakistani cable television on identity and culture discourse through the lens of three topics: identity and the new communication technologies; ideology, power and the new media; and new media and new popular cultures.

The inauguration of the Lee Myung-Bak administration in 2008 signaled a crisis for Internet-driven participatory, democratic public culture in South Korea. One of the most visible effects was immediately found in the administration's repressive media policy. A series of anti-democratic regulations,

Foreword

THE IMAGE, THE MESSAGE AND THE MEDIA: AN AFRICAN OVERVIEW

Partly based on Keynote Address to the Sixth Biennial Conference of African Broadcasters (AFRICAST 2006), Abuja, Nigeria, October 31–November 2, 2006.

The paper starts from the premise that what Africans learn about the rest of Africa comes disproportionately from Western sources of information. African media (electronic or print) cannot afford to have their own correspondents in major African capitals, or war reporters in Darfur and the Democratic Republic of the Congo. African media cannot send filming or taping crews of their own to meetings of the Pan African Parliament in South Africa or meetings of African foreign ministers in Banjul or Abidjan.

One of my favorite illustrative predicaments concerns my own country, Kenya. Kenya shares borders with five other African countries, each of which has had major convulsive breakdowns – Somalia, Ethiopia, Sudan, Uganda, and even Tanzania if we include the convulsions of Zanzibar.

Therefore, Kenya is surrounded by major news stories across five borders. And Kenya has the best African news media for its size outside Southern Africa. And yet, most of the big stories that Kenya's media have carried about its neighbors have relied disproportionately on Western sources.

Uganda's worst political breakdowns occurred during the reign of Idi Amin and the second administration of Milton Obote. It was Idi Amin's reign which drove me out of Uganda. A few years after I left Uganda, there was a big story about my old university in Kampala. The World Press reported that Amin's soldiers had invaded the Makerere campus, raped women students, looted the Halls of Residence and offices, and even mutilated the breasts of some of their female victims; some were killed.

Although I was already based in the United States at that time, I happened to be in Nairobi, Kenya, when the news broke. My wife in North America was visiting Canada. She broke down when she heard the story on the radio. She then called me in Nairobi to check how I was taking it.

I told her that although I was devastated, I was also bewildered that all the Kenyan newspapers and broadcasts were quoting British sources, especially the British Sunday paper, *The Observer*.

Why were the Kenya media relying almost exclusively on a British newspaper for an event which had occurred next door in broad daylight, in the capital city of Uganda, on the campus of the country's most distinguished educational institution?

Fortunately, I was scheduled to have lunch with Milton Obote in Dar es Salaam two days later. Obote, who had been overthrown by Idi Amin, was in exile in Tanzania.

At the lunch with Obote was the British correspondent responsible for the horrendous story, David Martin. I asked David Martin if he had been to Idi Amin's Uganda to check out the story. He had not. He said he got the story from students who had escaped across the border.

Then Obote challenged me to telephone my friends in Uganda to verify. I said such a phone call about such a subject might, if overheard, endanger my friends' lives under the Amin regime.

The story continued to be headline news in Britain, Africa and much of the rest of the world for another week or so.

We subsequently discovered that while Amin's troops had invaded the Makerere campus, their worst atrocities were slapping people, kicking people, and intimidating everyone in sight. There were no rapes or murders authoritatively reported and certainly no mutilation of female breasts. The enormity of the original blood-chilling story shrunk to one more incident of African soldiers misbehaving themselves.

But the world had been ready to believe any brutal story from Africa. And even neighboring African countries had no sources of confirmation independently of the West.

At that time, African media consisted primarily of print, radio, television, and social institutions like churches and mosques.

Today, the Internet would facilitate more cross-checking between African neighbors. African citizens can send e-mails to each other, and African governments are not yet sophisticated enough to keep track. Idi Amin would have been none the wiser in the age of the Internet.

But external stories covered in the African media continue to depend disproportionately on the messages and images transmitted between the West and the rest of us. And those messages and images are influenced by seven biases of perception across cultures.

The *racial* bias tends to ensure that a train accident which kills four white folks is a bigger story than an overturned bus which kills twenty Africans. The *gender* bias focuses more on the achievements of men than of women. The *elite* bias rewards the famous with additional fame – and regards the powerful as newsworthy. The *urban* bias attracts reporters to cities and urban centers – and pays far less attention to rural folks. The *generation* bias pays more attention to older folks than to the young – unless the young are at the center of exceptional events. The *exotic* bias is attracted to female circumcision, rather than high school graduation; to African witchcraft rather than African brain surgeons. As for the bias of *negativism*, this almost always ensures that real bad news is bigger news than happy news tends to be.

Let us return to the Makerere saga under Idi Amin. The *racial bias* made most of the world credulous about any story of barbarism coming out of Africa. The *gender bias* in the Makerere story was about women as passive creatures and preeminent victims of the ruthless lust of macho men. The *urban bias* was the setting of the story in the capital city of an African country and in broad daylight. The *elite bias* was the obvious linkage with a university, the premier educational institution of a postcolonial African country. The *exotic bias* was Africa's allegedly ruthless sexual drive – from promiscuity to rape. This time it allegedly involved Idi Amin's lust-activated warriors.

The *generation bias* in the Makerere story included negative and positive elements about young people. The Makerere students symbolized the youth as victims, but Idi Amin's soldiers were themselves disproportionately young. These symbolized the youth as alleged perpetrators. The one or two students who escaped across the border to break the story to David Martin would have been heroes had their story not been so ridiculously exaggerated. They must have been pro-Obote partisan witnesses.

Finally, the overarching *bias of negativism*. Massively cruel stories – true or false – make bigger news than heart-warming stories of human kindness. The Makerere saga encapsulated all the seven biases of messages and images between the West and the rest.

The Media and Seven Functions of Culture

The Internet in Africa operates in the context of thee civilizations – Africa's own indigenous legacy combined with the legacy of Islam, and both under the impact of Western culture. The *triple heritage* consists of Africanity, Islam, and the West.

The Internet and other media affect those three civilizations by influencing the seven functions of culture.

I. Culture as Perception

Culture is how we view the world and how we are viewed by others. The most objectionable portrayal of Muslim women in the world presents the women of Islam as *over-clothed*, with hijab and sometimes the veil.

The former British Foreign Minister has asked any Muslim visiting woman in his office to unveil: "We cannot meet face to face if one face is veiled." *France* wants Muslim women to abandon the hijab when going to school. *Turkey* does not want women members of Parliament to wear a head scarf in the House.

On the other hand, the most objectionable portrayal of indigenous African women (non-Islamized) is when the women are portrayed as *under-dressed* – targeting such dress codes as that of the Nuba, or of the Maasai or of the traditional Tiv.

There is a big debate now on the campus of Binghamton University concerning an exhibition of photographs of nude African women taken among under-clothed African cultures in the 1950s. On November 9, 2006, on campus, the particular curators of the photographic exhibition confronted their critics. One of the speakers was a Maasai post-doctoral fellow on campus.

The media in the West revels in debates about over-clothed Muslim women and under-clothed indigenous African women. Nakedness is in the eye of the beholder.

II. Culture as a Spring of Motivation

How can we motivate African rulers to be honest, fair-minded, incorruptible and democratically accountable? One answer suggests the Nile Valley as the cradle of African civilization. The latest civilizing gesture from the Nile Valley comes in 2006 from a Sudanese-Egyptian benefactor – offering billions of dollars to motivate African rulers to be just and incorruptible, in exchange for $5 million in retirement, and more money later.

In the past, we thought we would try to motivate good behavior in our rulers by threatening impeachment and trial. This was motivation by the *stick*. Now the Nile Valley seeks to motivate African rulers by the carrot – a financial reward.

The media's responsibility is to keep track of the performance of our rulers. The Internet's role is to facilitate greater debate.

III. Culture as Means of Communication

Are the media aggravating the marginalization of indigenous African languages by concentrating on the Euro-imperial languages, like English, French and Portuguese?

Is the Internet widening the linguistic gap between users of Euro-international languages and users of African indigenous languages?

The cellular phone as an oral tradition is multi-lingualizing users of telephones across the world, but the Internet as a written tradition is held hostage by the dominant languages of the world.

Even in Africa, the Internet friendly languages tend to be the big indigenous ones, such as Yoruba, Igbo, Hausa, Kiswahili, and, of course, the Arabic language.

IV. Culture as a Basis of Identity

The breakdown of culture as a basis of identity is most manifest in Africa's worst cases of instability.

The Somali people share the same language, the same religion, the same political culture based on clans and the same love of poetry. In the pre-colonial period, Somali governance was based on ordered anarchy. Then colonialism imposed on the Somali institutionalized order. Since then, the post-colonial period has eroded the artificial institutions and restored the old anarchy, but without the old order.

Was the absence of the mass media in Somalia one of the causes of the cultural breakdown? The Somali language remained unwritten until long after the colonial period. The Somalis could not make up their minds about the appropriate alphabet. Should it be the *Arabic* alphabet or the *Roman*, or a uniquely *indigenous* Somali alphabet which had already been invented?

A Somali military government subsequently chose the Roman or Latin alphabet in the 1970s. Would the Somali people have been nationally integrated better if they had had an alphabet for a mass media during their yesteryears as a people of ordered anarchy?

Even today, the Somali people are greater users of the *cell phone* (the oral tradition one to one) than of the *Internet* (the written tradition of collective accessibility). In Somalia, shared culture has continued to break down as a foundation of shared empathy.

There are other African examples of cultural breakdowns. Where in Africa has a shared language disastrously failed as a basis of national integration? One colossal example is Rwanda. The Hutu and Tutsi in Rwanda speak the same native language – Kinyarwanda. Yet Rwanda goes down in history as the country which had the fastest rate of genocide in recorded history. 800,000 people died within three months. Did the radio in Uganda serve as a trigger of anti-Tutsi massacres? Indeed, did a radio program trigger the anti-Igbo pogrom in Northern Nigeria in 1966?

Where else in Africa has a *shared language* spectacularly failed to hold a country together? In Rwanda's sister country, Burundi, with the same rival groups, Hutu and Tutsi, Kirundi has failed as a language of national concord. Have the media in Burundi fed into mutual ethnic massacres?

Where in Africa has a *shared religion* failed to cement a sense of nationhood? There is the example of *Darfur* in Sudan, where both sides are Muslims and yet many thousands have died in two or three years. Has the Western media served the positive role this time of being a whistle blower to avert the transition from ethnic cleansing to large-scale genocide in Darfur?

Or is the Western media selective where it blows the whistle? Why has it completely ignored two and a half decades of similar atrocities in Northern Uganda from 1986 to 2010? Is Northern Uganda a case of shared tribality failing to avert catastrophes? An Acholi civil war has raged, killing adults, imprisoning families, and depriving children of their childhood years.

Where in Africa once again has a shared *religion* failed to avert catastrophes? Examples include the Democratic Republic of the Congo (Christian vs. Christian), Algeria (Muslim vs. Muslim), as well as Rwanda and Burundi (Christian vs. Christian).

In those confrontations, the international media has a mixed record. In Rwanda it had been good in covering the final catastrophe (the genocide) and its aftermath, but not good at covering the events which led to the explosion.

In Darfur the international media started covering the calamity mid-stream. The media hopes to avert the ultimate catastrophe.

V. Culture as a Ladder of Stratification

Has the basic stratification changed from who *owns* what to who *knows* what? Are computer skills creating new rank-orders? Is Nigeria's North-South divide being digitized?

Although Northerners were historically the first to be *pen-literate* in Nigeria, are Southerners the first to be *computer-literate*? Are Igbo and Yoruba cultures more calculus friendly (numerophile) than Hausa culture? Are Indians of South Asia and Koreans of East Asia more calculus-friendly (numerophile) than Africans or African Americans? Are the Asians better at mathematics? Do the Bagarda enter the Internet culture more easily than do the Karamojong? Do the Kikuyu more than the Maasai?

Islam gets more blame for gender inequality than it gets credit for racial equality. Of the three Abrahamic religions (Judaism, Christianity and Islam), Islam has the best record in *race relations*.

But in modern history Islam may also have the worst record in gender relations. The U.S. doctrine, "Separate but Equal," lasted until 1954 [Brown vs. the Board of Education ruling]. Islam has an implicit doctrine of "Separate but Equal" for genders, rather than races. Gender segregation was potentially more feasible with the advent of the Internet. Women can surf and work at home and eventually make millions – surfing their way to freedom.

On the other hand, the hijab may gradually become more rare, as dating on the Internet becomes more common among the Muslim youth.

VI. Culture as a Pattern of Consumption and Mode of Production

Much of Africa has borrowed Western tastes without Western skills. It has borrowed consumption patterns without production techniques. It has promoted urbanization without industrialization and has learned capitalist greed without capitalist discipline.

The media has been a major instrument for advertising consumption patterns, but not of stimulating new production techniques. On the other hand, the media is an instrument for exposing corruption and tracking excessive greed. In the West the media have become indispensable for implementing democracy. In Africa the struggle to democratize continues with the help of the media worldwide.

Seven Pillars of Media Wisdom

As we confront the subject of democracy, we are stepping into the domain of modern pillars of wisdom. The first pillar of wisdom in contemporary Africa is the imperative of *liberation*. Kwame Nkrumah captured "liberation" at its most optimistic: "Seek ye first the political kingdom and all else will be added unto you."[1]

In some respects, liberation is the means towards democracy. In other respects, liberation is the whole purpose of democratization. Where do the media fit in? The African quest for liberty should influence the selection of stories to be covered and should be a major guide to editorial policy. But liberation does not really translate into democratization unless it is linked to the second modern pillar of wisdom – *accommodation*. The pursuit of liberty needs to be moderated by the principle of accommodating diverse and competitive interests. This second imperative is indeed *accommodation*. If African politics are ethnic-prone, how can African constitutions be ethnic-proof? The African media needs to address this dilemma in creative ways to ensure that liberation and accommodation are not antithetical.

The role of the media in African election campaigns is growing (especially the radio and newspapers and, less persuasive, television). Although the mass media have a long way to go before they become as decisive in African campaigns as they are in, say, U.S. elections, the African media have already become very important in African urban constituencies.

The third modern pillar of wisdom is the imperative of *communication*. Do the African media need to strike a balance between indigenous languages and Euro-imperial languages in their choice of story coverage and pictorial images? Should the media simply follow the priorities of the government in language policy for the country? Or should the media have its own linguistic priorities and act on them?

Many Africans who have learned to read only in an indigenous African language (e.g., Kiswahili in Tanzania) often need regular access to a newspaper in that language if they are to maintain their literacy at all over a long period. Minimally educated Africans can easily relapse back to illiteracy if there is no appropriate and interesting reading material to keep them reading day after day. The print media help to preserve the hard-earned ability to read.

As for ordinary folks who need to be kept informed about local, national, and international affairs during each week, there is nothing to compare with news bulletins in the local language on the local radio. All these elements are vital in promoting a relatively sophisticated electorate among the emerging African democracies.

The fourth pillar of wisdom in the media is *androgynization* in the sense of the imperative of *gender-balance*. This would require a constant effort to balance stories about men with stories about women and to balance male imagery with female.

Some years ago I was graciously invited by *The Guardian* newspaper in Lagos, Nigeria, to give their annual Anniversary Lecture at the Nigerian Institute of International Affairs. The newspaper assigned me a topic. I was to lecture on "The Black Woman." I agreed to give the lecture provided it was chaired by a woman. *The Guardian* agreed and also invited Mrs. Maryam Babangida, the First Lady of Nigeria, to be the Distinguished Guest at the lecture.

In my opening remarks I thanked *The Guardian* for the gender-balance.[2] I added these words: "Since this year you have invited a man to speak about 'The Black Woman,' I hope next year *The Guardian* will consider inviting a woman to speak about 'The Black Man.'" Until then no woman had ever been invited by *The Guardian* to give the annual Anniversary Distinguished Lecture. I would like to believe that my 1991 lecture helped to trigger a re-evaluation of the gender-balance in the editorial and public relations policies of *The Guardian*.

But the changes did not come fast enough to spare either *The Guardian* or myself from such Nigerian feminists as Dr. Molara Ogundipe-Leslie, who had at that time just resigned from the University of Ibadan to depart for a visiting professorship in the United States.[3] She launched an attack in an article entitled "Beyond Hearsay and Academic Journalism: The Black Woman and Ali Mazrui".

The fifth modern pillar of wisdom is *ecological balance*. The African media should be more sensitized to issues like desertification, population growth, the green-house effect and global warming, and the relationship between the ecology and the human condition. The electronic media may have already begun to respond to the distinction between the ethics and the aesthetics of environmental concerns.

Television documentaries about wild animals, the wonders of the sea-bed, the organisms of the deep, the wonders of the reefs, already abound. They concern the *beauty* of the environment (ecological aesthetics). The African media should close ranks with Al Gore, former U.S. Vice-President, in producing documentaries about global warming, the risks of population growth, deforestation and the depletion of the ozone layer (environmental ethics). Although television as a visual medium is more effective with such documentaries than either radio or the print media, there is already a lot of scope for environmental coverage in all the media.

The sixth modern pillar of wisdom is more purely economic, *balancing production, distribution, and consumption*. The African media has begun to be effective in covering economic calamities such as draught, famine, devastating floods, and large-scale fires. Progress is also being made in covering the work of economic organizations, such as the Economic Community of West African States (ECOWAS), the New Partnership for Africa's Development (NEPAD) and the emerging East African Community (ECA). But the media also needs to play a more active role in stimulating and guiding investment and supporting agricultural festivals and livestock exhibitions.

The seventh and final modern pillar of wisdom is eternal vigilance in pursuit and support of *the truth*. This would also include the verification and clarification of each story and a defense of honest *variant interpretations* of the truth.

Conclusion

The pursuit of the truth includes a readiness to blow the whistle when things are going seriously wrong, either in our own society or among our neighbors. We have mentioned earlier the increased readiness of the press in Africa to expose corruption in high places. But the African press should also be ready to tackle even higher crimes and misdemeanors, both within our countries and elsewhere in Africa. The African press has been much slower than the Western press in blowing the whistle about ethnic cleansing in Darfur, or the atrocities of the Lord Resistance Army in Northern Uganda, or the endless civil wars in the Democratic Republic of the Congo.

NEPAD and the African Union encourage *peer-review* between African states – checking on each other on economic performance and the ethics of governance. But there should be peer-review also at the level of the Fourth Estate. The press in each African country and the press between each country should help to raise the alarm about impending or unfolding calamities in our African region.

Ali A. Mazrui
Director, Institute of Global Cultural Studies and Albert Schweitzer Professor in Humanities,
State University of New York at Binghamton, USA

__Ali A. Mazrui,__ D.Phil (Oxon) - Albert Schweitzer Professor in the Humanities and Director of the Institute of Global Cultural Studies at Binghamton University, State University of New York. He is also Albert Luthuli Professor-at-Large at the University of Jos in Nigeria. He is Andrew D. White Professor-at-Large Emeritus and Senior Scholar in Africana Studies at Cornell University. Dr. Mazrui has also been appointed Chancellor of the Jomo Kenyatta University of Agriculture and Technology, Kenya. Mazrui is widely consulted on many issues including constitutional change and educational reform. Dr. Mazrui has been involved in a number of UN projects on matters which have ranged from human rights to nuclear proliferation. He is also internationally consulted on Islamic culture and Muslim history. He was also among the Eminent Personalities who advised on the transition from the OAU to the African Union (2002)

ENDNOTES

1 Nkrumah, Ghana: The Autobiography of Kwame Nkrumah (1957).

2 Mazrui, "The Eighth Annual Anniversary Lecture of The Guardian newspaper, delivered at the Nigerian Institute of International Affairs, Lagos, Nigeria, July 4, 1991.

3 See Professor Ogundipe-Leslie's angry article, "Beyond Hearsay and Academic Journalism: The Black Woman and Ali Mazrui," in Research in African Literature, Vol. 24, No. 1 (Spring, 1993), pp. 105-112.

Preface

Scholars from a wide range of fields continue to explore the social, political, economic, and cultural dimensions of new communication technologies (NCTs). A cursory look at the existing literature on NCTs shows that most studies have been focusing on NCTs potential to alter existing social-political categories that border on economic class, social status, race, gender, and ethnicity. While some of these studies illustrate how certain identity makers are potentially concealed by and through the traditional media, others examine the role of NCTs in bridging economic disparities, promoting social justice, and nurturing infrastructure conducive for democratization. But in as much as NCTs have revolutionized our representation capacities and capabilities, facilitated instantaneous transfer and exchange of information, and developed and expanded hybrid ways of communication, much that defines the traditional media is conspicuously manifested in new technologies. Little wonder that the aegis of "new" in NCTs has invited newer ways of examining whether these new technologies are changing the way we do things without altering the things we do.

But there is one area that has not been adequately examined: the intersection of NCTs and the ideological hegemony. In the following chapters, experts drawn from different disciplines engage in a debate that variously examine the extent to which NCTs are co-opted by the hegemony. This book is made up of chapters whose content is drawn from sub-Saharan Africa, North America, China, Pakistan, and South Korea. Specifically, the book seeks to:

a. Analyze the intersection of new communication technologies and culture, and its role in cultural development
b. Analyze the dynamics informing the techno-cultural landscape and its impact on political and ethnic structures
c. Examine the relationship between the use of new communication technologies and the existing ideological hegemony

There is a common motif among authors in their interrogation of the diagnostic degree to which the use of NCTs is demonstrative of the users' imaginary relations to imaginary reality, their thoughts and perceptions. As Ndirangu Wachanga, the editor of this volume demonstrates in chapter 1, NCTs technologies are seamlessly co-opted by the design of the hegemony; the use of new media technologies can be traced to causes rooted in both ideological and informational characteristics of a particular historical period. Wachanga illustrates how Kenya serves as an example of bottom-up model, which is a challenge to traditional media hegemonic practices, especially when placed in a historical-political framework. By analyzing 110 cellular phone messages circulated during the Kenyan 2007/2008 election violence,

Wachanga sheds light on the vulnerability of NCTs to manipulation and co-option by the dominant. He nonetheless presents NCTs as a conduit with potential to alter hegemonic designs.

Echoing Wachanga's sentiments are Ogechi and Ogechi who submit that NCTs have revolutionized the way of doing things in Kenya, including politics. Ogechi and Ogechi delineate how traditional communication technologies are being complemented by the new technologies, which are now taking over and which can be used to negotiate both narrow ethnic-blocks and national identities. In their chapter, Ogechi and Ogechi show that NCTs are not always used to portray and perpetuate narrow ethnic identities; rather, instances exist when the users want to project professional and national identities. Supported with data that focus on the Kenya's pre- and post-2007 elections violence, Ogechi and Ogechi argue that inasmuch as NCTs continue to be considered emancipatory, great caution should be exercised when using them since unchecked use might bring untold suffering to society.

But whether more caution is required in some societies than in others is a question addressed by Njoroge, Kimani, and Kikechi as they outline and discuss the intersection between NCTs and ethnicity in Kenya. In their chapter, they investigate ways in which the new media reinforced issues relating to ethnicity in Kenya prior to the 2007 presidential election. In demonstrating the nexus between new media and ethnicity, their chapter posits that the upsurge of ethnic animosity in Kenya before, during, and after the elections was, by and large, instigated by the influence of the new media. This chapter attempts to investigate how the new media covered ethnic topics. While the contributors admit that the new media have opened new spaces for debates formerly suppressed, they show their participation in churning predatory messages. The chapter concludes that there is an urgent need for the new media in Kenya to help the citizens to redefine the status of ethnic relationships through the recognition of ethnic differences and the re-discovery of equitable ways to accommodate them for; after all, there is more strength in these differences.

A chapter by Musambira and Muwanguzi departs from an examination of Kenyan politics to one on the role played by NCTs in relation to Uganda's central government and four of its kingdoms: Buganda, Bunyoro-Kitara, Busoga, and Toro. Their analysis is placed in the historical context, within which, each kingdom has pursued a strategy for preservation and development of its people's unique culture and ethnic aspirations. A convergence of cellular phone and FM radio designed by the Buganda kingdom to confront the central government with specific demands and the government's response are examined. The less combative use of NCTs by Bunyoro-Kitara, Busoga, and Toro kingdoms is described. An analysis of how each antagonist uses NCTs for a hegemonistic agenda is presented and cast against the application of NCTs for good governance and the prosperity of citizens.

Kang'ethe Iraki investigates the extent to which the cell phone technology has impacted the lives of young Kenyans, especially within the context of a university in Kenya. Noting that Kenya has had a tremendous technological leap from the 1990s where landlines were the main tools of communication, to mobile telephony where people can now communicate any time, any place, Iraki investigates the ramifications of such a new and stunning technology on a young nation like Kenya, especially if there are pernicious effects emanating from the novel technology. Iraki's chapter results from a study that involved 100 university students at the United States International University in Nairobi, Kenya. The questionnaires were randomly distributed and collected after a week. The findings point out that mobile telephony has become an indispensable tool for most youth providing them with a great sense of control and freedom. However, the technology is also rife with concerns about personal autonomy, dishonesty, breach of cultural norms, personal security, and security of others, especially when texting while driving.

Unlike most chapters in this book that focus specifically on one country, Onyebadi and Kalyango Jr. widen the scope of their focus to accommodate three East African countries: Kenya, Uganda, and Tanzania. In their co-authored chapter, Onyebadi and Kalyango Jr. set out to ascertain the use of and dependency on new media technology for political communication by voting- age citizens of these three East African countries. While they emphasize the importance of new media communication opportunities towards the struggle for liberal democracy, demonstrated by the unanimity of respondents' attitudes that their governments suppress political discourse, they are quick to note that the degree of use and dependence on these new media instruments are not uniform across the region. Gender and the level of education are two salient factors that create these differences in the use of and dependency on new media technology for political communication in East Africa. But their findings have no bearing on the authenticity of the messages sent through these new media technological devices, or the extent of divisiveness or unity that such messages might engender among citizens in the region in times of political crises. These scholars' primary contention is that such citizens now depend on these technological devices to serve their informational needs moreso when political and other forms of national emergency situations arise. This dependency phenomenon is partly the consequence of the existence of media laws enacted by governments in the region to hamstring mainstream media houses.

In this section, we include two chapters examining the impact of NCTs on the media landscape in Zimbabwe, a country whose press freedom and other civil liberties have deteriorated over the years. In his chapter titled "Whose TV Is It Anyway? An Examination of the Shift Towards Satellite Television in Zimbabwe," Rick Malleus argues that the rise in popularity of satellite television in Zimbabwe was not mainly driven by the capabilities of the new technology, but by dissatisfaction with Zimbabwe Television (ZTV). His chapter begins with a discussion problematizing the idea of what constitutes new technology in the African and Zimbabwean contexts. Malleus' focus then moves to the content broadcast locally and how a segment of the Zimbabwean population have turned away from ZTV to have their media needs and gratifications met from satellite TV. There is an extended discussion of propaganda and quality of programming, which are hypothesized as the driving factors for satellite TV's rise in popularity. A discussion of the cultural influences that satellite TV programming has on the Zimbabwean viewing public is included, and the chapter concludes with some thoughts on the future of satellite television in Zimbabwe if changes in local broadcast TV programming come about.

Still on Zimbabwe, Hayes Mabweazara uses a qualitative case study approach to critically examine the appropriations of the Internet by Zimbabwean mainstream print media journalists. In his chapter, Mabweazara draws on social constructivist approaches to technology and the sociology of journalism to highlight the impact of the Internet on practices and professionalism. He argues that the deployment of the Internet by Zimbabwean journalists (and indeed in Africa at large) is relative and contingent upon the 'internal' newsroom context(s) and the wider socio-political and economic circumstances in which the journalists operate. Among other functions, the technology shapes the mainstream press' news agenda as well as avails information often censored by government. This chapter concludes by demonstrating how the use of the Internet is replete with ethical and professional implications.

In the second section of this volume, we focus on the NCTs, language, and health communication. Mahiri Mwita examines theatrical and cultural texts that are performed through a Theatre for Development (TFD) rubric known as Magnet Theatre, which uses theatre-based outreaches to mobilize people in low–income communities into forums that discuss HIV/AIDS and how its problems manifest in their localities. Using examples from performances of four theatre groups that operate in Nakuru and Mombasa towns of Kenya, his chapter examines how the performances textualize, thematize, and theatize the main issues in HIV/AIDS as seen through the perspective of the performers and how the targeted audience reacts to these "AIDS performances." Beyond studying the theatrical outreaches, the research for this chapter surveyed communities in which these performances have taken place to further appreciate how the motifs discerned from the theatrical outreaches compare to realizations of the AIDS problem in the communities.

Just as worrying as the HIV/AIDS pandemic is the Malaria scourge, which kills more than 3,000 children in Africa on a daily basis. While most African governments and health institutions invest in protecting and nursing mothers and children, other demographics left rather vulnerable. Citing the case of Nigeria, Nwachukwu Egbunike examines how NCTs are gradually gaining ground as dependable means of meeting the communication needs of young Nigerians. His chapter discusses how the potential associated with the Internet and social networks can be tapped into and be adopted by the Insecticide Treated Nets (ITNs) to help Nigerian youth. Egbunike suggests solutions and recommendations based on the concept of participatory development communication.

If most of the contributors to this book point to the revolutionary potential of NCTs, to paraphrase Crystal (2001), then, that revolution is predominantly linguistic. This is partly demonstrated by Githinji, who examines the process of translanguaging in imagined communities as a form of style-shifting that reproduces hegemonic structures that dominate national discourse. Githinji examines ethnicities and identities as negotiated by Kenyans through the digital media. He argues that whether operating at national or local level, linguistic performance of positive and negative identity practices results in community members aligning themselves with those with whom they agree while alienating those with whom they disagree. While this alignment may at times validate existing social identities, it sometimes results in creating new identities that challenges the status quo. However, since these new identities are grafted on existing social categories, they are not radically different from the parent structures they seek to replace. He discusses the extent to which interlocutors who employ communication technology on the Internet or mobile phones constitute communities of practice through their linguistic practices. Although he draws parallels between broadcast media (especially community radio) and digital media, he pays more attention to NCTs. This, he argues, is because the Internet has become an important discourse site that competes with face-to-face interactions both in dissemination of information as well as in cementing bonds of social solidarity. But if NCTs comprise of leading edge in social interactions, they are increasingly serving as a central site for excavating the invisibility of covert ethnicity and race relations.

In her chapter, Harper investigates how covert whiteness (a tacit form of racialized consciousness), which does not manifest itself at the surface level in the same overt manner that extreme white cyber hate "imagined communities" do. By applying Critical Race Theory and Black Feminist methodology-based discursive analysis, Harper investigates performances of whiteness in a vegan/animal-rights-oriented website called Veganporn.com. As a progressive forum associated with social justice, this site provides a radically different environment in which to examine white supremacist ideologies; ideologies typically found in more overtly-racist, "extremist" online dialogues already examined by critical research.

Discourse analysis of a specific Veganporn.com forum topic revealed three major themes in the computer-mediated discussion: (1) discursive patrolling of epistemic borders to "protect" Standard English and colorblind expressions (whiteness) of veganism/animal rights from non-Standard English and non-white racialized expressions; (2) the use of blackface *cyber-minstrelsy* to reinforce the "superiority" of Standard English (whiteness) over the "inferiority" of speakers of Black English and Ebonics; (3) the premise among several white-identified Veganporn.com participants that a vegan- and animal-rights ideology is "colorblind" thus making invisible the current socio-historical implications of power structures created around white skin color.

Though Harper's thesis focuses on one discussion within a forum, the analysis of this event offers insight relevant to understanding whiteness as a system, an ideology, and a structure. Specifically, by employing certain theoretical components of critical race studies (racialized consciousness, social ontology of whiteness, and racial mapping), this analysis reveals how the World Wide Web can be an effective site for *cyber-ethnographers* focusing on "decoding" whiteness within progressive social justice movements.

Loosely tied to Harper's analysis is Mohochi and Wachanga's study, which analyzes the extent to which East African hip hop artists have managed to maintain a balance between educating and entertaining society at the local level, while keeping abreast with emerging global trends and influences. These scholars, firstly, strive to show how the young generation of East African musicians uses music to sensitize the public by serving as critics of the management of public affairs, and how this has contributed to political change. Secondly, they seek to investigate the effects these emerging practices have had on the use of African languages in the performing arts

The third section of this volume focuses on the impact of NCTs on the media in other part of the world. We include in this section a case study from China, an analysis from Pakistan, and another one from South Korea. An intriguing chapter in this section is on African Diaspora. Also, in this chapter, we include Adedayo Ladigbolu Abah, examination of the role of Nigerian video film in mediating the twin issues of culture and identity among African immigrants in the United States. For this chapter, Abah interviewed African immigrants, their children, and friends living in the Dallas/Fort Worth area of Texas, USA on the role of the nascent Nigerian video industry and the way these immigrants sustain and straddle their multiple identities and culture in the host nation. Abah found out that most of the immigrants view the videos as affirmation of the values they grew up with and with which they still subscribe to. This is in direct contradiction to the professed cultural denigration immigrants feel in their everyday professional lives in the United States. Most of the younger immigrants and first generation immigrants view these videos as a convenient way of accessing their Africanness as part of their multi-stranded identity and culture.

From a discussion on dynamics informing immigrants in the United States, we turn to James Schnell who addresses mass media and cultural identity dynamics that impact U.S. military policy in the execution of the Iraq war and related cross-cultural communication issues that have evolved during the course of the war. Schnell has completed primary levels of Professional Military Education (PME), and this context provides him with ample opportunity to become familiar with military doctrine and, after the start of the Iraq war in 2003, to observe how primary tenets of established military theory have been consistently ignored. He argues that the U.S. invasion/liberation of Iraq has been executed as a case study for future students to review and learn how not to address such a challenge. According to Schnell, the inability to consider, let alone plan for, cross-cultural ramifications has been a central communication failure that has proven tremendously problematic. The mass media, via traditional channels and NCTs, have been keen to report on these matters which, in turn, have modified public understanding of the matters being reported. Mass media messages have been vital in creating and impacting this phenomenon. This phenomenon has been equally true among traditional media channels and the new communication technologies

The last three chapters of this book are drawn from Asia. In their co-authored chapter titled "New Media and Hegemonic Discourse in Pakistan," Talib and Gardezi problematize the emancipatory promise of NCTs. They argue that the effects of the 'new media' may not be what we expect with respect to the democratization of society. Their analysis shows that ideological hegemony is not only alive and well, but might even thrive in these new media. To scrutinize these effects, their chapter focuses on the impact of Pakistani cable television on identity and culture discourse through the lens of three topics: identity and the new communication technologies; ideology, power, and the new media; and new media and new popular cultures

Taking the cue from Talib and Gardezi is Siho Nam who summarizes certain main debates regarding the role of the Internet in promoting or hindering democracy by paying close attention to recent spate of regulations regarding Internet content in Korea. Nam seeks to shed light on how the Internet is re-configured as a new site of cultural politics and democracy. This argument is picked by Jiafei Yin who investigates the social conditions and ways in which NCTs are transforming the politics, culture, and the society in China, the largest Internet user in the world. China had 384 million of its citizens connected to the information super-highway by the end of 2009. Yin analyzes the use of the Internet, different roles played by the traditional and the new media, cases that were catapulted to the national media spotlight by the online community, and government laws and regulations responding to the development of the Internet through contrasts with the roles new communication technologies play in western and African societies. Also, this chapter attempts to examine the factors that have made such transformations possible and the implications of these transformations

The diverse nature of this books' content is invaluable in the overall attempt to place NCTs in a context that recognizes their seamless co-option into the designs of politics, culture, ethnicity, and the ideological hegemony.

D. Ndirangu Wachanga
Wisconsin, August 2010

Acknowledgment

I am indebted to a number of persons and agencies that helped me along the way. I learned the value of journalism and media during my tenure as a student and later as an instructor at the Mayborn Graduate Institute of Journalism at the University of North Texas. The Mayborn director, Mitch Land, has been a mentor and a friend since my days at Eduardo Mondlane University, Mozambique. As I drafted the proposal for this book, received chapters, and edited the manuscript, I had numerous conversations with another of my mentor, Brian O'Connor. My conversations with Dr. O'Connor ranged from constant chats in his office when he served as my dissertation chair, to meetings in coffee shops, to discussions during the Document Academy meetings, to his response to my e-mails and text messages, to long telephone dialogues, especially after I left the University of North Texas. Just as our conversations have been diverse in form, so have they been in content: representation, human information seeking behavior, graffiti, clandestine radio broadcasting, poetry, and photography. It is my hope that the content of this book will provoke more conversations among the readers. My intellectual journey starts with the seeds planted by Emilia V. Ilieva during my undergraduate days at Egerton University.

I am grateful to Evan Mwangi, Northwestern University; Kiarie Wa'Njogu, Yale University; Kenneth Kimani, Tennessee State University; Henry Chakava and Kiarie Kamau of East African Educational Publishers; Milton Obote Joshua, University of Massachusetts at Amherst; Sangai Mohochi, Maseno University; Mahiri Mwita, Princeton University; Ali Mazrui, Binghamton University, Ravenna Narizanno, Binghamton University, Melinda Levin, University of North Texas, and Jurhee Curtis, University of North Texas.

I would, of course, like to register my great appreciation to my parents, Irene Njeri Wachanga and Simon Wachanga for their unwavering support despite of the geographical distance between us. My brother, Mwangi, and sister, Wangechi, encouraged me when it seemed difficult to tighten several looses of this volume. I would like to thank the readers in advance for taking their time to be part of various conversations contained herein.

D. Ndirangu Wachanga
University of Wisconsin-Whitewater, USA

Section 1

Emerging Media, Community, and Identity (Re)Construction

Chapter 1
Challenging Traditional Media Hegemonic Practices:
A Kenyan Case

D. Ndirangu Wachanga
University of Wisconsin-Whitewater, USA

ABSTRACT

The top-bottom model espoused by the traditional media structures is being problematized by the emerging technological change in the 21st century. Kenya presents an example of bottom-top model, which is a challenge to the hegemonic potential of top-down model. In the discussion that follows, we will establish how media in Kenya have been operating within a top-bottom model until recently. This top-bottom model has been problematized by the emergence of the new communication technologies (NCTs), which have allowed individuals to challenge dominant voices and myths, alter representation and meaning of symbols and vocabulary, and re-define politico-social structures around which the luminal rituals of the national have been interwoven for the sole purpose of fostering group cohesion. We argue in this discussion that the proliferation of NCTs and communication strategies have dismembered the nation by stimulating exponential multiplication of discursive regimes that may have been impossible previously when media technologies were confined to the control of a central, and oftentimes restrictive authority. It is against this background that we seek to probe the extent to which media technologies are ineffaceably co-opted into designs of the dominant ideology and various other hegemonic structures.

INTRODUCTION

In the 21st century, news media are like frogs before a thunderstorm and pose the question of whether we can hear anything on account of them. Unlike the 19th century with newspapers and the 20th with newspapers, radio, and television, 21st century media are numerous, novel, often participatory, and challenging. How are we to hear anything

DOI: 10.4018/978-1-60960-591-9.ch001

from them with increased novel convergence; a disruption of a predominantly top-bottom communication model?

The top-bottom model espoused by the traditional media structures is being problematized by the emerging technological changes. Kenya presents an example of bottom-top model, which is a challenge to the hegemonic potential of top-down model. In the discussion that follows, we will establish how media in Kenya have been operating within a top-bottom model until recently. This top-bottom model has been problematized by the emergence of new communication technologies (NCTs), which have allowed individuals to challenge dominant voices and myths, alter representation and meaning of symbols and vocabulary, and re-define politico-social structures around which the luminal rituals of the nation have been interwoven for the sole purpose of fostering group cohesion. We argue in this discussion that the proliferation of NCTs and communication strategies have dismembered the nation by stimulating exponential multiplication of discursive regimes that may have been impossible previously when media technologies were confined to the control of a central, and oftentimes restrictive, authority.

Historically, Kenya's restricted information flow and limited circulation of the print media formed an integral part of consolidating rather than challenging the spatial politics of the national. Little wonder that, until recently, Kenyans had became acquiesced to a discourse that presented the national both as a cognitive and administrative authority while (mis)presenting contesting and alternative voices as detrimental. Therein lay the success of the national communication model, which thrived on purging the self as a discursive and communicative space. We are not insinuating that this space was non-existent; rather, we are emphasizing how oppressive structures were put in place to render tolerant discursive spaces taciturn because the success of national politics thrived on its reticence. The burgeoning of NCTs is changing the laconic disposition of this space.

From a global standpoint, most African countries are still trying to get onto the information super highway despite the unprecedented technological advancement. Alemneh & Hastings (2006) have noted the increasing concerns on whether the Internet is assisting Africa to realize its development plans or, to the contrary, the Internet is merely intensifying the existing economic disparity between the poor and the rich. The Internet penetration in Africa is only 5.6% compared to 26.9% in the rest of the world (www.internetworldstats.com). By March 2009, it was estimated that more than 55 million Africans were accessing the Internet. This is only 3.4% of the global Internet access. In spite of these low figures, Africa is adopting new communication technologies at a notable rate. Kenya, with a population of 39 million, for instance, had an estimated 3 million Internet users by early 2009 (www.internetworldstats.com). This is 15 fold increase from 2000 when only 0.2 million people were connected to the Internet. Kenya is ranked fourth in sub-Saharan Africa behind Nigeria (10 million users), South Africa (4.6 million users) and Sudan (3.5 million users). Kenya has an 8% Internet penetration. But it is the mobile devices, especially cellular phones that have revolutionalized communication in Kenya because of their ability to bypass the necessity of elaborate infrastructure such as electricity installation, especially in rural areas. It is estimated that at least 12 million Kenyans, nearly a third of the population, have cellular phones. The current expanding mediascape in Kenya, increasing Internet penetration, and the rising number of cellular phone users present a potential for change in the media hegemony.

MEDIA TECHNOLOGIES, DOMINANT IDEOLOGY AND HEGEMONIC STRUCTURES

Across the globe, the media have generally been lauded for their positive contribution in democratization processes, as purveyors of civic education, gadflies of econo-political accountability, and as forums for cultural transformations and civic engagement. A case in point is the use of social networking sites by Iranian demonstrators protesting what they believed to be a rigged election in 2009. When the Iranian government limited internet access and restricted information flow, cellular phones became an alternative means of information dissemination. The popularity of what became Twitter in Tehran, led to the U.S. State Department requesting Twitter to postpone "network upgrade that would have shut down the site for some hours on June 15th and 16th" (Diane & Meier, 2009, p. 3). The social media - Facebook, e-mail, Twitter, YouTube - in North Africa played a major part in the domestic mobilization of protesters and in informing the rest of the world in 2011. The Hosni Mubarak and Muammar Gadhafi regimes in Egypt and Libya respectively, did try to control both the social media and the mass media (radio, television and the print media). But shutting down the Internet in Mubarak's Egypt only resulted in enlarging the crowds, which turned up in person at the Tahrir (Liberation) Square.

Such is the adulation for the uses of the emerging media technologies. But the traditional media have also been criticized and condemned for ethical irresponsibility and professional negligence; for being passionately profit and power-driven at the expense of offering community service, as an embarrassment to the integrity and sustenance of the democratic processes they purport to serve and promote; for peddling predatory messages that are, at times, socially antagonizing. Both the exaltation of celebratory and the intensification of counter-celebratory discourses have bolstered with the unprecedented burgeoning of information and communication technologies. Yet, as Grossberg, Wartella, Whitney, and Wise (2006) remind us, media technologies are often fashioned within, shaped by, and controlled by institutions involved in their production and use. Media technologies, therefore, develop within "specific economic and social frameworks, and from specific economic and social determinants" (Shaughnessy & Stadler, 2008, p. 6). In fact, most studies examining ideological implications of the emerging media conclude that "computer-mediated communication reproduces the social, political, and economic relations that exist in the real world" (Cook, 2004, p. 103). It is erroneous, therefore, to construct the Internet world "as something that exists in binary opposition to 'the real world.'" because "when it comes to questions of power, politics, and structural relations, cyberspace is as real as it gets" (Kolko et al. 2000, p. 4).

It is against this background that we seek to probe the extent to which media technologies are ineffaceably co-opted into designs of the dominant ideology and various other hegemonic structures. We will cite cases from ethnically stratified societies, with specific examples from Kenya, to show how media technologies disseminate messages that facilitate to reinforce patterns that are ideologically persuasive and compliant to hegemonic designs. Seeking to understand the use of media technologies' propagation and dissemination of messages may offer guidance in our extrapolation of contours informing such societies' ideological and hegemonic terrain. However, we will lace such propositions with attentiveness to media technologies instrumental value as much as their ideological value. Also, we will seek to probe the intersection of democratic processes, politics and the media by, firstly, extrapolating media dynamics in Africa in general, and secondly, paying particular attention to Kenya.

This discussion will be placed in a broader historical context before examining trends, informed by among other changes, information and technological revolution, processes geared toward expanding democratic and associational spaces, multiplicity of media voices, and diversity of media ownership, leading to, though not entirely responsible for, increased citizen participation, knowledge society and knowledge generation.

Broadly, we will be trying to examine how media technologies are seamlessly co-opted into the designs of the hegemonic order; that the use of new media technologies can be traced to causes rooted in both ideological and informational characteristics of a particular historical period. We maintain that any examination of new communication technologies needs to "be situated within the economic, social, and political contexts where the information and images they convey are consumed, circulated, and signified" (Cook, 2004. P. 104). Our justification for placing media technologies within such broad parameters is based on how emerging media technologies continue to alter and transform cultural, social, political and economic arrangements globally (See also Slack and Wise, 2005). But our justification recognizes that information, communication, and media technologies per se bring about neither development and democracy, nor social and political integration. We will, therefore, structure our discussion thus:

- First, provide a cursory overview of models associated with media in Africa including the modernization theory
- Second, illustrate the nexus between media and how their power is manifested in knowledge generation
- Third, use the Kenyan case in seeking to understand the extent to which traits of democracy (or its clamor) can be traced in historically based informational and media characteristics.
- Fourth, briefly examine the use of new communication technologies in Kenya in the 21st century by paying close attention to the 2007/2008 election violence, especially after the Kenyan government banned live television broadcasts.

Before extrapolating these four areas, there are certain terms we need to operationalize.

COMMUNICATION, TECHNOLOGY AND THE MEDIA

We will be alluding constantly to the terms information, communication, technology, and the media, and it is important to contextualize their usage. The terms information and communication in our case are conceived and conceptualized within a broader frame that is consistent with Hayes (1993) interpretation; that is, information as being more complex than dissemination of mere facts; a convergence of message and meaning, communication beyond the linear quantitative transmission model. By echoing Hayes' views, Bimber (2007, p. 11) notes that such broad conceptualization facilitates our understanding of information and communication's vitality "to democracy in myriad ways: in the processes by which citizens preferences are formed and aggregated, in the behaviors of citizens and elites, in formal procedures of representation, in acts of governmental decision making, in the administration of laws and regulations, and in the mechanism of accountability that freshen democracy and sustain its legitimacy." Information serves as an incentive that affects the cognitive structure on the part of the receiver. It is a complexity that is marked by the confluence of meaning and message. But we can receive a message and lack the requisite tools to decode the meaning. Information, therefore, does not engender interpretive processes; rather, it is informed by the ability to decode signs and symbols encoding a message. Interpretation requires the ability to extract meaning(s) from the data carrying the message; the construction of meaning from symbolic signification. It is possible to receive information without being informed, to receive a code and yet lack the requisite decoding skill or tools, or both; that availability of information does not equate to information accessibility. Still, we need information to communicate. Communication, simultaneously, involves the process of decoding the data, the content (e.g. in statements such as "we sent a communication), and the effect on the recipient (e.g. "Has the teacher communicated?"). Communication, therefore, alters the cognitive structure of the information recipient. Ascribing and associating meaning to the data is what we will refer to as understanding, a complexity of recognizing the information; its structure and its content (see also Hayes, 1993). Although information is, partly, communicated in language and, partly, interpreted through language (there are other numerous audio-visual ways of representation), the way it is mediated – through various forms of mass media including the NCTs – affect the interpretive process, echoing the McLuhanian dictum of the medium is the message. Mediation is more than conduction. The media are more than conduits; they're as interpretive and they are ideological. Lull (2002, p. 26) notes "that the technical form used to move information from one place to another communicates something itself that is just as important as, if not more important than, the medium's apparent "content."

We will refer to technology as the physical means of producing, processing, and packaging; distributing and disseminating, and of facilitating representation and retrieval of goods, services, and cultural products. We insist that technologies used in the production and processing, in the distribution and dissemination of information, are not independent of the society where they are devised and used. History of technology tells us thus. The dynamics informing the relationship between technologies and institutions are as contemporaneously conspicuous as they were in the Middle Ages, when the Roman Catholic Church sanctioned technologies of writing. During this period, it was only the clergy who taught how to write and who controlled the resources associated with the generation of knowledge and what was written and disseminated had to be endorsed by the church. When Johannes Gutenberg and Johann Fust fused the printing press with moving types in the 15th century, there was a major technological and ideological shift that accompanied the production of information. Now that individual letters of type could be moved around and reused, this innovation generated an institutional challenge: the power of the institution of the church. Even three centuries later – 18th century – nine out of ten books that appeared were not only written in Latin, but were also available almost exclusively to the intelligentsia. A major informational change was ushered through the book publishing industry in the 19th century that saw at least eight out of ten books written in vernacular (See Christians, Ferré & Facker, 1993). If the printing press facilitated the disengagement of technological and informational control from the institution of the church, it did not, however, escape the entanglement of other institutions. The printing press was "created within and became part of other institutions – medieval institutions such as guilds and later modern commercial institutions such as banks and mass media industries. And these, in their turn, controlled how the technology was used and what sort of things could be written, printed, and distributed" (Grossberg, Wartella, Whitney & Wise, 2006, p. 13).

But heavily influencing and determining the use of media technologies is a set of hegemonic, dominant taken-for-granted ideas and assumptions. Hegemony, a dominant ideological framework embodied, inscribed and often perceptively invisible via a multitude of symbolic practices of storytelling, artistic displays, rites of passage and many other acts, serves to construct the need and value of owning, consuming and using, and the convenience associated with various media technologies. However, because hegemony functions by acquiring disguised normative trappings; by passing nearly invisible to the technology users, media audiences potentially perpetuate the interests of the social, political, and economic class.

Having operationalized the terms information, communication, and technology, we realize how slippery the term media can become, especially because of the constant changing dynamics informing the mediascape. Yet, changes within the mediascape are as well significantly affected by changes in the technological, informational and communication realms. This multifarious intra and inter-mutative tendency of the media has been eloquently discussed by Appadurai (1990). Globally, Appadurai (1990, p. 299) demonstrates how mediascapes "tend to be image-centered, narrative-based account of strips of reality, and what they offer to those who experience and transform them is a series of elements (such as characters, plots and textual forms) out of which scripts can be formed of imagined lives, their own as well as those of others living in other places" (299). For the purposes of our discussion and that of contextualizing the term media, we will use Shaughnessy and Stadler's (2008, p. 4) six media traits:

- The media are human communication systems
- The media use processes of industrialized technology for producing message
- The media generally aim to reach large audiences or to be used through "mass production" leading to 'mass communication,' their success is often built on popularity.
- The media usually aim to facilitate communication in which the sender does not need to be present as the communication is recorded and then transmitted.
- The media are called "media" because they are literally in the middle, or are the middle chain, of this communication (*media* means 'middle' in Latin); they are the mechanism that connect the senders and the receivers of messages.
- The development of the media has been affected by commercial interests that recognize that the media are potentially highly profitable industries.

In the following section, we will examine how the above concepts ideologically tie to the general history of media in pre- and post- independent Africa. Scholars such as Nyamnjoh (2005), Hyden (2003), Hyden, Leslie, Ogundimu (2002) among other have eloquently discussed media history in Africa.

Ideological Inclination and the Media Landscape in Pre- and Post-Independent Africa

An overview of media history in pre- and post-independent Africa exudes expansive literature on communication outlets that thrived clandestinely in undermining autocratic and colonial regimes and in pushing for the expansion of democratic space.[1] But there is an academic hiatus in the examination of Africa's dependence on poorly operated and resource-restricted, though colonial-unsanctioned, means of mass communication when fighting for independence. From a sanctioning standpoint, which engenders history, politics, power and economics, most post-independent African states possess a Janusian[2] - double-faced - communication paradigm. Those who operated anti-colonial media later rose to power only to use the same media to perpetuate their ideas and ideal even as they criminalized cases, causes and courses of any form of opposition- real or imagined.[3]

Broadly, this paradigm exhibits two phases of information warfare that inflect information gathering and propagation, and information flow; a communication battle in two political settings – as first the critical alternative - often rebel and clandestine - source of information during the colonial struggle, and then, upon gaining self-governance, as the sanctioned source (and even sanctioned suppressor) of information. But, while both clandestine and sanctioned media outlets gather information, operate sanctioning mechanisms, and package information to suit their needs, there is variance in their techniques. One of the major reasons attributed to this difference is their informational and communication objective, which is mostly organized along competition for power. A deeper look at this paradigm reveals a three-dimensional typology of information organization:

a) Information input – processes involving information acquirement and processing
b) Information output – processes involving information dissemination and strategic propagating of messages
c) Information suppression – processes involving information suppression and criminalizing of countering unsanctioned messages. Propaganda and outright lies, censorship, intimidating and exterminating sources of information, fall in this category (See also Wachanga, 2007)

The Janusian character of this paradigm takes the form of a mirror image by reflecting itself; especially because elements characterizing each information category were not significantly altered by changes that punctuated the political dispensation (See also Wachanga, 2007). This historical and political situated-ness of the media in Africa reflects their anticipated roles in different African countries and different historico-political periods.

In his examination of the juxtaposition of the media and development, Zeleza (2009, p. 21) illustrates the multifaceted nature of these roles, which rest in the African media's fluidity, their intra-mutative tendencies as well as their changing relationship with other mutating aspects such as local, national and global politics and economy:

Politically, there is the question of the impact of the media on the democratization of society, as well as the impact of democratization on the media itself. Economically, at issue is the development of the media as a service industry. This would include growth and ownership patterns, as well as levels of profitability and competitiveness. Culturally, the media can contribute to the development of social capital – that is, relational networks involving social interactions, civic trust and normative behavior that can benefit or harm society – as well as to the articulation and performance of cultural identities. Technologically, the media have played a crucial role in the ways in which they introduce and incorporate technological innovation.

The intersection of politics, economic, culture and technology as illustrated by Zeleza calls for the need to differentiate between democratization of the media and democratization through the media. Media democratization, on one hand, underlies the commitment to augment media processes (gathering, storing, disseminating, securing, distributing and retrieving) and strengthen media institutions in ways that facilitate democratic values while allowing unrestrictive and informed participation in public discourses and decision-making. Democratization through the media, on the other hand, includes processes through which democratic ideals, however variously defined, are pursued through available means of communication. Media democratization emphasizes the relationship between changes in the mediascape and other socio-economic and political-cultural terrains. The relationship between democratization and media in contemporary Africa need, therefore, to be interrogated from various viewpoints - in terms of processes, paradigms and patterns.

But since different African countries have different histories, consequently espousing varying democratic agendas, communication visions and developmental directions, they inevitably followed different democratization routes at independence. And that has not significantly changed. Most countries' democratic orientation continues to be heavily influenced by their positions in relation to the West. Using this background, Zeleza (2009, p.22) presents the following models of democratization:

- Nativist model - This invokes the ethos of traditional or communal democracy.
- Liberal model – This is confined to multipartism, periodic electoral contests and good governance.

- Democratic model - Seeks the construction of distributionist democratic developmental states.
- Theoretical model - Envisions the creation of a new social and political order sanctioned by religion.
- Transnational model - Aspires to the construction of democratic regional federations and Pan-African dispensation.

These models, as proposed by Zeleza, presuppose the existence of structural means of information acquisition and dissemination. It is that presupposition that informs our interrogation of the nexus upon which ideology, democratization and media technologies oscillate. As Gumucion and Tufte (2006) have observed, any discussion about any form of development in Africa must be predicated on information acquisition and structural concerns, which include, among others, land ownership and human rights. This assertion depart from the tenets of modernization models, one of the earliest theoretical reasoning that sought to illustrate the contributory relation between media technologies and the democratization process (See Lerner, 1958; Schramm, 1964; Schiller, 1971). Modernization thinking served as one of the earliest robust discussions about development and communication in Africa in1950s and 1960s, as detailed in the following section.

MODERNIZATION REASONING

Modernization as an approach to development in Africa assumed that development processes witnessed in the global North would serve not only as a valuable guide, but also as a model worthy of emulation by the global South. At the core of the modernization debate was a more direct focus on socio-economic development rather than democracy. The modernization era assumed that "democracy would only be feasible once society had been modernized" (Hyden & Leslie, 2002, p.2). Its chief proponents argued for a convergence between indicators of modern and democratic societies. Others, even more persuasively, such as Lerner (1958), "replaced personal experiences as the font of new ideas" with mass media, which were articulated as vital link in modernization (Hyden & Leslie 2002, p.3). But these media would only grow, it was argued, if they had the support of a critical mass. This paradigm held urbanization and literacy as accessories to democracy and participatory governance, themselves assumed to be resultant features of the modernization process (See Lerner, 1958).

The value of Lerner's book, *The passing of the traditional society*, coincided with the United Nations General Assembly's campaign for the establishment of a program to facilitate the expansion of mass media in the developing world. In 1962, the UN General Assembly impressed its member governments on the necessity to integrate mass media expansion strategies with their economic development. Although the modernization paradigm recognized the vital role of the media in national development, it also largely conceived development "in terms of importing Western models, capital, technology and market institutions and dialogues" (Zeleza, 2009, p. 20). Still, the development question became more problematic, more so with the inclusion and involvement of governments in the development process. There emerged concerns of how much governments would be involved just at the same time as African countries were starting to express discontent with the mediocre performance and unrealized promises of the modernization paradigm itself.

According to Zeleza (2009, p. 20), this "growth without development, led many African states to join the broader Third World plea for a new international economic order. This call was rooted in the insights and implorations of dependency theory that ascribed underdevelopment in the global South and enduring inequities of the international division of labor." Ideologically, this new economic order was advanced as a necessary consequence because of its potential to attack global inequity. It was clear that development in Africa was far more socio-politically complex than modernist such as Lerner and Schramm had speculated. As a result of this realization, "The relatively linear and simple development model that they had advocated was gradually abandoned in favor of other approaches that highlighted different aspects of development, notably the redistribution of economic resources in favor of the poor" (Hyden & Leslie, 2004, p. 4). Modernization reasoning became tantamount to oversimplification of African traditions and their interrelationship with the western modernity.

However, one crucial feature of modernization that persisted in the 70s and 80s was the role governments played in developments. Governments had been perceived as sources of state solutions until early 1980s when the international community realized that governments were not necessarily a solution to challenges of national development. This modernization criticism was partly a upshot of transnational debate over the proposal at UNESCO and UN General Assembly for the establishment of what became the New World Information and Communication Order (NWICO), an international dialog generated by an imbalance in news, in both content and flow. NWICO sought to decolonize the information terrain while generating a new informational order in the communication field (See also Schiller, 1978).

During this period, African countries were learning from and encouraged by successes made by the Organization of Petroleum Exporting Countries (OPEC) in extracting higher prices for oil from industrialized countries. There was a rise a neocolonial developmental paradigm, which "constituted a fierce regime of free market fundamentalism, and gained ascendancy on the backs of a new wave of capitalist globalization as Keynesian welfare began to disintegrate, and the nationalist project in the global South collapsed under the weight of proxy wars and rising indebtedness" (Zeleza, 2009, p. 20). As the impact of modernization was withering in the 1990s, political-economic arrangements based on globalization theories were on the rise.

With the modernization theory becoming increasingly problematic, there was a scholarly shift in the media discourse toward an examination of the role of the media in expanding democratic participation through political plurality and increased political participation. According to UNDP:

Africa has the highest number of countries with democratic systems to date since the 1960s. A large majority of African countries now conduct regular elections at the national, regional and local levels, enabling populations to choose their political leaders and ensuring legitimacy of elected governments and officials. This is exemplified by the increasing number of peoples participating in elections (25.7 million or 80% in the DRC in 2006; 23 million or 77% in South Africa in 2009; and 8.6 million or 69.5% in Ghana in 2008).

http://www.undp.org/africa/governance.shtml

Media - emerging and traditional - democracy and development in Africa cannot be discussed without considerable attention to the freedom of the press. Available literature on the clamor for democracy in post-independent Africa – christened the second liberation – goes hand in hand with that of the freedom of the press. We will mention briefly that the debate on media freedom has failed to satisfactorily address the membership of media institutions as both information disseminating entities and commercial enterprises that are driven by the logic of capitalism. So, while freedom of the press has constantly been a bone of contention in Africa, it is ironic that most countries have entrenched in their constitutions the right to free expression, which among other provisions, reins in government infringement on human expression. But Christians, Ferre & Fackler (1993) are quick to remind us that political independence does not necessarily fend off successfully other media 'enemies.' These 'enemies' are rooted in economic, educational, scientific, ethnic-cultural and religious institutions that, out of either malice or sheer social complexity, dominate the news media's agenda. Beneath the alluring journalism patina of giving "truthful, comprehensive and intelligent account of the day's events in a context which gave them meaning," there are disguised interests of power wielders, and their ability to use their power to disguise their interest through the media (Hutchins Commission, 1947).

The kind of information that is disseminated by the media is not, therefore, innocent, or as Zaller (1991, p. 13) notes, "just information." Media are businesses, they are entities expected to make profits and operate within a devised legal system. But legal is different from ethical. It is not surprising that media companies address issues related to "quality control" because it "sounds better than 'social responsibility' to business ears simply because it involves not expense, but profit…"(Christians, Ferre & Fackler, 1993, p. vii). Informed by these mechanics, the media are vulnerable to manipulation by those with and in power. This susceptibility does not isolate the benefits of enlightening a society, say, through knowledge generation that requires intelligent and informed citizen participation. But there is an irony here: Let us, for instance, take democracy to be a political system that, arguably, permits unfettered citizen participation in processes of deciding who ascends to power. But meaningful participation in democratic or development activities presupposes knowledge acquisition. Yet knowledge results from the availability and accessibility of information, where availability of information does not necessary guarantee or translate to information accessibility. The media, therefore, play a vital role in knowledge generation, an area we turn to in the following section.

MEDIA, POWER, AND KNOWLEDGE GENERATION

The media role in knowledge generation is probably the clearest indicator of their power. Foucault (1980) has meticulously examined this tightly knit relationship. The profundity of this relationship can be perceived in Foucault's use of the "Bentham's Panopticon, a circular architectural structure in which cells are arranged around a central viewing tower in such a way as to ensure permanent surveillance, which ensures control of and discipline over the incarcerated bodies" (Karatzogianni, 2006, p.27). Foucault suggests an associative constitution between economies of knowledge and knowledge economies, not simply between the fields of knowledge and power. If understanding the media requires understanding power structure, then, we must as well examine how systems of knowledge are defined, and according to Foucault, how they are tied to and influenced by systems of power. And because different systems of power generate different types of knowledge, narratives constructed by the media are often tailored to generate and achieve certain meanings within a sequence of possibilities. Meaning-making, as a process is, therefore, not just a functional attempt at representation of the world using language, but also "a struggle for masterly in discourse – over which kind of social accenting is to prevail and to win credibility" (Hall, 1982, p.77), a social struggle.

This struggle emanates from and is facilitated by the polysemic nature of language – the ability to derive various meanings from a similar set of signifiers and sets of similar signifiers. The triumphant meaning, the one sanctioned and repeatedly enunciated by those with power, plays a significant role in the representation and in defining reality. The process leading to the emergence of a dominant accentuation of meaning serves as a formidable base in our formulation of knowledge and in our construction of a dominant ideology (See also Hall 1982, Althusser 1971, Gramsci, 1971,

Vološinov, 1973). The media are at the centre of constructing and disseminating meanings. The burgeoned Internet has allowed for multiple spaces to construct and negotiate meaning so powerful is the impact that "if the internet is a revolution…it is a likely to be a linguistic revolution" (Crystal, 2001, p. viii).

But the media use purposefully framed narratives whose intentions can potentially pass unnoticed among audiences. Such narratives become "meaningful, not because of the individual, but because of the discursive formation, providing subject-positions which the individual can occupy" (McNay 1994, p. 68 quoted in Karatzogianni, 2006, p. 27). The subtleties within which these narratives are crafted, interpellate media audiences through media technologies and other instruments associated with power.

The relation between meaning-making and power is more complex in ethnically stratified societies like Africa, especially where the process inevitably revolves around multiple inflectional lenses including linguistic, ethnic and political correctness. Besides these relations, there exist ethnic tensions and cultural struggles, which are indicative of the failure to achieve a cohesive hegemonic order in stratified societies, and a politicized and ethnically polarized search for nationalism or national cohesion.

Nowhere is this more evident than in Kenya. In the next few pages, we will attempt to demonstrate the extent to which characteristics of democratic processes can be traced to causes rooted in informational and media characteristics. We will do this by placing media development in Kenya in a generous historical-ideological bosom.

MEDIA IN COLONIAL KENYA

The media in Africa was first vibrant in the western part of the continent partly because of the "repatriation of freed slaves to Liberia, among them well educated people who embarked on journalism both as a business and as a means of advocacy" (Ibelema & Bosch, 2009, p. 294). The advocacy content of early publications served as an inspiration to freedom activists and nationalist leaders such as Kwame Nkrumah and Nnamdi Azikiwe who used the media to agitate for political freedom in Ghana and Nigeria respectively. There was, however, a different orientation both in media ownership and use among East Africans. As Ainslie (1966, p.99) notes, " whereas in West Africa, newspapers developed as a voice to express the protests of the ruled, in East Africa they were from the beginning vehicles for culture and concepts of the rulers, with the considerable resources of white capital at their command" (Quoted in Ibelema & Bosch, 2009, p. 295).

Although Bourgault (1995, p.164) has indicated how "Asian entrepreneurs working with the colonial government used their accumulated capital to establish newspapers that were essentially colonial in orientation" (in Ibelema & Bosch, 2009, p. 295), mass media made its debut in Kenya at the close of the 19th century in form of missionary newspapers, targeting Europeans both in Kenya and Britain. The first newspaper, *Taveta Chronicle,* was published by Reverend Albert Stegal of the Church Missionary Society in 1895 (See also Wa'Njogu, 2004). Almost five years later, the first English language weekly, *Leader of British East Africa and Uganda Mail*, was being circulated in Mombasa. Like its predecessor, this weekly served the interests of the settlers, even after its circulation base was relocated from the coastal town of Mombasa to Nairobi in 1908.

The coast region, which was the European entry point to East Africa, had a vibrant media compared to the hinterland. In Malindi, another coastal town to the north of Mombasa, *Observer* was published between 1901 and 1902. What is today the *East African Standard* was started in 1902 as the *African Standard.* Together with the *Daily Nation*, which was started in 1960 as a commercial entity by the Aga Khan of the Ismailia Islamic sect, they remain the two major newspaper outlets in East and Central Africa. These two media groups now have websites, FM radio stations, TV stations, and magazines associated with them.

As more foreigners settled in the interior, partly because of the completion of the Kenya-Uganda railway in 1901, which linked the interior to the outside world, more newspapers emerged in the hinterland. A case in point is the Kericho-published *Fort Ternan Times*, which targeted settlers and railway staff workers as well. In Nairobi, *Nairobi News, Times of East Africa*, and *Nairobi Star* - a weekly publication- were published around 1905. Also in Nairobi, the *Globe-Trotter*, a weekly newspaper, closed doors in 1906 after only three years of circulation.

Another notable publication in the interior of Kenya was *Wathiomo Mukinyu* (A True Friend), a Roman Catholic Church publication based in Central Kenya (Nyeri), whose content was religious and targeted the local Kikuyu community. Other religious and vernacular newspapers included the Church of Scotland Mission's *Kikuyu News* (1908-1957), the Anglican Church's *Lenga Juu* (Aim Higher) (1916). While most of these publications served colonial interests, Bougault (1995, p.164) notes how Asian, especially Indian, entrepreneurs "working with the colonial government used their accumulated capital to establish newspapers that were essentially colonial in orientation" (cited in Ibelema, 2009, p. 295). Intriguingly, it was the Indian community that started the first anti-colonial publications in Kenya. The first Indian newspapers targeting an African audience include *Samachar* (News), a Gujarati weekly, The Indian Voice, *Hindi Prakash* (Hindi Light), and *East African News*, which were published in both English and Gujarati. These publications, however, were not acerbic in their criticism of the colonial discriminatory policies as was the *East African Chronicle*, which was first published in 1919 by Manilal Ambalal Desai, an Indian industrialist and political activist.

Desai paved the way for political engagement using the media, a route that was taken by key Kenyan political activists and colonial liberation such as Jomo Kenyatta and Harry Thuku. "After World War 1, Africans and Asians in Kenya openly began to question and oppose policies and laws deliberately set to entrench white domination. In the 1920s and 30s, Thuku' *Tangazo,* Kikuyu Central Association's *Muigwithania* (edited by Kenyatta), and Owiti's Dholuo, *Luo Magazine* were part of a steadily growing alternative voice to the official colonial government's point of view" (Wa'Njogu, 2004, p. 58).

As colonial discomfort intensified among Kenyans, who continued to exert more pressure on the settlers so did the number of newspapers increase. In the 1950s, there were more than ten weekly publications that were predominantly anti-colonial. They included: *Mumenyereri* (The Care giver), *Sauti ya Mwafrika* (Voice of the African), *Uhuru wa Afrika* (Freedom for Africa), *African Leader, Kenya Mpya* (The New Kenya), *New Africa* among others (See also Wa'Njogu, 2004). The vibrancy of the anti-colonial weeklies did not sit well with the colonial government, which led to an amendment of the 1906 Newspaper and Books Ordinance, banning all anti-colonial media outlets.

This repressive legislative move was later relaxed to accommodate limited political discourse among Africans and Indians. Although anti-colonial media faced various challenges, it nonetheless continued to thrive. As Wa'Njogu (2004, p. 59) notes "the early African and Indian papers were popular: they challenged unpopular authority, restored the confidence of Africans in their capacity for political action, and gave African and Indian journalist a taste of their power to influence their destiny." Publications such as *Uhuru, Sauti ya Mwafrika, Mfanyakazi, East African News Review, Comment, Kenya Weekly News, Independent Comment* and *News Comment* were in circulation in 1950s. This was more than 30 years after the launching of the British African Broadcasting Limited relaying of radio broadcasts in English in 1928. This monopoly of the broadcasting language persisted until 1953, when a

30-mimute Kiswahili service was incorporated in the programming. The introduction of programs in Kiswahili, Arabic and other vernacular languages was part of a colonial propaganda initiative to insure obedience amongst the colonized, especially during the Mau Mau revolution.

POST-INDEPENDENT KENYAN MEDIASCAPE: AN OVERVIEW

In the 1960s – an era when most African countries gained independence – governments maintained a monopolistic control over the media. Even when there were no legalities preventing private media ownership, very few private citizens could compete against the government-owned outlet. When Kenya gained independence in 1963, there was only one state-run radio and TV station, Voice of Kenya. These two outlets served as mouth pieces of the Kenyan government and the then ruling political party, Kenya African National Union (KANU). This was during the reigns of presidents Jomo Kenyatta (1963- 1978) and never dissipated during Daniel arap Moi's era (1978- 2002). In early 1990s, there was intense pressure to accommodate political plurality. There is extensive literature detailing the dark days of Moi's dictatorship and how it suffocated development and manacled democracy (See Article 19's Censorship Reports, Wachanga, 2011, Ibelema, 2009, Wachanga, 2007, Ochieng, 1993, Wa'Njogu, 2004 etc).

When Kenya became a multi party democracy in 1991, media outlets burgeoned; there was a rise in media ownership, and diversification in both the target audience and content. By the time Moi was leaving power in 2002; Kenyan media was gaining vibrancy in terms of operations and programming, and was enjoying relatively less government interference.[4] Today, all major newspapers are accessible online; TV and Radio stations have digitized their programming; media outlets have converged and are meeting technological challenges of the 21st century, and programming has been expanded to meet diverging needs of diverse audiences – different gender, young and old, different social and economic classes, those at home and in Diaspora.

But even as we toast for these successes, it is important to heed Njogu's (2009, p. xi) caution on how communication among Africa urban elites is used as a generalization about the African mediascape. These elites, Njogu warns, are used as the norm, ignoring "the nature of the majority of the varied and ever-changing social groups and networks in both urban and rural Africa." Nonetheless, the vibrancy in the mediascape has played a key role in Kenya's developmental agenda by pushing for the expansion of the existing democratic space, enabling economic advancement through competitive commercial advertising, and by serving as a site for civic education. However, ideologically, like in most other African countries, Kenya's media have been assumed to be dedicated to forging a national identity. At the same time, political leaders have continued to rally their citizenry around a kind of ethnic, though masked as national, consciousness. And since these political leaders have also emerged as comprador bourgeoisies, their statuses are erroneously perceived as a representative of the hegemonic order. We can gain a better understanding by listening to Aseka (2009, p. 89).

During the first wave of independence movements in Africa, idioms of identity were encoded in nationalist constructs that were held to be integrally linked to notions of self-determination. Yet, these notions lacked any expression of ontological transcendence that could lead to the growth of supra-ethnic African identities. Expressed through the various nationalist media, idioms of identity became the basis for organizing and challenging the participatory experiences of those who contested the colonial order. As such, idioms of identity became the basis of the invention of the imagined political communities." Herein lays an intellectual lacuna in discussing ethnicity and its concomitant impact on media, and national development (89).

Other challenges include pressing global economic demands and raucous economic class disparities, which heavily determines the way in which the media carry out their duty. It is not surprising, for instance, that even the proliferation of FM radio stations and unprecedented use of cellular phones' short messaging service (SMS) in Kenya has not detribalized the country. As a matter of fact, these stations were heavily blamed for peddling ethnically predatory messages that antagonized communities during the 2007/2008 general elections, just as ethnically demeaning text messages were circulated, arguably fanning the post-election violence (Wachanga, 2011).[5] Associating FM stations with antagonistic ethnic sentiments was not without precedent. KASS FM was closed temporarily by the government in November 2005 during the run-up to a political plebiscite. During his tenure, President Daniel Arap Moi had threatened to close Kameme FM, which broadcasts in Gikuyu. While FM stations were hugely blamed, Wa-Mungai (2010, p. 220) attributes this vulnerability to the radio's high penetration rate in Kenya. He quips:

While the charges leveled against especially vernacular FM radio may be true up to a point, it is necessary to consider that radio broadcasts intersect in a complex ways with a whole range of other popular modes of communication – politicians' rhetoric in village baraza (public meetings), traditional and contemporary folklore, TV broadcasts and newspaper reports. Therefore, attempts at singling out for blame any of these modes of expression ignore various agentic subtleties (for instance ethnicity, kinship, economic status, religion and gender) that determine media consumption behavior.

While several indigenous radio stations – serving different ethnic communities – are in most cases owned by one proprietor (e.g. Royal Media), they are run to meet the needs of the target community even when those needs are couched in messages that marginalize and demonize other communities. Media owners ideologically belong to an ethnic community defined by economics as opposed to other ethnic indicators such as language. There are incidents where stations are not considered to authentically belong to a certain community, not because of the language used to broadcast, but because of the ethnic background of the proprietor. In Kenya, a renowned politician categorized Ramogi FM (which broadcasts in Dholuo) as a non-Luo radio station because its owner is not from the Luo community. According to this politician, a radio station authentically belongs to a community not solely on the language and content of its broadcasts. Ownership, an economic component in this politico-ethnic equation, becomes vital in determining means of knowledge generation. Therefore, an increase in media channels and emerging media technologies has not necessarily facilitated detribalization neither has it facilitated contestation for the disruption or alteration of the existing ideological terrain. However, it serves to mediate the changing political order, identity and citizenship in terms of alternate and self-centered discursive and communicative regime that NCTs inevitably help to midwife. We will use the Kenyan 2007/2008 election violence to further illustrate this point.

KENYAN ELECTION VIOLENCE: TECHNO-ETHNIC WARFARE

When the contentious 2007 presidential election results were announced in Kenya, the country degenerated into a socio-political abyss; a near genocide that claimed more than 1,133 lives and condemned more than 350,000 into medieval-like refugee camps. During this crisis, the government banned live television broadcasts, an act that led to an unprecedented use of new communication technologies (NCTs) by the citizens to receive and disseminate information.

Kenyan bloggers were on the rise, using Twitter to post items via Short Messaging Services (SMS). Others used the Yahoo-operated Flickr to upload pictures via the GPRS modem; cellular phones to capture videos that were later uploaded to YouTube; still others chronicled the events through live blogging 2.0 –CoverItLive; while others linked violence-hit areas to Google Earth and Google Maps. These digital devices provided alternative source of information just as they churned propagandistic and predatory messages that allegedly fanned ethnic passions and hatred.

With Kenya fast catching up with the new communication and information technologies, its public sphere was punctuated with an unprecedented plurality of voices. Yet, the context of this technological and polyphonic sphere needs to be understood within its economic-social-political landscape.

A BRIEF OVERVIEW OF THE 2007 KENYAN ELECTORAL VIOLENCE

When Kenya went to the polls in 2007, the country was more ethnically stratified than ever before in its political history. Two political forces were competing for power: the Orange Democratic Party (ODM) and the Party of National Unity (PNU). ODM was predominantly a coalition of two major ethnic communities: Luo and Kalenjin. Kikuyu, the most populous ethnic community in Kenya, predominantly identified with PNU. Other smaller communities aligned to one of these two parties. During the campaign period, both political parties mobilized their ethnic supporters and demonized their opponents. The content of campaign message was largely as ethnic as it was political. What followed was heightened propaganda that facilitated the construction of narratives of ethnic antagonism. When the presidential results were announced in favor of PNU, Kenya degenerated into an unprecedented period of political and ethnic violence leaving the country on a precipice of disaster. According to the Commission of Inquiry into Post-Election Violence (CIPEV, 2008), 1,133 people were killed and 350,000 internally displaced during this crisis. According CIPEV (2008, p. 306)

Many (witnesses who appeared before the Commission) recalled with horror, fear, and disgust the negative and inflammatory role of vernacular radio stations in their testimony and statements to the Commission. In particular, they singled out KASS FM as having contributed to a climate of hate, negative ethnicity, and having incited violence in the Rift Valley region. However, there were also similar complaints in other parts of the country. These included the vernacular music and negative ethnicity allegedly coming from Kikuyu FM stations including Kameme, Inooro, Coro, and others in other different parts of the country.

Indigenous radio stations were chastised for broadcasting vitriolic narratives that fanned ethnic antagonism. Given the political tension at the time, the context of deployment of broadcast messages framed their meaning and exacerbated their potency to rally listeners to acts of violence. Besides the indigenous radio stations, the mainstream media – television, radio and newspapers – took sides and gave prominence to news stories with total disregard to journalistic ethical and professional standards. "We became subjective, forgetting our professional tenet of objectivity and neutrality. In fact, this polarization was so bad in the newsrooms that some broadcast journalists refused to cover or read news that wasn't favorable to the candidate or party they supported,"[6] according to one Kenyan journalist. As violence escalated and the media took the blame, the Kenyan government banned live television broadcasts in December 2007.

But as the government banned live television broadcasts, the country was already witnessing an unprecedented use of NCTs and an emergence of citizenry dissemination of political, sometime acerbic, messages. The Kenyan digital landscape was abuzz with information activities.

During the Kenyan PEV, cellular phone subscribers started sending and/or receiving text messages whose content was ethnically antagonistic. Some of these messages were couched in political language that targeted each of the two dominant political camps- ODM and PNU. The ethno-political hostility at the time had rendered ODM synonymous with the Luo, Kalenjin and Raila Odinga,ODM presidential candidate who is Luo. PNU had become synonymous with Kikuyu and Mwai Kibaki, PNU presidential candidate who is a Kikuyu. Messages propagated and disseminated via SMS were at times simultaneously distributed via e-mail. Here is one such message in favor of ODM and Odinga:

He is bold, he is a team player, he is a liberator, and he is a panafricanist, a democrat with a national constituency. He is Raila Odinga and he has what it takes to deliver Kenyans to the Promised Land. Join other Kenyans in supporting this great son of Africa. Send this sms to 10 people.

The circumstances surrounding the propagation and dissemination of these text messages made it difficult to authenticate their originality because, in circulation, were several corrupted versions of similar messages. Here is a case in point where the above example was corrupted to demonize Odinga:

He is cold, he is a time player, he is a terminator, a destroyer, a demolist with an odd national constituency. He is Raila Odinga and he hasn't what it takes to deliver Kenyans to the Promised Land. Join other Kenyans in supporting this great son of Africa. Send this sms to 10 people.[7]

Some of the SMS in circulation accused the mainstream media of partisanship not only in their alleged biased news coverage, but also for facilitating belligerent activities. The Nation Media Group, one of the leading media groups in Kenya, was accused of sympathizing with PNU. There were SMS in circulation that accused this Media Group of providing vehicles to transport weaponry in war torn areas. The allegations were so far-reaching and deeply damaging that the Group issued a press release to clear its name.

The Nation Media Group's attention has been drawn to Short Text Messages (SMS) and e-mails linking the group and its executives to the political violence going on in Rift Valley Province and other parts of the country. We would like to inform Kenyans that there is no truth whatsoever in the messages, which are being circulated through mobile phones and the Internet. Nation Media Group is a law-abiding and responsible corporate citizen with a commitment to serving the public. We share the agony and distress of our fellow Kenyans in this difficult period and support every effort being made to resolve the national crisis. (http://politics.nationmedia.com/Blog/nation~speaks~blog.asp?Display=172&so=&srt=&stp=8)

The Kenyan government also added its voice by pleading with the citizens to stop spreading inflammatory and provocative text messages. This was an unprecedented move by the government to communicate to its citizenry via SMS. Initially, the government wanted to block the messaging service, but was advised otherwise by Michael Joseph, Safaricom, mobile phone service provider, director. Some bloggers have, however, claimed that "Bulk sms has been blocked by the government to prevent guys from sending inciteful (sic) messages." (http://www.kenyanpundit.com/2008/01/01/post-media-blackout-update-jan-1-430-pm/)

The government was also torn between the provocative and instructive roles the SMS were playing as exhibited in Bitange Ndemo's, permanent secretary in the ministry of information and communications, dilemma.

With more than 10 million SMSs sent every day, and an audience of nine million phone owners -- and growing fast -- the potential impact of SMS to incite violence in Kenya is huge. Sources reveal that pressure mounted on Dr Ndemo to block messages after banning live radio. But he didn't. "There were so many stranded people in the forests who were sending SMSs to their relatives," he said. "So, had we shut it down, we would have caused more damage than what we had intended to prevent." http://allafrica.com March 1, 2008

Besides SMS, the 2007 Kenyan pre- and post-election information warfare saw the rise of bloggers in an unprecedented way. In fact, bloggers started gaining prominence in Kenya in 2003. One of the earliest Kenyan blogs, Mentalacrobatics, for instance, had links to other different blogs on its site, establishing a Web-ring, which facilitated navigation to related Web-logs.

When the government banned live television broadcasts, it was Mentacrobatics the blogger who was among the first to blow the whistle while requesting other bloggers to tone down ethnic passions in their posts. (http://www.kenyanpundit.com)

All live broadcasts have been suspended by the government. The order was released as ODM was addressing their press conference. This is now officially a police state...I've had to moderate comments for now and will only allow comments that are sending information about what's going on the ground - too busy to focus on controlling the emotions of people on both sides. (Posted December 30, 2007).

Other bloggers, such as Juliana Rotich, chronicled the horror that visited the displaced, the pillage in local towns (e.g. Eldoret), and closure of transport system including local airports. Besides her blog posts, Rotich used the SMS to update her site, used Twitter even before it became popular, and uploaded her photos using the Yahoo-operated Flickr using the GPRS modem. While Rotich was compiling these reports from Eldoret, Mentalcrobatics, was chronicling events as they unfolded in Nairobi. This site had Flickr-linked images of opposition members of parliaments engaging in battles with the police in the streets of Nairobi, deserted streets of often-vibrant downtown Nairobi and pictures of displayed newspapers headlining their plea for political sanity.

After the election crisis, the Kenya Human Rights Commission (KHRC),[8] a non-governmental and non-profit human rights watchdog, put together a report on the root cause of the crisis. All of the cellular phone text messages used in this section were obtained from the material gathered by KHRC. For the purposes of this chapter, 110 text messages were analyzed. While most of the analyzed text messages are in English (some are ungrammatical and/or in shorthand English), others are in Kiswahili, Sheng' (a mixture of local languages and/or Kiswahili, and /or English), and Kikuyu - this researcher's first language.

Most of the circulated cellular phone messages were framed within the political slogans of ODM and PNU, and in ethnic stereotypes associated with communities allied to each of the political party. Other messages either glorified or demonized the two presidential candidates - Kibaki and Odinga. While some messages made implicit reference to their target party, candidate or ethnicity, others were outright vitriolic and contained personal invectives. Sample the following messages:

He is cold, he is a time player, he is a terminator, a destroyer, a demolist with an odd national constituency. He is Raila Odinga and he hasn't what it takes to deliver Kenyans to the Promised Land. Join other Kenyans in stopping this great son of Africa. Today the people have spoken. Send this sms to 10 people.

Some of the messages in vernacular were burlesque and alluded to ethnic stereotypes. In other cases, there was a poor fusion of ethnic and Biblical allusion. Here is one such example in Kikuyu:

Nake Ngai akiira Ibrahim atiri, 'amba uthii uruithie muruguo, ucoke umurehe ndimurathime"! Kwoguo kihii gitingiathana Kenya!!Pamoja tu-angamize kihii! (And God spoke to Abraham "go and circumcise your son, then bring him forth for my blessings!" Therefore, a boy (uncircumcised) cannot lead Kenya!! Together, let us do away with the boy)

The reference to circumcision results from its significance as a *rite de passage* among the Kikuyu. Circumcision is a significant rite of passage among the Kikuyu. It marks one's graduation from boyhood to manhood. Traditionally, one is bequeathed the secrets of the community and considered an adult. There is a belief among the Kikuyu that a circumcised man can never reveal a secret. One can be trusted with vital ethnic information and decisions after the rite. The Luo do not undergo the cut as a cultural rite. Among the Kikuyu, the Luo are therefore boys (*ihii*) who cannot be trusted. Among the Kikuyu, if one confides a secret in a boy, one is only relieved after the boy goes through the rite because he can now be trusted. Luo cannot, therefore, serve as a cognitive authority to the Kikuyu; especially given the significance of trust in matters pertaining to political leadership. The term *ihii* (boys) may be understood within the Gramscian hegemonic discourse, where language is used to construct reality. Such technologically mediated messages only helped to reproduce the social and political relations that exist between the Kikuyu and Luo in real life.

We used discourse analysis to decode the meaning of the text messages. The linguistic components (words and/or phrases), used to construct the cell phone messages, were examined as part of the discourse analysis process. The politics of ethnicity and ethnicity of politics in Kenya was considered in making sense of the content of these messages. The drive to participate in this information warfare is examined against Gramsci's notion of hegemony and Althusser's ideology. Ideology, as an organized thought, thrives on strategically selected ways of thinking by those who wield societal power. These ways of thinking are powerfully designed in systemic patterns that they become fundamentally the only perceivable avenues for subaltern to understand the world, the only way to make sense of their lives. As Lull (2002) notes, these patterns become "familiar and accepted because they are delivered to us systematically via the mass media, and are further circulates in the personal conversations we have with our families, co-workers, teachers, fellow students, neighbors, email correspondents, chat-room partners, and others."

The gathered messages were thematically placed in five broad categories.

- Construction of "the other"/ small world
- Sexuality, power and humor
- Call for peace
- Caricature and allusion
- Call to arms/ framing besiegement

We sought to gain an in-depth insight of the content of disseminated message by understanding the dispensers of these vitriolic messages within the ambit of the interpellated. Our analysis pointed how the senders of text messages confirmed their membership into identifiable collectives - ethnic, gender or class-which bestows upon them countless imaginary dividends. It does not matter whether this is true or not, neither does it matter that it is insensible; what matters is that the senders suspended without knowing it, their true relations to the reality and donned the imaginary relations to imaginary reality; a reality of the collective dominating others.

The Kenyan 2007 information warfare provides an insight in to the design of its dominant ideology around a political-social-ethnic axis, which the PEV revolved. The cell phone text messages analyzed in this section shed light on how the dominant ideological hegemony strategically gains and seeks to uphold credibility through structures that are "strengthened by an interlocking system of information-distributing agencies and taken-for-granted communication practices that permeate every corner of social and cultural reality" (Lull, 2002, p. 50).

In these messages, there is an embedded message of ethnic incompetencies of "the other." This raises the need to understand how citizenship and ethnic membership are negotiated in multiethnic and multicultural societies such as Kenya, especially when each ethnicity maintains its righteous claim to power and control. It is, therefore, plausible to argue that any attribution of normative to the vernacular and ethnically inclined violent tendencies must contend with the admission that what was witnessed in Kenya is not so much the actual circumstance of operationalized hegemony, but rather the enforcement of unipolar domination in the name of "nationalism" as a vehicle for achieving hegemony by a section that may even rely on repressive apparatus and other statutory instruments to enforce compliance.

CONCLUSION

This discussion has departed from the widely circulated argument that centers on the causally binary distinction of how media technologies affect culture/society to one that centers on media technology as an intricacy of the dominant ideology. We have not sought to provide a response to whether changes in media technologies and information and communication environments affect audiences' internal political knowledge, consequently impacting democratic processes, as parochially postulated by modernist theorists. We also note that there is a fundamental shift from 19th and 20th century media – cell phones use media channels as if they were utilities – essentially no content from the cell phone company, essentially all content from the very users – while there remains a strong top-down component in the media, bottom-up is now significant. This 21st century communicative change and discursive alteration continues to redefine geo-political influence which up till now reify national ideology.

REFERENCES

Alemneh, D., & Hastings, S. (2006). Developing the ICT infrastructure for Africa: Overview of barriers to harnessing the full power of the Internet. *Journal of Education for Library and Information Science*, *47*(1), 26–38. doi:10.2307/40324333

Althusser, L. (1971). Ideology and ideological state apparatuses. In *Lenin and Philosophy and other essays*. London, UK: New Left Books.

Appadurai, A. (1993). Disjuncture and difference. In During, S. (Ed.), *The cultural studies reader*. London, UK & New York, NY: Routledge.

Aseka, E. (2009). Media consumerism and cultural transformation. In Njogu, K., & Middleton, J. (Eds.), *Media and identity in Africa* (pp. 19–35). Edinburg University Press for the International African Institute.

Bomer, B. (2007). How information shapes political institutions. In D. Graber (Ed.), *Media power in politics*. (pp. 8- 18). Washington, DC: CQPress.

Bourgault, L. (1995). *The mass media in sub-Saharan Africa*. Bloomington/ Indianapolis, IN: Indiana University Press.

Carroll, W., & Hackett, R. (2006). Democratic media activism through the lens of social movement theory. *Media Culture & Society*, *28*(1), 83–104. doi:10.1177/0163443706059289

Clifford, J. (1992). Travelling cultures. In Grossberg, L., Nelson, C., & Treichler, P. (Eds.), *Cultural studies*. London, UK & New York, NY: Routledge.

Cook, S. (2004). New technologies and language change: Toward an anthropology of linguistic frontiers. *Annual Review of Anthropology*, *33*, 103–115. doi:10.1146/annurev.anthro.33.070203.143921

Crystal, D. (2001). *Language and the Internet*. Cambridge, UK: Cambridge University Press.

Diane, C., & Meier, P. (2009). *New technologies in emergencies and conflicts: The role of information and social networks*. Washington, DC & London, UK: UN Foundation-Vodafone Foundation Partnership.

Gramsci, A. (1971). *Selections from prison notebooks*. London, UK: Lawrence & Wishart.

Grossberg, L. (2006). *Media making: Mass media in a popular culture*. Thousand Oaks, CA: Sage.

Gumucio, D., & Tufte, T. (Eds.). (2006). *Communication for social change anthology*. New York, NY: Communication for Social Change Consortium.

Hall, S. (1982). The re-discovery of ideology. In Bennett, J. C., & Woollcott, J. (Eds.), *M. Gurevitch, T* (pp. 56–90). London, New York: Methuen.

Hyden, G. (Ed.). (2003). Media and democracy in Africa, Uppsalla: NordiskaAfrikainstitutet

Hyden, G., Leslie, M., & Ogundimu, F. (2002). *Media and democracy in Africa*. New Brunswick, NJ: Transaction Publisher.

Ibelema, M., & Bosch, T. (2009). Sub-saharan Africa. In De Beer, A., & Merrill, J. (Eds.), *Global journalism: Topical issues and media systems* (pp. 293–336).

Karatzogianni, A. (2006). *The politics of cyberconflict*. Routledge research in information technology and society.

Koloko, B., Nakamura, L., & Rodman, G. (2000). *Race in cyberspace*. New York, NY: Routledge.

Lull, J. (2002). *Media, communication, culture: A global approach*. New York, NY: Columbia University Press.

Njogu, K. (Ed.). (2009). *Media and identity in Africa*. Edinburg University Press for the International African Institute.

Nyamnjoh, F. (2005). *Africa's media, democracy and the politic of belonging*. London: Zed Books.

O'Shaughnessy, M., & Stadler, J. (2008). *Media and society*. Oxford University Press.

Ochieng, P. (1992). *I accuse the press: An insider's view of the media and politics in Africa*. Initiative Publishers.

Shepperson, A., & Tomaselli, K. (2009). Media in Africa: Political, cultural and theoretical trajectories in the global environment. *International Communication Gazette*, *71*(6), 473–489. doi:10.1177/1748048509339790

Volosinov, V. N. (1973). *Marxism and the philosophy of language*. New York, NY: Seminal Press.

Wa'Njogu, K. (2004). Language and multiparty democracy in a multiethnic Kenya. *Africa Today*, *3*(50), 55–73.

Wachanga, D. N. (2007). *Sanctioned and controlled message propagation in a restrictive information environment: The small world of clandestine radio broadcasting*. Unpublished doctoral dissertation. University of North Texas.

Wachanga, D. N. (2011). 'Kenya's indigenous radio stations and their use of metaphors in the 2007 election violence'. *Journal of African Media Studies*, *3*(1), 109–125.

Waki, P. (2008). *Report by the Commission of Inquiry into Post-Election Violence* (CIPEV). Wa-Mungai, M. (2010). "Soft power," popular culture and the 2007 elections. In Kanyinga, K., & Okello, D. (Eds.), *Tensions and Reversals in Democratic Transitions: The Kenya 2007 General Elections*. Kenya: Nairobi. Institute for Development Studies, University of Nairobi.

Zaller, J. (1991). *The nature and origins of mass opinion*. Cambridge University Press.

Zeleza, P. T. (2009). The media in social development in contemporary Africa. In Njogu, K., & Middleton, J. (Eds.), *Media and identity in Africa* (pp. 19–35). Edinburg University Press for the International African Institute.

WEBSITES

http://allafrica.com/stories/200802291070.html

http://politics.nationmedia.com/Blog/nation~speaks~blog.asp?Display=172&so=&srt=&stp=8

http://www.kenyanpundit.com/2007/12/30/media-blackout-announced/

http://www.kenyanpundit.com/2008/01/01/post-media-blackout-update-jan-1-430-pm/

http://www.khrc.or.ke/

www.cia.gov/index.html

www.internetworldstats.com

www.internetworldstats.com/stats1.htm

ENDNOTES

1 Most liberation movements in Africa realized the superseding significance of transmitting facilities – either clandestinely or granted airtime by stations of countries sympathetic to the cause. Research on the role played by clandestine media in Africa liberation struggle is lacking. Also see Wachanga, 2007

2 Janus was a Roman god who guarded the doors – both the entry and the exit – of Forum Romanum, the idyllic center for the Roman People. Janus was therefore portrayed as double-faced, looking back and forward, the beginning and the end. Most African political regimes assume a Janusian figure by being on both sides of the communication battle in a political setting.

3 Examples abound - from Nnamdi Azike of Nigerian, Kwame Nkrumah of Ghana, Jomo Kenyatta of Kenya, Robert Mugabe of Zimbabwe. They all ran media outlets when fighting colonialists. All rose to head their respective governments.

4 I am not unmindful that on May 2005 the Kenyan first lady, Lucy Kibaki, burst into the Nation Media Group offices and slapped a cameraman. Equally dramatic but more depressing was the 2006 attack of The East African Standard offices, the oldest Kenyan newspaper, and the Kenya Television Network (KTN). The government was accused of the raids.

5 On December 15, 2010, the International Criminal Court (ICC) mentioned six Kenyans suspected of committing crime against humanity during the 2007/2008 election crises in Kenya. One of the suspects, Joshua Arap Sang, a radio broadcaster, hosts a call-in program on Kass FM called "Lene Emet" (How is the country). According to ICC prosecutor, Luis Moreno-Ocampo, Sang used coded language disseminated through radio broadcasts to help coordinate ethnic attacks. Summonses to appear before ICC pre-chamber trial judges were March 2011.

6 http://www.ipsnews.net/news.asp?idnews=41049 A workshop titled "30 Days in Words and Pictures: Media Response

in Kenya During the Election Crisis," was organized by Internews in January 2008 to engage media professionals in a discussion on the role that was played by the local media during the post-election violence

7 These messages have not been edited for grammar, spelling or punctuation

8 More information about the KHRC's report on PEV, visit its Web site http://www.khrc.or.ke/

Chapter 2
Identity and the New Communication Technologies: Evidence from Kenya

Nathan Oyori Ogechi
Moi University, Kenya

Emily Bosire-Ogechi
Moi University, Kenya

ABSTRACT

This chapter discusses the use of short text messages on cell phones and e-mail conversations among social networks to negotiate identities in Kenya. It shows that these New Communication Technologies have revolutionised the way of doing things in Kenya, including politics. It comes out that Old Communication Technologies are being complemented by the new technologies which are now taking over and which can be used to negotiate both narrow ethnic-blocks and national identities. The chapter shows that the new technology is not always used to portray and perpetuate narrow ethnic identities; rather, instances exist when the users want to project professional and national identities. Supported with data that focus on the pre- and post-2007 elections, the chapter argues that inasmuch as the New Communication Technologies are good, great caution should be exercised when using them since unchecked use might bring untold suffering to society.

1. INTRODUCTION

The present chapter discusses the use of New Communication Technologies (NCTs) in negotiating and perpetuating various identities in Kenya. New Communication Technologies include computers, mobile phones, facsimile and the Internet. The NCTs are distinguished from Old Communication Technologies (OCTs) that comprise radio, television and newspapers which are also referred to as the old media (Danmole 2010).

DOI: 10.4018/978-1-60960-591-9.ch002

The paper focuses on the Internet and mobile phones (cell phones) for communication. Jensen (2001, p.98) has described the Internet as a product of the ICTs while Crystal (2006, p.3) views the Internet as "an association of computer networks with common standards which enable messages to be sent from any registered computer (or *host*) on one network to any host or any other". The Internet developed experimentally in the United States of America in the 1960s as a network which grew fast to include the military, federal, regional, university, business and personal users and is now the world's largest network (ibid). The Internet provides many services and enables many people to be in touch through electronic mail (e-mail), chart groups, virtual worlds, World Wide Web, instant messaging, blogging, etc. It is notable that more than a quarter of the world's population was using the Internet in 2009 whereby over a quarter of the world's population or 1.9 billion people had access to a computer at home (ITU, 2009). On the other hand, according to ITU (2009), the cell phone is the most rapidly adopted technology in history. It is the most popular and widespread personal technology in the world with an estimated 4.6 billion mobile broadband subscribers globally as of 2009 having overtaken the fixed broadband subscribers in 2008.

As already indicated, the present chapter focuses on the role of the Internet and the cell phone in negotiating identities. The New Communication Technologies vide mobile telephony and Internet services are presently widely available in Kenya so that sending short text messages (SMS) and e-mails is fashionable. The prevalence of mobile phones is a product of the liberalization and privatization of the telecommunications sector since the late 1990s when competition in the mobile telephone market came into being (Chepkonga 2002). Subsequently, mobile broadband subscribers increased from 15,000 in 1999 to 17.4 million by mid 2009 (RoK, 2009) through the four authorised mobile operators, namely Safaricom, Zain, Econet Wireless, and Telkom. During the same period, the number of cyber cafes also increased. Thus surfing the Internet to exchange e-mails and also to read and make comments on social networks appears to be widely in use among those who are computer literate. Internet has also gradually moved away from cyber cafes to mobile phones which have advanced facilities that allow one to surf the World Wide Web. Portable modems are also sold by mobile telecom companies and are ensuring that 'last mile' connectivity is achieved (Ratemo 2010).

As we discuss the role of NCTs, it should be noted that the NCTs are channels of communicating messages of different kinds including messages on individual or group identities using language which includes lexical choices. It is the lexical choices that the users make vide the NCTs that carry the message(s) which project(s) the identity(ies) of the sender(s)-receiver(s). Falling from this, the present chapter's argument is that identities in Kenya, just like elsewhere in the world, are dynamic; that is, they are negotiated and the process of negotiation is reflected in the way that language is used. Therefore, one can identify various identities such as those reflecting peer-groups, professionals and members of different ethnicities and political parties especially in line with claims of Tajfel's (1978) social identity theory. Closely related to Gumperz's (1982) categorization of 'we" and 'they' codes, the social identity theory distinguishes between in-group and out-group members whereby members of either group positively evaluate themselves vis-à-vis the other. Furthermore, in the context of the present chapter, the groups can be private individual(s), i.e. members of particular ethnic groups or impersonal bodies such as political parties, professional groups, etc. Thus the language uses done through the NCTs can be analyzed to yield the various identities some of which we have already mentioned. This chapter particularly seeks to unravel whether or not messages sent through the NCTs reflect identity(ies) that are deeply-seated and ones that usually start from ethnic ideological hegemony.

There are 42 or so ethnic communities in Kenya and it is possible to conservatively talk about over 40 ethnic identities. Furthermore, since the introduction of plural democracy in the country in the early 1990s, ethnicity seems to have entrenched itself in the country's politics. In this case, political parties are usually formed, financed and sustained by drawing their main support from certain regional turfs on the basis of ethnic identification. Whereas the pattern of political parties' formation has almost always been influenced by ethnicity, the trend seems to have been enhanced with the advent of New Communication Technologies. Against this backdrop, we interrogate and demonstrate how the short text messages through cell phones, on the one hand, and the group or social networks on the Internet, on the other hand, are used to negotiate and perpetuate the identities.

Therefore, we ask: What is identity in Kenya? Can one talk of a national identity? Do members of various ethnic groups in Kenya identify themselves as individual ethnic groups? Are there times and occasions when these people assume a national identity? How do they do it? Apart from ethnic identity, what other identities are seen in Kenya? Subsequently, is it possible to talk of multiple identities in Kenya? Do these identities manifest themselves in political life and political affiliations of Kenyans? How are these identities communicated amongst group members? What is it that New Communication Technologies have added to the communication of identities? What kinds of messages are sent and what identities do they convey? To what extent do the messages conveyed through the New Communication Technologies help to reinforce ethnicity or nationalism in Kenya? We attempt to answer the foregoing questions in a bid to show how mobile phones and the Internet have been used for negotiating identities in Kenya. We also assess whether New Communication Technologies are always used to communicate messages that help to reinforce patterns that are ethnically compliant or whether the messages also do the opposite of this.

The chapter is organised as follows. In section 2, we review views from the existing literature on New Communication Technologies, ethnicity and identity. Section 3 sketches the theoretical framework on which the arguments are based. This is followed in 4 by a description of the methodology we used to gather the data and analyses them. Finally, we offer our concluding remarks in 5.

2. BACKGROUND

As indicated in the foregoing section, two major categories of communication technologies exist, namely the Old Communication Technologies (OCTs) and the New Communication Technologies (NCTs). The distinction is based on the advances made in the field over the years and there appears to be fears that accompany the advent of any new communication technology (Crystal, 2006). For instance, the invention of printing in the 15th century was resisted by the Church that saw it as a satanic invention while the arrival of the telegraph was deemed a destroyer of the family institution and a promoter of crime. Later on, the invention of the telephone was seen as likely to undermine society while broadcasting was feared as a likely voice of propaganda.

In spite of these, the fears were overcome by the technological advancements. For example, the radio, though an OCT, has proved useful in generally disseminating information for development (e.g. Mwansoko, 1995; Makunja, 1995; Saleh & Kasisa, 1995; Rukiramakuba, 1995; Mpiranya, 1995). But it is also true that language use in radio has been instrumental in spreading and even "killing" the grammar of Kiswahili in Kenya (Mohochi, 2001) through incorrect use of the language. Mohochi (ibid) further contends that the preponderance of linguistic errors in both print and electronic media fails to not only promote Kiswahili but also passes wrong messages. Therefore, since the various types of old media are crucial in communicating to a wide audience, it is imperative that the media carefully consider their choice of language. Concerning identity, Mohochi claims that errors arising from mother tongue influence, especially on radio and TV, are grave since they expose the anchors' ethnic identities. To avoid this ethnic "betrayal" during the broadcasting, Said (2007) cautions media proprietors against using only paper certificates to hire anchors while ignoring their practical oral use of language. To him, it is possible to conceal ethnic identities in the media if careful and rigorous measures are put in place during the hiring process. The danger of exposing and foregrounding ethnic identity can also be reflected if possession of good translation skills is not emphasized during the hiring process (Bitugi, 2007). Unless one is well grounded in communication and translation skills, there is a danger of one's cultural background seeping through the translation that s/he does in a media house given that translation is a routine practice in the media industry.

Elsewhere, it has been shown that radio is an effective "weapon" in the fight against the enemies of development, namely, disease, ignorance and poverty in Kenya since it is cheap to purchase, service and use even in the most remote corners of the country (Ogechi, in press). In addition, it requires little or no literacy at all to operate. The radio has been used for edutainment in the communication of messages on family planning and the HIV/Aids pandemic in Kenya since the 1990s (Njogu, 2001). The FM radio stations are especially effective when they broadcast content on local issues and in local languages that listeners identify with (Kembo-Sure & Ogechi, 2006) as seen through the case of *Mang'eleta FM Radio* in Makueni District of Eastern Province which broadcasts from Makueni town and serves a small radius. Such local FM radio stations play a critical role not only in development issues but also in passing peace and reconciliation messages (Kawoya & Makokha, 2009; Ogechi & Sang, in press). Although the history of FM radio stations in Kenya only goes back to the freeing of airwaves in the early 1990s (Musau, 2000; Orao 2009), Kawoya & Makokha (2009, p. 22) sum up the profound effect of these stations as follows:

Today, FM stations, whether privately or state-sponsored, command large audiences with their customized programmes on news, entertainment, consumerism and especially political commentary. They have given voice to the masses in two ways. Firstly, in the spirit of free-market economic principle of demand and supply, the people have a greater say in the format, timing and even language of broadcasting. Failure to adhere to consumer tastes normally renders a radio station marginal and unpopular in the competitive field. Secondly, popular call-in interactive programmes such as the "People's Parliament" and the "ebimeeza" (interactive roundtable/call-in radio discussions) ensure greater engagement of the masses in topical and political issues affecting their localities and countries.

Besides radio, newspapers and magazines as forms of OCTs have been exploited in passing general messages of development and particularly messages on HIV/AIDs. Beck (2006), for example, explored the efficacy of HIV/AIDS communication or prevention campaigns through the Kiswahili comic in *Kingo* magazine and a UNICEF-ESARO sponsored comic *Sara* initiative in Kiswahili. The choice of Kiswahili medium was found to be very effective in the campaigns. Similarly, regional newspapers printed in Kiswahili under the sponsorship of the UNESCO were particularly useful in enhancing rural development when they existed in the 1980s (Luganda 2001). Indeed, newspapers have been central in identity and Kenyan politics in the way they choose their words in order to portray a liking and support or otherwise to politicians or political parties (Matu 2009, p.38). Van Dijk (1997) cited in Matu (2009, p. 38) has argued that lexical items may be selected not only because of the official criteria of decorum, but also because they effectively emphasize or de-emphasize political attitudes and opinions, garner support, manipulate public opinion, manufacture consent or legitimate political power. This lexical choice was seen in the Kenyan newspapers in the run-up to the 1997 general elections where individual politicians and political parties were disparagingly reported in the press. Skewed lexical choice was also used to enhance negative ethnicity and cause mayhem as happened in the run up to and after the disputed presidential elections of 2007.

It is also notable that at present the OCTs do not solely operate as they did in the past; rather there is convergence of the OCTs and NCTs. The call-in interactive programmes in radios mentioned by Kawoya & Makokha (ibid) reflect the convergence; in addition, journalists now make use of the Internet in their daily work (Reglitz, 2009). This development has both positive and negative ramifications (Danmole, 2010). Positively, the NCTs have assisted people to learn more about issues they are not knowledgeable in while negatively they have made it possible for people to read newspapers online or send text messages that could aggravate conflict as it happened among the Muslims in Nigeria who supported Ayatollah Khomeini's *fatwa* of a death sentence on Salman Rushdie following his publication of a book that was deemed offensive to Islam (Danmole, Ibid).

The NCTs and especially the Internet's distributed nature and the relatively low cost of digital production has made it a site of multiplicity and diversity since the mid-1990s (Wei & Kolko 2005, p.200). Many people spend many minutes and hours on the Internet and even do so as the first thing in the morning. The *Daily Nation* Tuesday June 1, 2010 reported of a Kenyan fiction writer who received news of her having been short listed for the 2010 Caine Prize for African Writing thus:

There was this e-mail from her, on a Friday morning late April, with the subject: "some good news from Wasafiri". I opened it while still in bed. My eyes blinked and my mind went into a spin. I remember thinking – this is one reason, Lily, why you should never ever log into your e-mail before taking a shower and drinking two mugs of strong tea to shepherd your brains into the real world.

In addition to e-mail usage craze, the computers and cell phones are now widely used by social networks such as face book and twitter. No wonder Hintereder (2009, p.3) in an editorial has said:

.... shame on you. After all everyone in this network and app page is always on the move and always online – day and night, of course. The brave new media world has also pretty unashamedly thrown the established theories of communication studies abroad, radically expanded old transmitter-receiver patterns, created a new and boundless cosmos of communicative freedom, brought to life new channels, new content and new formats.

Besides Internet on computers and mobile telephones, computer mediated communication extends to mobile phones in the form of short text services (Ondimu, 2006) (SMS) that cell phone owners send to each other with a profound effect on language choice and use. Posteguillo (2003) pioneered linguistic studies of SMS through his Netlinguistics Analytical Framework (NAF) whereby he focused on language analysis as used in the Internet. He concurred with Crystal (2006) that SMS is affecting language use and forms whereby increasingly new forms of spelling and shorthand-like characters are used. As further revealed in Ondimu's (2006) inquiry, the new linguistic forms used in the text messages, can be viewed as a new language variety complete with own lexicon, grammar, discourse and usage conditions. Indeed no single code is used in the SMS texts she studied. Instead a mixture of surface morphemes from different stable and unstable codes was seen. Elsewhere, however, single codes have been used on SMS as a New

Communication Technology for seeking and getting information in the provision of health, business and agricultural services to grassroots people in Kenya (Bosire 2010) in order to achieve enhanced development and alleviate poverty. Such SMS texts are auto responses to information seekers from the said departments. To the best of our knowledge, no studies have explored NCTs and the negotiation of identities. In addition, we have not come across works on whether or not the NCTs have been used to negotiate identities to perpetuate ethnic hegemonies, issues that this chapter examines.

Though we have already alluded to the term "identity", we now wish to discuss it in some depth. The term "identity" can be defined variously depending upon one's professional or academic background. In general, however, identity has something to do with the way people are categorized (identified) either by themselves and/or by others along a variety of dimensions, namely, race, nationality, ethnicity, culture, sex and gender (Ogechi, 2008, p. 76). Of all cultural elements, language is a powerful instrument of identity (and belonging). Kramsch (2001, p. 65) argues that:

It is widely believed that there is a natural connection between the language spoken by members of a social group and that group's identity. By their accent, their vocabulary, their discourse patterns, speakers identify themselves and are identified as members of this or that speech and discourse community.

But identity is not as easily defined vide language as such. It can be complex in heterogeneous linguistic communities and hybrid identities do exist (Maganda, 2006). Moreover, identities are not static but dynamic since they can be negotiated even when language is used to mark identity. For instance, while Gumperz (1982) talked of two identities in multilingual communities namely, "we" and "they" code identities, Kamwangamalu (1998) extended this to argue for a "code-in-between". The essence here was to provide for identities that are neutral even when a 'we-code' is for in-group solidarity while a 'they-code' is for distinguishing the out-group members. Therefore, the case of Kenya that is treated in this chapter is bound to present a complex scenario of identities – there exist 42 or so languages with English as the official language while Kiswahili is the national and co-official language. Heavy codeswitching involving these languages and the mother tongues including Sheng[1] do exist. This is not only seen in interaction on politics but also among professionals and peers vide New Communication Technologies. To understand how political discourse could display different ethnic identities, we discuss the term "ethnicity" and its role in the formation of political parties in Kenya.

The terms "ethnicity" and "ethnic group" are variously interpreted depending on one's discipline or political leaning. According to Jackson (1984), in Becher & Basedau (2008, p.6), ethnicity can be viewed as "embracing all kinds of socially attributed identities such as clan, community, ethnicity, faith (religion), gender, language, regional provenance, race, sect or tribe, and the like, all of which are the result of self-ascription by others but are usually inherited by birth and cannot easily be hanged by an act of individual will". This view is slightly different from sociolinguists who claim that ethnicity is "a set of descent based on cultural identifiers used to associate persons to social groupings that expand and contract in inverse relation to the scale of inclusiveness of membership" (Cohen, 1978, p.387). In spite of the little differences between the various disciplines, in this chapter, an ethnic group is distinguished from another by use of a common language, shared culture and descent from a common ancestry.

Based on language as an ethnic distinctive feature, Kenya is said to have 42 or so ethnic groups (Ntabo & Ogechi, 2009) divided into Bantu, Nilotic and Cushitic groups. Numerically, Bantus comprising Gusii, Gikuyu, Swahili, Kamba etc, are 60% followed by Nilotes (e.g.

Kalenjin, Luo and Maasai) and Cushites (such as Somali and Rendile) in that order. Formation of political parties has always been influenced by the coalition of different but like-minded ethnic groups (Elischer, 2008). As an illustration, in the run up to the national referendum on the proposed constitution in 2005, two groups emerged leading to the formation of the major competing political parties in the 2007 elections. There was the so-called Orange Group that opposed the proposed constitution. To a large extent, this group gravitated around the Luo, Kalenjin and Luhya. The group that supported the constitution namely, the Banana Group, was largely supported by the Gikuyu, Meru and Somali. Members of the other ethnic groups spread out their support in these two camps. Statements made by either group's politicians during the campaigns for the 2005 referendum used belittling, demeaning or derogatory terms towards the ethnic groups in the opposite side of the campaigns:

Context: Statement by a member of parliament supporting YES in relation to those in the NO camp.

1. They hate Kikuyus because we are hardworking. Luos just go fishing and fish is free and thereafter they ask the government for relief maize to make ugali. (KNCHR & KHRC 2006: 39)

Context: Statement by a member of parliament opposed to the then proposed constitution in relation those in the YES camp.

2. If YES wins, the Kikuyus should pack their bags and move out of Eldama Ravine. Kikuyus from Shauri, Maji Mazuri and Timboroa will not be issued with Title Deeds if they vote YES (KNCHR & KHRC 2006: 41)

These statements indicate that the relationship that was building up then was based on ethnicity. After the national referendum (November 2005), the two groups were transformed into political parties. The Orange Group was initially registered as the Orange Democratic Party of Kenya (ODM-K). However, about three months to the election date, it split into two, that is, ODM-K and Orange Democratic Movement (ODM). During the split, some members of the Orange Group (largely those in KANU[2]) joined the Banana Group that became the Party of National Unity (PNU). This group was associated with the then government, which started off as a NARC government on 30th December 2002. The three groups went into the national elections with distinct slogans that identified them and were widely used in their campaign messages. While PNU's slogan was *Kazi Iendelee!* (let work continue, i.e. let the existing government continue ruling), ODM-K waved greetings in the air while shoutng *Wiper!*. The ODM's slogan was *Kazi Ianze!* (let work begin!). These slogans were a major identity marker and in most cases they were associated with the party leaders and, by extension and remotely, the ethnic groups of the party leaders. Note that the identification was not that of individual ethnic groups but a collective of ethnicities associated with the group and later associated with a political party.

This chapter analyses SMS texts and internet messages to show how NCTs have been useful in enhancing collective identities. The data will illustrate political identities and political party stereotypes, on the one hand, and national identity, peace and reconciliation on the other. The e-mails analyzed reflect Kenyans identifying themselves as practitioners of a specific profession regardless of their ethnicity. Before we present our data, we wish to describe the theoretical framework on which the data were analysed.

3. SOCIAL IDENTITY THEORY

The data were analyzed on the basis of the social identity theory that is a construct of social psychology initially propounded by Tajfel (1978) and subsequently refined by his associates' studies. Two of the theory's tenets, namely categorization processes (Tajfel, 1978) and the ultimate attribution error (Pettigrew, 1979) are crucial to this chapter. On one hand, Tajfel (1978) claims that there exist categorization processes whereby individuals distinguish between in-group (read same ethnic or professional group) and out-group (read another ethnic or professional group) members. On the other hand, Turner (1987) argues that group behaviour is the behaviour of individuals acting on the basis of categorizing self and others as a social group. That is, behaviour of individuals within a group could reflect or represent the group. A central claim here is that individuals strive to achieve and/or maintain a positive social identity. In this case, there is negative evaluation of members of the out-group contrary to the positive evaluation of the in-group by members of the in-group. Negative evaluation results when there is emphasis and foregrounding of the differences between the in-group and out-group. For instance, by members of an ethnic group (in-group) stressing ethnic stereotypes about members of other ethnic groups (out-group), negative statements are likely to occur.

On the contrary, Pettigrew (1979 in Vrij, Van Schie & Cherryman, 1996), also a proponent of the social identity theory, asserts that the ultimate attribution error could yield both positive and negative assessment. The error refers to people's tendency to attribute (a) the positive behaviour of out-group members and the negative behaviour of in-group members to situational factors (external attribution) and (b) to attribute the negative behaviour of out-group members and the positive behaviour of in-group members to the member's personal dispositions (internal attribution). This argument could work well among members of a shared academic discipline and profession regardless of their ethnicity. That is, members view colleagues in the same profession who hail from another ethnic group positively and not negatively. They view them as comrades and not outsiders. The categorization into in-group-out-group alongside the attribution error was applied in the analysis of data for identity negotiation vide the SMS and e-mail messages in this chapter.

4. THE DATA AND ANALYSES

4.1 The Data

The data were collected between January and February 2010. The SMS comprised those that were sent out during campaigns in the run up to the 2007 general elections and soon after the eruption of post election violence in early 2008. The origin of the campaign SMS texts was anonymous but they could reach peoples' cell phones and circulate among people who knew each other. The circulation could be taken as part of the campaign or as a way of getting amusement from the messages. We requested our research assistant, a fourth year student at the university where we work, to ask her fellow students to forward to her cell phone the political SMS texts they received in the run up to the 2007 elections. Where they had deleted the texts, she requested the students to reconstruct them both in terms of content and even wording as much as they could recall. Other SMS texts were obtained from a colleague at Moi University who hosted a peace programme on *Biblia Husema Studios Radio* when the post-election violence was subsiding, whereby listeners were requested to send text messages to express their feelings about the post-election violence and what they felt should be the way forward.

As for the e-mail texts, we used data sent to our mail inboxes from networks of our professions. In particular, we used those messages circulating among former information science students at Moi University (one of this chapter's authors is an in-group member of the Infoscience Yahoogroup).

Before we collected the SMS, we assured our respondents that the data will be used strictly for academic purposes and that their identities will be concealed. With this assurance, the students volunteered to give the SMS or even reconstruct the SMS messages they had received during the 2007 election campaigns. We gave similar assurance to our email correspondents before we analyzed the data that we received.

The data were not always in one code; rather they were more often designed in a mixed code involving English, Kiswahili and Sheng lexemes. In the analysis, English lexemes are presented in normal font, Kiswahili words are in **bold** while Sheng words are in ***bold italics*** face. We tried to contextually situate the examples before subjecting them to the claims of the social identity theory upon which the analyses are premised.

4.2 Portrayal of Ethnicity and Stereotypes

A number of the SMS messages sent out during the 2007 election campaigns reflected negative ethnicity and skewed stereotypes of the supposed supporters and members of other parties while emphasizing solidarity among the in-group. As indicated earlier, it was very easy and possible to send out texts due to new developments concerning this New Communication Technology. The essence of the text messages sent out was to negatively portray out-group members (those of the rival political party) while glorifying the in-group members:

Context: The practice of conducting campaigns by different political parties.

3. While ODM men are already out of bed campaigning PNU men are still telling their women "*panua! kazi iendelee*" as ODM-K women are telling their men "*niko* wet *nipatie* wiper".

'While ODM politicians leave bed early to go out campaigning, PNU male politicians keep asking their female compatriots for more sex as do ODM-K male politicians whose ready-for-sex women ask for clothing to wipe their wet genitals'.

Context: Warning about the ODM presidential candidate.

4. Remember ODM means One Dangerous Man; One Deceased Man.

The foregoing two examples illustrate a revolutionised 2007 election in Kenya where New Communication Technologies were in use during election campaigns. There is no gainsaying the fact that the technology had the advantage of reaching many people at the same time compared to the use of public address systems at campaign rallies or using the Old Communication Technology through radio. In addition, unlike the old technology where it is not possible for a message to be retained for further consumption and possible spread at a later time to a potential voter, the SMS could be stored in the phone and forwarded to another potential voter's cell phone. However, the messages create long lasting negative evaluation of the rivals. Example (3) was most likely crafted by ODM supporters. As indicated earlier, the ethnic bedrock of ODM comprised the Luo, Kalenjin and Luhya while PNU's ethnic turf was the Kikuyu, Meru and Embu. ODM-K's ethnic following was largely Kamba. Therefore, (3) seems to be using the New Communication Technology to, first, negotiate ODM identity and solidarity and, secondly, negatively evaluate the ethnic groups that supported PNU and ODM-K as lazy sex maniacs who care little about achieving political power. They are portrayed as being leisure lovers compared to their ODM counterparts who are early risers and hardworking go-getters.

It should not be construed that it is only ODM that spewed negativities about their rivals. As (4) shows, even ODM's rivals were equally negative about ODM. While the acronym ODM refers to the Orange Democratic Movement, the supporters of ODM's rivals sought to portray the leader of ODM negatively as a dangerous person that should not be trusted with national leadership. They even misinterpreted the acronym to portray him as a sickly person knowing fully well that the constitution of Kenya does not allow a person that is medically unfit to lead. If stretched further, one can say that the negative portrayal of the ODM leader was not just meant to refer to him as an individual; rather his ODM-leaning ethnic groups were being portrayed as dangerous and sickly to lead. Such an SMS was meant to enhance existing stereotypes about the behaviour of ethnic groups by their out-group members.

The SMS texts analysed so far are those that were sent out during campaigns. It is clear that their focus was in negotiating identities of competing political parties and, by extension, either perceived positive or negative attributes of their leaders or the stereotypes attributed to the political party leader's ethnic group. Besides such SMS texts, there were those that were sent out as the post-election violence raged on in early 2008. These too were sent out by the two major opposing political groups and some of them were replete with veiled ethnic vitriol as seen in (5) and (6).

Context: Counting of votes and release of presidential results

5. When ODM votes were trickling in, Peter Marangi was seen painting State House orange but as soon as PNU votes started coming in Marangi was called to undo the mess.

Context: Reaction of some voters to the release of presidential results

6. PNU *ponyaka na ugali***,** ODM *okoa dunia maisha ianse*[3]*na* Pentagon; K***** *ni mwizi ashindwe*!!!

'PNU has run away with maize bread, ODM should save the world and life starts afresh with the Pentagon since K***** is a thief'.

The SMS text in (5) refers to the state of things as tallying and announcement of presidential votes was being done. Initially, the ODM presidential candidate was seen to have scored a win since votes from his ethnic bedrock were announced giving him a handsome lead. However, tables turned against him as votes from the PNU ethnic strongholds trickled in. It is also instructive to note that during the electioneering period a television advert by a paint manufacturing company was running using a model with a pseudonym, Peter Marangi, who came on to television screens painting walls. Thus in (5), by Peter Marangi painting the presidential residence (state house) orange (the official colour of the ODM) it signified an initial ODM "win". This win was however a "mess", according to the SMS writer, since it was nullified with the arrival of winning votes for the PNU candidate. The SMS must have been crafted by the PNU supporters to taunt and laugh at their ODM rivals for prematurely celebrating yet their candidate had not won an election.

What is pleasing is that the PNU SMS was merely taunting and not overtly hulling negative ethnicity contrary to (6). The SMS must have been sent out by an enraged ODM supporter who thought the ODM candidate had been robbed off his victory by PNU - PNU *ponyaka na ugali* (PNU has run away with food). S/he was pleading for good life to start afresh with the ODM team in power as seen through the statement - ODM *okoa dunia maisha ianse na* Pentagon (ODM save the situation and let life start afresh). The ODM Pentagon comprised of five ODM luminaries from different ethnic groups that spearheaded the national presidential campaign for the ODM leader. Example (6) illustrates the SMS writer's hurt feelings and expectations so that s/he even exaggerates her negative attribution of the PNU whose presidential candidate s/he calls a thief (*mwizi*) probably implying that the candidate rigged himself into office.

To demonstrate that identity is dynamic and is negotiated from time, there were SMS texts that struck a reconciliatory tone. (7) was sent out after the post-election violence when many internally displaced persons (IDP) were living in make shift camps:

Context: A male writer talks of his predicament after the outbreak of post-poll violence.

7. *Watoto wetu na wamama manaumia juu ya ODM na PNU na si' 'nanyeshewa na hata hakuna nyu'ba tena tunaishi kwa hema na makaratasi*[4].

'Our children and women are suffering because of ODM and PNU. They are being rained on and we have no houses since we live in tents and paper thatched houses'.

This is one of those examples that were collected from one of the presenters of a peace and reconciliation radio programme on *Biblia Husema Studios*. As mentioned earlier, the presenters invited people in the radio's North Rift catchment area to send SMS texts to express their feelings and what they thought was the solution to the crisis. Example 7 seems to be a disappointment from the internally displaced persons (IDPs). You could have expected him to blame the crisis on the leaders of a rival political party and the ethnic group to which that party's leader belonged. However, here we see a person that is striking a middle ground and blames both political parties for occasioning a crisis through their competition. With such a message, it is not possible to argue that new communication technologies are always used to enhance ethnic hegemonic identities. Rather, it appears that they can be used when talking about political issues to negotiate a semblance of a national identity.

Negotiating Profession-Based Identity

Groups of professionals who have studied the same discipline have also been seen to use New Communication Technologies to negotiate and sustain their identity regardless of their ethnic backgrounds. While the foregoing section showed how the cell phone is useful, this section draws from the Internet to demonstrate profession-based identity. With the advent of the Internet, different groups of people have established networks through which they keep touch on research developments and queries in their disciplines. For example, scholars who study codeswitching interact through the codeswitching group network. Similarly scholars whose studies were funded by a certain foundation usually keep in touch through an electronic network as do the German trained scholars funded by the Deutscher Akademischer Austausch Dienst (DAAD – the German Academic Exchange Service). Heavy use of the Internet is due to a changed environment concerning the New Communication Technologies in Kenya. It is notable that until the end of the 1990s, Internet services were uncommon and a preserve of a minority elite-computer-literate group in urban areas while cell phones were only for the high level state operatives plus their business collaborators and wheeler-dealers. The state had a monopoly over the provision of telecommunication services including phones and the Internet. The few Internet cafes that existed then were only in urban centres with power supply; Internet surfing was slow and very far out of the reach of the common man. However, from the late 1990s the state was forced to privatise the provision of mobile telephone services and with it followed a preponderance of cheap e-mail services.

In our study, we obtained data whereby former students of the School of Information Sciences at Moi University keep in touch through their Infosciences@yahoogroups.com. Through this New Communication Technology network, the subscribing graduates whose ethnic background is multifarious exchange views on many issues (current affairs, their new exploits, reminisce on their past without any hard feelings etc.). Intra-group distinctiveness is at times maintained through their use of a code that is only known to them. They do not necessarily have to harp on and perpetuate negative ethnicity as (8) shows. The excerpt comes from a hot issue involving a jazz maestro who had purportedly been running what the state called a cult like church. He had in custody at his luxurious residence a large number of his church members including a popular television anchor who quit her job to manage the maestro's music recording business. In the conversation excerpted here, five contributors from different ethnic groups gave their opinion on the maestro's presidential ambitions:

Context: A contribution on the Infoscience yahoogroup network about a cult like scandal involving a jazz maestro.

8. T1. I have watched half of this Gentleman's interview at K24 yesterday. Isn't he a sharp witted orator? *Jaluo Jeuri* and extremely bright! Or is he using this to cover ills we can't decipher with a naked eye? What are your opinions *Wangu watu*? Just methinks

(*Jaluo jeuri* – a crafty Luo, *Wangu watu* – my people)

T2. This guy has hypnotic power.. no vanity discernible from his speech at all. I think he's the real deal. ** just got part 1 though link slow ***mbaya!*** (terribly slow Internet link)

3. The real deal you say!!! On Presidential Campaign: "I am just about to sign a massive contract with PepsiCola........"I might just end up funding this thing single handedly" BUT DO I SAY!

He is hard hitting, words flow out of him with bewitching coherence. The problem is how to seamlessly fuse God and Kenyan Politics! He should speak of the latter more. If he took the latter and dedicated the stamina and wit he is exhibiting here, he could sweep the Kenyan young generation by surprise, The R***** of Kenya will wonder what hit them!! But he was right, this A***** Scandal has brought him out and he won't go back!!!

4. He has to decide what he wants to do. Is it jazz, gospel or politics? He has the political gait, but if he has to succeed, he needs to drop religion, jazz is no problem, in fact it might help if he can sing a little before addressing the campaign meetings. 2012, we are set for real drama!
5. Yes Mhe M******, That is exactly what I thought of him. Drop religion or just push it to his Runda Palace and lock it there, then paint the town red with Politics of 'Change is coming' sweetened with Jazzy melodies. I was even tempted to think of him debating the R**** of Kenya on issues of the day..... he will show them dust. And he was very clever when he talked of R**** and K*****.....mentioning the Govt's failure but not out-rightly condemning in the usual insult-laden version of Kenyan politics. Religion will be his toughest hurdle! Watch this Jazz Maestro!!

In T1, an interlocutor introduces the topic as s/he asks colleagues to give their views on the jazz maestro who had been caught in a scandal of keeping his members largely young girls in his house. He deflects the people's attention from the scandal to his purported presidential ambitions saying he will run in the next general elections scheduled for 2012. The writer of T1 starts by giving his opinion saying this is a wonderful orator of Luo origin.

T2's response ups the interest in the discussion as s/he too concurs with T1 although s/he says that s/he was not able to watch the maestro's entire talk. The contributor in T3 jokingly, it appears, offers to bank roll the maestro's campaign. S/he then concurs that the maestro is a strong candidate and could easily appeal to the youth and give the existing politicians a run for their money. This is what T4 takes up but cautions that the maestro has to choose between, religion, jazz and politics. In his opinion, he is more suited to politics than any of his other callings. He advises that the aspiring candidates could use his musical talent as an entertainment preamble during his campaigns. T5's contribution is an affirmation of the foregoing writers' views. While admiring the maestro, s/he argues that the maestro could use the theme of "change" and make scathing attacks on the current politicians. He says that, in his speech, the maestro's veiled political attack on the politicians is a pointer to his razor sharp intelligence that should be exploited.

On balance the infoscience yahoogroup network discourse analyzed here is about politics. However, it does not negotiate myopic political identity that is ethnic-based. Instead it is broader and is based on politics of generational identity. Though we have concealed the contributors' names (ethnic markers), they appear to be from diverse ethnic backgrounds yet in their contribution they castigate the older generation of politicians and see a generational change of politics in the maestro. This and other discussions by this network show that the group uses the New Communication Technology through the Internet to discuss diverse issues. The group also censures itself since it does not degenerate to identifying with their ethnic groups. Even when ethnicity is mentioned as seen in T1, the subsequent turns' writers ignore it and focus on the broader issue. By so doing, the network comes out as one with a mature appreciation of the significance of New Communication Technology in enhancing communication for meaningful development.

5. CONCLUSION

The present chapter sought to discuss how the cell phone text messages and social group networks as New Communication Technologies are used to negotiate identities in Kenya. This was done so as to illustrate whether or not the new technologies are being used to perpetuate hegemonic and narrow ethnic identities against the backdrop of the country's rampant ethnic-based political parties. On this basis, the chapter argued that identity is dynamic and not static. We contended that identity is negotiated from time to time and perceptions could change within a very short time span.

By analysing SMS texts in the run up to and after the disputed 2007 presidential elections, it has been shown that using the social identity theory, there is some evidence of the SMS texts portraying negotiation of blocks of ethnic identities. In particular, the data showed that before the elections ethnic identity and stereotypes of perceived political rivals were rampant. However, the outcome of the hyped up strong ethnicity feelings and identities drastically made the SMS writers to rethink and start renegotiating a national identity. This was clear when the writers started viewing themselves as innocent masses subjected to suffering through myopically following the political party leaders. The high noon of the broad national identity negotiation is seen in the use of New Communication Technologies by members of a profession through their networks. Though hailing from diverse ethnic backgrounds, the contributors to this network displayed mature use of the new technology.

On balance, the chapter has shown the pros and cons of identity negotiation vide the New Communication Technologies in Kenya. It has come out that though technological development is meant to ease communication and speed up development, it is possible to plunge society into untold suffering if not carefully utilized. It is true that access to information is a basic human right and a major prerequisite for development. For this reason, the Kenya government efforts to regulate the media and communication sectors in the last two years met strong opposition. But if unregulated and haphazard use of New Communication Technologies could plunge the young developing country into untold suffering, then there is need to support the move to regulate the use of New Communication Technology gadgets through registering one's cell phone line with the service providers by supplying one's bio data. This could guard against money laundering, terrorism, kidnapping and killing. After all, the regulation is already in place even in the developed democracies with the ability to counter terrorist attacks and kidnapping Kenya, which lacks the capacity to counter sophisticated terrorist attacks, cannot afford to expose itself to such a danger in the name of safeguarding the right to laissez-faire freedom even when the dangers of the same are known as seen when a self-confessed kidnapper-cum-serial killer was recently arrested.

ABBREVIATIONS USED IN THE CHAPTER

IDP: Internally Displaced Person
KANU: Kenya African National Union
KHRC: Kenya Human Rights Commission
KNCHR: Kenya National Commission on Human Rights
NAF: Netlinguistics Analytical Framework
NARC: National Rainbow Coalition
NTC: New Communication Technologies
OTC: Old Communication Technologies
RoK: Republic of Kenya
SMS: short message service

REFERENCES

Becher, A. & Basedau. M. (2008). *Promoting peace and democracy through party regulation? Ethnic party bans in Africa*. German Institute of Global and Area Studies (GIGA) Working Papers No. 66.

Beck, R. M. (2006). Popular media for HIV/AIDs prevention? Comparing two comics; *Kingo* and the *Sara* communication initiative. *The Journal of Modern African Studies*, *44*(4), 513–541.

Bitugi, E. M. (2007). Taaluma ya tafsiri katika vyombo vya habari nchini Kenya; matatizo, changamoto na mustakabali wake. In K. Njogu (Ed.), *Kiswahili na Elimu Nchini Kenya.* (pp. 158-170). Nairobi, Kenya: Twaweza Communications.

Bosire, E. K. (2010). *Access and use of Information Communication Technologies (ICTs) for rural poverty reduction in Uasin Gishu, Kenya.* (Unpublished PhD thesis). Edoret, Kenya: Moi University.

Chepkonga, S. (2002). *Helping Kenya extend access to ICTs to rural areas (Acacia II). Communications Commission of Kenya (CCK)*. Nairobi, Kenya: Kenya Bureau of Statistics.

Cohen, R. (1978). Ethnicity; problem and focus in anthropology. *Annual Review of Anthropology*, *7*, 379–403. doi:10.1146/annurev.an.07.100178.002115

Crystal, D. (2006). *Language and the Internet.* Cambridge, UK: Cambridge University Press. doi:10.1017/CBO9780511487002

Danmole, H. O. (2010). From the global to the local: The media and Islam in Nigeria, 1979-2007. In Adebayo, A. G., Adesina, O. C., & Olaniyi, R. O. (Eds.), *Marginality and crisis: Globalization and identity in Contemporary Africa* (pp. 89–196). New York, NY: Lexington Books.

Elischer, S. (2008). *Ethnic coalitions of convenience and commitment: Political parties and party systems in Kenya*. German Institute of Global and Area Studies (GIGA) Working Papers No. 68.

Gumperz, J. J. (1982). *Discourse strategies.* London, UK: Cambridge University Press. doi:10.1017/CBO9780511611834

Hintereder, P. (2009). Media moments. In *Magazin-deutschland; The New World of Media, 3.*

ITU. (2009). *Manual for measuring ICT access and use by household and individuals*. Geneva, Switzerland: ITU Telecom World.

Jensen, M. (2001). Afribox, telecenters, cybercafés: ICT in Africa. *ICT in Africa, 1*, 96–109.

Kamwangamalu, N. (1998). 'We-code', 'they-codes' and 'codes-in-between': Identities of English and codeswitching in post-apartheid South Africa. *Multilingua, 17*(2/3), 277–296. doi:10.1515/mult.1998.17.2-3.277

Kawoya, V., & Makokha, J. S. (2009). The case for Kiswahili as a regional broadcasting language in East Africa. *The Journal of Pan African Studies*, *2*(8), 11–35.

Kembo-Sure, & Ogechi, N. O. (2006). Language planning and language reform in Kenya. In Kembo-Sure, S. Waitiki & N. O. Ogechi (Eds.) *Language planning for development in Africa. (*pp. 37 – 54). Eldoret, Kenya: Moi University Press.

KNCHR & KHRC. (2006). *Behaving badly: Deception, chauvinism and waste during the referendum campaigns*. Nairobi, Kenya: KNCHR & KHRC.

Kramsch, C. (2001). *Language and culture*. Oxford, UK: Oxford University Press.

Luganda, M. M. (2001). Magazeti ya Kiswahili na maendeleo nchini Kenya. In Shitemi, N. L. (Eds.), *Kiswahili: A tool for development- a multidisciplinary approach* (pp. 197–205). Eldoret, Kenya: Moi University Press.

Maganda, F. (2006). A hybrid identity among the Sukuma in the Tanzanian Africa Inland Mission. *Southern African Review of Education (SARE) with Education with Production (EWP), 12*(1), 31-43.

Makunja, G. S. (1995). Sarufi na msamiati wa Kiswahili katika vyombo vya habari. In Mlacha, S. A. K. (Ed.), *Kiswahili na Vyombo vya Habari* (pp. 14–24). Dar es salaam, Tanzania: TUKI.

Matu, P. M. (2009). Newspaper reporting of the run-up to the 1997 Kenyan general election. *Journal of Intra-African Studies*, *1*, 37–53.

Mlacha, S. A. K. (Ed.). (1995). *Kiswahili na Vyombo vya Habari*. Dar es salaam, Tanzania: TUKI.

Mohochi, S. (2000). Kiswahili na vyombo vya habari nchini: Matatizo katika matumizi ya lugha. In Njogu, K. (Eds.), *Kiswahili katika Karne ya Ishirini na Moja (Kiswahili in the 21st Century)* (pp. 147–157). Cape Town, South Africa: Centre for Advanced Study of African Society.

Mpiranga, F. (1995). Kiswahili katika vyombo vya habari Rwanda. In Mlacha, S. A. K. (Ed.), *Kiswahili na Vyombo vya Habari* (pp. 93–123). Dar es salaam, Tanzania: TUKI.

Mukhongo, L. L. (2009). The Kenyan media: Confronting state terror and atrocities. *Journal of Communication and Language Arts. Special Issues on Media and Democracy*, *3*, 26–43.

Musau, P. M. (2001). Freeing the airwaves or cultural enslavement: The case of media in Kenya. In Shitemi, N. L. (Eds.), *Kiswahili: A tool for development- a multidisciplinary approach* (pp. 36–42). Eldoret, Kenya: Moi University Press.

Mwansoko, H. J. M. (1995). Dhima ya vyombo vya habari katika ukuzaji, usanifishaji na uenezaji wa istilahi. In Mlacha, S. A. K. (Ed.), *Kiswahili na Vyombo vya Habari* (pp. 1–13). Dar es salaam, Tanzania: TUKI.

Njogu, K. (2001). Discourse and serial drama in the promotion of health in Kenya. In Shitemi, N. L. (Eds.), *Kiswahili: A tool for development- a multidisciplinary approach* (pp. 181–187). Eldoret, Kenya: Moi University Press.

Ntabo, M. M., & Ogechi, N. O. (2009). *Siege politics: The underlying problem in the Kenyan crisis*. Paper presented at the Moi University-Indiana University Purdue University Indianapolis conference at Moi University, June 2009.

Ogechi, N. O. (2007). Building bridges through trichotomous youth identities in Kenya: Evidence from code-choice. In Njogu, K., & Oluoch-Olunya, G. (Eds.), *Cultural production and social change in Kenya: Building bridges* (pp. 129–150). Nairobi, Kenya: Twaweza Communications.

Ogechi, N. O. (2008). Sheng as a youth identity marker: Reality or misconception. In Njogu, K. (Ed.), *Culture, performance and identity: Paths of communication in Kenya* (pp. 75–92). Nairobi, Kenya: Twaweza Communications.

Ogechi, N. O. (in press). Linguistic human rights for development communication Kenya. *Frankfurter Afrikanistische Blaetter, 19.*

Ogechi, N. O., & Sang, J. K. (in press). Nonformal peace education through codeswitching and Sheng in FM radio for development in Kenya. *Africa Peace and Conflict Journal.*

Ondimu, J. K. (2006). *Linguistic analysis of computer-mediated text messages.* (Unpublished MPhil thesis). Eldoret, Kenya: Moi University.

Orao, J. (2009). The Kenyan indigenous languages and the mass media: Challenges and opportunities. *Stellenbosch Papers in Linguistics PLUS, 38*, 77–86.

Posteguillo, S. (2003). *Netlinguistics: Language, fiscourse and ideology in Internet.* Jaume, UK: Universitas Jaume.

Ratemo, J. (2010). Digital villages bring government to rural areas. Retrieved on June 16, 2010, from http://www.standardmedia.co.ke

Reglitz, E.(2009). Households online. *Magazine-Deutschland: The New World of Media, 48.*

RoK. (2009). *Economic survey.* Nairobi, Kenya: Government Printer.

Rukiramakuba, E. (1995). Mwanga na maendeleo ya Kiswahili, Rwanda. In Mlacha, S. A. K. (Ed.), *Kiswahili na Vyombo vya Habari* (pp. 83–92). Dar es salaam, Tanzania: TUKI.

Said, V. M. (2007). Matumizi ya Kiswahili nchini Kenya katika redio na runinga. In Njogu, K. (Ed.), *Kiswahili na Elimu Nchini Kenya* (pp. 20–27). Nairobi, Kenya: Twaweza Communications.

Saleh, N., & Kasisa, A. (1995). Usambazaji wa Kiswahili, Zaire. In Mlacha, S. A. K. (Ed.), *Kiswahili na Vyombo vya Habari* (pp. 63–71). Dar es salaam, Tanzania: TUKI.

Shitemi, N. L. (Eds.). (2001). *Kiswahili: A tool for development- a multidisciplinary approach.* Eldoret, Kenya: Moi University Press.

Tajfel, H. (1978). Social categorization, social identity, and social comparison. In H.

Tajfel, H., & Turner, J. C. (1986). The social identity theory of intergroup behaviour. In Worchel, S., & Austin, G. (Eds.), *Psychology of intergroup relations.* Chicago, IL: Nelson-Hall.

Tajfel (Ed.). *Differentiation between social groups: Studies in the social psychology of intergroup relations*, (61-76). New York, NY: Academic Press.

Turner, J. C. (1987). *Rediscovering the social groups: A self-categorization theory.* New York, NY: Basil Blackwell Inc.

Vrij, A., Van Schie, E., & Cherryman, J. (1996). Reducing ethnic prejudice through public communication programs: A social-psychological perspective. *The Journal of Psychology*, 130.

Wei, C. Y., & Kolko, B. (2005). Resistance to language and Internet diffusion patterns in Uzbekistan. *New Review of Hypermedia and Multimedia, 11*(2), 205–220. doi:10.1080/13614560500402817

ENDNOTES

1 An unstable code spoken in Kenya, previously attributed to the youth.

2 KANU refers to Kenya African National Union Party. This is the party that ruled Kenya from independence in 1963 until December 2002 when it was defeated by the National Rainbow Coalition (NARC). Some members of the ODM-K were influential politicians during the hay days of KANU.

3 This is ***ianze*** (it starts) in Standard Kiswahili. But the writer has a mother tongue influence that is reflected also in writing.

4 The statement is a reflection of mother tongue influence on Kiswahili.

Chapter 3
New Media in Kenya:
Putting Ethnicity in Perspective

Martin C. Njoroge
Kenyatta University, Kenya

Purity Kimani
Kenyatta University, Kenya

Bernard J. Kikech
Kenyatta University, Kenya

ABSTRACT

The way the media processes, frames, and passes on information either to the government or to the people affects the function of the political system. This chapter discusses the interaction between new media and ethnicity in Kenya, Africa. The chapter investigates ways in which the new media reinforced issues relating to ethnicity prior to Kenya's 2007 presidential election. In demonstrating the nexus between new media and ethnicity, the chapter argues that the upsurge of ethnic animosity was chiefly instigated by new media's influence. Prior to the election, politicians had mobilized their supporters along ethnic lines, and created a tinderbox situation. Thus, there is need for the new media in Kenya to help the citizens to redefine the status of ethnic relationships through the recognition of ethnic differences and the re-discovery of equitable ways to accommodate them; after all, there is more strength than weaknesses in these differences.

1. INTRODUCTION

The recent shift from traditional media to new media may not, on its own account, cause instant ripples in the media industry. However, to ignore this as a normal takeover would be to misread what is in no doubt a move indicative of more fundamental shifts in media. Digital technology such as Mobile phones and the Internet have simply not been around long enough in Kenya, Africa. However, incidents like the 2007-2008 crises in Kenya provide a flash of insight into the emerging power of these tools. It also puts into sharp focus the power new media technologies give citizens of developing nations to report news and organize responses to crisis situations. While Kenyan

DOI: 10.4018/978-1-60960-591-9.ch003

journalists and community organizers have a great deal to be proud of in their response to an electoral crisis and the concomitant ethnic violence, new media was also used both by the government and civilians to amplify tensions and coordinate violent attacks (see Goldstein and Rotich, 2008). Significantly, the ease in communication and dissemination of information provided by these digitally networked technologies is vital as digital tools can help promote transparency. Research findings reveal that new media commands 48 percent of the audience share dwarfing the old media (Synovate, 2009 cited in Ogola, 2009).

The term 'new media' is used in this chapter to mean the emergence of digital, computerized or networked information and communication technologies in the latter part of the 20th century. New media often have the characteristic of being manipulatable, networkable, dense, compressible and interactive.

The chapter identifies the specific ways in which new media contributed to negative ethnicity and particular emphasis is given to the role of the social media such as the internet and mobile computing which enables what Bayne (2008) refers to as 'different kinds of communication' than their analog antecedents. Bayne (2008: 43) writes, 'the first element is the shift from a hub-and-spoke architecture with unidirectional links to the end points in the mass media, to distributed architecture with multidirectional connections among all nodes in the networked information environment. The second is the practical elimination of communications costs as a barrier to speaking across associational boundaries.' In other words, digital technologies are tools that, in addition to allowing communication in the traditional one-to-one fashion, also allow us to become our own broadcasters and reach large numbers of people in unprecedented ways at trivial cost. Viewed through this lens, actual activities of the new media and ethnicity prior, during and after the election violence witnessed in Kenya are alluded to. Among the questions investigated are: what was the impact of digital technology on ethnicity prior to, during and after the 2007/2008 post election violence in the Kenyan context? To what extent did the new media contribute to fanning ethnicity in Kenya? What was the impact of this digital technology on ethnicity? The chapter also gives a theoretical and conceptual treatment of new media and ethnicity. The chapter rests on the premise that while digital tools can help promote and strengthen ethnicity, they can also increase the ease of promoting hate speech and ethnic divisions. It is also the argument of this chapter that access to new media in Kenya is critical. As our analysis shows, social media can be an alternative medium for citizen communication or participatory journalism.

2. RATIONALE OF THE STUDY

Recently in Kenya, there has been a massive expansion of Internet infrastructure and a mushrooming of digital technologies, all geared towards informing, entertaining and educating the citizenry. Mobile phones, the Internet and other digital avenues constitute an important outreach agent through which the public is sensitized on issues such as ethnicity, negative ethnicity and political ideologies. The numerous reports by national and international organizations that document the threatening atmosphere and ethnic violence before, during, and after the 2007 election all mention new media having fanned the flames of ethnic hatred, and the role of hate speech as a feature of the conflict (see Bayne 2008; European Union (EU) 2008; Kenya National Commission of Human Rights (KNCHR) 2007- 2008). The distressing conclusion of many of the reports is that some instances of ethnic based violent behavior were likely motivated by encounters with hate speech in the form of text messages, e-mails and blogs (new media). Many people relied on such information that was transmitted by these digital devices and often acted according to their interpretation of the same (KNHCR, 2008: 18).

It is, therefore, imperative that any information transmitted through this media should be accurate, moderate, adequate and comprehensible. It is thus of essence to examine claims made that blogs, e-mails, mobile telephony and text messages were used to inflame tension and incite ethnic hatred.

The speed with which Kenyans turned to digital technology prior to, during and in the aftermath of the 2007 elections is proof that new media (the Internet, mobile telephony and blogs) are increasingly shaping the ways in which crises play out in the developing world. However, the double-edged sword of technology extends beyond crisis situations. It is against this background that this chapter attempts to provide a rapid and succinct analysis of what happened in Kenya prior to, during and after the 2007 elections using the lens of new media devices as discussed later in the chapter, (see section 13) the 2007 election fiasco exposed the deliberate stoking of ethnic tension and animosity by new media, power-hungry elites, feeble democratic traditions and institutions in Kenya. This state of affairs threatened to consume the citizenry. The chapter strives to place the new media in Kenya under scrutiny by analyzing its role in the 2007 post-election violence and eventually suggesting the way forward.

3. STATEMENT OF ISSUE

Kenya, being a multi-ethnic country, has witnessed inter-ethnic conflicts experienced especially during the general elections carried out after every five years as witnessed in 1992, 1997, 2002 and the worst one being in 2007/2008 after the disputed presidential election. New media was mentioned in various post election commission reports as having contributed in instigating violence. For example, the Commission of Inquiry into Post-Election Violence (CIPEV) popularly known as the Waki Commission, Non-Governmental Organizations (NGO) such as Minority Rights Group International (MRG) and Kenya National Human Rights Commission (KNHRC) all report that the new media fanned tribal animosity and hatred. It is from this perspective that this chapter seeks to examine and analyze how the new media supposedly fuelled ethnic animosity leading to inter-ethnic wars. Using the lens of the 2007-2008 Kenyan presidential election crisis, this chapter illustrates how digitally networked technologies, specifically mobile phones and the Internet, were a catalyst to both 'predatory behavior', where behavior is driven by cynical, opportunistic and often violent norms such as ethnic-based mob violence and to 'civic behavior' where behavior is driven by the norms of toleration, accountability and equality such as citizen journalism. In particular, this chapter focuses on the significant ways that Kenyans used new media technology to coordinate action: Short Messaging Services (SMS) campaigns to promote and counter ethnic violence; blogs to challenge mainstream media narratives and online campaigns to promote awareness and transparency. It is argued that Kenyan politicians and new media resorted to the use of negative ethnicity to undermine the state through the exploitation of its most vulnerable fault-line ethnicity which exploded soon after the 2007 election leading to Kenyans butchering one another.

4. KENYA AT A GLIMPSE

Kenya's official name is 'The Republic of Kenya' which is a country in East Africa. The country is named after Mount Kenya, a significant landmark and the second highest mountain peak in Africa. The area covered by Kenya is 582,646 square kilometers. The population has grown rapidly in recent decades to nearly 40 million although a census conducted in 2009 has not been released for public use. However, sources indicate that it may portray a significant shift in the demographic shares of the different ethnic groups. Kenya achieved its independence in 1963 from Britain. It has now had three presidents. The capital city is Nairobi.

According to the 1999 population census about 97% of Kenya's people are Africans. In 2006, the World Bank estimated Kenya's literacy at 73.6%, with male literacy at 77.7% and female literacy at 70.2%. On religious matters the Kenyan population consists of protestant 45%, Roman Catholic 33%, indigenous beliefs 10%, Muslim 10%, and other religions 2% (1999, Census Report).

Politically, the country is a presidential representative democratic republic, whereby the President is both the head of state and head of Government, and of a multi-party system of government. Although ethnic violence has been a recurring ingredient of Kenya's general elections' process since 1963, the cutthroat nature of the competition for the presidency in 2007 propelled Kenya towards inter-tribal violence. The mega-ethnicity that was witnessed in 2007 was of a larger scale than had been evident in the previous elections. Kibaki's Party of National Unity (PNU) went beyond the horizons of the Gikuyu, Embu and Meru Association (GEMA) to incorporate parties that represented a grand-mega ethnic constituency in Kenya. Thus PNU had in its ranks the following political parties: Democratic Party of Kenya (DP) (with a strong base around the Gikuyu, Embu, Meru Association [GEMA] communities), NARC-Kenya (NARC-K) (mainly constituting the support of GEMA communities and from Mijikenda communities from the Coast and a few other ethnic groups from other regions), FORD-Kenya (FORD-K) (that was mainly made up of Bukusu sub-group of the Luhya of Western Kenya). On the other hand, Raila Odinga's Orange Democratic Movement (ODM), which was the main opposition party, was mainly made up of the Luo, the Kalenjin, and a number of other subgroups of the Luhya, apart from the Bukusu, and other ethnic groups from the rest of the country as those from the Coast and like-minded groups from North Eastern Province.

Prior to the 2007 elections opinion polls predicted a close contest between the two main parties namely: PNU and ODM (Steadman, 2007). The ODM party leader is from the Luo community, while the PNU party leader who has been in power for two terms is from the Kikuyu community. In Kenya executive power is exercised by the government and is vested on both the government and the national assembly. The judiciary is independent of the executive and the legislature. However, there was a growing concern especially during the last year of the first tenure of Mwai Kibaki that the executive was increasingly meddling in the affairs of the judiciary by appointing judges without consultation. Prior to the 2007 election, Kenya had maintained remarkable political stability despite changes in its political system and had not experienced political crises common in neighboring countries such as Somali and Sudan.

5. ETHNICITY AND ETHNIC GROUPINGS IN KENYA - A BANE OR A BLESSING?

Our knowledge on the various Kenyan ethnic communities in the pre-colonial and post-colonial period has been enriched by a multiplicity of sources, including oral traditions, archaeology, historical linguistic and cultural anthropology. Pioneer scholars, including Ogot (1967), Were (1967), Muriuki (1977), Ochieng (1974), Mwanzi (1977), Aseka (1989) among many others (cited in Onyango, 2008) have enriched our historical knowledge on individual ethnic communities that occupy present Kenya.

Their studies have emphasized the fact that the present day Kenya was already an ethnically complex region characterized by varied communal interactions by 1500 AD (Mwanzi, 1977). These interactions were constantly altering the social, economic and political entities of the communities in the region. Through the historical process of encounter and interaction, there were evolving ethnic communities that were neither definitive nor pure but hybridized in nature (Ochieng', 1974:44). This process of interaction was underway with the arrival of colonialists. What perhaps needs to be emphasized from the contributions of the pioneer historical studies is that, first, the evolutionary process in Kenya pre-dates the histories of the present day inhabitants. Secondly, that the ethnic composition of the present country Kenya is as a result of a crystallization of many centuries of interaction between the various peoples and ethnic groups.

According to the 1999 population census, the country has about 43 ethnic groups and each group has its own language and culture that defines it. The country is divided into three major linguistic groups. The largest being the Bantu, followed by Nilotes and the smallest group comprise the Cushites. The ethnic languages of Kenya feature prominently at the national level and each is important to the individual or community at the interpersonal and intrapersonal levels of communication and more so as a marker of ethnic identity. Ethnic languages symbolize an in-group identity for the users. This ethnic distinctiveness has sometimes led to clashes among ethnic groups.

The major ethnic groups in Kenya include: Kikuyu (Agĩkũyũ) 20.78%; Luhya (Bakhayo, Banyala, Banyore, Bukusu, Idakho, Isukha, Kabras, Kisa, Marachi, Maragoli, Marama, Samia, Tachoni, Tiriki, Wanga) 14.38%; Luo 12.38%; Kalenjin (Kipsigis, Marakwet, Nandi, Pokot, Sabaots, Sengwer, Terek, Turgen) 13.46%; Kamba 11.42%; Kisii 6.15%; Ameru (Achuka, Aigembe, Aimenti, Amuthambi, Amwimbi, Atharaka, Atigania) 5.07%; Maasai 1.76%; Turkana 1.52%; Embu 1.20%; Taita 0.95%; Swahili 0.60%; Samburu 0.50%; Somali 2.29% (1999, Census report). The percentage indicates the ethnic group's proportion of the population of Kenya. The country also has minority ethnic groups which include: the Aweer, Bajuni, Kore, Kuria, Miji Kenda, Ogiek, Orma, Pokomo, Rendille, Sengwer, Suba, Taveta, Watha, Yiaku, Dorobo, Elmolo, Malakote and Sanye. Ethnic minorities here are distinguished by the small size of their populations. Other non-Kenyan ethnic groups resident in Kenya for many generations are: Gujaratis, Baluchs, Punjabis and Goans from India, Britons, Hadhrami and Omani Arabs and Italians, plus a number of Africans who have become citizens of Kenya (see Makoloo and Ghai, 2008).

In the latest national population and housing census of 2009 though (not officially released), the government put forth a policy of not providing information on the ethnic breakdown of the peoples of Kenya. Instead it opts to provide information disaggregated by age, districts and gender. This action of the government, according to a Senior Demographic Officer with the Ministry of Planning and National Development, was due to the experience of the 1989 census. He stated that:

It was discovered that the ethnic figures were hijacked, abused and used for the wrong purposes mainly for political propaganda. This is why you found that in some cases senior politicians claimed that the figures of their communities had been doctored. This raised unnecessary tensions. The result of this, which is the other reason for the new decision, is that the debate having been taken over by politicians the main issues for which a census is done [sic] were not addressed in the national debate.[1]

According to this officer, the figures on ethnicity were collected but not processed for the reasons stated. The same might be the reason why the 2009 population and housing census results have not been officially released.

Ethnicity is, therefore, as old as humankind. People in every part of our modern world, just as in ancient times, belong to some kind of ethnic or tribal grouping that reinforces their sense of belonging, nationalism, patriotism, social values, political progress, and development. The languages we speak, the customs and traditions we cherish, the food we crave for, and the clothing we adorn all have some linkages to our ethnicities, whether as Blacks, Europeans, Asians, Icelanders, (Kukubor, 2006) or even as Kenyans. In this way, ethnicity reinforces our very beings as persons and nations in charting our destinies in this world in regard to national unity and progress. This is the good news.

Ethnicity should not be a hindrance to national unity and progress, or the source of the continuing ethnic violence and political instability in Kenya unless Kenyans, out of misguided individual egos used ethnicity for mischief, bordering on corruption, mismanagement, and greed for power. We do not categorically state that ethnicity was the cause of the 2007 post election violence. However, the bad news is that ethnicity was a resource in the hands of frivolous 'political entrepreneurs' who, because of corruption, mismanagement, and greed for power, manipulated ethnicity and used it to achieve their personal political agenda (Onyango, 2008). In the real world, as we know from either personal experience or through the media, cultural misunderstanding, ethnic conflict, prejudice, xenophobia, ethnocentrism, anti-Semitism and racism frequently characterize relations between groups that are somewhat different from each other (Bayne, 2008). This is especially the case when one group holds more power, has more privileges or more resources than the others and uses the difference (language or ethnicity) as a legitimation to dominate or marginalize others. Bayne (Ibid) indicates that many African nations have ignored the ethnic nature of their societies and chosen to pretend that the ethnic groups do not exist and have stubbornly refused to build their houses on the strong rudiments of ethnicity. This may sound rather perverse because it has always been argued that our strength lies in unity and that ethnicity is the bane of our nation.

This brings us immediately to the difference between ethnicity and ethnocentrism. Ethnicity is a noun from a Greek word 'ethnikos' meaning 'heathen', implicitly meaning the origin of a person. Ethnicity is the fact of the ethnic group. It poses no danger to the nation. Ethnocentrism on the other hand, is the danger; it is the misuse of the ethnic group, of ethnic sentiments against other ethnic groups (Cheeseman 2008; Yieke 2008). Ethnocentrism may refer to any action or attitude, conscious or unconscious, which subordinates an individual or a group based on origin (language, ethnic origin and culture). This action can be enacted individually or institutionally. This attitude or behavior is based on one's extreme viewpoint or loyalty to a tribe/ethnic or social group, ignorance, excessive pride in one's ethnicity, and/or intent to suppress and dominate others. Ethnocentrism is what Kenyans have been referring to as negative ethnicity (Wa Wamwere 2008). The way we see it then is that, ethnocentrism can be combated, whereas ethnicity is permanent. Wa Wamwere further notes that one can be ethnocentric, regardless of one's religion, intelligence, social status and benevolence.

The concept of 'Tribe' was a derogatory term developed in the 19th century by racist western scholars, and journalists to designate alien 'non-white' people as inferior or less civilized, and as having not yet evolved from a primal state. A lot of times, tribalism and ethnocentrism have been used interchangeably to convey the same thoughts and opinions (Yieke, 2008).

Tribal and ethnic issues are so fundamental in Kenyan society that they seem almost an integral component. These issues seem to have come up even more clearly in the run-up to the 2007 general election. And even though we know that tribalism has always been there, the level to which the Kenyan community had fallen prey to tribalism was and is still amazing. It is becoming a common phenomenon for some myopic thinking Kenyans to behave as if ethnic differences produce inherent superiority in people of some specific ethnic groups. In fact, such individuals respond to other Kenyans differently merely because of ethnic backgrounds. Individual tribalism occurs in our day-to-day activities at informal levels. This tendency is exhibited in daily conversations, jokes, and how we routinely relate to one another. At this level, the tribal behavior may be conscious or unconscious. The idea however, is to demean or lower one ethnic group in order to raise the profile of the one to which the ethnocentric belongs. An ethnocentric tendency at this level is implicit in behavior and can be identified by certain behavioral signs. This often is done in very subtle ways. For example, the belief that some ethnic groups are more adept in particular jobs or tasks, and the belief in differences in intelligence among certain ethnic groups (what we call stereotypes).

Furthermore, the ethnocentric have no insight into their own prejudice. They believe that their prejudice is based upon objective grounds that cannot be compromised. By this strong fixation, an ethnocentric is capable of fanning violence and other forms of crime towards members of what he or she views as the 'inferior' ethnic group. He or she could easily support the use of force to dogmatically maintain their 'superior' belief. Kenya as a Nation, witnessed this in the run-up to the general elections and even after these elections.

In Kenya, ethnicity has taken the form of the 'ethnification of political and economic processes', which means that people are made to treat ethnicity as increasingly relevant to their personal and collective choices in terms of choice of candidates during elections, investment, residence, and even social interactions such as marriages (Wodak 2001). According to Onyango (2008) political leaders and the people they lead are stimulated to consolidate, form or stabilize an ethnic identity; thus the political opportunity afforded by ethnic networks is easily exploited for political support.

Onyango (2008) further notes that most ethnic groups in Kenya associate the relative economic prosperity of some Kalenjins and Kikuyus to the real/imagined favors derived from the political advantages that accrued to them during the Moi - Kalenjin (1978-2002) and Kenyatta - Kikuyu (1963-1978) presidencies respectively. The same was and is now being said of the Kibaki presidency (2002-2012). Thus, there is common talk of those who have 'eaten'. In terms of political mobilization, therefore, political leaders call upon their ethnic communities to group together and fight for political office either as a way of ensuring their continued stay in power, so as to continue 'eating', or to gain political power as a way of finding an 'eating place'.

Mute (2008) argues that the ground was fertile for what ensued after the 2007 elections. Campaigning, particularly through new media, took an ugly bent, with ethnically prejudiced and stereotyped coverage. The Kenyan blogs and online sites, populated primarily by a youthful generation of Kenyans, many of whom lived abroad, did the same. Text messages circulated around the country, playing on angers and resentments arising from the material reality of historical and contemporary inequalities and injustices, but once again articulated and promoted as being ethnically-based, and experienced as such too. Given the ethnicities of the two presidential contenders - Kikuyu, in the case of Mwai Kibaki and Luo, in the case of Raila Odinga - perhaps insufficient attention was paid to similar angers and resentments in other ethnic communities, particularly among the Kalenjin, whose leading politicians had cast their lot in with Odinga's Orange Democratic Movement (ODM).

Recent political problems that threaten to tear Kenya apart require analysis that goes beyond ethnicity as portrayed in new media and current analyses that attempt to explain the situation. More correctly, emphasis and focus should be placed on the interpenetration of historical and current political developments whose origins can be traced in the early stages of state formation in Kenya. During this time, White settlers got 20% of Kenya's high-potential farmlands. As these settlers failed to provide enough state revenue and blocked African opportunity, the British increasingly encouraged African farming on the other 80%. So the second economic centre became Kikuyu-land: home to 20% of the population; located close to the capital, Nairobi; cool and attractive to missionaries, with more schools than the rest of the country. By geographical accident, then, the Kikuyu had a head start in making money (essential to advance political ambitions) and in acquiring modern managerial skill. Not surprisingly, the Kikuyu also bore the brunt of colonial capitalist dispossession and socialization, and were in the vanguard of the nationalist struggles that led to decolonization and they came to dominate the postcolonial state and economy. Capitalist development and centralization of power reinforced domination of the Kenyan economy by the Central Province and the Kikuyu, a process that withstood the twenty-four year reign of President Moi, a Kalenjin from the Rift Valley, and was reinvigorated under President Kibaki's administration. (See Mwolongo, 2008 and Hirsch, 2008)

In the new circumstances, other not-so-well-placed ethnic groups made the most of what they had. They were often driven by a local patriotism inspired by vernacular, mission-translated, Bibles that told of an enslaved people who became an ethnic nation. They embarked, in combinations of hope and desperation, on chain-migrations out of pauper peripheries (not unlike the Scots or Irish in comparable circumstances) to colonize particular niches of employment: on the railway; on white farms and plantations; in domestic service; or in the police and army. Yet others came to dominate the livestock trade.

Officials and employers exploited these various tendencies and stereotyped the supposed ethnic qualities of the group concerned. The British helped to harden ethnic divisions made greater by differing potentials for social mobility. The emergence of ethnic consciousness also arose from local debates about how the genders, generations, rich and poor should relate, as older inequalities were transformed into new differentiations less sensitive to existing moral audits of honor.

Central Province and Kikuyu dominance of Kenya's political economy bred resentment from other regions and ethnic groups. It fed into constitutional debates about presidential and political centralization of power, and the regional redistribution of resources that dominated Kenyan politics until 2005 when the draft constitution supported by the President and Parliament was rejected in a referendum. The ODM was born in the highly politicized maelstrom of the run up to the 2005 referendum.

This narrative tends to ignore an important qualifying fact, that not all Kikuyus are dominant and not all Luos are disempowered. Colonial, neo-colonial and neo-liberal capitalisms have bred class differentiations within communities as much as they have led to uneven development among regions. In other words, Kenyan elites have much more in common with each other than they do with their co-ethnics among peasants and workers who also have more in common with each other across ethnic boundaries than with their respective elites. This is a reality that both the elites and the masses strategically ignore during competitive national elections, because the former need to mobilize and manipulate their ethnic constituencies in intra-elite struggles for power, and the latter because elections offer one of the few moments to shake the elites for the crumbs of development for themselves and their areas.

In order to investigate the relationship between language and ethnic identity, or ethnicity, one has to first understand how the concept of ethnicity is indeed planted among an ethnic group. According to Wa Wamwere (2008), the study of ethnicity diverges into two different opinions, namely the "primordialism" and "instrumentalism" perspectives. In the primordial's view, ethnicity is regarded as 'constituting' a fundamental feature of society and that ethnic identity is natural and unalienable. In other words, the ethnicity of a group is defined by its cultural and biological heritage, and is territorially rooted. It is thus grounded by the group's primordial ties and bound by the ancestors' values, myths, languages, among other values.

On the other hand, instrumentalists argue that the primordial approach emphasizes too much on the objective nature of ethnicity, which stresses that ethnicity is a 'given' and one is born with it (Wa Wamwere, 2008: 52). They criticize the fact that the primordial approach cannot explain the evolution of ethnic groups over time. Instead of admitting solely to primordial ties, instrumentalists emphasize that ethnicity of a group should be understood in terms of its relationship to other groups. This simply means that the members of an ethnic group identify themselves subjectively in relation to other groups in order to maximize their social interest. Worsley (2007) argues that cultural traits are not absolute or simply intellectual categories, but are invoked to provide (ethnic) identities, which legitimize claims to rights. They are strategies or weapons in competitions over scarce social goods. Adopting the instrumental approach to ethnicity, the relationship between language and ethnic identity in new media will be much more transparent to us.

The instrumental view holds that ethnicity is a subjective way of interpreting a group's identity, often in the hope of maximizing the members' interest. However, in order to identify a group's separate and unique ethnicity, the members often have to in some way find themselves certain features which can distinguish them from the other ethnic groups. For instance, biological heritage, religious divergence and language difference are commonly cited as proofs of ethnicity.

It is, however, safe to argue that the very majority of our social life depends on the use of language: the use of different languages naturally separates people into different groups, each not being able to understand the others. Lacking channels of communication, we typically identify others as being 'different' from us. This is what makes language such a prominent objective factor in defining ethnicity. To say language is to say society (Yieke, 2007). Thus if one speaks a particular language, then one belongs to that particular society. The idea of language itself is, however, sometimes not an objective fact, but a matter of subjective interpretation and is often employed purposefully as an ethnic distinction. Linguistically, every regional dialect is more or less different from the neighboring dialects. Even though we may normally regard these dialects as dialects of the same language, it is so easy to take this difference as the evidence of independent ethnicity once the speakers find this parallel to their interest. Consequently, this flexibility (or otherwise ambiguity) in defining 'language' creates its link to ethnic identity in Kenya.

6. DIGITAL MEDIA IN KENYA

In Kenya, the traditional media (electronic and print media) has played significant roles in the mobilization of voters' as well as for political campaigns. The independent Kenyan government inherited the authoritarian colonial government's dominant perception of the media as a necessary evil that deserved close supervision and control. The colonial government feared a free and thriving nationalist media that acted as the mouthpiece for political independence. It enacted the Penal Code in 1930, to control alleged seditious nationalist publications such as Sauti ya Mwafrika (African voice), Uhuru wa Mwafrika (African freedom), African Leader, Inooro ria Agikuyu (The voice of the Gikuyu) among others. Similarly, the Kenyatta Government was averse to an independent and foreign owned media playing a watchdog role that could cause disaffection towards the young government. As a result similar patterns of media ownership and development continued as they were under the colonial rule. Consequently the Independent African government entered the shoes of the colonial rulers.

The traditional media on the other hand is not immune from governmental controls which take the form of political representations to the owners and threats to sue through courts of law, state raids as it happened with the Standard media Group in 2006 and the ban on live media reports as it happened on December 30th 2007.

During the 2007 Kenyan elections the new media technologies used by citizen journalists and community organizers were the same ones used by forces in the government who sought to rig the election, and agitators who attempted to expand ethnic violence. One lesson from the use of information technology in the Kenyan crisis is that the new media technology itself is neutral. It can be used powerfully to give citizens a voice in crisis situations, or used to aggravate those same crises and thus does not take sides. We therefore argue for both sides of the coin by putting into sharp focus the power of the new media technologies in the Kenyan context.

Understanding the role of new media in the elections crisis requires a brief history of Kenyan digital media as well. With an estimated 3 million Internet users, Kenya has one of the highest levels of Internet penetration in sub-Saharan Africa, at 7.9% (of major sub-Saharan African countries – discounting those with populations under a million – only South Africa has higher net penetration.) (http://www.internetworldstats.com/stats1.htm)

More than 12 million Kenyans – roughly 30% of the population – have mobile phones, as compared to a continent-wide penetration of 20%. Kenyan companies have been early adopters of mobile money transfer systems like M-PESA[2] and complex SMS-based systems like Kazi560[3] which matches jobseekers and employers via their phones. (http://www.boston.com/news/world/africa/articles/2008/06/09/kenya-mobile-phone-companys-ipo-get-warm-debut)

New technologies have done much to advance both the country's democracy and its economy. During the recent crisis, these technologies played a very different role, however.

Text messages spread hateful ethnic views rapidly and may have helped leaders to enact plans for ethnic violence on a large scale. The public was bombarded not just with hateful and derogatory depictions of politicians and of ethnic groups but also with the idea that it was acceptable to express hatred electronically and to share that expression much more widely than would be possible for an ordinary individual prior to the advent of this new medium. Some of the messages went beyond identifying groups and their interests to denigrating particular ethnicities by using familiar stereotypes of their qualities or behaviors, such as laziness, acquisitiveness, and callousness.

Charles Onyango-Obbo, the Nation Media Group's managing editor for convergence and new products, reported on Mashada blog on December 21st that "There were SMSs landing in his cell phone's inbox literally every 15 minutes in the last two weeks of the campaign, emanating from all sorts of support groups for the candidates… after the dispute over the outcome, the texts were arriving every five minutes, and they were meaner, nastier and more propaganda-filled." Pervasive texting of negative ethnic speech also allowed for the participation of Kenyans living in the Diaspora, who might have had less involvement in, and impact on, all aspects of the election.

Against this backdrop, it makes sense that Kenyans would emerge as early adopters of new media. Prominent Kenyan blogs, including Daudi Were's "Mental Acrobatics" have been online since early 2003. Starting in 2004, Kenya "Unlimited" has aggregated posts from individual blogs on a central site and provided a "web ring", a navigation mechanism that links related weblogs together. In 2006, a nationwide blogging contest – the Kenya Blog Awards or "Kay bees" – helped bring together individual Kenyan bloggers into a community. Afrigator, an African blog aggregator based in South Africa, cites two Kenyan blogs in its list of top twenty blogs, giving the country the second best representation on that list (after South Africa, which dominates).

Several Kenyan Bloggers took pains to document the 2007 election, but there was little indication from their posts that any anticipated the unusual events that would follow the election. Describing his voting experience, and the precautions taken by the Electoral Commission of Kenya (ECK) to prevent election fraud, Daudi Were observes:

One thing I noticed was that no one was wearing any political party merchandise and the conversations in the queue were distinctly non political and non ethnic. Rather than being divided, by queuing together to exercise our civic duty and responsibility, we were bound together in a sort of patriotic camaraderie. We all felt it was worthwhile to take part in the vote and that ultimately was what mattered. http://www.mentalacrobatics.com/think/archives/2007/12/voting-experience-kenya-election-2007-php.

The tone – and focus – of coverage changed sharply on December 30th, as it became clear that the disputed election would be declared in Kibaki's favor. The ban on live media reports particularly incensed Okolloh, who had been monitoring television, radio, the Internet, SMS and local gossip to produce several election updates per day. When the live coverage ban was announced, she declared: "All live broadcasts have been suspended by the government. The order was released as ODM was addressing their press conference. This is now officially a police state. So we have no idea what ODM is saying, and what the security situation is around the country. " http://www.kenyanpundit.com/2007/12/30/media-blackout-announced/

Onyango-Obbo, writing in the East African newspaper, argues that there were around 600 blogs around the elections: "Some of them spewing shocking ethnic vitriol... many African nations are already on the verge of being failed states, so we should be very afraid that at highly emotional moments like the recent Kenyan election, it's the subversive forces and hate-peddlers who are winning the SMS and blog wars." (http://www.kenyanpundit.com/2007/12/30/media-blackout-announced/) It is thus imperative that much of the Blogosphere in the country, as diverse as it is, can also be characterized as highly factionalized and often virulent in their content.

In the wake of a ban on live media, some Kenyan bloggers responded by redoubling their efforts as citizen journalists. Reeling from the violence in her native Eldoret residence, Juliana Rotich began posting brief bulletins on refugee movements, fuel shortages, road and airport closures. Some were posted via SMS using Twitter to disseminate messages to a wider audience; other bloggers featured photos and these were uploaded to Flickr using a GPRS modem. Were took to the streets on January 3rd 2008, following ODM activists as they attempted to march to Uhuru Park to attend a banned rally. His photos document the empty streets of the usually-bustling capital and the tense standoffs between activists and security forces, and provided insights on the confrontation hard to find in international media covering the confrontations. http://www.flickr.com/photos/afropicmusing/sets/72157603595279896.

As Kenyans and the international community try to come to terms with what happened after the 2007 elections, it would be useful to systematically think about the role played by new media which has been a key force for democratization in Kenya during the recent political upheavals like the 2002 elections. But as things unraveled after the 2007 elections, one could not help but wonder whether the new media could have done better or whether new media could have helped forestall the fallout. The new media in Kenya has played a central role in shaping Kenya's democracy. It had gained a reputation for acting as vigorous fora for public debate and it was seen as a guardian of the public interest against overweening state power. The recent record of the new media, according to many within it (see Rotich and Goldstein, 2008), is that new media has undermined as well as invigorated that democracy. As a result, development actors should be better engaged and more supportive of new media in the future. The argument being fore-grounded here then is that the problem facing Kenya's new media (citizen media) is not how technically sophisticated and globally connected it is but rather how neutral the technology itself was and how the new media was utilized positively or negatively in Kenya during the 2007 general elections.

7. THEORETICAL FRAMEWORK

The chapter has adopted an eclectic approach in its theoretical framework, De Fleur and Ball Rokeach's (1989) Functionalist Theory of Media, and Tan's (1984) Social Responsibility Theory and Ethnicity as a Colonial Formation conceptual framework guided the study.

The Functionalist theory posits that the more the audience is dependent on the media for information and the more a society is in a state of crisis or instability, then the more power the media are likely to have (or be credited with). The theory proposes an integral relationship among audiences, media and the larger social system. This theory predicts that we depend on media information to meet certain needs and achieve certain goals, like uses-and-gratifications. But we do not depend on all media equally. Two factors influence the degree of media dependence. First, one will become more dependent on media that meet a number of one's needs than on media that provide just a few. The second source of dependency is social stability. When social change and conflict are high, established institutions, beliefs, and practices are challenged, forcing one to re-evaluate and make new choices. At such times our reliance on the media for information will increase. At other, more stable times our dependency on media may go way down. One's needs are not always strictly personal but may be shaped by the culture or by various social conditions. In other words, individuals' needs, motives, and uses of media are contingent on outside factors that may not be in the individuals' control. These outside factors act as constraints on what and how media can be used and on the availability of other non-media alternatives. Furthermore, the more alternatives an individual has for gratifying needs, the less dependent he or she will become on any single medium. The number of functional alternatives, however, is not just a matter of individual choice or even of psychological traits but is limited also by factors such as availability of certain media. In 1989, Ball-Rokeach and DeFleur stated that as technology increases the way in which media can be delivered, its influence becomes even more powerful, and this assertion could not have been truer.

This functionalist theory is nearly only useful for considering questions of social integration. In a complex society, there will be a number of different ways for societies to achieve a required degree of control and consensus and new media are only one institution among several with overlapping tasks in this respect. The tenets of this theory are invoked in the analysis of the emergent patterns of whether the ethnic crisis in Kenya was as a result of the power credited to new media in Kenya. The theory also guided the researchers' conclusion that new media fell short of negative ethnicity control and consensus in Kenya.

Social responsibility theory proposed by Tan (1984) states that the media has a social responsibility that it is called upon to accept; in recognition of their essential role in political and social life. The media, according to this theory, has to provide a "full, truthful, comprehensive and intelligent account of the day's events in a context which gives them meaning. Secondly, the media should serve as a forum for the exchange of comment and criticism and be common carriers of the public expression. Thirdly, the media should give a representative picture of constituent groups in society and also present and clarify the goals and values of society. However, the social responsibility theory emphasizes responsibility of the media to the society it serves as opposed to freedom per se. The social responsibility theorists take the position that the new media need of necessity to assume both moral and legal responsibilities for all that they publish for the general good of the society. The researchers analyzed their data against these tenets to establish whether the data adhered to the tenets or deviated from them.

Ethnicity as a colonial formation conceptual framework is used to explain the division of Kenya into groups otherwise referred to as ethnic groups. The British divided the Kenyan territory along ethnic lines into eight provinces, creating a different majority in each; each province was subdivided into districts, often according to ethnic groups and subgroups. For example, the Luo are based mainly in Nyanza (though it is also the home to the Kisii, who have their own district); the Luhya, in Western Province; the Kikuyu, in Central Province; the Somali, in North-Eastern Province; and the Mijikenda, in the Coastal Province. The Rift Valley is dominated by the Kalenjin, but also contains Maasai, Turkana and Samburu districts. The Kamba share Eastern Province with Embu and Meru, among others. Nairobi is the most cosmopolitan province, with the Kikuyu forming a plurality. It has also often been noted that the structure of colonial administration crystallized the population into tribes creating them where they did not previously have meaning. This framework guided the researchers in data collection and analysis as the emerging patterns were analyzed, described and correlated with ethnicity. The colonial ethnic crystallization in turn created extremely new forms of ethnic competition. Today ethnicity in Kenya means politicized kinship more than it does anything else, a kind of overpowering identify informed by grievances of sense of people wronged of being under siege.

8. METHODOLOGY

The research employed purposeful sampling in collecting the data, drawn from the 2007 post election reports and new media. Excerpts of electronic mails and text messages circulated during the 2007 post election violence and those forwarded to the Waki commission of inquiry and major blogs updates prior to, during and after the 2007 elections were also analyzed through content analysis. The study used qualitative research methods where by the data were analyzed descriptively for meaning. The findings are derived from the facts collected and analyzed from oral and documentary evidence that was gathered by the researchers.

9. E-MAILS, TEXT MESSAGES, MOBILE TELEPHONY AND BLOGS

We collected e-mail messages and text messages that were circulated before, during and after the 2007 Kenyan elections for our analysis. The researchers made a public request through their Facebook, Twitter and My Space updates which forms part of new media for people to forward e-mails and text messages that were circulated prior, during and after the elections. Due to the sensitive nature of the content of information being circulated majority of the people declined to forward the e-mails and text messages. Some claimed that it might land them into trouble, as it was illegal to circulate confrontational information with a law on hate speech now in place. One update that captures this read "which sane and God fearing person will keep them." Other participants did not want to be associated with the issue claiming that we were collecting evidence to forward to the International Criminal Court (ICC). We were thus unable to collect the expected number of text messages for analysis since most people did not have handsets with a large memory to store them and thus had deleted them; however we analyzed the few that we collected which we believe are more than enough. We on the other hand collected 35 e-mail messages that were sampled purposively to get the e-mail messages analyzed for this study. Data from the Kenyan key Blogs such as Mashada, Thinker's Room, Ushahidi (witness) and What an African Woman Thinks were collected and analyzed. We did a content analysis of the e-mails, text messages and blogs specifically focusing on the use of derogatory terms, demeaning words, the nature of ethnic animosity expressed and, on a positive note, how these new digital technology media helped to counter negative ethnicity.

10. SUBJECTIVITY AND VALIDITY

As researchers, we filtered data through our own expertise and knowledge, which enabled us to derive meaning (Creswell, 1998; Denzin & Lincoln, 1994), yet this filter also likely colors our analysis and interpretations. We minimized possible negative effects of subjectivity by analyzing data from a variety of new media sources such as e-mails, text messages and blogs.

11. FINDINGS

Although the new media in Kenya portrayed the violence as purely political, its basis was clearly ethnic or tribal. Initially, the violence in Kenya was described as a revolt against disputed election results, but when observers began to identify ethnic patterns to the violence, the comparisons to the Rwandan genocide flowed swiftly through the international media. The Western media tore Kenya into shreds. For the two months, the hitherto peaceful Kenya found itself painted in the worst of adjectives, such as "genocide", "ethnic cleansing", savagery (or tribalism) and so on. For the entire two months after the 2007 December polls, Kenyans waited anxiously for the Western media to change tact and be compassionate. It never happened. For failing to cultivate its own niche, the Kenyan press which normally borrows heavily from the Western media--found itself torn between the sensational reporting of the Western media and the hard facts on the ground. It chose sensationalism. At a recent meeting under the aegis of The East African Editors Forum, top editors within the region accepted that they had played second fiddle to the Western media and engaged in a soul-searching of their true call. Adam Mynott, who reported on the crisis for the British Broadcasting Corporation (BBC) does not believe that the BBC exaggerated the scale of the violence, and he shares the criticisms made of those who compared the violence to the Rwandan genocide. "I don't think the BBC did compare the violence to the Rwandan genocide, although others did. It's a ludicrous parallel to draw," he says, "800,000 people died in Rwanda and 1,000 people died in Kenya, so it is a comparison that is odious." (http://commonco.typepad.com/8months/2008/02/reporting-on-co.html)

Every day, footage of the police beating protesters, houses burning, displaced persons, blocked roads, ad infinitum, was aired on online bulletin boards for all to revel in. Whereas telling-it-as-it-is is encouraged in the media, still it has to be done in a responsible and balanced way. But the slant of the Western media coverage was appalling. They put a spin on events, negating the very principles of a fair and just practice of the trade. They disproportionately made everyone to think that the whole country was aflame (http://www.e-ir.info/?p=475)

New media reports on the Kenyan elections especially blogs of the protests following the inauguration of President Kibaki almost invariably included the word 'tribal'; the reference is to 'tribes' and 'tribalism' as primordial identities untouched by history, as ancient hatreds immune to modernity, as pathological conditions peculiar to Africa. During the violence that surrounded Kenya's recent elections, for example, Mashada blog documented that "tribal war" had exploded between the Kikuyu and Luo with picture footage of "gangs going house to house, dragging people of certain tribes out of their homes and clubbing them to death". This was an "atavistic vein of tribal tension that always lay beneath the surface but up until now had not provoked widespread mayhem" (http://www.recesspa.org/index.php?option=com_content&view=article&id=98:a-framework-for-media&catid=44:april&Itemid=191).

As Oyugi (1997) correctly argues, when political contradictions are not adequately addressed in the media, they tend to have adverse consequences on inter-ethnic relations in the society. Conversely, Wa Wamwere (2008) contends that these clashes cannot properly be described as ethnically motivated, ethnicity is a positive distinction, and has nothing of the hatred here at work. Generally, Kenyans primarily identify themselves according to their tribe or ethnic group first, and then secondly as Kenyans.

Indeed, Kenyan media reports, during what many regarded as ethnic-divided elections, contributed to the violent ethnic clashes that killed over a thousand people and displaced more than a quarter of a million others (http://www.westminster.ac.uk/schools/media/news-and-events/events/2010/racism,-ethnicity-and-the-media-in-africa)

12. SHORT MESSAGE SERVICE (SMS), ELECTRONIC MAILS AND BLOGS

In Kenya, as in the rest of Africa, Short Message Service (SMS) is the most widely used digital application (Bayne 2008). Prior to, during and after the 2007 elections the new media and specifically mobile phones, e-mails, and blogs emerged as significant communication tools. This new technology was perceived to be efficient and a cost effective way of mobilization of voters by the Kenyan politicians prior to the elections. Mobile telephony and more specifically the Short Message Service (SMS) was spreading very fast like a virus. The vial nature of SMS was capitalized on by the political parties for political marketing and thereafter the elections as a tool for spreading ethnic hatred and organizing ethnic violence. The political marketing experts as well as independent mobile phone users, bloggers and the owners of the e-mail accounts, generated these messages. Significantly, most of the ethnic based jokes, humor and hate speech were written in Kenya's local vernaculars. Very few were written in English while the rest were in Swahili. The reason for using the vernaculars was, first, that some of the messages lose their weight and meaning when translated into Swahili or English. Secondly, ethnicity is very central in the Kenyan election process. Political language is translated and interpreted using ethnic background and history as such any political communication must be codified into relevant vernaculars for it to leave an impression in people's mind. On the one hand, these messages focused intense attention on the competing presidential candidates from both the Party of National Unity (PNU) and the Orange Democratic Movement (ODM) in a way that promoted a healthy debate through providing new space relevant for democratic strengthening never witnessed before. On the other hand, these new media devices led to negative political campaigns that also became extremely ethnic based after the elections.

However, if the ethnic discourse of disparagement was covert in the public, in private it was more overt and marked outright disparagement (Waki, 2008). Largely, the SMS in mobile phones was a remarkable medium. There were so many messages used to campaign prior to the elections but we just use the sampled examples to illustrate our point. On the part of ODM supporters, one message that was circulated read:

A Deadly Mountain Flu known as PNU, which affects the brain, has been reported in Central Kenya. The region is under quarantine. People in other parts of Kenya are advised to take ODM pills. One full orange for 3 months to avoid infection. (http://inanafricanminute.blogspot.com/2007/04/dark-sideof-mobilization.html.)

This was coded ethnic discourse. In the first place, PNU was seen as a Central Kenya affair. Therefore, PNU was mainly seen in terms of the Kikuyu ethnic group who are the leading inhabitants of Central Kenya. Secondly, PNU was also seen as a negative thing (flu or disease) that is associated with Mount Kenya region. In fact, the supposed sacred cows in President Kibaki's first five-year tenure were referred to as "Mount Kenya Mafia" (see Onyango 2008). This was a referential strategy that associated PNU stronghold with a disease rooted mainly in the Gikuyu, Embu and Meru Association (GEMA) - mega-ethnicity that had its stronghold in the Mount Kenya region. To many Mount Kenya out-groups any time the mountain is mentioned, the bell that rings is that of GEMA, because all the conglomerate ethnic groups of GEMA come from around Mount Kenya. This message of disparagement ended with a fallacy cushioned in persuasion, in line with the spirit of campaign. It persuaded people to take a full orange for three months to avoid the flu. The implicit message here was that people who were targeted by the message were to remain steadfast in ODM during the whole period of the campaigns (shown by the three months). The issue of quarantine was also a persuasive strategy that appealed to the audience of the message to keep off PNU. The idea of full orange was very precise to ODM's party symbol that was a full orange.

On the part of the PNU supporters, one short message against the Luo ethnic group read thus:

Do you want to be ruled by Luo to take us back to joblessness? Safeguard the Kingdom. Let us ALL come out and give all the votes to Kibaki so that we are not ruled by an uncircumcised man who will make us wear shorts and plunder all our wealth. It's your vote that will prevent our country from going back to Egypt. May our God bless you. (http://allafrica.com/stories/200802291070.html)

In this text message, an ethnic stereotype that was coined against the Luo during the colonial period is revisited. During the colonial period, careful social engineering came up with ethnocentric labels that labeled the Kikuyu as cheeky, the Luo: genetically lazy and the Maasai as trustworthy albeit trustworthy natives (Ochieng, 1975). Thus, this is the right place to place "the Luo to take us to joblessness" referential strategy in the above discourse. In the question of the Luo not being circumcised implies not being ripe to lead. The Luo do not traditionally circumcise their male, the inner groups that supported PNU traditionally circumcise their male. The question of male circumcision has been a very intriguing point in ethnic discourse that is associated with top leadership. Although Ochieng (2001) reckons that the circumcision trivia on the Luo is remarkably central Kenya discourse, it is, however, true that this discourse is common rhetoric among the male circumcising groups of Kenya when referring to the Luo in negative terms. In this line of thought, among the people of Coast, the Luo are called "watu wazima", that is complete persons, because they have not been cut. Similarly, the Bukusu refer to the Luo as "omusinde", the uncircumcised. Actually, in the National Referendum campaign on the new Constitution in 2005, Honorable Simon Nyachae, who hails from Nyanza with Raila Odinga but from the male circumcising Kisii group also expressed the uninitiated boy sentiments when referring to Raila Odinga. The idea of "putting on shorts" can be seen in the context of "boyhood."

Although Kenya is a tropical country and therefore putting on of shorts would be good for adults, generally, it is boys and not men who are associated with shorts. This is understood in full in relation to our discussion on circumcision above. "Going back to Egypt" is taken from the Bible, a Holy Book. Egypt is associated with the suffering of God's chosen people. In this text, therefore, we see a fallacious campaign maneuver that was sought to associate ODM leadership with suffering. The conclusion of the text was "may our God Bless you", that was, in a nutshell, a false appeal to revered authority. However, whereas we know that God is not discriminatory, in this message the appeal for God's blessings was for those who were to heed the message of not voting for a Luo and God is reduced to a specific group; that is, the senders of the message and their targeted ethnic audience seen in, "our God". The issue of revering PNU in this text was seen in the urge to safeguard the "kingdom", which was in fact a mega- ethnic kingdom, no more no less. The use of ALL in capital letters was populist and a persuasive appeal. This is the important background to the discursive analysis of new media and ethnicity in Kenya prior to, during and after the 2007 elections.

This epithet also seems to combine gender discrimination with ethnocentrism. Gender discrimination occurs in the sense that the uncircumcised are viewed to be feminine and thus unqualified to rule. Specifically, the comment on the suitability of a political candidate, depending on whether he had been circumcised or not addresses the politics of ethnicity and gender in relation to power. When people are criticized for their lack of circumcision as evidence of their demeaned ethnicity, the insult is as much targeting masculinity as ethnicity. Mwolongo (2008) asserts that, the history of sexist use of language makes that terrain of insult more readily available and makes certain such an ethnic slur doubly demeaning.

Yieke (2008) documents how utterances that urged Kenyans from particular ethnic groups settled in various parts of the country to return to their region of origin expressed a "politics of inclusion and exclusion" that related directly to longstanding land disputes and the movement of ethnic groups. Certain references were not newly invented for the election violence but rather resembled statements made about people from surrounding nations who had entered Kenya as refugees or economic migrants and were no longer welcome. These individuals castigated as "foreigners" and "visitors" were clearly positioned as having no entitlement to Kenyan residence. Using such terms in reference to Kenyan citizens has the effect of suggesting that they too can be disenfranchised from civil rights, land, residence, and even identity as Kenyans. Note here that seemingly ordinary terms such as foreigner and visitor gain hurtful power when used in a context of ethnic mobilization.

Those who engaged in negative ethnicity during the 2007 election violence voiced not just the well-worn and familiar stereotypes of commonly circulating ethnic jokes about lazy or lascivious politicians or ethnic groups, but they also used cold, dehumanizing language. They called people "spots" or "weeds" that needed to be cleansed or pulled out. They referred to fellow Kenyans as animals (e.g., mongoose, snakes) or insects (see, Mwalongo 2008). The following blog on Mashada tells it all, "Kikuyu are like mongoose which is ready to eat chickens. All other tribes, i.e. Luo, Kisii, Luhyas are all animals in the forest. They cannot be able to lead this country like Kikuyus." The use of dehumanizing epithets, as well as the refusal to speak of certain groups of people as humans, is a key indicator that groups in conflict may have become locked in axiological opposition (Rothbart and Bartlett 2007). One other e-mail referring to a Kibaki minister with his portrait read:

Rattlesnake Me-Chuki (hatred). This man is the most famous rigging mastermind Kikuyu in Kenya. Known for his 'rattlesnake' temper, he has a tongue that dates back from colonial administration. Look at his eyes, cunning and foxy. He is the power-behind-the-throne. He is the shadow president of Kenya.[4]

The rationale implicit in the use of dehumanizing, value-laden language is: "We are the good and they are the evil. We are the people and they are the animals. We are the worthy and they are the killable." Such dualistic rhetoric tends to be an indicator that significant violence is possible. Some uses of dehumanizing language may operate as fighting words that incite ethnic violence in self defense by those who experience insult and threat. Relatedly, these words also offer an ideological justification for those - either speaker or hearer - who might engage in physical violence. As an aside, for words to cause or justify ethnic violence, that is, to have a specific effect, technologies of violence must also exist. Words alone may heighten emotions, but when people decide to engage in violence they must have the tools (even rudimentary) and the organization (even haphazard) to do so.

From the foregoing discussion of ethnic conflict in Kenya, we can infer that ethnic discourse depicts ethnic conflict. The argumentation scheme in ethnic discourse is one that explicates positive self-presentation and negative other-presentation. It is in the question of positive self-presentation and negative other-presentation that we locate the issue of discrimination in ethnic discourse. In this respect, we argue that ethnic discourse decries, discredits, debases, degrades, defames and so on, members of the competing out-groups.

Referential strategies are strategies in which one constructs and represents social actors, in-groups, out-groups by ways of reference tropes, biological, naturalizing, depersonalizing, metaphors, metonymies as well as synecdoche. Predication strategies are, for example, the stereotypical, evaluative attributions of negative and positive traits in the linguistic form of implicit or explicit predicates. They are very close to referential strategies. In Kenya, there are a number of ethnocentric labels that embed referential and predicative strategies that are used concerning various ethnic groups. For example, in the text message, "…uproot the "sangari", 'shake off the soil', 'gather it together' and 'burn it'" (Waki 2008: 218). The Kikuyu here are labeled as 'sangari' – weed by the Kalenjin ethnic group while in other text messages they were labeled as Kenyambi (Kikuyu grass, a weed, by the Gusii), the Kalenjin on the other hand are called warriors.

In the deep heat of the campaigns before the 2007 general elections, disparaging ethnic discourse was evident across the partisan divide. For example, there were e-mail messages that openly called Raila Odinga of ODM "One Dangerous Man" (Hirsch 2008: 8). This referential strategy depicts pars pro toto (part standing for the whole) (Reisigl and Wodak 2001:45 cited in Onyango, 2007), where Raila was taken as a part to represent the Luo community or the whole of ODM. Just one year before the 2007 general elections, Odinga's biography had been released that was interpreted by strong PNU adherents as having implied that he actively participated in the abortive coup of 1982, during Moi's regime. It is important to note that during this abortive coup, most of the top brass coup leaders who were later executed were the Luo (Raila's ethnic group). Thus, in the text of "One dangerous Man" (a caricature of ODM, so to speak), was the implication of danger implied in a dangerous coup plotter who should not be entrusted with the legitimate reins of power. On the other hand, ODM supporters presented a counter argument to PNU's "One Dangerous Man" text. To them the counter text was "One Daring Man" that depicted Odinga as an unrivalled political rights crusader in Kenya.

The use of mobile phones to pass negative campaign and propaganda message proved to be very infectious. As the Kenyan crisis unfolded, many cell phone owners received SMS messages that urged them to drive neighbors from their houses: "If your neighbor is a Kikuyu, just kick him or her out of that house. No one is going to ask you anything." Messages included expressions of ethnic hatred and warnings that one ethnic group would attack another. Opposition sympathizers who translated such messages into their respective vernaculars and passed them on to their contacts personalized these messages that were centrally produced by activists. The mobile phone users actively participated and aggressively changed the course of the campaign prior to the 2007 elections and ethnic based violence after the elections. In addition, the mobile phone users aware of the limitation of SMS characters and aware of negative and limited coverage given to opposition parties by the state owned media dedicatedly invented lethal negative attacks and rumors that were targeted at PNU candidates. Although the SMS targeted the people with access to mobile phones, such messages clearly appealed to them to pass on the message and to ensure that as many people as possible accessed them.

Regarding the escalating violence in Kenya, the European Union (EU) spokesman in the country, one Bernard Barret is quoted as saying that rumors were being spread by mobile phone text messages predicting imminent attacks by one ethnic group or another and that this is heightening tensions. It is difficult to attach a positive or negative value to these messages collectively. If they are true, then they serve as a useful warning, enabling those who are due to be attacked to protect themselves or to flee. If they are not true, on the other hand, they cause unnecessary panic and might lead to those receiving them planning and executing attacks of their own in order to pre-empt the attack of the perceived enemy. Such messages according to a post on a Kenyan blog "What an African Woman Thinks," were less sinister but just as dangerous. The author draws a parallel between the calls to genocide on Radio Rwanda in 1994 and the violent SMS ethnic campaigns in Kenya:

It is one thing to broadcast subversive messages on Radio as was the case in Rwanda, and is alarmingly the case with some vernacular radio stations in Kenya. It is an entirely different thing to send these messages to a carefully selected list of people on your contact list that will in turn send them on to their own select list of people so that the message spreads like a virus but catches only people who answer to certain 'characteristics.'[5]

And if you think this is farfetched, then take another look at the Akiwumi (1994) Report according to which some people defended their acts of aggression by saying that they had received word that they were due to be attacked and that, therefore, they were merely being offensive in their own defense. It was more dangerous because there was more stealth to it. It was not done in the open; it was done in secret, making it harder to put an end to it. In addition, the dissemination instrument was not situated in one central place that could be clamped down on easily as compared to old media. Rather, every mobile phone within and without the country or blogger was a potential dissemination instrument, making it almost impossible to crackdown on the proliferators of these messages. Observers were afraid that mobile phones would be for Kenya what Radio was for Rwanda.

Unfortunately in such a state of unrest, it was difficult to distinguish between fact and falsehood in the flood of text messages that were filling people's phones each day. There were also messages spreading rumors that implicated Kenyan companies and institutions in promoting violence. The Nation Media Group had to officially deny claims, spread primarily through text messages, that their vans were being used to ferry guns to far flung parts of the country. Half truths, untruths and propaganda spread like wildfire. One person sends it to five, those five send it to twenty, those twenty send it to one hundred, and so it spreads like wildfire.

Worse though, were text messages and e-mails that were unrepentantly filled with hatred and subversion. These became increasingly more frenetic in the days leading up to the 2007 general election, and reached a climax after the Electoral Commission of Kenya (ECK) botched up the tallying of votes and a disputed government was hurriedly sworn in. On January 1, 2008, Kenyans started to receive frightening text messages that urged readers to express their frustrations with the election outcome by attacking other ethnic groups. One such message reads, "Fellow Kenyans, the Kikuyu's have stolen our children's future...we must deal with them in a way they understand... violence." (Hirsch, 2008: 11). In reaction, another reads, "No more innocent Kikuyu blood will be shed. We will slaughter them right here in the capital city. For justice, compile a list of Luo's, Luhyas and Kalenjins you know...we will give you numbers to text this information." (Hirsch, 2008:11). Mass SMS tools were remarkably useful for organizing this type of explicit, systematic, and publicly organized campaign of mob ethnic violence. These messages are part of a troubling trend in East Africa. In April 2007, three Ugandans died in Kampala when violent acts were organized via SMS to protest the Government of Uganda's sale of the Mabira Forest to Kakira Sugar Works.

These were the text messages that preached a radical and dangerous message. These were the messages that tell an individual all that is good about him/her on account of ethnic identity and all that is wrong with the Other on account of their ethnic roots. These were the sometimes hysterical text messages that justify hardliner stances and violence visited upon the Other simply because they are Other (ethnically). These text messages called on the recipient to act in a certain way on the basis of their ethnicity, and further, to regard the Other or act upon the Other in a certain way because of their ethnicity. What made these subversive messages spreading by mobile phone most sinister though, was the ability to select the audience.

According Yieke (2008) mobile phones were also used to coordinate riots and attacks on various ethnic groups in Kenya. Text messages inciting ethnic violence started to spread as early as January 1, 2008, urging Kenyans to "deal with them (incumbent president Mwai Kibaki's tribe, the Kikuyu) the way they understand...violence." (Yieke, 2008:27) Messages in response called on the Kikuyu to "slaughter them (members of opposition candidate Raila Odinga's tribe, the Luo) right here in the capital city." (Yieke, 2008: 27)

There were also messages with a stern warning on members of the Kikuyu community to leave Coast region. One text message read:

Onyo kali nyinyi Wakikuyu Wakamba muliofika tayari na wale ambao tuko na habari wanakuja tutawauwa nyote. Hayo magari mlioleta tutayachoma yote. Kwa hivyo tunawapa masaa sabini na mbili muondoke. Muondoee magari yenu kabla ya muda kumalizika. Ama damu na majivu ya mali yenu itamwagika from Kaya Revolution Council. Similani Midzichenda. (Human Rights Watch 2008: 54)

Translation:

You Kikuyus, Kambas, who have already arrived here and those we have information that you are coming. We shall kill all of you. We shall burn all those vehicles you have brought. Therefore, we are giving you 72 hours to leave. Get your vehicles out of here before this time runs out. Or your blood and ashes of your property will be poured. From Kaya Revolution Council. Similani Midzichenda (sic).

In retaliation other ethnic groups such as the Luo, Luhya, Kalenjin and to a lesser extent the Kisii communities living in Central Province which is home to the Kikuyu ethnic group were targeted during the post-election violence. Text messaging and e-mailing ordering non-Kikuyu ethnic groups to leave Central Province were done with precision. The perpetrators of these evictions appeared to have known exactly where non-Kikuyu people lived, their mobile phone numbers and their e-mails. The messages were sent to non-Kikuyu residents giving them an ultimatum within which they should have left. As a result Short Messaging Services (SMS) were used to organize, rally and galvanize the Kikuyu ethnic community against other ethnic communities.

Yieke (2008) estimates that, "... according to satellite mapping of the violence in the Rift Valley, 95 percent of the recent clashes in that area have occurred on land affected by (settlement) schemes." Much of the violence in this region is motivated by majimboism, which is a Swahili term referring to the aspiration of a type of federalism composed of semi-independent regions organized by ethnic groups. To many in the Rift Valley, however, majimboism legitimizes violence against Kikuyu's who are seen as encroaching on the ancestral land of other ethnic groups. It is also worth noting that there are signs that those who attempted to seek majimboism through targeted violence largely achieved their goals. Human Rights Watch (2008: 111) reports that:

The events of the first months of 2008 have dramatically altered the ethnic makeup of many parts of Kenya. Scores of communities across the Rift Valley, including most of Eldoret itself, are no longer home to any Kikuyu residents. The rural areas outside of Naivasha, Nakuru, and Molo are similarly emptying of Kikuyu while Kalenjin and Luo are leaving the urban areas. In Central Province, few non-Kikuyu remain. The slums of Mathare, Kibera and others in Nairobi have been carved into enclaves where vigilantes from one ethnic group or another patrol 'their' areas. (See http://www.knchr.org/dmdocuments/KPTJ_Final_Press_Release.pdf)

However, since new media, unlike old media, is a multi-directional tool, there is also hope that voices of moderation can make themselves heard. In 1994 in Rwanda, radio was used to mobilize the genocide, and moderate voices were unable to respond. In Kenya, as hateful messages extended their reach into the Kenyan population, Michael Joseph, the Chief Executive Officer (CEO) of Safaricom, Kenya's largest mobile phone provider, was approached by a government official who was considering shutting down the SMS system. Joseph convinced the government not to shut down the SMS system, and instead to allow SMS providers to send out messages of peace and calm, which Safaricom did to all nine million of its customers. Kenyan mobile phone operators also cooperated with the Kibaki government to send messages to subscribers, urging them not to send or forward inflammatory messages. Juliana Rotich reported on the Mashada blog receiving the following message on her mobile phone in Eldoret: "The ministry Of Internal security urges you to please desist from sending or forwarding any SMS that may cause public unrest. This may lead to your prosecution".

Further, it is quite easy for governments and companies to identify and track individuals that promote hate speech. In the aftermath of the violence, contact information for the individuals who allegedly promoted mob violence was forwarded to the Government of Kenya. The ministry of information may have been premature in threatening prosecution for forwarding messages that incited violence. The Daily Nation newspaper (old media) reported on March 1, 2008 that the government had compiled a list of 1700 people who had forwarded text messages that incited ethnic violence. However no action was taken as Kenya did not yet have an applicable law to prosecute SMS, e-mail and blog-based hate speech; a debate was instituted in Parliament to create such a law (on electronic hate speech) which is already in operation and is being used to persecute hate speech peddlers in the forth coming constitutional referendum.

Negative ethnicity was not limited to mobile telephony. Blogs and e-mails were instrumental in documenting post-election ethnic violence in Kenya and in sharing this information with a global audience during an otherwise total media blackout instituted by the government. The Kenyan online forum "Mashada" became so filled with violent messages that its creator, David Kobia, shut it down on January 29, 2008 for a cooling-off period noting that it had been overwhelmed with hostile and divisive messages. A screenshot taken of the forum that day reads, "While we feel that people need a space to interact; the majority of interaction on Mashada.com has began to reflect the negative aspects of what is happening in Kenya":

Mashada will be back shortly...

While we feel that people need space to interact, the majority of interaction on Mashada.com has begun to reflect the negative aspects of what is happening in Kenya.

In the meantime, anyone who feels that they still want an avenue to express their views should either visit the blogs, or better still their own blog.

We still welcome any suggestions, comments or questions. Contact us.

Let us promote Peace, Love and Unity in our country. Pray for our beloved nation **Kenya.**

In an interview several months after the elections, Kobia noted that the media blackout caused many Kenyans to look for information online. He called the freedom of expression offered by Mashada a "double-edged sword," admitting that it allowed people to share "imagined account[s] of the truth" and that for those with "the intent to spread ethnic vitriol on the website this is a microphone" (www.mashada.com)

However, a few days later, Mashada's site administrator David Kobia launched "I Have No Tribe", a site explicitly centered on constructive dialogue among Kenyans. As Ethan Zuckerman writes: (I Have No Tribe) showed posts from Kenyans around the country and around the world wrestling with the statement, "I have no tribe… I am Kenyan." Kobia redirected the Mashada site to the new site, and it rapidly filled with comments - combative as well as supportive, as well poems and prayers. Kobia re-opened (Mashada) on February 14th, having elegantly demonstrated that one possible response to destructive speech online is to encourage constructive speech.

The examples of SMS and online bulletin boards illustrate the tension inherent in new many-to-many digital communications tools. In the Kenyan context, this architecture is a new space where the predatory impulse to deepen existing cultural divides meets head on with the civic impulse for constructive and healing dialogue. Just as human rights activists used digital tools to amplify their voice and lower their cost of operations, so too did citizen journalists, who used blogs, SMS, and e-mails to challenge the narrative presented by mainstream Kenyan media and the government.

14. CONCLUSION

There is solid evidence from our analysis for the proposition that new media and the World Wide Web offers a nascent public sphere in which a national conversation about ethnicity that has yet to happen might somehow be forged. It as well emerges that new media has a significant role to play in uniting or dividing a society with a variety of ethnic groups. On the contrary, this chapter also observes that ethnic based violent behaviors were largely motivated by encounters with hate speech in new media. The ethnic violence that characterized Kenya's elections appears to have been the result of deliberate manipulation and instigation by new media. From our analysis, 85% of the data analyzed indicated that the new media took a central role in fanning the post-election violence.

The chapter contends that the upsurge of ethnic violence instigated by new media has resulted into more pronounced ethnic consciousness and that Kenyan politics hinges primarily on ethnicity and not ideology. This, therefore, demonstrates that ethnic violence has had far-reaching implications for the country. Our analysis has also shown that the intervention - though late - by the mobile telephone companies and blog owners decision to counter the hate speech with messages of tolerance (for example, Kenya: One Nation, One People) and compelling local leaders to denounce egregious utterances - might have been effective in pre-empting the violence and in handling future instances. Our analysis has also shown that "some new media became sensational and unnecessarily alarmed their audiences and inflamed their passions".

Further, the chapter concludes that language is instrumental in accomplishing new media roles and new media entrepreneurs must wake up to the reality that all is not well. There is an urgent need for new media to redefine a prescriptive strategy based on historical reality and the current status of ethnic relationships in Kenya. This should encompass the promotion of ethnic enlightenment through the recognition of ethnic differences and the discovery of equitable ways to accommodate them. It is our contention that with responsible journalism, good governance, good media laws and high literacy levels, such cases as witnessed in the 2007 post Election violence in Kenya would be avoided.

The chapter's overall conclusion is that the role of the new media in Kenya's ethnic and political crisis was entirely preventable and controllable, and had it been prevented, the violence and the ethnic animosity itself may well have been much more limited. These were unambiguous criminal acts that demanded government intervention, including through Section 96 of the Kenyan Penal Code which outlaws (among other acts) language calculated to bring death or injury to any person or community of persons. New media's role in the future may be critical in the reconciliation and restoration of democratic and ethnic legitimacy in the months and years ahead.

15. SOLUTIONS AND RECOMMENDATIONS

The government and non governmental bodies, using new media, should design activities that could broaden the public space for discussing ethnicity in productive ways. Dialogues, media presentations, and the recounting of histories of inter-ethnic cooperation could all be considered. In the wake of the violence, none of these were easy to pursue and might not have been possible until more time has passed. National dialogue on ethnic relations could be a useful parallel to addressing root causes as described above. Relatedly, efforts towards post violence remedies, including reconciliation among groups, should include an explicit discussion of the relations among language, power, ethnic violence, new media and the context and histories behind the recent expressions of hate speech in new media. Kenya, Africa is not alone in needing such a conversation; the United States and other nations would benefit from attention to the power of hateful language to effect violence in various forms.

ACKNOWLEDGMENT

We would like to express our gratitude to Solomon Aunga and Meshack Nyanamba for their constructive input and meaningful discussions during the conceptual stages of this chapter. We further thank Marissa Furaha and Charles Kebaya for their constructive comments.

REFERENCES

Akiwumi, A. (1998). *Land clashes commission report*. Nairobi.

Aseka, E. M. (1989). *A political economy of Baluyia, 1900-1964*. Unpublished Ph.D Thesis, Kenyatta University, Kenya.

Bayne, S. (2008). *Post-election violence in Kenya: An assessment for the UK government*. London, UK: DFID Kenya-UK Government.

Cheeseman, N. (2008). *The Kenyan elections of 2007: An introduction. Journal of Eastern African Studies, 2(2)*. London, UK: Routledge.

Creswell, J. W. (1998). *Qualitative inquiry and research design: Choosing among five traditions*. Thousand Oaks, CA: Sage Publications.

DeFleur, M. L., & Ball-Rokeach, S. (1989). *Theories of mass communication* (5th ed.). White Plains, NY: Longman.

Denzin, N. K., & Lincoln, Y. S. (Eds.). (1994). *Handbook of qualitative research*. Thousand Oaks, CA: Sage.

Dialogue Kenya. (2007). *The Kenya national human rights commission report on post election violence*. Retrieved December 2, 2009 from http://www.dialoguekenya.org/docs/Agreement-commission-on-post-Election-violence.pdf

European Union, Election Observation Mission. (2008). *Kenya: Final report general elections*. 27 December 2007. European Union.

Ferguson, R. (2004). *The media in question*. New York, NY: Arnold Publishers.

Fleur, M. L. (1973). *Theories of mass communication* (2nd ed.). Seattle, WA: Washington State University.

Goldstein, J., & Rotich, J. (2008). *Digitally networked technology in Kenya's 2007 post election crisis*. Harvard, MA: Berkman Publishers.

Hirsch, F. S. (2008). *Putting hate speech in context: Observations on speech, power, and violence in Kenya.* Paper prepared by George Mason University.

Human Rights Watch. (2008). *Ballots to bullets: Organized political violence and Kenya's crisis of government*. Retrieved from http://hrw.org/reports/2008/Kenya0308/kenya0308web.pdf 37

Kenya National Commission on Human Rights. (2007). *Still behaving badly: Second periodic report of the election-monitoring project*. Nairobi, Kenya: Kenya National Commission on Human Rights.

Kenya Population Census. (1999). Retrieved January 16, 2010 from http://www.minority-Rights.org/download.php?id=147/html

Kiai, M. (2008). Speech, power, and violence: Hate speech and the political crisis in Kenya. In *Proceedings of the International Professional Communication Conference*. Nairobi, Kenya: Kenya National Commission on Human Rights.

Kukubor, K. B. (2006). *Ethnicity, discrimination and national integration*. Accra, Ghana: Daily Graphic.

Lynch, G. (2006). Negotiating ethnicity: Identity politics in contemporary Kenya. *Review of African Political Economy*, *33*(107), 49–65. doi:10.1080/03056240600671282

Makoloo, M., & Ghai, Y. (2008). *Kenya: Minorities, indigenous peoples and ethnic diversity*. Nairobi, Kenya: Minority Rights Group International.

Muriuki, G. (1977). *A history of the Agikuyu*. Nairobi, Kenya: Oxford University Press.

Mute, L. (2008). Kenya: Legislation, hate speech, and freedom of expression. Retrieved from http://allafrica.com/stories/200810240331.html

Mwalongo, R. (2008, January 26). Spreading the word of hate in Kenya. *Guardian*.

Mwanzi, H. A. (1977). *A history of the Kipsigis*. Nairobi, Kenya: East African Literature Bureau.

Ochieng, E. (2001). Tribalism and national unity: The Kenyan case. In Ojuka, A., & Ochieng, W. R. (Eds.), *Politics and leadership in Africa*. Nairobi, Kenya: East African Literature Bureau.

Ochieng, W. R. (1974). *An outline history of Nyanza up to 1914. Essays in memory of Professor G.S. Were*. Nairobi, Kenya: EAEP.

Ogola, G. (2009). *Why sands are shifting in the media.* (p. 4). Retrieved October 15, 2009, from www.multimediamarsgroupkenya.org/storyid265079

Ogot, B. A. (1967). *A history of the Luo*. Nairobi, Kenya: EAPH.

Onyango, J. O. (2008). *Ethnic conflict and ethnic discourse in Kenya: Advanced researchers fellowship report*. Dakar, Senegal: CODESRIA.

Oyugi, E. (1997). *Ethnic politics: An ideological framework for antagonistic social mobility in Africa. Bujumbura workshop*. Bujumbura: CODESRIA.

Reuters. (2008, January 13). *Kenya death toll hits 693: News for South Africa and the World.* Retrieved January 4, 2010 from http://www.101.co.za/index.phpset-id=nw2008011303/

Reuters. (2008). *Up to 1,000 killed in Kenya crisis*. Retrieved January 18, 2010, from http://www.reuters.com/article/homepage crisis/idus-lo743589-CH.2400

Rieder, R. (2007). *Counting the spoons in the American journalism review of February/ March 2007 issue*.

Rothbart, D., & Bartlett, T. (2007). Rwandan radio broadcasts and Hutu/Tutsi positioning. In Moghaddam, F. M., Harre, R., & Lee, N. (Eds.), *Global conflict resolution through positioning analysis*. New York, NY: Springer.

Tan, A. S. (1984). *Mass communication theories and research.* Columbus.

Wa-Wamwere, K. (2008). *Kenya: Loving each other won't cure negative ethnicity*. Retrieved January 5, 2010, from www.allafrican.com/stories/2011002010256.html

Waki, P. (2008). *Commission report of the Kenyan 2007 post-election violence*. Retrieved December 2, 2009, from www.dialoguekenya.org/docs/pev%report.pdf

Were, G. S. (1967). [Nairobi, Kenya: E.A.P.H.]. *History of the Abaluyia of Western Kenya, C*, 1500–1930.

Wodak, R. (2001). The discourse historical approach. In Wodak, R., & Meyer, M. (Eds.), *Methods of critical discourse analysis*. London, UK: Sage.

Worsley, K. J., & Taylor, J. E. (2007). Detecting sparse signals in random fields, with an application of brain mapping. *Journal of the American Statistical Association, 102*, 913–928. doi:10.1198/016214507000000815

Yieke, F. A. (2008). *The discursive construction of ethnicity: The case of the 2007 Kenyan general election*. Paper read at governing the African public sphere, Yaounde, Cameroon.

WEBSITES

http://www.allafrica.com/stories/200802291070.html.

http://www.boston.com/news/world/africa/articles/2008/06/09/kenya-mobile-phone-companys-ipo-get-warm-debut

http://www.Common.co.typepad.com/8months.2008/02/reporting.on.co.html.

http://www.e-ir.info/?p=475

http://www.flickr.com/photos/afropicmusing/sets/72157603595279896

http:www.inanafricaminut.blogspot.com/2007/04/darksideof-mobilization.html.

http://www.internetworldstats.com/stats1.htm

http://www.kenyanpundit.com/2007/12/30/media-blackout-announced/

http://www.mentalacrobatics.com/think/archives/2007/12/voting-experience-kenya-election-2007-php

http://www.recesspa.org/index.php?option=com_content&view=article&id=98:a-framework-for-media&catid=44:april&Itemid=191

http://www.westminster.ac.uk/schools/media/news-and-events/events/2010/racism,-ethnicity-and-the-media-in-africa

KEY TERMS AND DEFINITIONS

Ethnic Conflict: Cleavages between groups based on differentiations in ethnic identities.

Ethnic Group: Group of humans who share the same primordial characteristics such as common ancestry, language and culture.

Ethnicity: Behavior and feeling (about oneself and others) that supposedly emanates from membership of an ethnic group.

Hate Speech: Bigoted speech attacking or disparaging a social or ethnic group or a member of such a group.

New Media: Emergence of digital computerized or networked information and communication technologies in the latter part of the 20th century which are manipulatable, networkable, dense, compressible and interactive.

ENDNOTES

1 See, for example, David Anderson, "Majimboism: the troubled history of an idea," in Daniel Branch & Nic Cheeseman (eds.), *Our Turn to Eat! Politics in Kenya since 1950* (Berlin: Lit Verlag, 2008).

2 M-pesa is a popular safe, reliable and fast electronic money transfer system which allows Kenyans to bank and transfer money for daily business transactions using their mobile phones.

3 KAZI560 is a Kenyan job alert service that matches subscribed job seekers to their career jobs through mobile phones.

4 Make Text Not War, Humanitarian.info, http://www.humanitarian.info/2008/04/24/make-text-not-war/. of the 2007 Kenya's general elections

5 Ofeibea Quist-Arcton, "Text Messages Used To Incite Violence in Kenya," *NPR News*, http://www.npr.org/templates/story/story.php?storyId=19188853

Chapter 4
The Role of New Information and Communication Technologies (NICTs) in the Relations between the Central Government and Four Major Kingdoms in Uganda

George Musambira
University of Central Florida, USA

Samuel Muwanguzi
University of North Texas, USA

ABSTRACT

The role NICTs are playing in the relations between Uganda's central government and four of the kingdoms in the country is analyzed and placed in the historical context within which each kingdom has pursued a strategy for preservation and development of its people's unique culture and ethnic aspirations. A convergence of cellular phone and FM radio designed by the Buganda kingdom to confront the central government with specific demands and the government's response are examined. The less combative use of NICTs by Bunyoro-Kitara, Busoga, and Toro kingdoms is described. An analysis of how each antagonist uses NICTs for a hegemonistic agenda is presented and cast against the application of NICTs for good governance and the prosperity of citizens.

DOI: 10.4018/978-1-60960-591-9.ch004

INTRODUCTION

Uganda is no stranger to political strife. From its colonial experience to the present, the young nation has endured more than its fair share of political violence. However, what is novel to Uganda's tumultuous political experience lately is the pivotal role New Information and Communication Technologies (NICTs) have played in the ongoing contentious relations between the administration of Buganda's kingdom (also known as "Mengo" in reference to the location of its headquarters) and the central government.

The strained relations exist in contrast with those between the central government and the country's three other kingdoms: Bunyoro-Kitara, Busoga, and Toro. This chapter examines several ways through which each antagonist has employed NICTs to perpetuate its hegemonic interests. The convergence of FM radio and cell phones, conceptualized in this chapter, as new media in the African context, is a potent mix that has enabled Mengo to mobilize its loyalists within Buganda and abroad to publicly, sometimes with emotional disdain and ominous rhetoric, agitate for the restoration of a federal form of government for Buganda, that would purportedly restore its past political, economic, and cultural glory.

The role the new media are playing in the Buganda kingdom's agitation is distinctive when compared with the impact the same NICTs have had in the other kingdoms in Uganda such as Bunyoro-Kitara, Busoga, and Toro where relations with the Uganda central government are less contentious. This chapter illustrates how, through the use of new media, the common person in Buganda has been a target of manipulation on the one hand by the resurgent and lately recalcitrant if not combative Buganda kingdom, and on the other by the central government with its populist but, in many cases, hollow rhetoric. The chapter shows how each of these antagonists is pursuing a hegemonistic agenda into which the common person in Buganda has been inadvertently drawn as a pawn in the political drama. In particular, the common person's access and use of the cell phone to engage in live radio talk shows and also to participate in the so-called "Bimeza" or public forums relayed live on the Buganda kingdom radio station, the Central Broadcasting Service (CBS), is most illuminated in this political drama.

We contend that the convergence of new media; the FM radio and the cell phone, a popular and easily accessible platform on which calls into FM radio broadcasts can be made, has helped to galvanize and embolden Mengo's efforts to rekindle Buganda's relentless agitation for her glorious past and future political agenda. This analysis of new media occurs against a backdrop of the running conflict, recently turned bloody, but not without historical precedent, between Mengo and the central government.

The chapter suggests that rather than deploy NICTs for parochial and selfish political interests that inflate the egos of key antagonists, they should instead be used for civic education that fosters constructive dialogue, ethnic relativism, national unity, and development. The chapter also points out that the global winds of change that are primarily driven by NICTs demand that central government and traditional leaders must, as of necessity, adapt to new global changes that emphasize, to a large extent, horizontal collaboration, communication, and mutual respect rather than the fading command and control structure that dominated the world stage prior to the ubiquitous access and use of NICTs.

BACKGROUND

In order to put the contentious relations between Mengo and the central government in perspective, a historical review of the genesis of this conflict is apropos. The conflict dates back to Uganda's pre-colonial era when Buganda kingdom emerged as one of the most powerful states in East and

Central Africa, with a highly organized and superior system of government.

The kingdom's political, military, cultural, and economic structures revolved around the kabaka (king). A highly revered cultural and political leader of the Baganda, the kabaka is looked upon as an embodiment of the community's welfare, aspirations, and symbol of its existence and survival (Kiwanuka, 1971). It is this inextricably linked cultural and political institution, an entrenched handwork of centuries of evolution, that, to the kabakas and their loyal subjects, is a priceless source of identity and belonging that must be preserved and perpetuated at any cost. Indeed, while echoing Apter (1967), Mulira (2009) stated that historically, the title of kabaka is a political one that refers to an absolute ruler of Buganda and only changed to a "titular ruler" of Buganda in the constitutions of 1955, 1961 and 1962. Mulira adds that it is therefore inconceivable in the Baganda mindset to imagine a ruler whether absolute or titular who does not wield political power.

The History of Buganda in Uganda

Thus, the current conflict is rooted in Buganda's historical and strategic maneuvers to preserve its territorial integrity, identity, and the kabaka as sole arbiter of her people's cultural, political, and economic destiny. The pre-colonial regime of Kabaka Mutesa 1 is a reasonable starting point for examining and understanding this geo-political dynamic. Long before the British declared Uganda a protectorate in 1890, Kabaka Mutesa 1 had made a strategic move to get into the good graces of Arab ivory traders in the early 1850s, by readily opening his kingdom to commercial links with them. The move assured Buganda of a steady flow of arms and ammunitions to protect the kingdom from external aggression and preserve her sovereignty especially from the expansionist tendencies of the rival Bunyoro-Kitara kingdom.

Kabaka Mutesa 1 also proactively invited British Protestant missionaries, the Christian Missionary Society (CMS), who arrived in the kingdom in 1877. He quickly countered their influence in his Kingdom by inviting French Catholic missionaries and in 1879, Roman Catholic priests from the Society of Missionaries of Africa (White Fathers), and a French religious order, arrived in Buganda. Moreover, through his earlier contacts with Arab traders, Mutesa 1 had also courted an Islamic presence in his kingdom. Through guile, the kabaka managed the tripartite religious scenario well enough to ensure that none of the foreign forces was any more powerful than him.

By the time of his death in 1884, both Protestant and Catholic missionaries had established their presence in his palatial court. The Protestant and Catholic missionaries competed for converts in the kingdom and became embroiled in local politics that culminated in the Buganda religious wars that pitted the Protestant converts armed by the British against Catholic converts supported by the Germans, and later against Muslim converts. Following the death of Kabaka Muteesa 1, the Buganda throne was assumed by heirs who could not match his foreign affairs adroitness. For example, when Kabaka Mwanga assumed the throne, he switched religious allegiances so haphazardly and failed to moderate the influence of the three religions over his subjects. In fact, the renowned Uganda Martyrs including Bishop Hannington were his bidding in trying to show his supremacy in the kingdom. By 1888, the religious rivalries had reached a point of no return and a series of religious wars erupted in the kingdom with an initial Muslim victory resulting in the deportation of Kabaka Mwanga.

Byrnes (1992) and Apter (1967) state that allied Christians converts soon rallied forces and overthrew the Muslims and, while their alliance lasted, installed several puppet rulers. The alliance of Protestant and Catholic converts was short-lived as in January 1892, fighting broke out between Protestant and Catholic converts. Initially, Catholic converts emerged victorious but Capt. Lugard of the British colonial force intervened on behalf of the Protestants, defeated the Catholics and reinstated Kabaka Mwanga as a puppet king. According to Byrnes (1992), Cohen (1970), Apter (1967) and Kiwanuka (1971), unable to stand the humiliation in his kingdom, Mwanga escaped from the palace in July 1897and launched a resistance against the British. He later linked with Omukama (king) Kabalega of Bunyoro and continued the resistance until they were both defeated and captured in 1899.

Following the defeat of Mwanga, his 3-year old son Prince Daudi Chwa 11 ascended the Buganda throne and the Baganda Christian allies of the British continued to support the colonialists in their wars to conquer other territories that form the modern state of Uganda. As a reward for their formidable military support and collaboration, the British negotiated and signed a separate treaty; the Uganda (Buganda) Agreement of 1900 in which Buganda was granted semi-autonomy and self-government within the protectorate and rewarded with half of Bunyoro's conquered territory that doubled the size of Buganda from ten to twenty counties (Low & Pratt, 1960; Byrnes, 1992; Apter, 1967). However, the "lost counties" of Bunyoro were to remain a sore spot between Buganda and Bunyoro kingdoms to the present day.

Even during the colonial era, Buganda's efforts to preserve her identity amidst all the privileges she enjoyed over the other kingdoms were relentless. For example, from the late 1920s through the 1930s, the Kabaka was a symbol of resistance against the colonial plans to introduce Swahili as the country's lingua franca. The Kabaka's efforts, according to Mulira (2009), led to the rapid development of the Luganda and other local languages' authography as a counter move against the introduction of Kiswahili.

When the country was moving closer to independence in 1962, Buganda pursued her interests independently while the rest of the country's interests were managed via evolving legislative and executive councils; a development that awkwardly resulted in the granting of independence to Buganda on October 8, 1962 and to Uganda on October 9, 1962 with Buganda as a special federal state within Uganda.

Thus, at independence, the Buganda kingship controversy was the most sensitive and important political question—how much control over Buganda should the central government have and how much power should the Kabaka wield in a unitary government? When Mengo boycotted the 1961 elections and Democratic Party (DP) won the elections in Buganda, Uganda People's Congress (UPC) and Kabaka Yekka (KY) led by the kabaka concluded a fragile and convenient political alliance between the two erstwhile ineffectual parties to deny the more popular DP majority seats in parliament.

As a result, at independence in 1962, UPC's Obote became the prime minister and Kabaka Muteesa 11, became president. As prime minister, Obote was responsible for holding together the loose coalition of the diverse interests both within his party and with KY. According to Low and Pratt (1960) and Byrnes (1992), Obote called for a referendum over the "lost counties" of Bunyoro that the British had donated to Buganda.

Although the Kabaka opposed the plebiscite, Obote jammed the motion through parliament and when the referendum was held in Bunyoro, the "lost counties" voted overwhelmingly to return to Bunyoro and parliament duly endorsed the result. Nothing worse could have strained the relationship between Obote and the Kabaka than the loss of the annexed counties. Buganda's members of the legislative council and Obote's opponents in his ruling UPC, acting on evidence from a bank where Colonel Idi Amin, Obote's military confidant, had deposited an unusual huge amount of cash from sales of smuggled gold and ivory from the Congo, moved and passed a motion of no confidence in the prime minister on charges of corruption on February 4, 1966, while Obote was on a trip to West Nile, a remote territory in the north of the country (Low & Pratt, 1960; Byrnes, 1992).

Rather than resign his position as prime minister following the no confidence vote, Obote turned to Col. Idi Amin and the military and effectively staged a coup against his own government. He suspended the constitution, arrested the offending cabinet ministers, forced a new constitution through parliament without debate, abolished the federal powers of all kingdoms most notably the internal autonomy enjoyed by Buganda, and concentrated all the presidential powers in the office of the prime minister. Buganda objected and the Mengo leadership rhetorically demanded that Obote remove his illegal government from Buganda soil (Byrnes, 1992). Moving swiftly, Obote directed his military ally, Colonel Idi Amin to attack the kabaka's palace in Mengo. The palace was overrun but the kabaka managed to escape and flee into exile in London where he died poor and miserable in 1969.

The new constitution introduced by Obote in 1967 declared Uganda a republic, completely abolished all Kingdoms, cultural and hereditary leaders, and martial law was slapped over Buganda—a new development that introduced military intervention and domination in Uganda's post-colonial politics (Low & Pratt, 1960; Byrnes, 1992). After the 1967 constitution's abolition of all kings, the Ugandan army turned the kabaka's palace into their barracks and the Buganda parliament building into its headquarters. After Idi Amin overthrew the Obote 1 regime, he returned the body of Kabaka Muteesa 11 who had died in exile in London but neither restored the monarchy nor returned the seat of the kingdom to the Baganda.

When President Museveni's National Resistance Movement shot its way to power in 1986, according to Byrnes (1992), it was quite ambivalent in its response to the emotion-laden issue of the restoration of the Buganda monarchy. Until its final year, President Museveni's guerrilla struggle to remove Obote had been conducted entirely on Buganda soil and involved a large number of Baganda fighters, capitalizing heavily on the revulsion most Baganda felt for Obote and his UPC government. The fact that Museveni was reported to have promised in public, towards the end of the guerrilla struggle, to restore the kingship and to permit Ronald Muwenda Mutebi, the heir apparent, to become king did not help diffuse the controversy. Besides, there were many other Ugandans opposed to the restoration just as strongly, primarily for fear of the political advantages it would bestow upon Buganda. Eventually, in 1993, owing to the political pressure from Buganda loyalists, the NRM government enacted a law that restored traditional leaders paving the way for the coronation of Prince Ronald Muwenda Mutebi as kabaka of Buganda as a cultural king without political power. This was interpreted by the Buganda kingdom faithful as hot air or "Byoya bya Nswa," literally meaning useless fluffy ant feathers.

Besides the foregoing historical raison d'être, other reasons why Mengo believes Buganda holds special status vis-avis other kingdoms and regions include its geographic location in the centre of the country. Buganda is the most strategic region in Uganda where the nation's commercial and political capital is located. In addition, Buganda is the most populous region with a culturally cohesive ethnic community whose political and cultural systems are not just inextricably linked to but also revolve around the kabaka, a larger than life figure who doubles as the political and cultural head of the Baganda under their well-entrenched clan system. When the Odoki Commission began collecting views from all over the country to draft the new 1995 constitution, over 90% of the people in Buganda demanded that a federal form of government be restored in the kingdom (Odoki Commission, 1993). In the subsequent Ssempebwa Commission of 2003 (as cited in Namutebi, 2009), 68% of all Ugandans demanded that the country be governed under a federal system of government.

However, the Museveni government used its majorities both in the Constituent Assembly in 1995 and in the Uganda Parliament in 2002, to defeat the Buganda-spearheaded relentless clamor for a federal type of government. The failure to grant a federal status (special treatment) to Buganda on both occasions, was interpreted by Buganda loyalists as an anti-Buganda agenda harbored by the Museveni government. The two developments started to strain the hitherto cordial relationships between President Museveni's NRM government and Mengo. While the restored Buganda kingdom in 1993 was to operate as a cultural institution with no role in politics, several of the Buganda palatial ministers openly engaged in politics—and that would be acceptable if their involvement was not viewed by the Museveni regime as an alliance with opposition political parties to undermine government programs. Matters came to a head in 2007 when the NRM government introduced the Land Amendment Bill that sought to criminalize the eviction of tenants by landlords (Bako, 2009; Tangri & Mwenda, 2010).

As the largest landowner in the country, the kabaka of Buganda and landlords in Buganda, many of whom held top positions in Mengo, viewed the bill as a ploy by the government to grab Buganda's land, destroy the kingdom's economic and cultural foundation, donate their land to foreigners, impoverish the Baganda, and emasculate the kabaka. The Mengo establishment openly and fiercely opposed the proposed legislation and reignited her calls for a federal form of government for Buganda to ensure her survival and destiny (Tangri & Mwenda, 2010). According to Mulira (2009), since 2005, when a group of radical Baganda politicians were appointed to leadership positions in Mengo, they orchestrated a fiercely defiant and anti-government campaign that fermented a hostile mood towards the Museveni regime, publicly opposing any interface between the kabaka and the president. Indeed, as a sign of opposition to the proposed legislation on land, the kabaka rejected all invitations to meet with President Museveni and declined to answer his phone calls for over two years.

It is this historically privileged status, wrung deep inside its identity, that has given Mengo the impetus to stand up to Uganda's central government with bloody consequences first during the so called kabaka Crisis of 1966, and then now with the current imbroglio over the NRM government land legislation and the refusal to restore Buganda's federal status. It is within this context that NICTs are being engaged by the kingdom to restore its lost status.

The Other Three Kingdoms

Because there has been little, if any, agitation for federal status from the other three kingdoms, new media in these monarchies have not been put to use for purposes of asserting political autonomy and ethnic nationalism as has been the case of defiant Buganda through its mouthpiece; CBS. The radio was closed following the September 11, 2009 riots in Kampala allegedly incited by CBS itself, which left nearly 30 people killed, hundreds injured, and properties worth millions of shillings destroyed. After keeping CBS shut down for one year, and in what some saw as a calculated move to appease Baganda voters ahead of the imminent national elections, President Museveni's government allowed it back on the air (Nalugo & Mugerwa, 2010).

The case for Bunyoro-Kitara kingdom. The Bunyoro-Kitara kingdom reached its peak of power and glory in the pre-colonial era after Omukama Kabalega ascended the throne in 1867. According to the official Bunyoro-Kitara Kingdom Web site (2007) Omukama Kabalega transformed the culturally declining kingdom into a regional trading center exchanging salt from Lakes Katwe, Kasenyi and Kibiro and ivory for guns and ammunitions to consolidate his internal power base and defense against external aggression from the neighboring Buganda kingdom and the advancing British colonialists. The kingdom's blacksmiths made spears, pans and other commodities for export.

Omukama Kabalega, who ascended the throne of the declining kingdom from his aging father

(Kiwanuka, 1968), quickly built a strong army and constructed a 10 mile underground bunker at Kakerenge with grain enough to see his army through the worst famine in time of war. And war indeed came in the form of British colonialists who sought to take control of the region and forestall the French who were moving southwards from the Sudan.

Omukama Kabalega resisted colonial rule by launching guerrilla warfare against the British for eight years from 1891 to 1899. Before his defeat, he linked up with kabaka Mwanga of Buganda (as mentioned earlier) fighting together gallantly until they were captured on April 4, 1899 in Lango, northern Uganda, then exiled to Seychelles Islands where Mwanga died while Kabalega was allowed to return but died in Jinja in 1926 (Bunyoro-Kitara Kingdom, 2007; My Uganda, 2010a). To this day, Kabalega remains an icon of resilience and resistance towards colonial rule in Africa.

The fall of Kabalega had marked the fall of Bunyoro and British colonization effectively killed the kingdom. The British donated huge chunks of Bunyoro kingdom land to Baganda collaborators who had allied with them to defeat Kabalega (Roberts, 1962). It is estimated that nearly 2 million Banyoro were killed during the Kabalega resistance and the Banyoro have since taken the British government to court demanding war reparation (World History Archives, 2004). When Uganda attained independence in 1962 with Bunyoro under a semi-federal system of governance, hopes were raised that the kingdom would emerge from the neglect it had endured during the colonial era. Although the lost counties were returned to Bunyoro during the Obote 1 regime in 1964, ownership of the land still remained in the hands of absentee Baganda landlords to this date. In addition, the semi-federal status of kingdoms was abolished by the Obote I administration in 1967 which dashed any hopes of the region emerging from the fangs of poverty and helplessness.

After the restoration of traditional institutions in 1993 by the NRM government, rival princes; John Mpuuga Rukidi and Solomon Iguru fought protracted succession battles in courts of law and wrangled over the kingdom's assets, which according to Israel Ndahura (My Uganda, 2010a), the Bunyoro-Kitara kingdom prime minister, further paralyzed the development of the area. Eventually, Prince Gafabusa Iguru won the case and on June 11, 1994, he was crowned the 27th Omukama in the Babiito dynasty of Bunyoro-Kitara kingdom.

Today, Bunyoro kingdom comprises four districts; Hoima, Masindi, Buliisa, and Kibaale with widespread poverty as the commonest denominator among them. Since his coronation in 1994, Omukama Iguru has been a strong but subtle supporter of President Yoweri Museveni's NRM government. However, the kingdom is still dogged by widespread poverty, lack of infrastructure, denial of rights to indigenous lands, a lingering sense of apathy, and neglect. Whether Bunyoro kingdom can rise to the standard that Kabalega had lifted it and once again lay claim to greatness are the questions the Banyoro confront as they embark on translating the recently discovered huge commercial oil deposits into real economic gains for the region.

However, the oil deposits and the contentious land conflicts pitting the indigenous Banyoro against the largely Bakiga immigrants now commonly referred to as the Bafuruki, the alleged involvement of Libyan President Muammer Gaddafi in Uganda kingdom affairs; he has promised setting up a radio station for Bunyoro kingdom, could engender a new course in the otherwise delicate relations between the once-upon-a-time powerful Bunyoro-Kitara kingdom of Omukama Kabalega and the central government. Whether the Banyoro will muster the courage to optimize their enhanced status as seen through the potential oil-boom and demand their pound of flesh from the central government, is a matter that could be resolved if committed to the new-found platform offered by NICTs. As noted above, the kingdom does not own a radio station. However, whenever kingdom officials appear on stations owned by individuals in the kingdom, in addition to taking care of their official media tasks, they often spare a moment to emphasize their support and partnership with the NRM government to implement development programs for their respective regions.

However, the recently-released Bunyoro-Kitara Kingdom proposal to government on the sharing of the oil revenue that was largely informed and shaped by the extensive use of NICTs in the collection of data to back their demands for a constitutional share of the oil revenue, points to an emerging new dynamic in the NRM-Bunyoro oil relationship. According to a news report (Ultimate Media, 2006), on top of elaborate consultations with all local stakeholders, the kingdom extensively used local and international media including newspapers, radio, television, and the Internet as sources of valuable information that offered a basis for the kingdom to develop a proposal to government to ensure equitable distribution of oil revenue.

The implications in the emphasis on the effective use of NICTs to inform and shape Bunyoro kingdom's proposal to the Uganda government demanding specific, fair, and constitutional distribution of oil revenues should not be lost on us. The same NICTs could be deployed to launch a popular campaign to resist any moves by the government to short-change the currently obliging Banyoro on the sharing of oil revenues and other attendant benefits, to which, they strongly believe, have a God-given birth right to receive and use.

The royal succession rivalries in Busoga kingdom. The kingdom of Busoga, located in eastern Uganda bordering Buganda to the east, is situated east of the River Nile, north of Lake Victoria, south of Lake Kyoga and bordering River Mpologoma to the East, comprises eleven counties that were administered by hereditary rulers as independent principalities in the pre-colonial era (Byrnes, 1992). A predominantly agricultural area, the principalities of Busoga were, in 1896, together with the kingdoms of Toro, Ankole, and Bunyoro, added to the British Protectorate that originally comprised Buganda (Cohen, 1970). Cohen further states that in 1906, the British Protectorate consolidated all the diverse principalities of Busoga into one administrative entity with the central authority granted to the Isebantu, the kyabazinga of Busoga as an elected king from five of the eleven princes on a rotational basis.

When Uganda attained independence in 1962, with the kabaka of Buganda Edward Muteesa 11 as president, observes Byrnes (1992), the kyabazinga of Busoga, Sir William Wilberforce Nadhiope, with political leanings to the UPC party of Milton Obote, became the first vice president of the republic of Uganda. However, the post of kyabazinga, like those of all other cultural and hereditary rulers was abolished by the Obote 1 government in 1966 followed by the promulgation of the republican constitution in 1967.

The Busoga Kingdom official Web site (Busoga Kingdom, 2007) states that when cultural and hereditary leaders were restored by the 1993 Act and entrenched by the 1995 constitution, Isebantu Henry Wako Muloki, who had ascended to the Busoga throne from Nadhiope, was reinstated as the Kyabazinga on February 11, 1996 (Musoke & Bita, 2008). Muloki's reign as king of Busoga, with 8 percent of Uganda's almost 25 millionpopulation o (Busoga Kingdom, 2004; Otiso, 2006), was marked by unreserved support to the ruling NRM government with a particularly cordial relationship established between the kyabazinga and President Yoweri Museveni until the former's death on September 1, 2008 (Muyita & Bareebe, 2008).

The bitter internal struggles for succession to the Busoga throne among princes from the royal clans in Busoga, could, however, potentially disrupt the otherwise amicable relations with the central government. And, such rivalries could offer cannon fodder to the burgeoning new media in Busoga as politicians from various political parties criss-cross the region in search for votes in the forthcoming 2011 elections. While Busoga kingdom does not own a radio station, rival princes that subscribe to different political parties could be lured by politicians to deploy the new media to advance their interests in the name of fighting for Busoga's needs.

Toro's frail and caged history. Founded in 1836, Toro kingdom seceded from the declining Bunyoro-Kitara empire by Prince Olimi Kaboyo, the oldest son of the aging Omukama Kyebambe Nyamutukula as a revolt against the cultural adulteration caused by the excessive settlement of foreign immigrants in the empire (Kiwanuka, 1968; My Uganda, 2010b).

The breakaway of Toro, with the richest and fertile counties, home to the largest salt deposits in Lake Katwe and the bulk of cattle herds, denied the empire her greatest source of revenue. In the ensuing years, the newly founded kingdom soon became a contested battleground between the resolute Omukama Kabalega who ascended the Bunyoro-Kitara throne in 1867 on one hand, the colonial government's Capt. Fredrick Lugard with his intimidating Maxim gun and Buganda kingdom armies on the other, who took turns to protect and reinstate the kings of the nascent and fragile Toro kingdom (Low & Pratt, 1960; Roberts, 1962).

Not until Omukama Kabalega was defeated and exiled by the British colonial army in 1899, did Toro witness some semblance of calm. According to Cohen (1970), years that followed the British Protectorate over the kingdom from 1896 through the colonial era to the post-independence 1966 Uganda crisis when Prime Minister Obote abolished cultural institutions in Uganda, Toro's economy flourished through the exploitation of her lucrative salt, fisheries, and copper resources in addition to the tea and tourism industries.

The young Omukama Patrick Olimi Kaboyo, who had just succeeded his father, Omukama Rukidi Kamurasi in March 1966, was forced into exile (Byrnes, 1992). Not surprisingly, the young King's elder sister, Princess Elizabeth Bagaya, strongly backed Yoweri Museveni's National Resistance Army (NRA) guerrillas in their five-year bush war against the Obote 11 regime leading to capture of power in 1986. Before the traditional leaders were restored by an Act of parliament in 1993 and entrenched by the 1995 constitution, Princess Bagaya and her brother Omukama Patrick Kaboyo served in Yoweri Museveni's NRM administration as ambassadors in the United States and Cuba, and Tanzania respectively.

Omukama Patrick Kaboyo was reinstated to the Toro throne following the enactment of the 1993 law that restored traditional and cultural leaders in Uganda. However, two years later, in 1995, he died and was succeeded on September 12, 1995 by his 3-year old son; Prince Oyo Nyimba Kabamba Iguru who only turned 18 in April this year. In an interesting repeat of history, President Museveni was appointed as one of the three regents of the infant King; acting up the combined roles of Lugard and Mutesa I—a role he has ably played to enhance his political fortunes in the region.

While the kingdom of Toro owns and operates Voice of Toro—one of the first kingdom FM Radio stations in the country, the mouthpiece has consistently toyed the official line and has steered clear of any controversy with the central government. Even in the face of apparent marginalization of the region by the NRM government especially in the wake of overwhelming support its residents give the ruling government in presidential, parliamentary, and local government elections, no voices of dissent have emerged from the kingdom save for occasional fringe complaints by journalists and disgruntled NRM party members.

As for the demand for a federal status, the kingdom has not taken up the mantle of agitation. Except for some private FM radio stations in the region that frequently have run-ins with the Queen Mother and local politicians and complaints about corruption, the demand for federal status in this kingdom is as good as dead or at most, in hibernation. However, there are fears coming out of the kingdom that external forces are up to their notorious historical role in Toro—offering royal guidance and protection to the frail kingdom to rump up its demand for a federal status.

Whether the kingdom's Voice of Toro will be deployed in the agitation for the federal form of government during the forthcoming elections in 2011, is a matter of conjecture and goes against any historical precedent in the kingdom. But given the dynamics of new media, it would be an exercise in futility to determine whether changing circumstances, if indeed massively popular, could not lead to the deployment of NICTs to advance the kingdom's interests.

Therefore, the emerging trend of NICT use in the three kingdoms discussed in the foregoing sections might witness their dedication to pursuing parochial interests if the central government does not address some of their discontents ahead of the 2011 general elections. Only time will tell.

BUGANDA KINGDOM (CBS) RADIO CAMPAIGN AND UGANDA GOVERNMENT'S RESPONSE

The conflict came to a head in 2007 when the government announced its initiative to amend the land act with the aim of protecting peasants, on both privately owned land or "mailo land' and public land, who are vulnerable to evictions by landlords (Kiwere, 2008). In response, the Buganda kingdom officials engaged the airwaves via the kingdom owned CBS FM radio to resist and rally the kingdom's faithful to rise up against the proposed land proposals (Olupot, 2008; Tangri & Mwenda, 2010). According to CBS (2008), the "government had a hidden agenda to grab kabaka's land and leave all the Baganda landless." The intensive participatory media campaign aired on the kingdom's CBS FM radio, drummed up open disdain for the government among the Baganda and turned into an instant victory for Mengo.

The broadcasts were in Luganda, the most widely used local language in Uganda, and the radio stations' live talk shows employed a listener-participatory format involving live-phone-in protocol on the two frequencies; 88.8 and 89.2 FM. Ssemujju (2008a) reported that this move enabled the tapping of all the echelons of the Buganda kingdom across economic, gender, rural versus urban, or literacy demarcations.

Moreover, the FM broadcasts were augmented by Web casting to engage the economically and politically vibrant Baganda in the Diaspora. In Sssemuju's (2008a) view "In return, the Baganda in the diaspora also weighed in by calling directly into the programs. The programs accomplished a phenomenal listener response in both Uganda and in the diaspora expressing listener opinions" (p. 1).

As such, the FM radio broadcasts and cell phones converged to heighten radio program participation to extraordinary levels. Talk show moderators engaged an innovative platform commonly dubbed "Bimeeza" or "Round Table." Armed with a cell phone, a talk show host carried radio to the people by dropping in at any public space. He or she would invite an audience to contribute to a current debate which would then be relayed live to the studio and then aired in real time. Interestingly, it was not unusual for government supporters, officials, ministers, and even President Museveni himself or his confidants to call in to offer their side of the story. For Nnyago (2008), "The *Bimeeza* by CBS radio have become the most popular and effective way of participation and resistance to unpopular government policies especially the land amendment bill. Government attempted to ban them but due to fear of a backlash, the bid was dropped" (p. 12).

When the Buganda Lukiiko (parliament) proposal to the central government to withdraw the land amendment bill was rebuffed, Mengo decided to boost its media campaign by launching a Central Civic Education Committee (CCEC) composed of politicians opposed to the central government (Kiwere, 2008). The CCEC was charged with the mandate "to educate the people of Buganda through radio about the Kingdom's position and the threat to their cultural inheritance and the only economic source of livelihood," (Ssemujju, 2008d, p. 1). CBS radio programs were interspersed with an incessant barrage of *promos* such as: *Ettaka lyaffe ligenda* or "Our land is going;" "Abagwira bagenda kubba *ettaka Lya Kabaka"* or "The King's land is being stolen by foreigners;" *Muveeyo tukuume ettaka lyaffe* or "Come out and we fight to protect our land from grabbers and foreigners," (CBS, 2008).

All the while, the nation's print media teemed with banner headlines quoting guest speaker after guest speaker on the kingdom CBS radio excoriating the central government for its perfidious manner of handling the land question in the country. The present authors identified a total of 160 front page and national news stories on the land impasse in the three major Ugandan newspapers, the *Monitor*, the *New Vision*, and the *Ugandan Observer* during the period January 2 and June 15, 2008.

Underscoring the resounding "success" of the CBS radio campaign in the power struggle between the central government and the Buganda kingdom regarding the land question, Nyago (2008) opined that:

All the major newspapers in the country; the New Vision newspaper, the Monitor newspaper, and the Uganda Observer have variously reported that the kabaka of Buganda, the Lukiiko (parliament) of Buganda, the 52 Buganda clan leaders, the association of bankers, the Catholic Church—nearly 50 percent of all Ugandans are Catholics, the Uganda Joint Christian Council, land economists association, all opposition political parties, the Uganda Law Society, parliamentary caucuses from other regions of Uganda such as Acholi, Lango, Karamoja, and Teso have rejected the land amendment bill because CBS radio has made it too risky to support and alienate themselves from their constituents." (p. 12)

According to Obbo (2010), the political passion with which ordinary people participated in the CBS orchestrated "Bimeza" was a politically dangerous development in the eyes of the ruling NRM government because the response was similar to creation of a popular mass movement never witnessed in Uganda before. Obbo speculates that because the Bimezas were too successful for CBS existence, the radio may not open until after the 2016 elections.

Generally, the broadcasts were marked by defiance, a recalcitrant stance on the land problem, and incessant agitation from all corners of the kingdom for Buganda kingdom's federal status. In some cases, albeit rare, some speakers even preached a total split of the Buganda kingdom from the rest of Uganda. Dismaying to the entire nation were prophecies of doom stressed in the broadcasts drawing parallels between the fate of Uganda's land deadlock and the Rwandan genocide of 1994 or the Kenyan ethnic bloodbath following the controversial 2007 presidential elections (Akaki, 2008).

Attuned to the persistent assault on his person and the authority of his government, President Yoweri Museveni took the kabaka and the CBS radio head on. In a hard-hitting letter published in the print media, he accused Mengo of all kinds of vices the least of which were intrigue, bad faith, sedition, and acting as avenues for opposition politicians to incite the public through the kingdom's radio sabotaging government initiatives, especially the land amendment bill. He impugned the motive and veracity behind the slogan *Ettaka lyaffe ligenda* (our land is going) by mocking: *Ligenda wa*? or "Where is it going?" Moreover, the potency of the newly formed Bimeeza (round table broadcasts) was not lost on President Museveni as he scolded those who "waste their time abusing him on Bimeeza and chide[d] them to stop wasting time and go and engage in agriculture to produce more food and take advantage of the rising food prices on the global market," (Muwanga, 2008).

In an attempt to deescalate the clash, the president called for a summit with the kabaka but to no avail. The kabaka snubbed the invitation insisting that the meeting should only occur at his bidding (Ssemujju, 2008c; Ssenkabirwa, 2008). Offended by Buganda and kabaka's parochial and recalcitrant attitude, President Museveni unleashed his security apparatus on Mengo.

First, the police criminal investigation department investigated "four Buganda government officials over allegations of inciting violence and promoting hatred against the government in the stand off between Mengo and the central government over land matters. The officials included "the kingdom's Attorney General, the CBS general manager, and two CBS radio presenters," (Mwanje & Ssenkabirwa, 2008). Second, a female opposition member of parliament representing a constituency within Buganda was investigated and later charged with sedition and fomenting ethnic hatred.

During his televised state of the nation address, President Museveni made a point to belittle the African monarchy as "traditional chiefs" (a put down) who are "irrelevant to modern African leadership because they had failed to defend the continent from colonialism" (Muhumuza, 2008). He also castigated Buganda's clamor for land on CBS radio as wrong headed and irresponsible (Among & Tebajjukira, 2008).

Not to be outdone, Buganda kingdom officials fired back through the kingdom's radio accusing the president of using them as distraction from his government's failure to provide even the most basic social services to the people, and to mask a record as the most corrupt of all regimes since the country's independence (Mwanje, 2008). One kingdom official even retorted that "Traditional chiefs were not corrupt. According to Mwanje, they did not mortgage the country to foreign interests as this government has done. We will not accept to be smeared. We will continue demanding what is due to us. Nobody will intimidate us into submission. We will not stop unless you suck our blood."

The government then further legally couched its actions against CBS radio and other "irresponsible media" by convening a cabinet retreat to set up a sub-committee on the media charged with: "[studying] all aspects of media management, including the ownership, media laws, funding, man-power, training and qualifications and then advise cabinet to later come out with a government position," (Musoke, 2008). But this move stirred the ire of all kinds of popular media in the country, captured in headlines such as: "a crackdown on the media," "muzzling the critical media," "the death of democracy in Uganda," "central government loses battle for hearts and minds," "cabinet to delete "press freedom" from constitution", and "press freedom remains elusive in Uganda" (Butagira, 2008; Matsiko, 2008; Ssengoba, 2008, Sserwanga, 2008; Ssemujju, 2008d).

Hegemony and the Political Relations between Uganda's Central Government and the Country's Four Major Kingdoms

The concept of hegemony treats cultural processes as central to understanding the role communication plays in facilitating power relations between social entities (Park, 1998). We follow with a review of the concept and a delineation of its implications for analyzing the political relations between Uganda's central government and the country's four major kingdoms.

The concept of hegemony. Gramsci (1971) originally formulated this concept to encapsulate how, in capitalist societies, a dominant class attains control of a less powerful one politically, culturally, ideologically, economically and morally. Ever since, proponents have expanded the concept's application to cover virtually any asymmetrical power relationship. The control or dominance is exercised via an amount of force or consent that is commensurate with how much members of the dominated entity are enamored by their dominant counterparts. For Gramsci, in capitalist societies these entities are part of the superstructure rather than the base or infrastructure that includes the means of production.

The superstructure, then, involves civil society or the private sphere and political society consisting of the state machinery. In a hegemonic relationship, civil society is subjected (though as emphasized later, not as a passive party) to coercion or persuasion by political society in order to uphold particular meaning structures or practices. This implies that the more democratic the context is the more the control is likely to be secured by consent of the dominated entity. Likewise, the more autocratic the context is the more the control is likely to be achieved by coercive means.

However, Anderson (1964) extols the cultural basis of hegemonic control by arguing that this "dominance of one social bloc over another [is] not simply [secured] by means of force or wealth, [but] by total social authority whose ultimate sanction and expression is a profound *cultural supremacy*" (p. 37 emphasis ours). To the extent that such cultural supremacy is in place, domination will transpire without the dominant entity having to resort to coercion in order to safeguard its privileged status. However, when the cultural supremacy is challenged or threatened use of force to secure the hegemonic relationship becomes a more viable option.

Hegemonic relations are viewed as not constant or given but malleable or dynamic and in the hands of both the subjugated and the dominant parties. It is recognized that the relationship fluctuates due to changes (economic, political, cultural, etc.) in either of the parties or both. Thus, the subjugated party is not viewed as passive in a hegemonic relationship. It is just that the meanings, practices, and interests that the subjugated party engages at any given time are the ones that tend to favor or privilege the dominant entity and the ones it holds. Hegemony, then, constitutes an exploitative dominant ideology but it also implies the possibility of alternative ideologies to which the parties can subscribe.

The above review of the hegemony concept has important implications for understanding the role NICTs have played and continue to play in the cultural and political dynamics involving Uganda's central government and the Uganda's four major kingdoms. By pursuing policies of a liberalized economy since the 1980s, the government of Uganda set the country on the path to a capitalist society (Ochieng, 1991). Therefore, based on Gramsci's conceptualization of hegemony the Ugandan political context is apropos for analysis using the concept.

Political society in the case of Uganda is constituted not only by institutions of the central government (ruling National Resistance Movement Party and its caucus in parliament, government ministries, the armed forces, the police, prisons, resident district commissioners etc.) but also the institutions of the monarchy in each of the four major kingdoms (ministries and other bodies created by the monarchy). Although chapter sixteen (article 246 sub-section 3e) of Uganda's 1995 constitution (Constitution of the Republic of Uganda, 1995) expressly forbids monarchies from engaging in partisan politics, this has not stopped them from doing so. As emphasized earlier, they have been political institutions historically, so totally divorcing themselves from political matters of the day, such as those concerning land rights, tends to be difficult.

The Buganda kingdom's political clash with the central government provides obvious testimony to this claim. Less obvious, but nevertheless political, have been pro central government statements made by leaders in the other three kingdoms. As for civil society, it is primarily the common citizenry whose subordination or cooperation is sought by central government or the kingdoms either by the consent or force if consent fails. Both the central government and the kingdoms employ communication tools to woo the citizenry and gain their support of the agenda proposed. For the Buganda kingdom, this has meant resisting the new land bill and agitating for federal status whereas for the central government registering the citizenry's support of the land bill and the regional tier have been priorities. Beyond politics, each party has sought to establish itself as the protector and promoter of the citizenry's cultural values.

REQUIREMENTS, CIRCUMSTANCES, AND ROLE OF ICTS IN HUMAN DEVELOPMENT

According to Labelle (2005), the role of ICTs in the development of a society should be analyzed in terms of their appropriateness to the circumstances in the society and the associated needs of its people. This requires taking stock of the circumstances and associated human needs in the society. Labelle insists that "user needs assessments are essential in planning the introduction of ICTs to communities" regardless of their development status (p. 15). As he intimates:

"An electronic bulletin board system for sharing information about job opportunities may be useful to those who are connected to the Internet, but useless to people who are not. On the other hand, in societies where the oral tradition is important, accessing the Internet to inform a community radio station may be a way of communicating valuable information to computer illiterate users in rural locations" (p. 15).

Labelle notes that largely due to the harsh economic conditions in developing countries, the diffusion of ICTs and the acquisition of the necessary technological skills are relatively limited. Adesida (2001) points out that although President Yoweri Museveni's Uganda is usually showcased in the popular international press as one of the rare "success stories" of good governance in Africa, the sober reality is that even this so-called success story (as well as others) has not lived up to the promises it bandies in the effort to win national elections as well as international acclaim. As a result, despite some impressive macro-economic indicators such as a high economic growth rate (7% average since 2000) and the laudable fight against the spread of HIV/AIDS in the country; high levels of corruption, poor infrastructure, a runaway population growth and marginal service delivery have stymied the government's ability to fulfill the promises it makes to its people (Department for International Development, 2010).

Power supply for running technologies is often unreliable in many developing countries. Even basic connectivity tends to be an issue because the majority of people in these countries live in rural areas that lack the necessary technological or telecommunication infrastructure, which limits access to ICTs (with the exception of radio). Mwesige (2004) corroborates this digital divide within Uganda in this respect by noting a pattern of Internet cafes being concentrated in urban centers at the expense of rural centers.

Labelle (2005) adds that war and other forms of civil strife such as intractable conflict or recovery from such strife are staple experiences in many of these countries. In fact, Adesida (2001) and Carmody (2009) note that it is not unusual for ICTs to be employed for purposes of intensifying or perpetuating civil strife (e.g., inflaming ethnic or religious intolerance). Adesida (2001) argues that absent effective but fair regulatory and policy conditions to monitor talk shows on radios, ethnic hostilities that were previously held at bay may accrue. Adesida also observes that corruption is often endemic in many developing countries, and violation of basic human rights is rife let alone adulteration of the environment.

Labelle (2005) emphasizes that disproportionate barriers against women and widespread illiteracy are also usually an issue.

Given such untenable conditions, Adesida (2001) argues that fulfilling basic human needs of food, safe housing, clean water, or electricity, remains a challenge in many African nations. Added to this scenario is the need for sufficient public participation in political debates and decision-making in order to facilitate the people's effective involvement in policy formation and accountability of public officials. The public's right to cultural identity expression is yet another important need to consider.

Although ICTs do not by themselves deliver development, they have a proven record of facilitating transformation by enhancing information transactions (Adesida, 2001; Robbin, Courtright, & Davis, 2004; Labelle, 2005). ICTs can enhance the management of the needs discussed above in several ways including electronic-business/commerce (e-business/commerce), electronic-government (e-government), electronic-health (e-health), and electronic-learning (e-learning). Through e-business/commerce, ICTs can help developing countries such as Uganda to improve their economic performance which then boosts their ability to address the public's basic necessities of life. Among several real examples of e-business that Adesida provides is how the nation of Mauritania streamlined the process of doing business in it by instituting an "ICT-based trade management system" (p. 14). Labelle (2005) also indicates how satellite imagery has aided fishermen, herdsmen, and farmers to become more productive and profitable by providing them with strategic information.

For e-government, it is a means of making government services more efficient, accessible, and transparent (Labelle). An example of e-government is the Electronic Graft Management System (EGM) program in Kenya instituted to monitor corruption in government (Adesida, 2001). Another case of e-government presented by the author is the assistance of Ugandan female parliamentarians with internet skills and information provided by the Forum for Women in Democracy (FOWODE) organization to enhance their participation in parliamentary proceedings.

Other scholars on the role of NICTs in Africa (e.g., Nyamnjoh, 2005; Van der Veur, 2003) averred that, owing to their relatively low cost, FM radio and mobile phones would flood the African political landscape and transform it by empowering people to rise against authoritarian rule. This view is consistent with postulations by Friedman (2006), Nissanke and Thorbecke (2006) and Zembylas and Vrasidas (2005) that radio and cell phones in liberalized market environments would likely revolutionize politics let alone countering the emergence of cultural homogeneity.

The literature on how NICTs affect governance and democracy invariably posits that because media conglomerates and the government had monopolized political discourse, they were reluctant to initiate synchronous communications to enrich popular participation in politics and governance (Ebo, 2002; Norris, 2001; Ogunsola, 2005; Parayil, 2005; Robbin, Courtright, & Davis, 2004; Warschauer, 2003a, 2003b). Researchers (e.g., Giddens, 1990) also postulated that due to the dominance by corporations in harnessing the global information revolution, countries and conventional institutions would politically, culturally, and economically homogenize in the face of the overwhelming tidal wave of globalization.

Meanwhile, e-health is a purposeful means of enhancing the management or delivery of healthcare such as the Healthnet project which involves the collaboration of medical personnel from a number of African countries including Uganda. Regarding e-learning, the aim is to make education accessible virtually to people in remote areas in particular (Zembylas & Vrasidas, 2005). The African Virtual University sponsored by the World Bank is an example (Adesida, 2001).

The Hegemonistic Designs of Mengo

Mengo's bid for federal status is hegemonistic because the claim that restoration of its federal status would genuinely empower the 5.5 million Baganda to seize their destiny is illusory, judging by the kingdom's own record regarding the basic human needs outlined earlier. The demand for the restoration of a federal status for Buganda entails granting political power to the Kabaka as a constitutional monarch, the return of the 9000 square miles of land and forests to the kingdom confiscated by the Obote 1 central government when all the kingdoms were abolished in 1967, the return of all the 20 county and over 50 sub county headquarters in Buganda to Mengo, and mandating the kingdom administration to levy taxes on all businesses located in Buganda, is simply a ploy to harness the nests of the closest of royal courtiers and unlikely to engender development of the majority of the common Baganda peasants historically referred to as the Bakopi.

Mengo's fierce advocacy to uphold the status quo (before new land law) indicates a willingness to perpetuate the exploitative landlord-peasant relationship that denies peasants security of both tenure and occupancy. Yet, going by the overwhelming support ordinary Baganda have extended to Mengo's opposition to the land law and the demand for a federal status through the popular and participatory CBS FM radio programs, the enthusiastic willingness into subservience to the Mengo establishment as highlighted in the discussion on the concept of hegemony is quite apparent. That several of youthful Buganda loyalists voluntarily put their lives in the line of fire and fought running battles with armed police during the Kampala riots on September 11, 2009 to defend the interests of Buganda and the integrity of the Kabaka—the embodiment of their welfare and aspirations and the survival of the kingdom, as was continuously broadcast live on CBS radio, are evidence of a well entrenched and self-perpetuating hegemonistic institution.

Central Government's Populist and Opportunistic Rhetoric

While the government's populist rhetoric of fostering modernization and ensuring security for all citizens with particular focus on the historically exploited and vulnerable peasants in Buganda; holders of unsecured tracts of land, may appear progressive on the surface, the populist and opportunistic rhetoric in which it is shrouded, on close scrutiny, betrays its hollowness and exposes its equally hegemonistic designs to perpetuate its stranglehold on power. It is a far cry from fulfillment of the basic human needs of the population articulated earlier.

The 24 year record of the NRM administration, touted by its leader Yoweri Museveni as a fundamental change, has equally degenerated into a hegemonistic institution bent on self preservation. Although widely supported and embraced by a majority of Ugandans especially in the southern and western parts of the country, the high levels of seemingly state-inspired corruption, the manipulation of the constitution to perpetuate itself in power, and the failure to improve the delivery of social services has rendered its much hyped fundamental change hogwash. The 1995 constitution, the land amendment billct, and the recently proposed regional tier bill are widely viewed as legal schemes designed by the central government to consolidate its hold onto power and subdue Mengo. The 2005 Constitutional Amendment Act that lifted the presidential term limits amidst allegations of widespread bribery of members of parliament by government in addition to the damning Transparency International and World Bank reports on corruption in government illustrate the duplicity in the government's rhetoric.

Overall, President Yoweri Museveni, Kabaka Ronald Muwenda Mutebi, and their protégés have together massively exploited the new media to advance their respective agendas and perpetuated or are seeking to regain their hegemonic hold over the people of Buganda and Uganda in general.

SOLUTIONS AND RECOMMENDATIONS

As mentioned earlier, the primary use of NICTs should be to enhance good and participatory governance, fighting corruption, eradication of the many indicators of underdevelopment not only in Uganda but also in other developing countries to realize the UN Millennium Development Goals (MDG) whose target of 2015 is fast approaching. NICTs should not be used to incite violence or genocide as the case was with Radio Rwanda and Radio Télévision Libre des Mille Collines (RTLM) in 1994.

Rather, the convergence of NICTs should primarily promote cultural co-existence, civilized dialogue, observance of individual, minority, and the rights of majorities. More importantly, as effective tools of seamless communication and collaboration, NICTs should be deployed to promote democracy—a conscious effort to advance a culture of amicable resolution of conflict; ethnic relativism; and mutual respect for divergent opinions. NICTs should also be used to educate and empower present and future generations with functional knowledge to uphold and adopt appropriate social value competencies to avoid being held hostage to cultural ethnocentrism, political manipulation, and hegemonic tendencies from either cultural or political forces masquerading as their sophisticated guardian angels.

As popular platforms that are easily accessible and usable, the convergence of, for example, the cell phone and FM Radios in Uganda and indeed Africa, should be deployed to facilitate the predominantly rural populations engaged in subsistence agriculture find markets that offer competitive prices for their produce, transfer knowledge, skills, and best practices from urban areas to improve the livelihood of the rural folk. Worthy noting too, is that as tools of communication, NICTs have the potential to build but also to destroy communities.

Proprietors, regulators, and users have a duty to harmonize their interests in NICTs to maximize their benefits to society. Uganda, for example, needs to re-consider the ownership, management, and use of NICTs, notably the popular platforms offered by the convergence of FM radio and the ubiquitous cell phone. We suggest, as an interim measure, that various stakeholders in the NICT sector come together, as equal partners, to thrash out a mutually agreeable modus operandi that would ensure a fair and just broadcast regime to all stakeholders without fear or favor.

Without a win-win solution to the ownership, operation, and use of NICTs in Uganda, there is likely no end in sight, at least in the near future, to the political strife that the country has, for far too long, painfully endured. All stakeholders owe it to present and future generations by rising above their egos and privileged positions to ensure that bloody clashes similar to those that took place on September 11, 2009 are not repeated.

Change is inevitable. To Uganda's political and cultural leaders, including the protagonists in the Buganda kingdom-central government media warfare and those who have assigned themselves the role of expert commentators and the others that are apportioning blame to either side of the divide, the solution is to stop pointing fingers. None of the protagonist is beyond blemish. Both the principal actors and cheer leaders in the central government and at Mengo take a portion of responsibility in the acrimonious struggle for hegemony. But appropriating blame is beyond the scope of this chapter. Suffice to state, inter alia, that leaders in Uganda, from whichever color of the rainbow, should transform themselves into genuine statesmen and embrace democracy and good governance. Democracy is about communication and integrative conflict resolution and not about the right of might to coerce the seemingly weak into submission. The use of NICTs should not be a monopoly of one point of view. Either side to the vitriolic hegemonic warfare should step back and re-think their strategy and future of both Buganda and Uganda in general.

There is no short-cut to change and adaptation to new ideas and the ever evolving global dynamics. Rigidity in politics is not only a recipe for disaster but also a pointer to a failure in leadership. Neither does cultural nostalgia nor inflexibility reflect a mark of competence and expectation. Taken together, failures by the political and cultural leadership in Uganda to spearhead some positive change could necessitate a fresh Ugandan renascence to re-discover the country's bearings and chart a new course. But all is not lost. The antagonists may find the following lines not only a premonition but also a sobering food for thought: "Change is hard. Change is hardest on those caught by surprise. Change is hardest on those who have difficulty changing too. But change is natural; change is not new; change is important. We simply must adapt and first survive then thrive," (Friedman, 2006, p. 11).

FUTURE RESEARCH DIRECTIONS

NICTs present a unique opportunity for scholars interested in the examination of the intersection between culture and modernity especially as NICTs continue to play a catalytic role in the globalization process. The dynamic and complex nature of NICTs, fostered by the rapid technological developments and changes in their use, are having significant and unpredictable impact on societies as never witnessed before.

In Africa, for example, where, according to the World Bank (2009), the fastest growing market for the cellular phone is located, the potential for the emergence of cutting edge scholarly and professional research that could have far-reaching implications is enormous. On the political scene, it may not be long before the use of NICTs by populations to result in massive political movements that could shift the post-colonial paradigm of election rigging and militarization of conflicts by incumbent regimes to pave way for popular democratic civil action. For Uganda in particular, the contested political ground, could, during the forthcoming 2011 presidential, parliamentary, and local elections, offer a turning point in studies on the use of NICTs in political campaigns especially coming against a backdrop of electoral irregularities confirmed by the country's supreme court in the 2001 and 2006 elections. Efforts by the active political parties and civil societies to upgrade their NICTs ahead of the elections point to interesting future case studies in NICT use. Such likely scenarios, not only limited to the political arena, could be replicated in the economic, social, and cultural domains that will offer opportunities for future research both in Uganda and the proposed East Africa federation.

CONCLUSION

This chapter has examined the complex ways in which NICTs are being employed in Uganda in a delicate and sometimes tumultuous political gamesmanship that has played out in the relations between the central government and Uganda's four major kingdoms. Specifically, the chapter conceptualized the FM Radio and the cell phone as new media in the African context primarily because other forms of NICTs are less accessible and least used in Africa, Uganda in particular.

Whereas each kingdom represents the unique culture and ethnic aspirations of its peoples, our analysis indicates that the historical context within which the kingdoms have co-existed with the central government helps to account for the differences in which NICTs have been deployed in each kingdom's dealings with the central government for the sake of self-preservation and self-determination. This reinforces arguments by scholars that NICTs are mainly tools that are socially constructed in the way they are engaged by users. The above analysis shows that the Buganda Kingdom has historically chosen to assert itself in ways that it believes can preserve and even extend its privileged status in Uganda while the other three kingdoms, two of which do not own radio stations, have chosen to employ NICTs in ways that eschew confrontations with the central government.

Our assessment suggests that despite the participatory nature of the CBS radio campaign that innovatively tapped cell phones to accomplish a level of audience participation never seen before in Uganda's media history, the intransigence with which both the Buganda establishment and central government behaved regarding their showdown on the federal and land questions, betrays an unwillingness to learn from the ugly history between the two institutions that has proved counterproductive. As argued previously, the quicker the antagonists realize that NICTs are resources that should be harnessed for genuinely advancing the lives of their people the better the outcomes will be for Uganda.

The central government and each of the kingdoms discussed above have the proven potential to deliver for their people economically, politically, culturally, and socially if they can hold their negative tendencies in check and deploy NICTs to capitalize on the positive potential. If the protagonists do not heed to the global winds of change that are blowing through the continent and indeed Uganda, and re-position themselves as renewed statesmen, forewarns the chapter, citizens are likely to deploy the increasingly accessible NICTs to engender a shift in the political, economic, and cultural paradigms for the betterment of the disenfranchised population.

The chapter, it is envisaged, will offer a fresh perspective to the use of NICTs in politics especially with regard to overt hegemonic agendas and their subtle sub-texts that manifest during political and cultural interactions.

REFERENCES

Adesida, O. (2001). *Governance in Africa: The role for information and communication Technologies*. Retrieved June 18, 2008, from http://www.comminit.com/en/ node/221626/307

Akaki, S. (January 20, 2008). Will 'Balaalo' and land define the 2011 elections? *The Sunday Monitor Online*. Retrieved January 20, 2008 from http://monitor.co.ug

Among, B., & Tebajjukira, M. (March 3, 2008). Museveni says 9,000 sq miles claim false. Retrieved March 3, 2008, from http://www.newvision.co.ug/D/8/13/614763

Anderson, P. (1964). Origins of the present crisis. *New Left Review*, *23*, 24–53.

Apter, D. (1967). *The political kingdom in Uganda* (2nd ed.). Princeton, NJ: Princeton University Press.

Bako, P. (2009). Walking the talk: Are land evictions in Uganda in line with human rights standards? Masters thesis. Retrieved March 13, 2011 from http://upetd.up.ac.za/thesis/available/ etd-08032010-145221/unrestricted/dissertation.pdf

Buganda Kingdom. (2004). *Buganda culture*. Retrieved June 12, 2008, from http://www.buganda.com/culture.htm

Bunyoro-Kitara Kingdom. (2007). *The Bunyoro-Kitara kingdom*. Retrieved December 14, 2009 from http://www.bunyoro-kitara.com/history.html

Busoga Kingdom. (2004). *Obwa kyabazinga bwa Busoga*. Retrieved June 10, 2010, from http://www.busoga.com/facts.php

Busoga Kingdom. (2007). *Evolution of obwa kyabazinga bwa Busoga*. Retrieved February 12, 2009 from http://www.busoga.com/history.php

Butagira, T. (May 25, 2008). Democracy, media under threat. Retrieved May 25, 2008 from http://allafrica.com/stories/200805270901.html

Byrnes, R. M. (1992). The society and its environment. In R. M. Brynes (Ed.), *Uganda: A country study* (chapter 2). Retrieved from http://lcweb2.loc.gov/cgi-bin/query2 /r?frd/cstdy:@field(DOCID+ug0025)

Carmody, P. (2009). *A new socio-economy in Africa: Thintegration and the mobile phone revolution.* Retrieved February 4, 2010 from https://www.tcd.ie/iiis/documents/ discussion/ pdfs/iiisdp279.pdf

Central Broadcasting Services (CBS). (2008). *Ettaka Lyaffe Ligenda.* Retrieved June 12, 2008 from http://www.radiocbsfm.com

Cohen, D. W. (1970). A survey of interlacustrine chronology. *Journal of African History, 11*(2), 177–202. doi:10.1017/S0021853700009932

Constitution of the Republic of Uganda. (1995). Retrieved November 31, 2009, from http://www. ugandaonlinelawlibrary.com/files/ constitution/ constitution_1995.pdf

Department for International Development. (2010). *Debt relief in Uganda.* Retrieved February 26, 2010, from http://www.dfid.gov.uk/ Media-Room/News-Stories/ 2010/Debt-relief-in-Uganda/

Ebo, B. M. P. (2002). *Old wine in new wine bottle: The Internet and the techolonization of Africa.* Retrieved February 20, 2008, from http://www. arts.uwa.edu.au/ MotsPluriels/MP2002be.html

Friedman, T. L. (2006). *The world is flat: A brief history of the twenty-first century* (3rd, ed.). New York: Farrar, Straus and Giroux.

Giddens, A. (1990). *The Consequences of Modernity.* Cambridge, UK: Polity.

Gramsci, A. (1971). *Selections from the prison notebooks* (Hoare, Q., & Smith, G. N. (Trans. Eds.)). New York, NY: International Publishers.

International Monetary Fund (IMF). (March, 2008). *Statement by IMF Staff Mission to Uganda.* Press Release No. 08/58, March 21. Retrieved June 18, 2008, from http://www.imf.org/external/ np/ sec/pr/2008/pr0858.htm

Kiwanuka, M. S. M. (1968). Bunyoro and the British: A reappraisal of the causes for the decline and fall of an African kingdom. *Journal of African History, 9*, 603–619. doi:10.1017/ S0021853700009063

Kiwanuka, M. S. M. (1971). *The kings of Buganda.* Nairobi, Kenya: East African Press.

Kiwere, N. (January 10, 2008). Land saga brewing widespread agitation in Uganda. Retrieved on March 13, 2011 from http://www.ugandanlink. com/politics/land-saga-brewing- widespread-agitation-in-uganda.html

Labelle, R. (2005). *ICT policy formulation and e-strategy development: A comprehensive guidebook.* Retrieved June 8, 2010 from http://www. apdip.net/publications/ict4d/ict4dlabelle.pdf

Low, D. A., & Pratt, R. C. (1960). *Buganda and British overrule, 1900-1955: Two studies.* Nairobi, Kenya: Oxford University Press.

March 07, 2008, from http://www.ugandaobserver. com/new/index.php

Maseruka, J., & Musoke, C. (September 30, 2009). *Kabaka Mutebi, Museveni meet.* Retrieved October 5, 2009, from http://www.newvision. co.ug/D/8/12/696342

Matsiko, G. (May 2, 2008) Press freedom remains elusive in Uganda. Retrieved May 2, 2008, from http://allafrica.com/stories/ 200805010675.html

Muhumuza, R. (May 28, 2008). Traditional institutions have no place – Museveni. Retrieved May 28, 2008, from http://allafrica.com/stories/ 200805280331.html

Mulira, P. (November 6, 2006). Demarcation of Uganda led to present problems. *The New Vision Online.* Retrieved November 6, 2006, from http:// www.newvision.co.ug/D/8/20/530625

Mulira, P. (October 7, 2009). Buganda must outgrow the culture of nostalgia. Retrieved December 14, 2009 from http://www.newvision.co.ug/D/8/20/697029

Muloki dies at 87. Retrieved December 12, 2008, from http://majimbokenya.com/home/2008/09/01/ rest-in-peace-kyabazinga-wako-muloki-dies-at-87/

Musoke, C. (May 21, 2008). Cabinet discusses state of media. Retrieved May 21, 2008, from http://www.newvision.co.ug/D/8/12/629236

Musoke, C., & Bita, G. (August 25, 2008). Museveni visits Kyabazinga in hospital. Retrieved March 13, 2011 from http://www.newvision.co.ug/ D/8/12/646353

Muwanga, D. (June 18, 2008). Uganda gets tractor assembly plant. Retrieved June 18, 2008, from http://www.newvision.co.ug/D/ 8/13/634339

Muyita, S. & Bareebe, G. (September 1, 2008). Rest in peace: Kyabazinga Wako

Mwanje, R. (2008, June 15). Buganda refuses to be smeared. Retrieved June 15, 2008 from http://allafrica.com/stories/ 200806160517.html

Mwanje, R., & Senkabirwa, A. M. (January 5, 2008). Police grills Buganda officials over land debate. *The Monitor Online*. Retrieved January 5, 2008 from http://monitor.co.ug/

Mwesige, P. G. (2004). Cyber elites: A survey of Internet cafe users in Uganda. *Telematics and Informatics*, *21*(1), 83–101. doi:10.1016/S0736-5853(03)00024-8

My Uganda. (2010a). *Monarchies-Bunyoro kingdom.* Retrieved January 20, 2010, from http://www.myuganda.co.ug/about/peoples/ monarchies/bunyoro.php

My Uganda. (2010b). Monarchies-Tooro kingdom. Retrieved January 12, 2010, from http://www.myuganda.co.ug/about/peoples/ monarchies/toro.php

Nalugo, M., & Mugerwa, Y. (October 24, 2010). Museveni orders re-opening of CBS radio station. Retrieved on March 13, 2011 from http://allafrica.com/stories/ 201011010005.html

Namutebi, J. (December 11, 2009). Regional government bill tabled. *The New Vision*. Retrieved June 10, 2010, from http://allafrica.com/stories/ 200912141161.html

Nissanke, M., & Thorbecke, E. (2006). Channels and policy debate in the globalization–inequality–poverty nexus. *World Development*, *34*, 1338–1360. doi:10.1016/j.worlddev.2005.10.008

Nnyago, U. K. (February 15, 2008). The land amendment bill: Buganda's blessing in disguise. Retrieved February 15, 2008 from http://allafrica.com/stories/ 200802141011.html

Norris, P. (2001). *Digital divide: Civic engagement, information poverty, and the Internet worldwide*. Cambridge, MA: Cambridge University Press.

Nyamnjoh, F. B. (2005). *Africa's media: Democracy and the politics of belonging*. London, UK: Zed Books.

Obbo, O. C. (February 24, 2010). Why CBS might be off air until 2016. Retrieved February 25, 2010 from http://www.monitor.co.ug/OpEd/OpEdColumnists/ CharlesOnyangoObbo/-/878504/867528/-/xgnuus/-/index.html

Ochieng, E. O. (1991). Economic adjustment programmes in Uganda, 1985-8. In Hansen, H. B., & Twaddle, M. (Eds.), *Changing Uganda: Between decay and development* (pp. 43–60). Oxford, UK: James Currey.

Odoki Commission. (1993). *The report of the Uganda constitutional commission: Analysis and recommendations*. Entebbe, Uganda: Government Printer.

Ogunsola, L. A. (2005). Information and communication technologies and the effects of globalization: Twenty-first century "Digital Slavery" for developing countries-myth or reality? *Electronic Journal of Academic and Special Librarianship, 6,* 1-11. Retrieved June 4, 2008 from http://southernlibrarianship.icaap.org/ content/v06n01/ogunsola_I01.htm

Olupot, M. (July 23, 2008). Mengo, govt conflicts not new. Retrieved March 13, 2011 from http://newvision.co.ug/D/8/426/640785/Mengo

Otiso, K. (2006). *Culture and customs of Uganda*. Westport, CT: Greenwood press.

Parayil, G. (2005). Digital divide and increasing returns: Contradictions of informational capitalism. *The Information Society, 21*(1), 41–51. doi:10.1080/01972240590895900

Park, H. (1998). A Gramscian approach to interpreting international communication. *The Journal of Communication, 48*(4), 79–99. doi:10.1111/j.1460-2466.1998.tb02771.x

Robbin, A., Courtright, C., & Davis, L. (2004). ICTs and political life. In B. Cronin (Ed.), *Annual Review of Information Science and Technology (ASIS&T), 38*, 411-482. Medford, NJ: Information Today Inc.

Roberts, A. (1962). The sub-imperialism of the Baganda. *Journal of African History, 8*, 435–450. doi:10.1017/S0021853700003340

Ssejjoba, E., & Lukwago, J. (April 13, 2008). Archbishop Lwanga warns oninsulting Kabaka. Retrieved April 13, 2008 from http://www.newvision.co.ug/ D/8/12/622091

Ssemujju, I. N. (January 3, 2008a). Why Kabaka refused to meet Museveni. *The Uganda Observer Online.* Retrieved January 3, 2008 from http://www.ugandaobserver.com/new/index.php

Ssemujju, I. N. (January 6, 2008c). Kabaka sets terms to meet Museveni. *The Uganda Observer Online.* Retrieved January 6, 2008, from http://www.ugandaobserver.com/new/index.php

Ssemujju, I. N. (May 29, 2008d). Cabinet to delete "press freedom" from constitution. *The Uganda Observer Online.* Retrieved May 29, 2008, from http://www.observer.ug/index.php?option=com_content&view=article&id=103:ssemujju- ibrahim-nganda&catid=34:news&Itemid=59

Ssemujju, I. N. (March 7, 2008b). Swords out. *The Uganda Observer Online.* Retrieved

Ssengoba, N. (May 27, 2008). Press crackdown; NRM loses battle for hearts and minds. Retrieved May 27, 2008, from http://allafrica.com/stories/200805270747.html

Ssenkabirwa, A. M. (January 2, 2008). Mengo replies Museveni's letter to Kabaka. Retrieved January 2, 2008, from http://allafrica.com/stories/200801030314.html

Sserwanga, M. (May 27, 2008). Media crackdown? Bye-bye to democracy. *The Monitor Online.* Retrieved May 27, 2008, from http://allafrica.com/stories/200805270222.html

Tangri, R., & Mwenda, A. (2010). President Museveni and the politics of presidential tenure in Uganda. *Journal of Contemporary African Studies, 28*(1), 31–49..doi:10.1080/02589000903542574

The Observer. (October 4, 2009). *Untold story of Museveni Kabaka meet.* Retrieved on October 5, 2009, from http://www.observer.ug/index.php?option=com_ content&view=article&id=5390:untold-story-of-museveni- kabaka-meet&catid=78:topstories&Itemid=59

The Sunday Vision Online. (February 9, 2008). *President Museveni to sue MP Beti Kamya.* Retrieved February 9, 2008, from http://allafrica.com/stories/2008021 Toro Kingdom. (2008). King Oyo Nyimba Kabamba Iguru Rukidi IV. Retrieved January 22, 2010, from http://www.torokingdom.org/Oyo.htm

Ultimate Media. (2006). Uganda business news: Bunyoro minister and kingdom press secretary resign over oil. Retrieved March 13, 2011 from http://www.ugpulse.com/articles/daily/ news.asp?about=Bunyoro+Minister+ and+Kingdom +Press+Secretary+resign+over+oil&ID=2393

United Nations Information and Communication Technologies Task Force. (2002). *Definition of information communication and technologies (ICTs).* Retrieved June 12, 2008, from http://www.unicttaskforce.org/

Van der Veur, P. R. (2003). *Media and democracy in Africa, broadcasting and political reform.* New Brunswick, NJ: Transaction Publishers.

Warschauer, M. (2003a). *Technology and social inclusion: Rethinking the digital divide.* Cambridge, MA: MIT Press.

Warschauer, M. (2003b). Dissecting the "digital divide": A case study in Egypt. [from http://olt-bta.hrdc-drhc.gc.ca/ download/ oltdualdivide.pdf]. *The Information Society, 19,* 297–304. Retrieved June 14, 2008. doi:10.1080/01972240309490

World Bank. (2008). *Facts and figures about Uganda.* Retrieved June 18, 2008, from http://www.worldbank.org/afr/ug

World Bank. (2009). *Information and communications for development (IC4D): Extending reach and increasing impact of ICTs.* Oxford, Uk: Oxford University Press. Retrieved from http://www.worldbank.org/ ic4d World History Archives. (March 2004). *The contemporary political history of the Kingdom of Bunyoro-Kitara.* Retrieved January 22, 2010, from http://www.hartford-hwp.com/archives/ 36/index-ijb.html

Zembylas, M., & Vrasidas, C. (2005). Globalization, information and communication technologies and the prospect of a global village: Promises of inclusion or electronic colonization? *Journal of Curriculum Studies, 37,* 65–83. Retrieved June 16, 2008, from http://www.courses.unt.edu/efiga/ 6880_Globalization/zembylas_vrasidas_2005_global_village.pdf

ADDITIONAL READING

Avirgan, T. & Martha H. (1982). *War in Uganda: The Legacy of Idi Amin.*

Davidson, B. (1969). *A history of East and Central Africa to the late 19th century.* Garden City, New York: Anchor Books.

Dunbar, A. R. (1969). *A History of Bunyoro-Kitara.* (Rev. ed.) Nairobi: Oxford University Press on behalf of Makerere Institute of Social Research.

Furley, O. W. (1989). Britain and Uganda from Amin to Museveni: Blind eye diplomacy. In Rupesinghe, K. (Ed.), *Conflict Resolution in Uganda, pp. 275-94 Athens.* Ohio University Press.

Hansen, H. (1984). *Mission, church, and state in a colonial setting: Uganda, 1890-1925.* New York: St. Martin's Press.

Hansen, H. Michael Twaddle (eds.) (1988). *Uganda now: Between decay and development.* Athens: Ohio University Press.

Ibingira, G. S. K. (1973). *The forging of an African nation: The political and constitutional evolution of Uganda from colonial rule to independence, 1894-1962.* New York: Viking Press.

Ingham, K. (1983). *The making of modern Uganda.* Westport, Connecticut: Greenwood Press.

Jorgensen, J. (1981). *Uganda: A modern history.* New York: St. Martin's Press.

Karugire, S. R. (1980). *A political history of Uganda*. Exeter, New Hampshire: Heinemann Educational Books.

Kasfir, N. (1976). *The shrinking political arena: Participation and ethnicity in Africa politics*. Berkeley: University Press.

Westport, Connecticut: Hill Baker, S. W. *(1969). A Narrative of the expedition to Central Africa*. London: Macmillan1874. Reprint. New York: Negro Universities Press.

KEY TERMS AND DEFINITIONS

Cell Phone: Mobile phone

Central Government: National government

Convergence: Combination of different media to achieve a goal

Federal Status: Constitutionally autonomous kingdoms linked to weak central government

FM Radio: Frequency modulated air waves

Hegemonistic Agenda: Interests of the powerful entity

New Information and Communication Technologies (NICTs): FM radio, cell phones, and Internet in the African context

Chapter 5
The Cultural, Economic and Political Implications of New Media:
A Case Study on Mobile Telephony among University Students in Kenya

Fredrick Kang 'ethe Iraki
United States International University, Kenya

ABSTRACT

This study investigates the extent to which the cell phone technology has impacted on the lives of young Kenyans, especially within the context of a university in Kenya. Kenya has had a tremendous technological leap from the 1990s where landlines were the main tools of communication, to mobile telephony where people can now communicate any time, any place. The ramifications of such a new and stunning technology in Kenya are very important, especially if there are pernicious effects emanating from the novel technology.

This study involved 100 university students at the United States International University in Nairobi, Kenya. The questionnaires were randomly distributed and collected after a week. The entire survey took one month.

The findings point out that mobile telephony has become an indispensable tool for most youth providing them with a great sense of control and freedom. However, the technology is also rife with concerns about personal autonomy, dishonesty, breach of cultural norms, personal security and that of others, especially when texting while driving, et cetera.

DOI: 10.4018/978-1-60960-591-9.ch005

1. INTRODUCTION TO THE STUDY

The mobile telephony in Kenya has been a veritable technological as well as a cultural revolution. It is simply mind-boggling how a technology that was hardly known in the early 1990s has now become too popular to the point of becoming banal. Similarly, the incredible ubiquity of the mobile technology has also confounded economic analysts since the technology had been deemed too onerous for ordinary Kenyans who purportedly live under one US dollar per day. Research by KENET (2009) reveals that the mobile technology has the all-high penetration of more than 6 million Kenyans spread across the entire country. Safaricom alone has 3 million subscribers, almost a tenth of Kenya's entire population. This penetration does not seem to care about the glaring economic disparities in the various parts of Kenya.

In Kenya today, owning a mobile phone is no longer trendy but functional. It is less a status symbol than a tool whose existence has become quasi-indispensable. True, some units (iPods, Blackberry, top Nokia series, etc) are still loud statements of economic class. Nevertheless, the technology is no longer novel and its ubiquity across the country is amazing.

A glance at some of the previous studies conducted in other parts of the world indicates that the new media has been associated with some obvious cultural transformations. For instance, Weinberg (2009) notes that in Syria, relatives call before making a courtesy call, a clear departure from the culture that greatly values family relations. She further notes that Syrian women feel liberated by the cell phone as they can now freely talk with men without the family restrictions and surveillance associated with the landline.

Drolet (2009) notes with concern how the cell phone can impact social behavior. For instance, some people respond to cell phone calls even while in the bathroom, a behavior that is normally reprehensible. In addition, the habit of talking loudly and for long spells of time on the cell phone appears bizarre as it stands in stark contrast with the conventional image of "talking to someone visible". Talking to oneself – as it appears – is a sign of dementia in most cultures.

In Uzbekistan, Wei (2005) describes three reasons for owning a phone, namely prestige, necessity for work, and a gift from parents. A preliminary survey among university students in Kenya echoes the same reasons. However, most of the students do not consider prestige to be a significant issue; for them the need to "connect with buddies" seems most important.

Bhuiyan (2006) makes a case for the convergence of media where the mobile phone becomes a tool for social communication, business, surfing the internet, blogging, vlogging etc. With such potency, the new media revolutionizes the way a society configures its information origination, delivery and commentary.

The versatility and wide usage of the cell phone technology creates an urgency not to mention immediacy in the transmission and reception of information. In Kenya today, it's almost unacceptable not to return a missed call; neither is it forgivable to forget a cell phone at home. The unsaid rule is *one should be reachable at all times, and anywhere*; a scenario reminiscent of Big Brother in George Orwell's *1984*.

The current study focussed on the acquisition, usage and impact of cell phone technology among university students – undergraduate and graduate in Kenya. In particular, the survey showed how the technology has transformed the way the students generate/access and comment information within the context of media convergence. Of particular interest was how the students tapped into the now popular electronic money transfer system (M-Pesa) for daily business transactions. Will the mobile telephony change the way Kenyans do business? Are the students already engaging in business transactions? These and other questions were addressed in the study.

Culture has a dialectical relationship with language. As the cell phone transforms the manner we generate/access, comment and relay messages, there is a concomitant mutation in the way messages are encoded. For instance, *c u b4 wak* could replace the elegant *I will see you before work.* This new mode of encryption, we argue, may or may not have repercussions on the manner students communicate in written English. The study considered the linguistic material produced by students and analyze the results. Do the constraints of *speed and accuracy* compel the encoder to sacrifice syntactical elegance? Will this new shorthand insinuate itself in the writing assignments? Will it be a good or a bad thing?

In Kenya, politics is the most important news item for many Kenyans. The Report on Media Survey of 2009 explains how the mobile phones were used extensively to communicate the election results in the 2002 General Elections. This innovative practice minimized vote rigging and theft. In the same vein, the usage of the same technology in the 2007 General Elections was associated with bigotry and ethnic hate speech. These two contrasting and contradictory roles of the new technology call for re-examination especially when it is understood that cell phone communication is fast and immediate. Indeed, it is now a criminal offence in Kenya to send hate messages; cybercrime is now a reality in the country.

Further, in the just-concluded Referendum of 4 August 2010 on the Draft Constitution, the mobile phones were employed by the Interim Independent Electoral Commission (IIEC) to relay the results, especially from areas in Kenya where there is no electricity.

The current study examined to what extent the university students at the United States International University (USIU) engaged with the new media with a view to participating in the democratization of the political space in Kenya. Do they blog or vlog on political issues or are their phones for social purposes only? Currently, there are many offers to use the cell phones for internet services such as emailing, blogging or vlogging. Students are incessantly advised to wean themselves from twitter and facebook as they spend an inordinate amount of time on them at the expense of studying. Could these blogs or vlogs be a good opportunity for the students to learn new things and ways of doing things? Is the academy afraid it might lose control of knowledge if the students are constantly exposed to alternative sources and types of knowledge?

The study was primarily conducted via a questionnaire administered to USIU students of ages 19-24 years. In addition, an oral interview was done with 10 American Students on a study-abroad program at USIU.

2. KENYA AT A GLANCE

There is no gainsaying the fact that mobile telephony has revolutionized the way Kenyans communicate with each other as well as the manner in which they relate with one another. Long gone are the days when people queued outside poorly-maintained and unsightly telephone booths to dial a call. Similarly, the booming private call-booths in the 1990s simply exited from the scene as Kenyans acquired more and more mobile phones, thus enabling them make calls from wherever they pleased and at any time. Such a technological mutation cannot be unaccompanied by cultural, social and behavioral change, to say the least.

Kenya has had mixed political and economic fortunes since her independence from Great Britain in 1963. During the period when she groaned under Jomo Kenyatta, 1963-1978, Kenya's GDP leapt dramatically despite the slow-down occasioned by the Oil Crisis of 1973 (Okello & Owino, 2004). Despite this economic lurch, the real growth in terms of improvement of living standards left a lot to be desired. Politically, Kenyatta's was a regime of utter repression of basic freedoms notably, the freedom of free thought. The State controlled every facette of Kenya.

Moi's regime (1978-2002) was a continuation of Kenyatta's decrepit leadership albeit with more acuity. Moi ransacked the economy and transformed the country into his personal estate. *L'Etat c'est Moi*[1] (The State is Moi) would have been an apt description of the syncretism between Moi and the Kenyan State. Kenyans evolved under terror at a period when all means of communication, including telephony, was in the hands of the State. Tapping of phones and torture were all too common under the Moi administration.

The Kibaki dispensation ushered in a new wave of freedom and economic recovery. Indeed, the GDP rose rapidly to an impressive 7% between 2003 and 2005; then Kibaki's star began to wax and wane. Politically, Kibaki was unable to marshal his political experience under Kenyatta (as Finance Minister) and Moi (as Vice-President and Minister for Health) to tame a rowdy rebel team under Raila Odinga. As a result, the economy was thrown into a tail-spin tumbling to near-zero in the wake of the Post-Election Violence in January 2008. Despite the political turmoil and the economic downturn, Kenya enjoys unequalled freedom of expression, thought and enquiry not witnessed in her entire history. Within this context, new forms of communication and doing business have evolved and grown at a mind-boggling rate. These include the internet services (online journals, magazines, blogs, vlogs, facebook, Twitter, etc) and mobile telephony and mobile banking, among other innovations. The new electronic environment in Kenya has already made impact on the way Kenyans interact with one another.

In the 2002 General Election, the mobile phone, which by then was still an expensive tool, was employed extensively to relay election results to radio and Television stations, thus avoiding manipulation of results by corrupt officers. Politically, the mobile phone established itself as a formidable force in transforming Kenya.

Similarly, although this time in an ignoble sense, in the 2007 General Election, the mobile phone was again remarkable. This time round, it was employed to relay hate messages across the country in a bid to stoke the fires of negative ethnicity. Misinformation, panic, hatred, suspicion and negative sentiments shot across the country making national healing a great challenge.[2]

The advent of mobile banking, notably M-Pesa, has made banking accessible to millions of Kenyans who previously could not afford a bank account. Moreover, e-banking via the mobile phone has been praised to be safe, reliable, fast and extremely practical in money transactions (Mbogo 2010).

In sum, the mobile telephone has been a critical factor in politics (minimally), economics and socio-cultural dynamics in Kenya. We now turn to a brief history of mobile telephony in Kenya.

3. MOBILE TELEPHONY IN KENYA: A BRIEF OVERVIEW

According to the Communications Commission of Kenya (CCK)[3], Mobile telephony made its debut in Kenya in 1992 under the analogue system known as Extended Total Access Communication System (ETACS). This system was commercially launched in 1993. The handset was a status symbol with a price-tag of Kshs 250,000 (USD 3,300) with charges including making and receiving calls. At a time when Kenya was grappling with the World Bank-imposed Structural Adjustment Programs and frozen IMF funding, to own a mobile phone at such a prohibitive cost could only be termed as obscene ostentation. Ordinary Kenyans were wallowing in abject poverty due to grand corruption and gross mismanagement of State resources by the Moi regime. In view of the prohibitive cost of the handset, the industry registered a mere 20,000 subscribers for the 7-year period, 1993-1999.

The mobile telephony entered the arena of communication where the State ruled supreme[4]. The Kenya Communications Act of 1998 ended the State monopoly of telephony with the licensing of two private entities, namely Safaricom Limited and KenCell Communications[5] (now known as Airtel). Subsequently, two other mobile phone operators were licensed: Telkom Kenya (operating as Orange Mobile) and Econet Wireless. In 2008, YU also joined the other 4 operators.

According to the Report by the CCK, the mobile companies have had phenomenal subscriptions with Safaricom and Zain accounting for 12.9 million subscribers by June 2008. In addition, the network capacity of the two leading companies has shot up from 640,000 in June 2001 to 25,964,700 in June 2008 (see Table 1)

Kaloki (2010), claims that the Kenyan mobile telephony industry accounts for 7% of mobile phone subscribers in Africa. He further notes that Kenya has 17.4 million mobile phone subscribers making it the third largest mobile telephony industry after Nigeria (26%) and South Africa (19%). The figures for Kenya are expected to rise to 29.28 million accounting for 66.7% penetration by the year 2013.

The decade from 1999 was to become a veritable revolution in the history of mobile telephony in Kenya. The price-tag of the handset tumbled dramatically thus providing a large cross-section of Kenyans with an opportunity to own a phone. Currently, there are basic handsets at Kshs 1,000 (USD 13) and *haut de gamme* handsets retailing at upwards of Kshs 90,000 (USD 1200)[6].

Economic Implications

Nyabiage (2009) notes that the subscriber base for mobile telephony rose from round 6,000 in 2000 to 18 million in 2009 with a coverage of 80% of the entire Kenyan population. The latter is estimated at 40 million. The extensive coverage has economic ramifications. Nyabiage observes that for the decade 1999-2009, the information and communications industry was amassing about Kshs 100 billion[7] yearly with a projection of Kshs 150 billion in 2013.

Thanks to low-denomination calling cards and reliability of services[8] (except Fridays due to network congestion) the mobile industry has registered a formidable penetration of 45.9% as at June 2009 (Communications Commission of Kenya – CCK). Nyabiega cites Mr. Dearbhla McHenry, an analyst at Pyramid Research, a telecoms research firm, who argues that:

The mobile data is "telecoms sector's fastest-growing revenue stream, increasing from $62 million (Sh4.7 billion) in 2008 to $224 million (Sh17 billion) in 2013, partly due to the launch of 3G services, and partly due to the explosive growth of low-tech, low-margin mobile data services, particularly mobile money transfers."

Mbogo (2010) observes that "There is appeal and utility of mobile banking and mobile payment services across the country as there are probably more people with mobile handsets than with bank accounts." The introduction of Mteja-Pesa (M-Pesa) – mobile Money transfer – by Safaricom has had a tremendous positive impact on micro-business as attested by Mbogo (2010). She notes that the service is safe (money can be stored in the phone and passworded), convenient (money can be transferred very fast) and reliable

Table 1. Mobile network growth indicators

	1999/00	200/01	2001/02	2002/03	2003/2004	2004/2005	2005/2006	2006/2007	2007/2008
Capacity					**3,935,000**	**6,800,000**	**10,600,000**	**18,200,000**	**25,964,700**
Connections	114,000	585,131	1,325,222	1,590,785	2,546,157	4,611,970	6,484,791	**9,304,818**	**12,933,653**
BTS					-		**1,519**	**1,924**	**2,658**
SMS					-	-	**201,445,683**	**315,557,601**	**287,145,378**

Source: Communications Commission of Kenya

(the network is very stable with infrequent disruptions). The electronic facility has made money exchange very efficient in terms of speed and safety thus transforming the manner in which ordinary Kenyans relate financially with one another. A similar service has been introduced by the Postal Corporation of Kenya but it pales in comparison with the Safaricom market.

Gathanju (2007) cites a Vodafon research:

The report also examined mobile telephony's impact on GDP growth in African nations and found that between 1996 and 2003, a country with an average of 10 or more mobile phones per 100 people enjoyed 0.6 percent greater GDP growth per capita than otherwise identical countries that lacked such mobile telephony penetration. The report said that mobile telephony had twice the impact on GDP growth for developing African nations as for developed countries.

We now turn to the political arena.

Political Impact

The mobile phone was praised for ensuring that the 2002 General Elections in Kenya were free from rigging. The cell phones were employed to relay election results from the poll-counting stations to national Television and Radio stations in Nairobi. The strategy had several advantages. Firstly, it ensured that the tallying officers had no time to meddle with the results. Secondly, the results were relayed electronically thus circumventing the delays occasioned by bad roads, rains and (more importantly) poll theft.

Otieno (2005) notes:

The contribution of mobile telephony to the defeat of KANU, which had dominated politics first as the only legal party since 1963 up to and including the multi party era that started in 1991, was phenomenal. Yet, such contribution can only be appreciated within the context of the wave of democratization process more than technological revolution. Mobile telephony contributed to the liberalization of the voters as individuals and as communities. Furthermore, this new media brought politics closer to the people, allowed them to interrogate both the political class as well as the political process. Lastly, mobile telephony in the context of the 2002 allowed space for participation in the election process by as many voters as possible especially and significantly including in marginalized areas in the country and this contributed to the ongoing consolidation of democracy.

Otieno further notes that the mobile phone encouraged people to exhort each other to vote, hence contributing to wider voter participation in the democratization process.

Despite the noble role of the cell phone in the 2002 General Elections, in the 2007 General Elections and the ensuing poll violence in 2008, the same tool was cited as playing the ignominious role of promoting ethnic hatred and bigotry among Kenyans. To address the problem, the Kenya Government criminalized hate speech via sms (Short Messages Service) in 2009. Kenyans relaying hate messages of any kind would be committing a crime that is actionable.

4. THE RESEARCH OBJECTIVE

The current study aimed at documenting some aspects of how university students, aged 19-24, interacted with mobile telephony and the effects (if any) of the latter on their behavior or culture. Further, the study sought to understand the extent to which the students were using the mobile phones for economic and political reasons.

5. STATEMENT OF THE PROBLEM

In view of the rapid expansion and availability of cell phones in Kenya, there have been dramatic changes in the way Kenyans do business or interact with one another. The mobile phone technology as noted above has revolutionize the manner in which Kenyans think; hence the mutation (relative) of cultural practices and behaviors. The extent of the change in behavior or culture deserves investigation in order to ascertain how the mobile technology is affecting Kenya, especially among the youth who account for 65% of the population.

6. SIGNIFICANCE OF THE STUDY

The current investigation is a miniature research that could trigger off a larger study. It is important to document how Kenyans relate with their cell phones; and how the technology impacts their social behavior. Historically, technology is known to change how people do things and eventually how they think. For instance, the advent of the motor car, the airplane, bullet trains, etc. was accompanied by rapid social and economic transformations of the people.

7. METHODOLOGY

The study involved 100 students from the United States International University (USIU) in Nairobi. The data was collected via a questionnaire prepared by the researcher and a student-research group that he mentors. The instrument was further refined by the Institutional Research department. The questionnaires were distributed to students irrespective of their majors; most of the students were undergraduate students. The exercise took one month. The data collected was then analyzed using the SPSS statistical package for both the qualitative and quantitative aspects.

8. ASSUMPTIONS

The fundamental presumption was that the data collected from the university students would enable the researcher to ascertain the impact of cell telephony on the youth in Kenya, and by extension the Kenyan people. Coupled with this assumption was the limitation due to the number of students involved. USIU has a population of 4689 students; therefore to analyze 100 and then draw general conclusions about USIU (or Kenya for that matter) would be grossly preposterous. The investigation was a mere exploration of a subject that warrants a substantive survey.

9. RESEARCH FINDINGS

Demographics

The sample was made up of students between 19-24 years (77.4%) and those between 25-30 years (7.1%); a further 15.5% did not indicate their gender on the questionnaire. The female students accounted for 76% of the sample; the male students 24%. The sample was 61% Kenyan, 4.8% Tanzanian, 2.4% Ugandan; the rest was shared (1.2% each) between Pakistani, Mozambican, French, Afghani, Indian, Rwandan, Zimbabwean and British. In terms of working, 94% of the students were not working and 6% were gainfully employed.

Acquisition & Purpose of Mobile Phone

On the issue of how the cell phone was acquired, the respondents mainly got their phones from the parents (54.8%); from personal savings (14.3%); a gift from relatives (11.9); and at times the cell phone was a hand-down from an older sibling/ parent (1.2%).

The purpose for acquiring the cell phone was primarily for communication (66.7%); peer pressure (since other friends had one) – 4.8%; the cell phone was deemed a necessity (2.4%): and 1.2% each for security reasons[9], socialization and chatting with friends.

Table 2 indicates the age at which the cell phones were acquired by the respondents.

It appears that most of the students acquire their phones during teenage years; from age 13 there is a visible rise in acquisition culminating at age 17 (see Figure 1). These are also the years when youngsters are in High School. At USIU, the students enroll mainly from age 19.

Mobile Phone Expenses

On the question of how many cell phones the respondents had, the majority had one unit (59.2%); those with two (33.8%); three phones (5.6%) and four (1.4%). The main reason for having more than one phone was the possibility to take advantage of the different tariffs offered by the mobile phone companies.

The type (model) of phone was very important to the respondents: 72.8% said it mattered a great while 28.2% disagreed. The former group said that the model should have as many features as possible (internet, music, beauty, design, etc). The latter argued that as long as it could be used for calling and texting, the rest was immaterial. The preferred models had internet (21.4%) and music/radio (9.5%)

Airtime expenses per day were mainly between Kshs 50 (USD 0.7) and Kshs 100 (USD 1.4) – accounting for 60.7%; and the main source was the parents (48.7%) – with mothers giving more than the fathers; the rest came from relatives, boyfriends and girlfriends.

Indispensability of the Cell Phone

On the question of whether the respondents knew of any student on campus without a cell phone, 72% of the respondents answered in the negative while 28% were affirmative. Most of the respondents said the cell phone is very important to their lives. When asked whether they would go back for the phone if they had forgotten it at home or at the office, 79% responded in the affirmative while 21% were negative. Further queried whether they

Table 2. At what age did you get your first mobile phone?

Age		Frequency	Percent	Valid Percent	Cumulative Percent
Valid	9	1	1.2	1.4	1.4
	10	1	1.2	1.4	2.9
	12	2	3.4	2.9	5.8
	13	8	9.5	11.6	17.4
	14	2	2.4	2.9	20.3
	15	9	10.7	13.0	33.3
	16	8	9.5	11.6	44.9
	17	14	16.7	20.3	65.2
	18	13	15.5	18.8	84.1
	19	7	8.3	10.1	94.2
	20	3	3.6	4.3	98.6
	21	1	1.2	1.4	100.0
	Total	69	82.1	100	
Missing	System	15	17.9		
Total		84	100		

Figure 1. Graph of data from Table 1

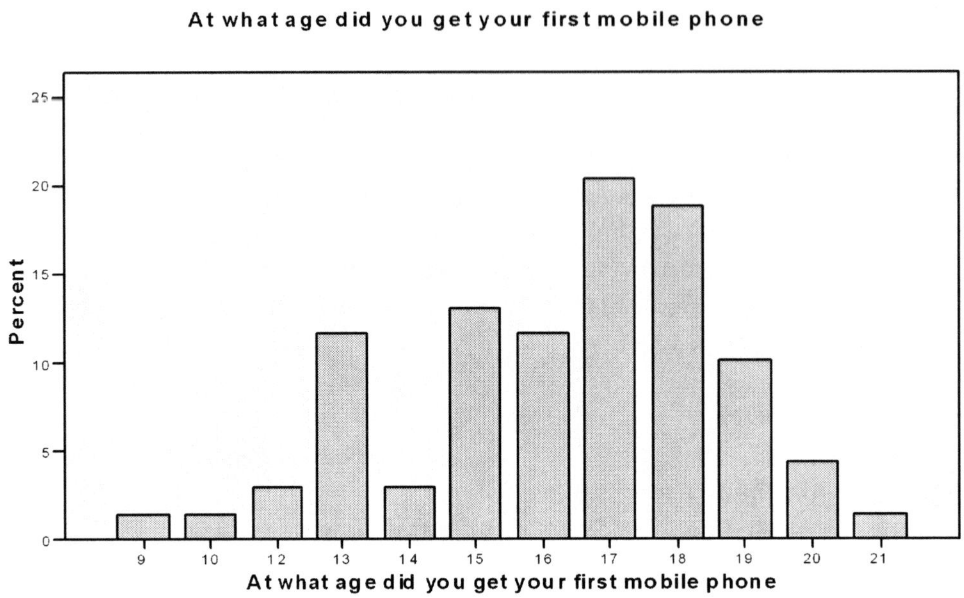

would replace a stolen cell phone *immediately* or not, 71% confirmed they would while 29% said they would replace it later.

Cell Phone and Social Behavior

The respondents were asked whether they picked up their phones in various places and the responses are indicated in Table 3.

On the issue of whether they answered the phone while driving, most the respondents (71.9%) were negative while 28.1% were positive. The latter admitted to picking the handset up (88.2%) rather than using the free-hand mode (11.8%). On the question of whether mobile phone usage contributed to absentmindedness while driving, 48.5% were affirmative while 51.5% were negative.

The respondents were interviewed on texting and whether they did it in certain places. The responses are found in Table 4.

In a social setting (pub, restaurant, etc), the respondents preferred starting up a conversation with a stranger (56.9%) to playing with the cell phone (texting, music, calling other people) – 43.1%. Further, fewer respondents felt that the mobile phone had affected their interaction skills (43.8%) compared to 56.2% who thought it had.

When asked whether mobile phones can promote dishonesty and lying, most of the respondents answered in the affirmative (76.1%) while

Table 3.

Place	Yes %	No%
Bathroom	49.3	50.7
Classroom	14.5	85.5
Meeting	17.4	82.6

Table 4.

Place	Yes %	No %
While walking	92.9	7.1
While driving	20.6	79.4
While in a group	94.1	5.9
While in church	33.8	66.2
While in class	67.6	32.4

the rest thought otherwise. The main reason was that it is easier to lie when not in a face-to-face situation. Some respondents argued that at times lying is inevitable.

Has the mobile phone affected/changed your cultural values? To this question, the respondents were divided right in the middle, i.e. 50% affirmative and 50% negative.

Cell Phone and Individual Freedom

The mobile phone was seen as a great liberator in the communication field. Indeed, 92.8% of the respondents claimed that the cell phone gave them a great sense of freedom to communication with whomever they pleased; 6.2% thought in the contrary. The respondents also said they did not mind if their close partners (boyfriend/girlfriend) had access to their phones (69%) claiming that they had nothing to hide. However, 31% were concerned claiming that a cell phone is very personal and private.

Closely related with the question of privacy was the issue of passwording access to the cell phone. The respondents were 42% to have installed a password, while 58% hadn't. The former argued that the phone was a private and confidential tool while the latter said they had nothing to hide. On whether, the respondents had coded some of the contacts, 27.3% were affirmative while 72.7 weren't.

Inconveniences of the Cell Phone

The respondents were asked to list any conveniences associated with the mobile phones. The results are in Table 5.

Evidently, the cell phone becomes a burden when one is trying to relax or unwind. Topping the list are flashers[10], stalkers[11] and people who call all the time.

Asked whether they deliberately switch off their mobile phones, the respondents were 80.3% in the affirmative citing the need to be alone and disconnected from the world. The reasons provided are collected in Table 6.

On whether the respondents had any reason to change their cell numbers, 43% were in the affirmative. The main reason for the change of number was to avoid "stalkers".

Some respondents (27.1%) also confirmed that they concealed their cell number while making a call. They argued that a surprise can be fun to a friend. Further, in order to reach a debtor who knows your number, the best way is to conceal the identity of your number. Another category said they enjoyed communicating secretly to some people for various reasons. The other respondents (72.9%) thought it was bad manners to conceal the cell phone number and they would not want someone to treat them in like manner. Transparency in communication is very crucial, they argued.

Cell Phone Etiquette, Internet Access

Many respondents (69%) found it important to call back when they found missed calls because they wanted to find out the reason for the call. Some argued that it could be a matter of life and death. They also said it was courteous to call back. The remainder (31%) preferred not to call back arguing that if the message was important, the caller will call again. They also argued that it depended on who had made the call; some people are just a nuisance and there is no need to call back.

A good number of respondents (64%) thought it courteous to excuse themselves before making or receiving a call. The rest (36%) did not see the need arguing that it was their business.

Some mobile phone habits are irksome to the public. Indeed, 84% of respondents said they were offended by the manner in which some people spoke on the phone. The main problem was speaking loudly and to a lesser extent use of socially offensive words, e.g. four letter words.

Asked whether there are text messages that they deleted immediately, the respondents were many

Table 5. In what ways do you find the mobile phone inconveniencing to you?

		Frequency	Percent	Valid Percent	Cumulative Percent
Valid		45	53.6	53.6	53.6
	Carrying it and keeping it safe	1	1.2	1.2	54.8
	Charging the phone	4	4.8	4.8	59.5
	Air time expenses	3	3.6	3.6	63.1
	Credit is expensive	1	1.2	1.2	64.3
	Facebook is time consuming	1	1.2	1.2	65.5
	Flashers	3	3.6	3.6	69.0
	If the network is down	1	1.2	1.2	70.2
	It saves time spent looking for someone	1	1.2	1.2	71.4
	It's addictive	2	2.4	2.4	73.8
	It's a major distraction	1	1.2	1.2	75.0
	It's expensive	1	1.2	1.2	76.2
	It's hard to avoid people since they have your contact	2	2.4	2.4	82.1
	Some people use the Internet wrongly	1	1.2	1.2	79.8
	Stalkers can get to you	2	2.4	2.4	82.1
	Thieves	1	1.2	1.2	83.3
	Too many people calling	1	1.2	1.2	84.5
	When called by unknown people	1	1.2	1.2	85.7
	When I need some time alone and people are disturbing	8	9.5	9.5	95.2
	When someone asks for your number and you don't want to give it to them	1	1.2	1.2	96.4
	When there is no network	1	1.2	1.2	97.6
	When you need to reach someone urgently and they are not available	1	1.2	1.2	98.8
	When you want to avoid some calls	1	1.2	1.2	100.0
	Total	84	100.0	100.0	

in the affirmative (76%) and 24% in the negative. The various reasons are tabulated in Table 7.

The study then sought to ascertain how the respondents employed the cell phone. Between calling and texting, 64% of the respondents preferred calling while 36% were for texting. The reasons are listed in Table 8.

Further 60% preferred calling back to texting back after receiving a call.

Concerning the internet, 78% of the respondents had access compared to 22% whose phones were not internet-enabled. The most frequented sites and frequency of visits are compiled in Table 9.

The frequency of the web visits was also investigated. Results are shown in Table 10.

Mobile Telephony and Language

The languages used for texting/calling are English (71.7%), Kiswahili (7.5%) and Sheng[12] (20.8%). The respondents argued that these languages were easier for them to use. The default language selected for the phone was mainly English and

Table 6. Reasons for switching off mobile phone

		Frequency	Percent	Valid Percent	Cumulative Percent
Valid		36	42.9	42.9	42.9
	At night	1	1.2	1.2	44.0
	During examinations	1	1.2	1.2	45.2
	Flashers	1	1.2	1.2	46.4
	I don't see the need	1	1.2	1.2	47.6
	I don't want to be reached	4	4.8	4.8	52.4
	If I am avoiding someone	2	2.4	2.4	54.8
	I don't want any disruptions	1	1.2	1.2	56.0
	In class and church	1	1.2	1.2	57.1
	It depends on the mood	1	1.2	1.2	58.3
	It's rude	1	1.2	1.2	59.5
	Stalkers	4	4.8	4.8	63.3
	To be alone for a while	5	6.0	6.0	69.3
	To sleep	1	1.2	1.2	70.5
	When busy	1	1.2	1.2	71.7
	When disturbed	1	1.2	1.2	73.9
	When I am feeling low	1	1.2	1.2	75.1
	When I don't	1	1.2	1.2	76.3
	When I need peace of mind	10	11.9	11.9	88.2
	When I need some time alone	5	6.0	6.0	94.2
	When I need some time out to relax	1	1.2	1.2	95.4
	When I want to save power	1	1.2	1.2	96.6
	When in a meeting	2	2.4	2.4	99.0
	Total	84	100.0	100.0	

55.2% of the respondents thought that the use of English on their phones helped them enhance their English.

On how often the respondents texted in a day we had the results presented in Table 11.

But we also sought to know the style of the texting. Do they employ short cut language (c u) or full language (see you)? Most the respondents employed the short-cut language (88%), while 12% used full language.

On whether the texting style affected their language, 48% of the respondents were in the affirmative while 52% declined. Some of the reasons are outlined in Table 12.

Cell Phone, Politics, Business, and the Media

Do the respondents use their phones to receive and send political messages? 94% of the respondents answered in the negative while 6% were in the affirmative. On whether the cell phone was used to transact business via electronic money transfers (M-Pesa, Sokotele, Zap, etc), 62.3% of respondents said they used it while 37.7% did not use the services. The reasons for employing the electronic money transfers via mobile phone are shown in Table 13.

Table 7. Reasons for deleting/not deleting some messages

		Frequency	Percent	Valid Percent	Cumulative Percent
Valid		40	47.6	47.6	47.6
	Am the only one with access	1	1.2	1.2	48.8
	Confidentiality	10	11.9	11.9	60.7
	Explicit courting messages	5	6.0	6.0	66.7
	Flash backs	5	6.0	6.0	72.6
	From stalkers	1	1.2	1.2	73.8
	I have no secrets	1	1.2	1.2	75.0
	I have nothing to hide	2	2.4	2.4	77.4
	Intimate	2	2.4	2.4	79.8
	It's private	1	1.2	1.2	81.0
	My phone is my phone	1	1.2	1.2	82.1
	Not important	3	3.6	3.6	85.7
	People who are not close to me	1	1.2	1.2	86.9
	Private business	1	1.2	1.2	88.1
	Some are not important	1	1.2	1.2	89.3
	Messages from Safaricom are not important	1	1.2	1.2	90.5
	They might offend someone	1	1.2	1.2	91.7
	To avoid unnecessary drama	3	3.6	3.6	95.2
	Unless the inbox is full	4	4.8	4.8	100.0
	Total	84	100.0	100.0	

The respondents were then asked whether they subscribed to instant press messages from media houses in Kenya (e.g. 411). Most of them (90%) said they didn't while only a paltry 10% did.

10. DISCUSSION OF THE FINDINGS

A detailed discussion of the findings from this survey would surpass the confines of this chapter. However, a couple of trends emerge from the research. The majority of the respondents are university students, aged 19-24 years, who are still under the tutelage of their parents. This explains why most of them got their phones and air time money from their parents; the latter may have invested in the cell phone in order to keep tabs on their progeny.

The importance and indispensability of the cell phone comes out forcefully from the survey. In addition, the prevalence of the cell phone in terms of usage (church, classroom, while driving, etc) cannot be overemphasized. The fact that the handset is deemed indispensable, personal and private, a tool for expressing freedom of communication is a powerful statement of how the cell phone is central in the lives of the studied, and probably most Kenyans.

Human communication has a code of conduct and cell telephony is not exempt from it. Many respondents thought it courteous to call back missed calls, excuse oneself before taking/making a call, talk softly, use decent language, and avoid harassing people. In the same token, human language is employed for lying and the cell phone has become a formidable tool in promoting

Table 8. Reasons for preferring either texting or calling

		Frequency	Percent	Valid Percent	Cumulative Percent
Valid		22	26.2	26.2	26.2
	Both	2	2.4	2.4	28.6
	Both are convenient	1	1.2	1.2	29.8
	Can do both if necessary	1	1.2	1.2	31.0
	Cheaper	1	1.2	1.2	32.1
	Cheaper and you can say anything	1	1.2	1.2	33.3
	Depends on communication	1	1.2	1.2	34.5
	For clarification	3	3.6	3.6	38.1
	For future reference	1	1.2	1.2	39.3
	Get response	1	1.2	1.2	40.5
	I am more of a writer than a talker	1	1.2	1.2	41.7
	I don't like talking	1	1.2	1.2	42.9
	I find it more convenient	1	1.2	1.2	44.0
	I have a lot to say when I text	1	1.2	1.2	45.2
	It is cheaper	12	14.3	14.3	59.5
	It depends on the situation	2	2.4	2.4	61.9
	It is easier and convenient	1	1.2	1.2	63.1
	It lasts	1	1.2	1.2	64.3
	It is convenient	1	1.2	1.2	65.5
	It is easier and personal	1	1.2	1.2	66.7
	It is easier and there is no evidence	1	1.2	1.2	67.9
	It is easier to communicate	3	3.6	3.6	71.4
	It is faster	9	10.7	10.7	82.1
	It is more expensive	1	1.2	1.2	83.3
	It is friendlier	1	1.2	1.2	84.5
	It is much easier to understand	1	1.2	1.2	85.7
	It is quicker	1	1.2	1.2	86.9
	It is more direct	1	1.2	1.2	88.1
	No evidence and it's cheap	1	1.2	1.2	89.3
	Personal	1	1.2	1.2	90.5
	Prompt response	1	1.2	1.2	91.7
	Record	1	1.2	1.2	92.9
	Spend less if I have to explain something	1	1.2	1.2	94.0
	Take less time	1	1.2	1.2	95.2
	Texting takes longer	1	1.2	1.2	96.4
	Texts are limiting the information you pass	1	1.2	1.2	97.6
	You can think about what you want to say	1	1.2	1.2	98.8
	You say more when you write	1	1.2	1.2	100.0
	Total	84	100.0	100.0	

Table 9. Which are your most frequented sites?

		Frequency	Percent	Valid Percent	Cumulative Percent
Valid		37	44.0	44.0	44.0
	Facebook	19	22.6	22.6	66.7
	Facebook BBC sports	2	2.4	2.4	69.0
	Facebook Gmail	1	1.2	1.2	70.2
	Facebook, google, Gmail	7	8.3	8.3	78.6
	Facebook, twitter, ebuddy	1	1.2	1.2	79.8
	Facebook, Google, Yahoo	8	9.5	9.5	89.3
	Faceboook.com	1	1.2	1.2	90.5
	Gmail Waptruck	2	2.4	2.4	92.9
	Google	2	2.4	2.4	95.2
	Nokia	1	1.2	1.2	96.4
	Yahoo	1	1.2	1.2	97.6
	Yahoo mail	2	2.4	2.4	100.0
	Total	84	100.0	100.0	

this human vice mainly because it allows one to conceal their whereabouts. Some of the reasons advanced for preferring texting to calling was that texting allows someone to lie at ease since the text is devoid of emotions normally associated with a lying speech.

It is notable that the cell phone can also be a source of discomfort. The fact that one is accessible to all can be a destabilizing factor especially when one craves solitude and silence. In other words, the incessant calls and text messages can deluge the mind to the extent that only the switch off button can salvage the receiver. This is particularly the case among young female students who get many calls from unrelenting suitors whom they refer to as “stalkers”.

Table 10. How often do you visit these sites?

		Frequency	Percent	Valid Percent	Cumulative Percent
Valid		47	56.0	56.0	56.0
	3 times a day	3	3.6	3.6	59.5
	5 times a day	1	1.2	1.2	60.7
	5 times a week	1	1.2	1.2	61.9
	As often as I can	1	1.2	1.2	63.1
	Daily	5	6.0	6.0	69.0
	Every hour	7	8.3	8.3	77.4
	Everyday	7	8.3	8.3	85.7
	Often	9	10.7	10.7	96.4
	Once a week	2	2.4	2.4	98.8
	Rarely	1	1.2	1.2	100.0
	Total	84	100.0	100.0	

Table 11. On average, how often do you text in a day?

		Frequency	Percent	Valid Percent	Cumulative Percent
Valid		27	32.1	32.1	32.1
	10 times	14	16.7	16.7	48.8
	100 times	1	1.2	1.2	50.0
	15 times	1	1.2	1.2	51.2
	20 times	6	7.1	7.1	58.3
	3 times	4	4.8	4.8	63.1
	30 times	3	3.6	3.6	66.7
	34 times	1	1.2	1.2	67.9
	35 times	1	1.2	1.2	69.0
	36 times	1	1.2	1.2	70.2
	40 times	2	2.4	2.4	72.6
	5 times	11	13.1	13.1	85.7
	50 times	1	1.2	1.2	86.9
	7 times	1	1.2	1.2	88.1
	It depends	1	1.2	1.2	89.3
	Once	5	6.0	6.0	95.2
	Rarely	1	1.2	1.2	96.4
	Twice	3	3.6	3.6	100.0
	Total	84	100.0	100.0	

Table 12. Explanations

		Frequency	Percent	Valid Percent	Cumulative Percent
Valid		54	64.3	64.3	64.3
	Am always keen	1	1.2	1.2	65.5
	Am used to short forms	1	1.2	1.2	66.7
	I am able to switch	1	1.2	1.2	67.9
	I can differentiate	7	8.4	8.4	76.3
	I forget the way of writing correct English	1	1.2	1.2	77.5
	I have forgotten to spell some words	1	1.2	1.2	78.7
	I know where I am writing	1	1.2	1.2	79.9
	I tend to shorten my sentences at times	12	14.3	14.3	94.2
	I write in full language	2	2.4	2.4	96.6
	I write short forms in class	1	1.2	1.2	97.8
	I write things in slang	1	1.2	1.2	99
	It's confusing at times	1	1.2	1.2	100.0
	Total	84.0	100.0	100.0	

Table 13. Reasons for mobile money transfers

		Frequency	Percent	Valid Percent	Cumulative Percent
Valid		57	67.9	67.9	67.9
	I no longer bank cash because M-pesa is cheaper	1	1.2	1.2	69.0
	Access of money	3	3.6	3.6	72.6
	Convenient	1	1.2	1.2	73.8
	Easy payments of bills	3	3.6	3.6	77.4
	Effective and efficient money transfer	11	13.1	13.1	90.5
	Faster	1	1.2	1.2	91.7
	For the better	1	1.2	1.2	92.9
	Improved	1	1.2	1.2	94.0
	Instant cash	1	1.2	1.2	95.2
	It's easier doing business	1	1.2	1.2	96.4
	It's easy to receive and send money	1	1.2	1.2	97.6
	Made it more efficient	1	1.2	1.2	98.8
	It helps when one is in need	1	1.2	1.2	100.0
	Total	84.0	100.0	100.0	

Further, the fact that texting is done while walking, driving, while in church, etc presents a new phenomenon in the Kenyan culture. This phenomenon has far-reaching consequences in terms of personal security and the manner in which Kenyans have come to regard church sermons, classroom lessons, etc. One would expect respondents to shut off the cell phone while praying in church or while following a lecture. Texting while driving can lead to serious road accidents (Wambugu 2010).

The finding that some students receive calls without the hands-free facility points to the great danger that such a practice poses for road users. In fact, since 2009, it's now an offence punishable in law to use the cell phone while driving in Kenya. Court fines are as high as Kshs 8,000.

In terms of language, it is surprising that young university students in Nairobi should prefer English then Sheng before Kiswahili, especially since the main reason given for the language choice was ease of use. We would have expected the order of preference to be Sheng, Kiswahili and then English since English is a foreign language and Kiswahili is considered more challenging than Sheng. However, the fact that English is Kenya's official language and it's the default language on the cell phones in Kenya may have contributed to making English the preferred language. We presume that in non-urban setting or among older Kenyans, Kiswahili would have taken the place of Sheng. The order would then have been English, Kiswahili with little, if any, Sheng.

With regard to news from media houses, the students do not subscribe to instant news services since they prefer browsing the internet for the same, and for free. In terms of political engagement, Kenyan universities are no longer the conscience of the nation as they used to be under the old system of education. Since 1985, the university students are very passive politically. This may explain why the students surveyed did not use cell phones to relay or receive political messages.

The use of the electronic money transfers via cell phones is not prevalent among students for the reason that most of them are not working gainfully. However, those that are working found it convenient, safe and fast.

12. CONCLUSION

The current survey sought to document, though modestly, the manner in which university students relate with their mobile phones. We sought to understand how this novel technology was impacting on the lives of young Kenyans. The findings provided some expected (trivial) trends such as the ubiquity, privacy, indispensability and power of the mobile phone. Nonetheless, there are non-trivial trends such as use of cell phone in situations that could raise concern: toilet, while driving, while in class, while in church, etc. Similarly, the emphasis on calling rather than texting is surprising since the latter is cheaper. Michael Joseph, Safaricom's CEO is quoted as saying that "Kenyans have peculiar calling habits" (Wachira 2010: 36-38). This means that Kenyans call more than they text, and secondly they mainly call in the mornings (6 a.m. – 8 a.m) and evening (5 p.m. – 8 p.m.).

The fact that the cell telephony is seen as a catalytic agent for providing misinformation, e.g. lying about one's whereabouts, is a serious issue in a country that claims to be very religious.

Of particular interest was the issue of whether the respondents felt whether cell telephony had affected their culture. Most of them did not feel so and yet other questions in the survey confirmed that the cell phone had changed the manner in which way people do their business today in Kenya. This discrepancy may have been due to a misunderstanding of the meaning of the word *culture*.

In fine, the study provides a prima facie investigation into the impact of mobile telephony on the Kenyan people. Although our sample was small and far from representative of Kenyan universities, let alone Kenyan society, the findings provide a small window to some of the effects of the cell phones on the Kenyan people. We hope that the preliminary findings will trigger off other studies, substantive and expansive, to unearth the raft of effects of the mobile phone technology on the way Kenyans think and do their daily business.

ACKNOWLEDGMENT

We would like to express our gratitude to Mrs Beatrice Gichohi, Head of Institutional Research at the United States International University, Nairobi, for her expertise and diligence in coding and analyzing the data. We further thank the Student Research Group under my mentorship for compiling all the questions for this survey. These included Margaret Otieno, Jim-Dennis Okore, Martha Waithera Mbote, Khalif Mohammed Abdi and Franck Asseko.

REFERENCES

Bhuiyan, S. I. (2006, September 5). Impact of new media technology on society. *Daily Star*.

Drolet, K. (2009). *Cell phone culture*. Retrieved October 7, 2009, from http://media.www.thecurrentonline.com/home

Gathanju, D. (2007). *Safaricom introduces mobile money transfer*.

Kaloki, T. (2010). *Kenya accounts for 7% of Africa mobile industry*. Retrieved March 17, 2010, from http://bizcommunity.co.ke/Article/111/78/43445.html

Mbogo, M. (2010). The impact of mobile payments on the success and growth of micro-business: The case of M-Pesa in Kenya. In Iraki, F. K. (Ed.), *Journal of Language, Technology & Entrepreneurship in Africa (JOLTE), 2(1)*.

Nyabiaegea, J. (2009). *Mobile was a gift horse from the heavens*. Retrieved March 17, 2010, from http://www.nation.co.ke/news/-/1056/830852/-/vona75/-/index.html

Okello, D., & Owino, K. (2004). Socio-economic context of governance in Kenya. In Bujra, A. (Ed.), *Democratic transition in Kenya: The struggle from liberal to social democracy* (pp. 57–103). Nairobi & Addis Ababa, Kenya: African Centre for Economic Growth & Development Policy Management.

Otieno, I. O. (2005). *Mobile telephony and democratic struggles: A case of 2002 elections in Kenya*. Retrieved on 17 March, 2010, from mokk.bme.hu/centre/conferences/reactivism/FP/fpOtieno.

Wachira, N. (2010, 14 August). He built Shs25bn empire from scratch. *Saturday Nation*.

Wambugu, S. (2010, 23 January). Why that SMS behind the wheels kills. *Daily Nation*.

Wei, C., & Kolko, B. E. (2005). Studying mobile phone use in context: Cultural, political, and economic dimensions of mobile phone use. In *Proceedings of the International Professional Communication Conference*.

Weinberg, J. (2009). *A language of its own: The Syrian mobile phone culture*.

ADDITIONAL READING

Communications Commission of Kenya report, 2009.

Kenya Education Network (KENET) report 2009

Report on Media Survey in Kenya, 2009.

ENDNOTES

1 A game of words from Louis IV's statement in French meaning "I am the State". Moi (*me* in French) and Moi (Kenya's former president).

2 It is now a criminal offence in Kenya to send hate messages. This is a measure to curb abuse of mobile telephony and to promote national cohesion.

3 http://www.cck.go.ke/market_information_telecommunications accessed on 28 January 2010.

4 In fact, the first mobile phone owners were government operatives or people who had benefited from the Moi Regime.

5 The company changed its name to Celtel following a 60% buy out by Celtel International.

6 These handsets (Blackberry, Nokia, etc) are actual mini-computers with highly sophisticated features such fast Internet connections, email, etc. They are still relatively cheaper in relation to the prohibitive prices of the 1990s. These models are strong status symbols for the affluent Kenyans.

7 Currently the exchange rate is 1USD = 75 Kshs.

8 Mr. Michael Joseph, the CEO of Safaricom noted that Kenyans talk a great deal, and that is good business for his company. On Fridays, the networks are congested as Kenyans make arrangements for evening socialization and weekend outings.

9 A sample of 7 American students cited security as the main reason why their parents had bought them cell phones. The parents can track their children to know whether they were fine.

10 *Flashing* in Kenyan parlance is the act (or is art?) of calling people on the cell phone and then disconnecting immediately so that the intended recipient can call them back, of course at his or her expense. The practice is

quite irksome as can be noted in the results above.

[11] *Stalkers* is a term employed by young people to mean people, especially men, who keep calling someone in pursuit of sexual favors.

[12] Sheng is an idiom with a basic syntax of Kiswahili and words drawn from local languages in Kenya and English. It is the main language of the youth in urban centers. The medium is greatly vilified by the older generation, language purists and the educational system apparently for assaulting the dignity of Kiswahili and English.

Chapter 6
New Media and Gender in East Africa:
Assessing Media Dependency and Public Attitudes

Uche Onyebadi
Southern Illinois University Carbondale, USA

Yusuf Kalyango Jr.
Ohio University, USA

ABSTRACT

This study set out to ascertain the use of and dependency on new media technology for political communication by voting- age citizens of the three main East African countries, namely Kenya, Uganda, and Tanzania. We learn the importance of new media communication opportunities towards the struggle for liberal democracy, which is demonstrated by the unanimity of respondents' attitudes that their governments suppress political discourse. But the degree of use and dependence on these new media instruments are not uniform across the region. Gender and level of education are two salient factors that create these differences in the use of and dependency on new media technology for political communication in East Africa. But our findings have no bearing on the authenticity of the messages sent through these new media technological devices, or the extent of divisiveness or unity that such messages might engender among citizens in the region in times of political crises. Our primary contention is that such citizens now depend on these technological devices to serve their informational needs moreso when political and other forms of national emergency situations arise. This dependency phenomenon is partly the consequence of the existence of media laws enacted by governments in the region to hamstring mainstream media houses.

DOI: 10.4018/978-1-60960-591-9.ch006

INTRODUCTION

This chapter investigates the use of new media for political communication by voting-age citizens across three major East African countries: Kenya, Tanzania and Uganda, accounting for gender differences and levels of education, with primary focus on women in the broader scheme of new media use and dependency. One of the major reasons for focusing on these countries is because of their dominant presence in the region, which dates back to 1967 when they formed what is arguably the nucleus of the East African Community (EAC). The EAC now embraces five countries in the region.

Developments in new media technology have had significant impact on the content and delivery of messages in political communication, especially in periods of electioneering campaigns. King (1997) argues that elections are tripartite affairs between the contestants who are encoders of political messages, the electorate or decoders, and the media which serve as the bridge between both groups. However, the reality in modern times is that the institutionalized media's monopoly of the "bridge" is no longer tenable. Also, the flow of messages has ceased to be vertical or uni-dimensional.

With the increasingly pervasive use of new media technologies, message source in political communication has been decentralized (Anduiza, 2009), with the electorate being transformed from mere "passive receivers of a message constructed and controlled by established institutions" (Turiera-Puigbo, 2009, p. 14), to people who simultaneously produce, disseminate and consume political information.

This opportunity to construct and share information by the electorate poses new challenges to political parties and candidates in their role of mobilizing people for political participation (Anduiza, 2009; Montero, 2009). It was in response to this challenge that campaign strategists for John McCain and Barack Obama extensively used the internet and other media platforms to reach the electorate in the 2008 U.S. presidential election (Kreiss, 2009; Turiera-Puigbo, 2009).

The impact of new communication technology in political communication is also substantial in other parts of the world such as South Korea (Lee, 2009) and the Philippines (Karan, Gimeno & Tandoc, 2009). In Europe for instance, the influence of political parties is on the decline due to the communication technology revolution (Zittel, 2009) and in China, the use of cellular phones is now an important factor in communication (Latham, 2007).

How about Africa, especially East Africa, which is the focus of this chapter? We acknowledge that the three main countries in that region – Kenya, Tanzania and Uganda – have always had political parties in their practice of democracy. However, the opposition in those countries have generally been marginalized or incapacitated by autocratic regimes since the countries gained independence. In reality, each of these nations is a *de facto* one-party state where government direct or subterranean control of communication channels has been prevalent. But, the advent of new communication technology now appears to offer citizens an alternative to controlled government sources of information.

For instance, the figures below show the level of cellular phone and internet opportunities available to the citizens of the three nations. According to the *World Factbook*, cell phone use in Kenya and Uganda in 2008 was 16.2 million and 8.6 million respectively, while Tanzania had 14.7 million users in 2009. Internet access in Uganda was 2 million in 2007 and 400,000 in Tanzania in the same year. Kenya's figure for 2008 was 3 million. The foregoing data show a modest explosion in the availability and use of new media communication channels in the East African region for mass mobilization, political contestation, and communication.

With regard to political participation and the extent of women's engagement in politics in the region, these figures are salient to this study. Uganda has a population of 12.5 million men and 11.9 women (2002 census), but according to the 2009 report by the Inter Parliamentary Union (IPU), an organization for national parliaments all over the world, women only constitute 30.7 percent of members of parliament (MPs) in the country. In Tanzania, women (17.6 million) also outnumber men (16.8 million) according to its 2007 population figure. Yet, the IPU report shows that only 30.4 percent of MPs in the country are women. The IPU figure for women MPs in Kenya is 9.8 percent, yet the country's 2007 population statistics show that Kenya had 18.5 million women and 17.7 million men.

While this number of female MPs in all three countries might be low, previous studies show a relatively high level of political participation by women given that factors such as culture, level of education and financial stability generally weigh against women in the region (see Kalyango & Winfield, 2009). We observe that in spite of these modest gains by women in East African politics, and the expanded new media opportunities in the region, no known studies have specifically addressed the issue of gender, political communication and new media in East Africa. What is available is a handful of studies about women and affirmative action and gender equality attributes in the media (Kalyango & Winfield, 2009; Tamale, 1999; Tripp, 2000). We, hence, investigate and evaluate the use of new media for political communication in the region, with emphasis on women, gender differences and levels of education.

This study draws from the media dependency theory developed by Ball-Rokeach and DeFleur (1976). The scholars suggest that media dependency is pervasive in modern societies, and define dependency as "a relationship in which the satisfaction of needs or the attainment of goals by one party is contingent upon the resources of another party" (p.6). We examine media dependency within the context of the growing use of new media communications facilities in East Africa, by specifically looking at trends in political communication in the region. We attempt to provide empirical evidence to these questions: In what ways do women in East Africa use the mobile cell phones and the internet to learn about politics and to express their political views? How strongly do women in the region regard text messaging as a quick and immediate source of political information vis-à-vis other media? Which means of communication do women in the three countries rely upon as their first choice of information source in times of national emergencies? Do public attitudes towards new media and national politics vary between users based on their gender within provinces in each country and across East Africa when we account for education levels?

MEDIA DEPENDENCY: THEORY AND LITERATURE

The Media Systems Dependency (MSD) theory as developed by Ball-Rokeach and DeFleur (1976) was not simply an intellectual effort to formulate and add another layer to the existing repertoire of media effects theories. The scholars specifically asserted that media dependency is largely heightened and sustained by two factors: When the need for information is central to decision making, and "when there is a high degree of structural instability in the society due to conflict and change" (Ball-Rokeach & DeFleur, 1976, p. 7). Natural disasters such as the tsunami that hit south-east Asia in 2004, Hurricane Katrina that ravaged New Orleans in 2005 and the earthquake that devastated Haiti in 2010 equally fuel a sense of dependency on media messages especially for families of the victims of the disasters.

This assertion implicitly recognizes that the degree and sustenance of media dependency vary from person to person and groups (Salwen, 1987). Skumanich and Kintsfather (1998) also add that one medium might satisfy multiple dependency goals. On the other hand, several individuals may use one medium to satisfy their collective dependency.

MSD theory is anchored on individuals' goals and the resources available to the media to facilitate the attainment of those goals (DeFleur & Ball-Rokeach, 1989; Grant, Guthrie & Ball-Rokeach, 1991). As such, the theory is predicated on a relational paradigm. It goes beyond an individual's exposure to media messages (Becker & Whitney, 1980), to linking his or her goals to the capacity of the media to satisfy the goals. As enunciated by Ball-Rokeach (1985), "dependency" is about goals set by individuals and the extent to which such goals are fulfilled by the media. Dependency therefore structurally weakens as the media's inability to facilitate the realization of those goals increases, and vice versa.

These individual goals are compartmentalized into three: play, orientation and understanding (Ball-Rokeach, 1985; DeFleur & Ball- Rokeach, 1989; Loges & Ball-Rokeach, 1993). Each dimension is further sub-divided into two, with understanding referring to the goals of "self" discovery or knowing about oneself, or about going beyond self to be acquainted with one's community or "social" environment; orientation is about an "action" such as deciding what dress to purchase and where to buy it, and the "interaction" involved in reaching out to learn how to handle situations; and play or engagement involves issues such as "solitary" play in relaxing with a beer or "social" play which involves group activity such as partying with friends.

Further refinement of MSD expands it beyond its original, micro, individualist focus to a broader and macro, ecological perspective. This entails the examination of the tripartite relationship between the audience, media systems and society at large. The premise is that human society is like an ecological system which is an amalgamation of individuals, groups, associations etc. These sub-sets are structurally interlinked and their dependencies are reciprocal. The media are concerned with information gathering, processing and dissemination but they depend on other members of society's ecological system to perform their functions. In other words, media power is contingent upon dependency relations with other organizations, and this has implications for media system dependency for the individual (DeFleur & Ball- Rokeach, 1989; Ball-Roakeach, Power, Guthrie & Waring, 1990; Ball-Rokeach, 1998).

Although "MSD is a theory in progress" with micro and macro manifestations (Ball-Roakeach et al 1990, p. 249), its application to research has been substantial. In a cross-cultural study about mobile phone dependency, Wei (2009) found that Chinese high-school students depended more on their mobile telephones for orientation and play than their American counterparts. Another study in China (Tai & Sun, 2007) underscored the heightened nature of media dependence in crisis situations, as in the case of the SARS epidemic in 2003.

Much as these studies show great latitude in the use of MSD theory, the advent of the internet and other forms of social media still require some critical reexamination of the application of the theory, particularly in Africa. Ball-Rokeach (1998) has a robust appraisal of this new information regime. She concedes that MSD is not a "revolutionary" but an "evolutionary" theory. It is our contention that there is need to expand the theory to the new realities brought about by new information technologies, where they seem to play a major role in East Africa for political mobilization. The traditional notion of dependence on formal media systems must be reappraised to account for dependence on other social media in East Africa.

POLITICAL COMMUNICATION, NEW MEDIA AND MSD

Political communication is a multi-disciplinary field that straddles across mass communication, political science, psychology, sociology, philosophy, all of which study communication in political settings largely at three main pedestals of analysis: the individual, institutions and systems (Perloff, 1998). This multi-dimensional character of the discipline does not predispose it to having a generally acceptable conceptual definition.

For instance, Denton and Woodward (1990) see political communication as discussions about resource allocation – with authority to govern and how sanctions are to be imposed in a political system. McNair (2007) and Graber (1981) conceptualize it as purposeful communication about politics that incorporates verbal, written and visual forms. Truman (2008) notes that it is "the discursive process by which political information is shared and promotes awareness, ignorance, manipulation, consent, dissent, action, or passivity" in the political environment (p. 8).

While these definitions add value to the understanding of the nature of political communication, this study will adopt the definition offered by Chaffee (1975) and adapted by Kaid (2004). Chafee suggests that political communication is about the "role of communication in the political process" (p.15). This suggestion is premised on the fact that participatory democracy entails the flow of communication between and among the various segments of a society, from individuals to organizations (Gastil, 2008). Such communication is pivotal to having an informed citizenry and winning elections, and it comes in various forms and across multiple media platforms. In the East African context therefore, we see the role and means of sharing information about politics and the political process amongst citizens, as the central theme in political communication.

Modern political communication is no longer confined to the traditional provision of information by contestants and institutions for consumption by the electorate. It is now a public dialogue between political office seekers and potential voters (Jackson & Lilleker, 2009). Thus the one-to-many or candidate-to-electorate model of political communication is increasingly making way for the more interactive many-to-many system brought about by new communication technologies (Karan, Tandoc & Gimeno, 2009).

The huge diversification of these means or platforms for information sharing as a result of modern communications and technological development, has effectively challenged the orthodoxy of formal media organizations, which had monopolized information gathering, processing and dissemination in electoral campaign periods in East Africa. Consequently, new media technology has expanded the scope of media dependency on the part of the electorate. As Tedesco (2004) observes, "the Internet appeared boundless to limits of individual expression, provided hope for lateral communication structure, and offered promise to sever the hierarchical public dependence on the established media" (p. 510).

In this regard, the primary source of political communication among the electorate in the region appears to be the mobile phone. Some scholars have argued that its portability and affordability has made it one of the most sought-after (political) communication devices in most parts of the world (Karan et al., 2009), and especially in Uganda, Kenya, and Tanzania where other electronic communication devices remain largely unaffordable.

The "swift emergence of the internet as a communication tool" (Tai & Sun, p.993) and the burgeoning use of new social media in political communication are not symptomatic of the death of traditional forms of communication. Rife, Lacy and Varouhakis (2008) examined studies on new media technology and concluded that the Internet's capacity to displace traditional media may not be as realistic as some scholars predict. Anduiza (2009) is more succinct as she presents the Internet hype as no more than "inflated utopias." This may still be the case in Africa despite the remarkable growth of the information technology sector; because majority of citizens in rural areas of East Africa still depend on other traditional media such as radio for all aspects of political communication (Kalyango 2009, Kalyango & Eckler, 2010). Even among residents in the urban areas, the availability of disposal income might limit ownership of the device. As documented in the *World Fact Book*, the 2009 estimated Gross Domestic Product (GDP) per capita in the three East African nations was $1,600 in Kenya, $1,300 in Uganda, and $1,400 in Tanzania. These low income levels suggest a restricted affordability of the device in the region.

Nevertheless, what is hardly in contention is that the Internet and new media technology have changed the landscape of political communication in the three countries examined here. The challenge in this study therefore is to examine how this change affects media dependency and public attitudes in the region as a whole, plus within and across 15 provinces in the region.

EAST AFRICA: MEDIA POLICIES AND POLITICAL COMMUNICATION

East Africa is located on the eastern coast of sub-Saharan Africa and includes the three sovereign states of Kenya, Tanzania, and Uganda. Some geographers place Rwanda and Ethiopia in the greater East Africa, but this study is only concerned with the political triad of the former communal East Africa. Kenya, Tanzania and Uganda share common borders, markets, legislature, Lake Victoria, the Kiswahili language, and a common passport under the administrative umbrella of the East African Community (EAC).

Despite the existence of the EAC as a regional intergovernmental organization and symbol of unity among Kenya, Tanzania and Uganda, each of these countries has a unique political culture and ideologically unique regime as some of the factors which differentiate them, and which sometimes contradict the common political, economic and administrative goals of regional cooperation (Kalyango & Eckler, 2010). Consequently, the assumptions here are that the major challenges facing the future of the East African political federation include the variations in characteristics of political participation, contestation by citizens based on gender, and the fragility of news media advocacy to educate citizens on their political rights and civic duties.

The political independence gained by Uganda (1962), Kenya (1963), and the Union of Tanzania (1964) did not result in jettisoning all administrative and legal relics of the ousted colonial regimes in these countries. The government-media relationship is one area where the incoming indigenous governments were not in a rush to discard the legacy of their predecessors.

Ng'weno (1969) contends that while censorship did not officially exist in the region, the East African governments had enormous powers to curb press freedom in those early post-colonial years. Kenya, Uganda and Tanzania all had restrictive sedition laws that were inherited from the British and were kept in the statute books almost unchanged. He further asserts that the governments had preventive detention acts which empowered them to detain without trial those responsible for acts and statements (including newspaper reports) considered by the governments to threaten their national interests.

What Ng'weno observed over forty years ago is still a largely legitimate comment about the nature of press freedom in contemporary East Africa, and other parts of Africa. But, quite curiously, Skjerdal (2008) opines that what is prevalent in Africa is not official censorship but self-censorship. The author describes self-censorship as something triggered by "perceived external pressure which leads journalists to alter media content in order to avoid personal reprimands and stay clear of accusations of inciting social and political tensions" (p.185). Such an argument tends to ignore the reality that a journalist's self-censorship is merely symptomatic of a much greater debilitating precursor and malaise – the official censorship.

Evidence of official censorship abounds in the East African region. As recently as 2007, the

Ugandan government introduced a controversial draft bill – the Interception of Communications Bill - daubed by the media as the phone-tapping bill. This bill grants the Ugandan government powers to monitor telephone conversations for national security purposes. In 2008, the Kenyan government also introduced its latest media gag bill. The British Broadcasting Corporation reported on its web site (January 2, 2009) that the "Kenyan President Mwai Kibaki has signed into law a media bill that journalists say will curtail press freedom. The law gives the Kenyan authorities the power to raid media offices, tap phones and control broadcast content on grounds of national security." In the case of Tanzania, Mulama (2008) reported that the government has laws in place that empower it to confiscate broadcasting equipment and muzzle the press.

While it is relatively easy for the governments of Kenya, Tanzania, and Uganda to attempt to shackle institutionalized media houses, the advent of new media technology is posing a threat to their control of news and dissemination of current affairs in the region. Kenya is a good example of this official dilemma. In the wake of the 2007 presidential election in the country, the government exercised it legal muscle and prohibited live news broadcasts in the country amidst allegations of election rigging by the Kibaki administration. The unintended consequence of this action was the soaring of politically inflammatory messages through mobile phone texting or SMS (short message service) and the blogosphere among Kenyans.

In a policy briefing paper sponsored by the British Broadcasting Corporation's (BBC) World Service Trust, Abdi and Deane (2008) contended that in the confusion following the disputed Kenyan election, the government unsuccessfully tried to shut down the SMS facility provided by Safaricom, the country's leading provider of cell phone services. The authors quote the company's former general manager, Michael Joseph, as saying that his organization came under intense pressure from government to suspend its SMS service on December 31, 2007 (four days after the election) but they resisted the interference "because we felt that panic might ensue" (p.10).

This Kenyan experience imposed unprecedented and unanticipated limitations on government's ability to control the content of information that gets to the public. On the other hand, it flung open new vistas in political communication among citizens. In this regard, Mwesige (2009) and Kalyango (2009) argue that talk radio, in the case of Uganda, has expanded the platform for the expression of alternative opinion by the opposition, major stakeholders and citizens in the country. Kalyango (2009) specifically found considerable use of radio for political information, but no direct influence on support for democracy. He also determined that greater use of radio to get political news in East Africa was strongly related to measures of political interest; and that political interest was a moderating factor for democratic support. Our contention here is that the penetration of new media technology, especially the cellular phones, has even quadrupled opportunities for this expression and exchange of political information among citizens of the East African region.

The downside of this new opportunity is obvious: It can be used for negative purposes, as was the case in Kenya where new media technology was used to circulate messages that fuelled the electoral crisis. While this is true, and efforts should be made to curtail such technology abuse, what is of greater significance is that new media technology is challenging information gathering and dissemination and government control of information in East Africa. The negative attributes, however, do not diminish the benefits of mobilizing citizens to engage in political discourse or protests during national emergencies. With rapid innovations in information technology such as the nternet and mobile electronic devices, citizens can no longer be starved of political information merely because government has clamped down on established media organizations (McMillan, 2006). They can originate and share such information through the use of their mobile phones' text messaging facility and other new technology devices.

DATA AND RESEARCH DESIGN

Data for this study are based on a survey of public attitudes on some key models of media dependency and media use in Kenya, Tanzania and Uganda. A cross-sectional survey of East Africans was conducted between June 2 and July 30, 2007. A total of 1,395 surveys in Tanzania (n=465), Kenya (n=448), and Uganda (n=482) were returned out of the 1,500 surveys that were randomly handed out. Citizens were selected using a stratified multi-stage cluster random sampling. A "citizen" of East Africa is defined as an individual over 18 years of age who resides in one of the three countries, and who can articulate his/her interests and attitudes based on reasoning and consciousness.

The media (radio, television, newspapers, online news, mobile phone texting or SMS messaging) are assessed here as monolithic concepts of traditional media and new media. The new media in this context are the online news, mobile phone texting or SMS messaging. During data collection, we considered the fact that radio is very popular among both rural and urban dwellers and that other media such as television are restricted largely to urban dwellers who can afford TV sets and in areas with electricity (Kalyango, 2009). We also considered that mobile phones are available to both literate and illiterate citizens in urban areas who can afford them (Kalyango & Eckler, 2010). Households and individuals who are exposed and use some or all of these media were surveyed.

A total of 28 field survey coordinators were recruited from East Africa to administer these surveys from house to house, at workplaces, and also at academic and public/state institutions. The survey instrument was given to respondents in hard copy (paper format) as a face-to-face survey. Respondents filled out either an English version or a translated version of the survey in Kiswahili, the most widely spoken language in the region. A face-to-face survey was administered instead of other data collection methods because more than 80 percent of the population has no access to a conventional land telephone line and an estimated 60 percent of residences or citizens in East Africa do not have access to a physical-address infrastructure where a mail-in or a self-administered survey can be effectively sent.

The survey instrument was first administered in a pilot study to four Africans studying at a major Midwestern university in the United States, as well as to five people resident in Tanzania, five in Uganda, and five in Kenya, to ascertain the reliability of the items that were used to construct the scales, before they were used. Field survey coordinators were matched with a province of their ethnicity or residence where their presence would create trust and comfort to respondents in order to minimize sensitivity, inaccuracy, and biased responses. Forcomparative purposes, the same survey instrument and questions were used in all three countries.

Items on the instrument utilized attitude measurements of media systems dependency (MSD), gender, levels of education, national politics, national security, and good governance to assess the citizens' degree of strong agreement or strong disagreement. Other items sought opinion about news coverage and political information on traditional media vs. the new media, and about the government and the presidency. Respondents were asked on some items to check the response that best corresponded with their opinion on point values assigned for items such that the higher the score, the more positive the response was. MSD was evaluated based – first of all, on *media consumption* (use of various media), *orientation* (degree of their attitudes), and *engagement* (reaction to or participation). MSD was predicted to have occurred based on the assumption that the higher an index of consumption towards a particular medium, and orientation or engagement towards an issue or national emergency, the more respondents have been impacted and the more they are likely to depend on the medium. The scale was scored by summing the point values of the responses, and then dividing them by the total number of responses. All items were coded such that a more positive (accepting) attitude toward the outcome variable received the higher response value.

The hierarchical linear modeling (HLM) statistical technique was employed to properly reflect the structure of the dataset with citizens grouped within their respective provinces in East Africa. HLM is a class of techniques used to analyze data that have a nested structure (Kang & Kwak, 2003; Raudenbush & Bryk, 2002). This method estimates how attitudes towards new media and national politics vary across 15 provinces of East Africa. The provinces are central, northern, eastern, southern, and western in Kenya, Tanzania, and Uganda. The 15 provinces (also geopolitically referred to as "regions" in Uganda) make up the major populated parts in the East African region. Sample sizes averaged about 92 citizens per province. Provinces are considered here because they account for the nature of geopolitics, ethnic fragmentation, and because of the assumption that national political issues are likely to influence provincial (or local) politics.

HLM was employed to answer the following question: do public attitudes towards new media and national political discourse vary between users based on their gender within each country and across East Africa when we account for education levels? The HLM application consisted of two sets of linear regression equations for predictor variables at two levels. The level 1 model assessed individuals within provinces and then nested at level 2 across all provinces of East Africa.

The model estimated how the consumption of new media led to public attitudes within provinces and across provinces in times of national emergencies and anxiety, accounting for levels of education and engagement. Attention was restricted to three predictor variables (national emergencies) $1j$ + (women) $2j$, and (education levels) $3j$. Province-level predictors include (mean-national emergencies) $1w$ + (mean-gender) $2w$, and (mean-education levels) $3w$, to assess media systems dependency.

RESULTS

Descriptive Findings

The survey yielded 1,395 complete responses in Tanzania (n=465), Kenya (n=448), and Uganda (n=482). Of the total sample, 712 (51%) were women and 683 (49%) were men. There were 210 (46.9%) men and 238 (53.1%) women in Kenya, 239 (49.6%) men and 243 (50.4%) women respondents in Uganda, and 234 (50.4%) men and 231 (49.6%) women in Tanzania. Among all respondents, 799 (57.3%) had completed high school and only 388 (27.8%) had completed a bachelors degree; and the rest of the respondents (14.9%) reported that they had not completed a high school diploma.

Participants also described their dependency on the media, namely: newspapers, radio, television, and Internet news, and mobile text messaging for information and current affairs. In Uganda, 74.5% (n=482) of respondents indicated that they listen to the news on radio four or more days a week, compared to 66% (n=448) in Kenya and 78% (n=465) in Tanzania. In Uganda, 56.1% (n=481) watch television four or more days in a typical week compared to 58.3% (n=448) in Kenya and 49.3% (n=465) respondents in Tanzania.

Data indicate that more East African respondents depend on phone text messaging than on newspapers as a source of information during emergencies. A slight majority of respondents throughout East Africa also indicated that they depend on phone text messaging than television as a source of information during emergencies. The Internet and newspapers are the least dependable source of information during emergencies. See

Table 1 for details on media use, also categorized by gender differences.

Overall, respondents who completed a high school education and beyond in Uganda (n=266), Kenya (n=287) and Tanzania (n=246) were more likely to get news from radio and television than from the Internet. However, respondents who completed a high school education and beyond in Kenya and Uganda depend more on the Internet than on newspapers as the main source of dependable information about politics, but not in Tanzania, χ^2 *(df = 3) 19.16, p <.009*. Those without a high school education in Uganda (n = 216), Kenya (n=186) and Tanzania (n=194) were more likely to get news from newspapers than from the internet. Data also indicate that all Tanzanian respondents, regardless of their level of education are more likely to depend on texting as their dependable source political information than their Kenyan and Ugandan counterparts who use that medium. Respondents in Kenya and Uganda are more likely to depend more on the Internet as the main source of dependable information about politics than Tanzanian respondents. See Table 2 for details on media use about politics, also categorized by education differences.

Respondents in Kenya and Uganda trust political information in their newspapers more than

Table 1. Dependable source of information during emergencies, by gender category

	KENYA		TANZANIA		UGANDA	
	Women	Men	Women	Men	Women	Men
Radio	99	85	97	98	103	94
Television	59	54	59	56	58	62
Newspapers	9	8	11	13	2	12
Internet	7	6	3	4	8	4
Texting	64	57	61	63	72	67
Total	**238***	**210***	**231***	**234***	**243***	**239***

Notes: The difference between the sources of information is significant, χ^2 (df=4) 36.14, p <.001. For mean differences in dependable media outlets between countries: * = p<.05; *** = p<.001 Gender differences on dependable sources between countries, not statically significant, p>.05

Table 2. Main source of dependable information about politics, by education category

	KENYA		TANZANIA		UGANDA	
	Higher Education	No High School	Higher Education	No High School	Higher Education	No High School
Newspapers	15%	10%	19%	17%	14%	6%
Radio	33%	36%	38%	42%	34%	45%
Internet	19%	3%	16%	3%	20%	2%
Texting	5%	24%	10%	14%	4%	12%
Television	27%	27%	17%	24%	29%	35%
	χ^2 *(df = 3) 19.16, p <.009*		χ^2 *(df = 3) 16.79, p =.019*		χ^2 *(df = 3) 21.13, p =.017*	

Notes: The "Higher Education" column refers to respondents who had completed high school, tertiary education diploma, as well as respondents with a college or advanced; The percentages are based on the following data: Respondents who had completed high school and beyond in Kenya (n=287) and those without a high school diploma (n=186). In Tanzania, respondents who had completed high school and beyond (n=246) and those without a high school diploma (n=194). In Uganda, the respondents with high school diplomas and beyond were (n=266) and no high school diploma (n=216).

in Tanzania, but there is a statistically significant difference in public attitudes towards newspapers in the three countries. Of the five media types examined here, respondents in all three countries trust political information on the Internet more than they trust political information on radio, television, newspapers, and mobile phone texting. Furthermore, the Internet was the only online media outlet in which positive public attitudes on that question of 'trust' were not statistically significant between the three countries (at the 99% alpha level).

Nearly all respondents in Kenya and Uganda strongly agreed that the Internet and phone texting are good for democracy (Kenya = 7.0 and Uganda = 6.31) and that the majority of respondents who have used cell phones also indicated that they have called and sent text messages to radio talk shows about politics (Kenya = 6.89, Tanzania = 6.91, and Uganda = 6.76). As Table 3 shows, the levels of strong disagreement are nearly identical in all countries (Kenya = 1.13, Tanzania = 1.31, Uganda, 1.07) that governments should regulate mobile phone texting of political information. Respondents in all three countries also agree or strongly agree that the government wire taps their mobile phones and the government often suppresses political discourse. See Table 3 for details on public attitudes towards the media and national political discourse between the three countries.

HLM was employed to estimate the extent to which public attitudes towards new media and national politics vary between users based on their gender within each country and across East Africa when we account for education levels. The level 1 estimates exhibit results from two multilevel logistic regression models on citizens' media dependency for three predictor variables. The *P*-value for ß*01* (national emergencies) at level-1 of the One-way ANOVA with random effects was 0.036, which implies that there is a significant between-gender variance in media dependency during national emergencies. The ICC (intra-class correlation) is.047, according to the formula from Raudenbush & Bryk (2002). In other words, about 5 percent of the variance in the citizen's dependency (use) of the media during national emergencies could be attributed to

Table 3. Public attitudes towards the media and national political discourse

	Kenya	Tanzania	Uganda
Trust political information in newspapers	5.97	4.08	4.93***
Trust political information on radio	4.74	5.19	5.87*
Trust political information on television	4.60	4.33	3.98*
Trust political information on internet	6.13	6.28	6.40
Emergency phone texting messages necessary	6.88	6.91	6.93
Government should regulate information texting	1.13	1.31	1.07
Have texted a message to inform or alert others	6.89	6.91	6.76
Have phoned/texted talk-radio about politics	6.33	6.16	6.59
Have called/texted TV talk shows about politics	6.11	5.83	6.39*
Written letters to newspaper editors about politics	1.29	1.21	2.29*
Used or would use the internet to discuss politics	6.83	6.34	6.49
Internet/phone texting is good for democracy	7.0	5.01	6.31***
The government wire taps my mobile phone	5.78	3.39	5.21*
Government often suppresses politics discourse	6.01	5.11	6.22**

Notes: For mean differences between countries: * = $p < .05$; *** = $p < .001$; Scale: (1= low, 7=high)

the gender differences. That correlation gives us confidence to proceed with multilevel modeling since some social science disciplines, such as sociology and psychology, prefer an ICC which is higher than 30% to provide substantial benefits for the HLM analysis.

Model 2 added all level-1 predictors to the model, which is the Means-as-Outcome Regression. This model allows levels of education and gender variation in media dependency, but fixes all demographic effects to be the same across gender in the three countries. In general, women, men, highly educated citizens with access to all or a few media outlets or interest in political information are all accounted for in the estimations. Model 3 entered all predictor variables. Gender closeness is positively related to newspaper use. Women are more likely to use the media for political discourse when they are more educated beyond a high school completion. Most interestingly, the dependency on media (as a whole) by women is positively associated with both less educated and highly educated citizens in Kenya and Uganda ($\gamma 03 = .041$, $p < .01$) while less educated women (without a high school diploma) are negatively associated with media dependency in Tanzania ($\gamma 04 = -.027$, $p < .01$).

The estimated variance on public attitudes towards the new media (Internet and mobile phone texting) and national political discourse across all provinces was statistically significant in Uganda ($\gamma 10 = 0.041$, SE = 0.011; $p < 0.05$) and not significant across all provinces in Kenya ($\gamma 10 = 0.033$, SE = 0.008; $p > 0.05$) and Tanzania ($\gamma 10 = 0.044$, SE = 0.019; $p > 0.05$), when gender and education were added to the model. Overall, the new media did not influence citizens across all 15 provinces in East Africa on issues of political information when we account for the citizens' gender and education levels across the region ($\gamma 10 = 0.054$, SE = 0.032; $p > 0.05$). This implies that individual East Africans, who are influenced by the media on matters of political information, do not collectively report the Internet and mobile phone texting as their main source of dependable information about politics across the region regardless of their level of education.

DISCUSSION

This study set out to ascertain the use of and dependency on new media technology for political communication by voting- age citizens of the three main East African countries, namely Kenya, Uganda and Tanzania. We also investigated how citizens use new media technology by considering any differences in gender and level of education to inform their political views. These are factors that might indicate public attitudes toward sources of communication and information sharing in the region.

Our findings suggest that the advent of new media communication technologies is good for the development of democracy in the East African region. This includes Uganda where a previous study had found no radio influence on support for democracy without other intervening factors such as interest in politics (Kalyango, 2009). The importance of this finding is further underscored by the fact that respondents who engage in political participation through call-ins to radio and television programs acknowledge doing so through the use of their mobile phones and SMS texting facilities.

We learn the importance of new media communication opportunities towards the struggle for liberal democracy, which is demonstrated by the unanimity of respondents' attitudes that their governments suppress political discourse (see Table 3). Given that governments in Kenya, Uganda and Tanzania have various anti-press laws as posited earlier in this study, new communications technology therefore provide needed avenues for citizens to voice and challenge their governments' repressive regulations on information dissemination in the system.

While citizens in all three countries gather information from a variety of media available to them, our findings indicate that they *trust* information on the Internet more than any other source. This is a welcome development, particularly in a region where the use of new technologies such as the Internet is still very limited in major towns and community centres in all three countries. But the question of access to the Internet cannot be ignored with regard to this finding. The issue of minimal Internet access is perhaps a plausible explanation as to why the respondents in the three countries also said they were more likely to get their news from the radio and television than the Internet.

However, it is noteworthy that when Internet access is available, respondents in Kenya and Uganda would prefer using it as a news source than newspapers. As Table 2 illustrates, Tanzanian respondents did not indicate this. In addition, this situation in Kenya and Uganda is largely attributable to respondents having a minimum of high school education, as Table 1 illustrates. Moreover, it is obvious that a modest level of education is required to effectively use the Internet. The *World Fact Book* shows that the literacy level in Kenya is 85%. In Uganda, it is 66%, while in Tanzania it is 69%. Education also plays a role in the source of news about politics in particular as respondents with less than high school education in all three countries are more likely to rely on newspapers for such news. Yet, quite curiously, despite their level of education, respondents in Tanzania showed a preference for text messaging as their main source of political information.

Notwithstanding their levels of education and sources of information, survey respondents across Kenya, Uganda and Tanzania acknowledge a measure of dependency on the media for news and political communication. Ball-Rokeach and DeFleur (1976) postulate that media dependency is triggered by the need for information for decision-making, and the ability of the media to satisfy this need. Our findings suggest that the suppression of unfavorable political discourse and the enactment of anti-press laws by governments in the East African region heighten the yearnings for information by citizens in all three countries. The findings also show that the use of new media outlets, especially the Internet and mobile phone text messaging, satisfies this need.

Equally prominent in MSD literature is the contention that structural instability in society also breeds media dependency (Ball-Rokeach, 1985; DeFleur & Ball-Rokeach, 1989). Accordingly, respondents in all 15 provinces not only depend on the media for information during national emergencies but rely more on mobile phone text messaging in such periods than newspapers and even the television.

Our conceptualization of national emergencies is not limited to natural catastrophes. Kenya's post presidential election violence in December 2007 was a major national emergency in the country. Earlier on, we had indicated how the SMS text messaging facility and the blogs became vital channels for exchanging political messages among Kenyans in that post election period when their government imposed a ban on live broadcasts. The people had to depend on new media sources to acquaint themselves with developments in that crisis period to satisfy their informational needs and to decide on their mode of engagement or response to the unfolding political violence. Kenya's president Mwai Kibaki, in early 2007 had supported Article 19 of the Kenya Media Bill in parliament, which sought to regulate "reckless and irresponsible journalism." The use of traditional media such as radio and new media (mobile phones) between citizens and media commentators mobilized the masses to reject its enactment, and ultimately forced parliament to reject the repressive clauses in late 2007 (Kalyango & Eckler, 2010). This was a critical victory and a good indication that the emergence of new media technologies with a fairly critical independent media, which exposes the shortcomings of the Kenyan government, can succeed in mobilizing the masses against repressive government measures.

However, we do not make the argument that the new media facilities completely displaced the newspapers, radio and television in that period. We agree with Rife et al (2008) that new media technology may not have the capacity to completely displace traditional forms of communication. In the Kenyan example as well as in Tanzania, the new platforms are only quickly filling a void in an environment where government intervened and suppressed information gathering and dissemination by the established media. In Tanzania, the government and the ruling party, CCM, still own and control a national radio station, *Radio Tanzania Dar es Salaam*, a national television station, *Televisheni ya Taifa*, and two national newspapers: the English-language *Daily News* and the Swahili-language *Uhuru* (Tripp, 2000). Yet, the use of mobile phones by Tanzanians to interact with and engage in the national political discourse by calling radio and television talk shows makes the people more assertive, visible, and at the same time still dependent on traditional broadcast media.

Our focus on new media technology should also not be misconstrued that they are the only sources of information during national emergencies. We recognize that other forms of communication, including inter-personal and inter-group channels, also play a role in such periods. We are also mindful that external or non-media agents influence messages disseminated through new media technology because human society is more like an ecological system where the various components in it relate to and impact one another (DeFleur & Ball-Rokeach, 1989; Ball-Rokeach, 1998).

The practical implication of this cross-fertilization of influences is that by trusting, depending on and using new media technology, recipients of messages from new media sources get a variety of opinions that have implications for their levels of media dependency. On the other hand, governments realize that such messages might well emanate from opposition groups, hence, as in the case of Kenya, the Kibaki administration unsuccessfully tried to disrupt SMS text messaging in the post election period. Or, in Uganda where Museveni's government introduced the phone-tapping bill against the backdrop of the forthcoming (now held) presidential election in 2011.

The continued use and dependence on the new media technology by people in East Africa will however be a function of the extent to which information gained from them satisfies peoples' needs for, as Ball-Rokeach (1985) argues, media dependency weakens when the media are no longer able to satisfy the goals set by consumers of media messages.

CONCLUSION

This study has demonstrated new media use and media dependency in East Africa from a holistic perspective. First, the nested HLM analysis shows gender disparity in the use and dependence on new media technology for political communication when levels of education are factored in. The findings are illustrated in Table 3. However, this window of disparity which is narrow at five percent is understandably so. One plausible explanation is that attitudes toward the use of new technology for people who have attained substantial educational standards and for whom access to new media technology is not an impediment are not likely to vary a great deal.

The within-provinces findings however differentiate women in East Africa on their use of new media technology for political discourse on the basis of their levels of educational attainment. In addition, while media dependency was found to cut across women of all education levels in Kenya and Uganda, in Tanzania women without high school qualification were not so dependent on the media. We did not specifically investigate the reasons for this, but we can reasonably conjecture that people in such low-level education group are expectedly not literate enough to take advantage of new technological devices nor can they indeed afford them.

Future studies can specifically examine that issue.

We conclude this analysis by reasserting that new media technology, especially the mobile phone text messaging and internet facilities, is gaining ground as a means of political communication among East African citizens. But our findings have no bearing on the authenticity of the messages sent through these new media technological devices, or the extent of divisiveness or unity that such messages might engender among citizens in the region in times of political crises.

Our primary contention is that such citizens now depend on these technological devices to serve their informational needs moreso when political and other forms of national emergency situations arise. This dependency phenomenon is partly the consequence of the existence of media laws enacted by governments in the region to hamstring mainstream media houses. But the degree of use and dependence on these new media instruments are not uniform across the region. Gender and the level of education are two salient factors that create these differences in the use of and dependency on new media technology for political communication in East Africa.

REFERENCES

Abdi, J., & Deane, J. (2008, April). *The Kenyan 2007 elections and their aftermath: The role of media and communication.* BBC World Service Trust. Retrieved February 17, 2010 from http://downloads.bbc.co.uk/worldservice/trust/pdf/kenya_policy_briefing_08.pdf

Anduzia, A. (2009). The Internet, election campaigns and citizens: State of affairs. *Quaderns del CAC, 33*, 5–12.

Ball-Rokeach, S. J. (1985). The origins of individual media-system dependency: A sociological framework. *Communication Research, 12*(4), 485–510. doi:10.1177/009365085012004003

Ball-Rokeach, S. J. (1998). A theory of media power and a theory of media use: Different stories, questions, and ways of thinking. *Mass Communication & Society, 1*(1/2), 5–40. doi:10.1207/s15327825mcs0101&2_2

Ball-Rokeach, S. J., & Defleur, M. L. (1976). A dependency model of mass-media effects. *Communication Research, 3*, 3–21. doi:10.1177/009365027600300101

Ball-Rokeach, S. J., Power, G. J., Guthrie, K. K., & Waring, H. R. (1990). Value-framing abortion in the United States: An application of media system dependency theory. *International Journal of Public Opinion Research, 2*(3), 249–273. doi:10.1093/ijpor/2.3.249

Becker, L. B., & Whitney, D. C. (1980). Effects of media dependencies: Audience assessment of government. *Communication Research, 7*(1), 95–120. doi:10.1177/009365028000700105

Chaffee, S. H. (Ed.). (1975). *Political communication.* Beverly Hills, CA: Sage Publications.

DeFleur, M. L., & Ball-Rokeach, S. J. (1989). *Theories of mass communication* (5th ed.). White Plains, NY: Longman.

Denton, R. E., & Woodward, G. C. (1990). *Political communication in America.* New York, NY: Praeger.

Gastil, J. (2008). *Political communication and deliberation.* Los Angeles, CA: Sage Publications.

Graber, D. A. (1981). Political language. In Nimmo, D., & Sanders, K. R. (Eds.), *Handbook of political communication* (pp. 195–223). Beverly Hills, CA: Sage.

Grant, A. E., Guthrie, K. K., & Ball-Rokeach, S. (1991). Television shopping: A media system dependency perspective. *Communication Research, 18*(6), 773–798. doi:10.1177/009365091018006004

International Parliamentary Union. (2009). *Women in national parliaments*. Retrieved October 25, 2009, from http://www.ipu.org/wmn-e/classif.htm

Jackson, N. A., & Lilleker, D. G. (2009). Building an architecture of participation? Political parties and Web. 2.0 in Britain. *Journal of Information Technology & Politics, 6*, 232–250. doi:10.1080/19331680903028438

Kaid, L. L. (2004). Introduction and overview of the field. In Kaid, L. L. (Ed.), *Handbook ofpolitical communication research* (pp. xiii–xviii). Mahwah, NJ: Lawrence Erlbaum Associates, Publishers.

Kalyango, Y. Jr. (2009). Political news use and democratic support: A study of Uganda's radio impact. *Journal of Radio and Audio Media, 16*(2), 200–215. doi:10.1080/19376520903279399

Kalyango, Y. Jr, & Eckler, P. (2010). Media performance, agenda building, and democratization in East Africa. *Communication Yearbook, 34*, 355–389.

Kalyango, Y., Jr., & Winfield, B. (2009). Rhetorical media framing of two First Lady candidates across cultures. *Global Media Journal, 8*(15), 1-16. American Edition.

Kang, N., & Kwak, N. (2003). A multilevel approach to civic participation. *Communication Research, 30*(1), 80–106. doi:10.1177/0093650202239028

Karan, K., Gimeno, J. D. M., & Tandoc, E. (2009). The Internet and mobile technologies in election campaigns: The Gabriella Women's Party during the 2007 Philippine elections. *Journal of Information Technology & Politics, 6*, 326–339. doi:10.1080/19331680903047420

Karan, K., Tandoc, E., & Gimeno, J. D. M. (2009). The case of party-list winners and new media technologies in the 2007 Philippines elections. *Quaderns del CAC, 33*, 87–94.

Kenya Information Guide. (2009). *Kenya population*. Retrieved October 25, 2009, from http://www.kenya-information-guide.com/kenya-population.html

King, P. (1997). The press, candidate images, and voter perceptions. In McCombs, M., Shaw, D. L., & Weaver, D. (Eds.), *Communication and democracy: Exploring the intellectual frontiers in agenda-setting theory* (pp. 29–40). Mahwah, NJ: Lawrence Erlbaum Associates.

Kreiss, D. (2009). Developing the good citizen: Digital artifacts, peer networks, and formal organizations during the 2003-2004 Howard Dean campaign. *Journal of Information Technology & Politics, 6*, 281–297. doi:10.1080/19331680903035441

Latham, K. (2007). SMS, communication, and citizenship in China's information society. *Critical Asian Studies, 39*(2), 295–314. doi:10.1080/14672710701339493

Lee, Y. (2009). Internet election 2.0? Culture, institutions, and technology in the Korean presidential elections of 2002 and 2007. *Journal of Information Technology & Politics, 6*, 312–325. doi:10.1080/19331680903050085

Loges, W. E., & Ball-Rokeach, S. J. (1993). Dependency relations and newspaper readership. *The Journalism Quarterly, 70*(3), 602–614.

McMillan, S. J. (2006). Coming of age with the internet: A qualitative exploration of how the Internet has become an integral part of young people's lives. *New Media & Society, 8*(1), 73–95. doi:10.1177/1461444806059871

McNair, B. (2007). *An introduction to political communication* (4th ed.). New York, NY: Routledge.

Montero, M. D. (2009). Political e-mobilisation and participation in the election campaigns of Ségolène Royal (2007) and Barack Obama (2008). *Quaderns del CAC, 33*, 27–34.

Mulama, J. (2008, December 24). *Kenya press freedom: Going, going, gone.* Retrieved February 14, 2010, from http://ipsnews.net/news.asp?idnews=45214

Mwesige, P. G. (2009). The democratic functions and dysfunctions of political talk radio: The case of Uganda. *Journal of African Media Studies, 1*(2), 221–245. doi:10.1386/jams.1.2.221_1

National Bureau of Statistics Tanzania. (2009). *Census Population*. Retrieved October 25, 2009, from http://www.nbs.go.tz/population.htm

News, B. B. C. (2009, January 2). *Kenya 'media gag' law approved.* Retrieved February 14, 2010, from http://news.bbc.co.uk/go/pr/fr/-/2/hi/africa/7808815.stm

Ng'weno, H. B. (1969). The nature of the threat to press freedom in East Africa. *Africa Today, 16*(3), 1–4.

Perloff, R. M. (1998). *Political communication: Politics, press, and public in America*. Mahwah, NJ: Lawrence Earlbaum Associates, Publishers.

Raudenbush, S., & Bryk, A. (2002). *Hierarchical linear models: Applications and data analysis methods* (2nd ed.). Thousand Oaks, CA: Sage Publications, Inc.

Riffe, D., Lacy, S., & Varouhakis, M. (2008, January). Media system dependency theory and using Internet for in-depth, specialized information. *Web Journal of Mass Communication Research.* Retrieved January 15, 2010 from http://www.scripps.ohiou.edu/wjmcr/vol11/11-b.html

Salwen, M. B. (1987). Mass media issue dependency and agenda-setting. *Communication Research Reports, 4*(1), 26–31.

Skjerdal, T. S. (2008). Self-censorship among news journalists in the Ethiopian state media. *African Communication Research, 1*(2), 185–205.

Skumanich, S. A., & Kintsfather, D. P. (1998). Individual media dependency relations within television shopping programming: A causal model reviewed and revised. *Communication Research, 25*(2), 200–219. doi:10.1177/009365098025002004

Tai, Z., & Sun, T. (2007). Media dependencies in a changing media environment: The case of the 2003 SARS epidemic in China. *New Media & Society, 9*(6), 987–1009. doi:10.1177/1461444807082691

Tamale, S. (1999). Towards legitimate governance in Africa: The case of affirmative action and parliamentary politics in Uganda. In Quashigah, E. K., & Okafor, O. C. (Eds.), *Legitimate governance in Africa: International and domestic legal perspective* (pp. 235–262). The Netherlands: Kluwer Law International.

Tedesco, J. C. (2004). Changing the channel: Use of the Internet for communicating about politics. In Kaid, L. L. (Ed.), *Handbook of political communication research* (pp. 507–532). Mahwah, NJ: Lawrence Erlbaum Associates, Publishers.

The World Factbook. (2009). *Kenya*. Retrieved October 20, 2009, from https://www.cia.gov/library/publications/the-world-factbook/geos/ke.html

The World Factbook. (2009). *Tanzania*. Retrieved October 20, 2009 from https://www.cia.gov/library/publications/the-world-factbook/geos/tz.html

The World Factbook. (2009). *Uganda*. Retrieved October 20, 2009, from https://www.cia.gov/library/publications/the-world-factbook/geos/ug.html

Tripp, A. M. (2000). *Women and politics in Uganda*. Madison, WI: University of Wisconsin Press.

Truman, J. S. (2008). *Political communication in American campaigns*. Los Angeles, CA: Sage Publications.

Turiera-Puigbo, T. (2009). How is Internet use changing the way in which politics is carried out and communicated? *Quaderns del CAC*, *33*, 13–19.

Uganda Bureau of Statistics. (2003). *2002 population report*. Retrieved October 25, 2009, from http://www.ubos.org/index.php

Wei, R. (2009). The influence of mobile phone advertising on dependency: A cross-cultural study of mobile phone use between American and Chinese youth. In Golan, G. J., Johnson, T. J., & Wanta, W. (Eds.), *International media communication in a global age* (pp. 401–421). New York, NY: Routledge.

Zittel, T. (2009). Lost in technology? Political parties and the online campaigns of constituency candidates in Germany's mixed member electoral system. *Journal of Information Technology & Politics*, *6*, 298–311. doi:10.1080/19331680903048832

Chapter 7
Whose TV is it Anyway?
An Examination of the Shift towards Satellite Television in Zimbabwe

Rick Malleus
Seattle University, USA

ABSTRACT

The thesis of this chapter is that the rise in popularity of satellite television in Zimbabwe was not mainly driven by the capabilities of the new technology, but by dissatisfaction with Zimbabwe Television (ZTV). The chapter will begin with a discussion that problematises the idea of what constitutes new technology in the African and Zimbabwean contexts. The focus then moves to the content broadcast locally and how a segment of the Zimbabwean population have turned away from ZTV to have their media needs and gratifications met from satellite TV. There is an extended discussion of propaganda and quality of programming, which are hypothesized as the driving factors for satellite TV's rise in popularity. A discussion of the cultural influences that satellite TV programming has on the Zimbabwean viewing public is included, and the chapter will conclude with some thoughts on the future of satellite television in Zimbabwe if changes in local broadcast TV programming come about.

INTRODUCTION

In 2010, most people accept the idea that we are living in a globalized, interconnected world. One consequence of globalization is the divide that exists between developed and developing countries in information and communication technologies (ICTs). The divide surrounding access to and available types of ICTs has become pronounced. This book is premised on the idea of interrogating new ICTs. This chapter's relevance to this discussion is premised on the notion that "new technology"

DOI: 10.4018/978-1-60960-591-9.ch007

is a relative term, with some technologies that are considered older in developed countries rightly being explored as new technologies in developing countries such as Zimbabwe.

The interconnection that ICTs allow provide a much easier and freer flowing media stream of informative, educational and entertaining programming than existed in the past. In Zimbabwe, however, there is one local television station (ZTV) that is the majority of the local population's only televised source of news and entertainment. A government appointed board controls the content of programming on ZTV. In a diverse, global world, what population would feel satisfied with access to one TV station with a limited and controlled program set?

When the technology of satellite TV was introduced into that media landscape, it allowed the Zimbabwean audience more choice in programming. As the major approved purveyor of satellite television in Zimbabwe, DStv, represented by Multichoice Zimbabwe, became the provider of choice for Zimbabweans who had the money to pay for the necessary equipment as well as the monthly charge in foreign currency. Having one main satellite provider of choice is a common experience on the African continent and depends largely on where you live. According to Wright (2008) "Most African countries obtain their satellite services from South Africa-based Multichoice's DSTV or France's Canale 5" (p. 44).

The major objective of this chapter is to provide the reader with an appreciation for the TV media landscape that exists in Zimbabwe and how satellite TV, which is a relatively new ICT for much of the population, has grown in popularity for a complex variety of reasons. A secondary objective it to make clear that Zimbabwe has experienced an extremely difficult decade and that it is within this *current context* that the issues explored need to be understood. Readers should realize that any exploration of the importance of satellite TV, local TV and cultural influence is extremely complex and cannot be divorced from historical context. Finally, readers should grasp the notion that if the media space in Zimbabwe changes, then the impact on DStv's audience and Zimbabwean television's audience is unknown and will have to be explored in that new context.

BACKGROUND

It is important to remember that in Zimbabwe, as in many African countries, a large portion of the population get their information and entertainment from radio. This chapter however focuses on satellite TV, but readers should note that there is a larger radio than TV audience in Zimbabwe.

Having made that distinction clear, it is important now to consider the idea of what constitutes a "new ICT" in the Zimbabwean context. Why? The term *new ICT* is often used as if the definition is clear and universal. What needs to be recognized is that the term is neither clear nor universal and it is crucial to understand this in order to locate Zimbabweans on the spectrum of ICT exposure. It is this understanding that underpins this chapter's inclusion in a text on new ICTs.

While perhaps a little overstated, Polikanov and Abramova's (2003) point that "...high-brow analysts should not forget that many Africans have never seen a TV set..." (p. 50) is germane. *Africa Business'* Tom Nevin makes a similar point when he argues that the "continent strives under a paucity of ICT infrastructure with many people having never even made a telephone call" (p24). The idea that to some, a telephone would be a new technology, while to others a TV would be a new technology is often not one that is considered when new ICTs are discussed. Most discussions of new ICTs focus on the latest technological developments in that realm, which most often occur in developed countries with strong and widespread ICT infrastructures. While this is not wrong to do, it is also not an accurate global interrogation of the notion of what a new ICT is.

Context is key in doing this. As Odendaal (2006) argues "...technology appropriation is highly contextual; culture, social systems, ascribed norms and networks..." (p32) are important considerations. While examinations of ICTs often look at them as tools or boosters of economic development, it is important to think about issues of freedom, social inclusion and participation (Zheng and Walsham, 2008). One way to look beyond ICTs as tools is to reconsider them in context relative to the exposure of specific populations to an ICT and the impact it has on that specific population.

While many discussions about new ICTs focus on the internet and telecommunications, in the Zimbabwean context, satellite TV can play an important part in helping develop an information society in Zimbabwe, one of the Southern African Development Community's goals (Mutula, 2004). An information society can be defined in part as one where "people know and appreciate what information they need, where to get it, how to get the information, and in the end, how to use it" (Mutula, 2004, p. 145). It can be argued that satellite TV, as a new ICT to many in the Zimbabwean population, can help achieve this end.

Many ICT researchers agree that unless African countries "become full actors in the global information revolution, the gap between the haves and the have-nots will widen" (Alemneh and Hastings, 2006 p. 5). While this new ICT debate is often framed around the internet, in the Zimbabwean context, *DStv is an important element of the ICT infrastructure* that can help reduce that gap.

Kirsty Brien, Marketing and Public Relations Manager for Multichoice Zimbabwe, confirms the growth of DStv's subscriber base in a newspaper interview; "...'Year-on-year, the subscriber base has grown and there are also seasonal trends that also affect subscriptions, soccer tournaments are particularly popular with seasonal subscribers.'" (Mokwetsi, 2005)

A drive through both high and low density suburbs of Zimbabwean cities and towns will also provide evidence of people being consumers of satellite TV as satellite dishes hang from rooftops of mansions, middle class houses, blocks of flats and one room homes. It is also safe to assume that in the collective culture that Zimbabwe is, not only the owners of satellite TVs are watching the programming as neighbors, friends and extended family share viewing time.

So, why is satellite TV popular? There are a complex set of reasons for the rise of satellite TV. Zimbabwe has been through a period of extremely intense social and political upheaval in the last decade and all aspects of Zimbabwean life have been affected by that upheaval. ZTV has been no exception. A new political party emerged to challenge the entrenched political leadership, an agrarian land reform policy was implemented with national and international consequences, a hyperinflationary economy made living and working in Zimbabwe extremely difficult, and the social fabric of the country was torn.

There are multiple sides and perspectives to the events that led Zimbabwe down this path; that discussion is not the focus of this chapter. However, it is important to understand this context, because the role that satellite TV played and plays in Zimbabwe must be understood within the framework of this wrenching social, political and economic environment.

The sentiments expressed below by a Zimbabwean school teacher echo the thoughts and feelings of many Zimbabweans, and anecdotal evidence points to the fact that her opinions coincide with the reasons many Zimbabweans have turned off ZTV, have looked for alternate television fare, and have found it in satellite TV.

"I can't stand ZTV any more. There is too much propaganda, hate language, repeat programming, poor picture quality and low broadcasting standards. Most of the presenters and newsreaders seem to have been picked on the streets and thrown behind the camera without any proper coaching" says Busi, a school teacher. "That's why I sacrificed to save my meager salary in or-

der to buy a satellite dish and enjoy quality and variety." (Masuku, 2009)

The following section of the chapter will examine specific issues that have made DStv popular in Zimbabwe. It is hypothesized that the rise in popularity of satellite television in Zimbabwe *was not mainly driven by the capabilities* of the new technology, but *by dissatisfaction with ZTV*. Three specific elements will be examined: propaganda, quality and escapism.

Propaganda

Jay Black (2001) points out that political propaganda can be "spread through the mass media to get the public to accept some political or economic structure or to participate in some action…" (p. 125). Perhaps the most important reason that satellite TV has become very popular is that ZTV came to be seen as a propaganda tool for the government of the day.

Zimbabweans complain about the propaganda they hear and see on ZTV, but what might they mean by that term? Merrill and Lowenstein (1971) looked at many definitions of propaganda and some of the core ideas of those definitions they found seem to fit the Zimbabwean context: manipulation, purposeful management, preconceived plan, reinforcement of biases, irrational appeals and arousal to action. The propaganda messages sent out by ZTV may have been the major factor that contributed to viewers seeking programming elsewhere. Research analysts suggest that state media, like ZTV, became "virtual organs of the ruling party" (Bratton, Chikwana and Sithole, p. 104).

An *extended series of examples* related to just ***one*** focus of ZTV's propaganda is warranted here to illustrate the points made above. Over the last decade, the Zimbabwean government and the ruling party, ZANU-PF (which have through all government controlled media, including ZTV, come to be rhetorically recognized as the same) have increasingly used *rhetoric that frames the country as being at war*, when in fact Zimbabwe was not involved in a war.

ZTV, as a publicly funded institution, was used through *purposeful management*, to *implement a preconceived plan* that *reinforced biases* of the population based in Zimbabwe's colonial history to make *irrational appeals* that would *lead to actions* that government supported or discourage actions that the government did not support. The rhetorical implication of the use of "war talk" was that the government was at war, *not with all the people of Zimbabwe*, but certainly *with some of the people of Zimbabwe*. ZTV also broadcast the government view that they were under attack by and therefore at war with, foreign enemies.

As a public broadcaster, ZTV sent these messages from the government to both "friends" and "enemies" of government within the country. ZTV's audience was subjected to a rhetorical frame set by government and not rigorously interrogated nor objectively investigated by those broadcasting the messages. One impact on the audience is to then narrow the agenda focused on as it is framed by the government. (Bratton, Chikwana and Sithole, 2005, p. 102) In other words the message to the audience is 'think about the war we are fighting, not your other troubles or concerns'.

In a war you have to have enemies to fight against, so who were painted as the enemy within Zimbabwe on ZTV over the last decade?

- Those that did not support Zimbabwe, defined by ZTV as any who opposed government policy in any way.
- Those that openly supported opposition political formations, mainly the Movement for Democratic Change (MDC) and a grouping who wanted a new constitution for the country, the National Constitutional Assembly (NCA).
- Labor unions that were not favored by government.

- Women of Zimbabwe Arise (WOZA), a women's group who advocated for women's and children's rights to health, education and affordable living conditions.
- White farmers who were resisting the government land reform program.
- Black farm labor and their families who sided with white farmers whose land was being expropriated under the land reform program.
- Local Non-Governmental Organizations providing aid of various kinds to Zimbabweans.
- White Zimbabweans as an undifferentiated conglomeration.

Just as internal groups of people were painted as enemies on ZTV, the propaganda also painted external "enemies" for the government and population to fight against:

- The British Government-especially Prime Minister Tony Blair.
- The American Government.
- The European Union.
- The Commonwealth.
- Select African political or prominent figures critical of Zimbabwean government policy.
- "The West".
- International Non-Governmental Organization providing aid of various kinds to Zimbabwe.

When a country and culture is experiencing crises, racial and class hatred are vulnerable divisions to be exploited through propaganda. Enemies can be invented, but often historical existing hates and fears will be exploited (Gustainis, 1990). From the list above, it is clear that the "enemies" ZTV told Zimbabweans about often fell into the category of existing historical hate and fear.

For example, because Zimbabwe was a British colony, ruled by a white minority government with racist policies that took a war of liberation to dislodge and achieve independence, the population has an historical memory of "them" as being the enemy. The "them" would include white Zimbabweans, and "the British." In addition, any black Zimbabwean individual or group that could in some way be tied rhetorically to "them" could also be painted as the enemy.

Specific segments of the Zimbabwean population were selected for labeling as the enemy during specific events: during elections or while government programs were ongoing. For example, people living in Mbare, a poor suburb of Harare, were viewed as supportive of the MDC and as such were labeled by government as people without totems (Ranger, 2004) and this denigration was not interrogated on ZTV. To the Shona people totems are an important mark of social identity, belonging and kinship, so this label was extremely insulting but never challenged by ZTV.

Pfukwa (2008) suggests "Names that people call each other are powerful barometers of social relations…" (p. 38). ZTV reporters would use the language of the government in their reporting to paint the enemy without restraint or interrogation about whether those terms were warranted, true or ethical to use in their broadcasting. The MDC and other black Zimbabwean political leadership who were not affiliated with or supportive of ZANU-PF were labeled puppets, sellouts, uncle toms, unpatriotic, and traitors. White farmers were labeled settlers, colonialists, Rhodesians, British, and the English on TV with all the historical baggage and attitudes that came with those labels.

Any visual images that could reinforce the language that was being used were also shown repeatedly on ZTV. For example, as the MDC was gaining in popularity and raising money, a ZTV crew filmed the MDC leader having checks to support the party coffers given to him by some white farmers at an MDC rally. This visual made "fact" the government's claim that the MDC were puppets. What was ignored by ZTV in their reporting and repetitive use of this image for years was the fact that ZANU-PF also accepted money from white farmers to support their party.

Gruesome images of the violent war of liberation were shown on ZTV decades after the events. With limited archival footage, the same images of bombings and attacks by the Rhodesian forces were shown repeatedly on various different ZTV programs and advertisements. It is important to remember a country's history, but large sections of ZTV's audience wondered why these images were repeatedly being shown now. The images were used because they once again provided a visual image of who the enemies were and still are.

In addition to images of the enemy, there was also a limited number of decades' old images of liberation war veterans singing Chimurenga (liberation struggle) songs, training and fighting that were used by ZTV to drive home the point that the government was for the people and that war veterans who were prominent in the land reform farm expropriations were still leading the struggle. In context both sets of images are useful as tools to learn about and recall Zimbabwe's past, to tell the culture's stories. In this fractured and divisive decade, the images were used by ZTV as an adjunct to the language of a war that was not really a war at all but was only framed as such.

One further example of how ZTV was viewed by the audience as an uncritical propaganda mouthpiece for the government can be seen in the "war talk" labels that government used to frame various policy programs that they implemented in the last decade. The agrarian land reform program that the government undertook was known as the Third Chimurenga. The label and ideological implications of this label were clear: the First and Second Chimurenga's had been real wars against the colonial powers and Rhodesians that resulted in Zimbabwe's independence. "Hondo ye Minda" which loosely translates to 'war over controlling the fields' was the slogan that accompanied the government program and was continually broadcast on ZTV. It is a noted irony that sometimes "Music originally used to resist oppression was used in the service of oppression" (Thram, 2006, p. 86)

A major reason the Second Chimurenga was fought was about equitable distribution of Zimbabwe's land, that is not in question, what was not ever thoughtfully and critically done by ZVT news or current affairs programs was to ask questions about this "war", the Third Chimurenga. Why was it being "fought" twenty years after independence? Why had the government waited so long to take firm action? Who was the government waging this war against as the land question involved all Zimbabweans? This is what bothered the audience and one place people went to get a more thoughtful analysis without inflammatory labels, rhetoric and visual images was to alternate news sources on DStv.

Most major government programs during this past decade became "operations" a term more appropriately associated with war than with social policy plans. Operation Murumbatsvina (Shona) or Operation Restore Order (English) was one such operation. The government policy was introduced to "reassert formal planning procedures, by-laws and local state institutions" (Fontein, 2009, p. 3). The focus was on urban areas where housing structures had been built over decades without official planning permission and often without meeting legal building code requirements. A skeptical population did not accept the officially stated reason for this operation: urban centers had voted overwhelmingly for the MDC in national and local elections. Most people saw the policy as punishment for exercising their franchise. The manner in which people's houses were demolished "by council officials, police and the military" (Fontein, 2009, p. 3) seemed brutal, indiscriminate and often outside of other Zimbabwean laws. The coverage of this operation on ZTV was once again one-sided and without rigorous investigation. People turned to foreign satellite TV sources to gain information about what was really happening around the country as the operation unfolded.

The label for the operation in English sounds reasonable, however, a loose translation of "murumbatsvina" is much less benign; sweep away the dirt, reject the rubbish (Fontein, 2009). Were the illegally built houses the dirt to be swept away or the people? Both the connotative and denotative meanings of the operation's label needed to be considered by ZTV and were not. Other news sources found on DStv channels did a more effective job of sorting through the complexities of both the labeling of the operation and the operation itself.

People also resisted ZTV's rhetoric by labeling the operation the 'tsunami' (Fontein, 2009) connoting massive destruction with no connection to the supposed 'order out of chaos logic' the operation was predicated on, but actually connoting the exact opposite. Interestingly, people came up with their own label for the operation based in large measure on the extensive coverage of the destruction brought about by the tragic Asian tsunami that many had seen unfold on their satellite TV's.

With the reality on the ground in Zimbabwe often not being reflected in the reporting on ZTV, the audience saw the un-interrogated war rhetoric as propaganda attempts to hide what was actually going on and perhaps more importantly to avoid critical discussion and analysis of the need for such "operations" to be carried out in the first place. The views expressed below further emphasize this point:

"In Operation Maguta, agricultural production was put under the command of the military, while Operation Chikorokodza Chapera (stamping out illegal exploitation of resources) was mainly directed at asserting state control (by military means) of the Marange and Chiadzwa diamond fields in the eastern province of Manicaland." (Chikomo, Gokova, Matambanadzo, Stewart and Wakatama, 2008.)

Clearly, a hunger for "unbiased" news, or sources of news on TV that represented multiple perspectives on the lived reality of Zimbabweans was a substantive reason for the rise in the popularity and consumption of DStv programming. It can certainly be argued that all sources of news will have an ideological bias, but the blatantly obvious propagandized version of reality that the Zimbabwean audience was being subjected to crossed the line for many. A narrow and skewed agenda setting of what the important issues of the decade were, did not sit well with many.

Quality

A second important factor that led people to DStv was the poor quality of programming offered by ZTV. Local production standards are now often low and the audience is not prepared to watch programs whose standards fail to meet their expectations. The Zimbabwean television audience knows what quality programming is based on previous decades' worth of higher quality Zimbabwean TV and on their exposure to international quality productions that used to be shown on ZTV.

When tuning in to a ZTV news broadcast or current affairs show, the audience is not satisfied with seeing news clips that have an audio track that does not match the visual images, not satisfied with audio failures where sound simply disappears, not satisfied with unprofessionally written scripts, mispronounced words and clumsily edited pieces. Occasional lapses are allowed for, but repeated instances of poor production quality discourage viewers from tuning in. Masuku (2009) considers some of the reasons for this:

A few years ago ZBC was forced to adhere to seventy-five percent local content as prescribed in the controversial broadcasting laws. That was the beginning of a massive exodus of viewers to other channels. Such a quota of local content is extremely ambitious considering that ZBC itself and many independent film producers don't have the required funding and equipment to produce the desired local films. (Masuku, 2009)

This government requirement that local content fill most of ZTV air time could be seen in normal times as a positive step toward building a viable production and distribution network for local directors, scriptwriters and actors. However, at the time the new content requirements came into force, the times were not normal in Zimbabwe. The government perceived themselves to be under "attack" from "enemies" within Zimbabwe and from outside Zimbabwe. This move toward local content was not received by the majority of the population as a well-considered strategy to protect and build local TV programming and a way to tell Zimbabwean cultural stories from a Zimbabwean perspective, it was perceived as a purely political move meant to stifle alternative voices and limit flows of information into the country from abroad that might contradict the government line. As the Media Monitoring Project Zimbabwe group asserted "…it is cynical and unfair for ZBH to claim that it is the country's "permanent choice" of radio and television station; the people have no choice." (MMPZ, 2007)

In addition, if arguments about motives are set aside, there simply was not enough local content-of any sort of quality-to fill the broadcast time on ZTV which had moved to being a 24 hour channel. Logical planning would dictate that a phased in local content requirement be developed so that the production capacity could match the need. What resulted from the hasty implementation of the new 75% local content policy were a number of magazine format programs, current affairs programs and a proliferation of local music videos, many of which had poor production quality and were often not of the most popular artists whose messages might not have been to the government's liking.

What also resulted from the local content requirement was that some international programs (mostly of American or British origin) that were popular with viewers were no longer able to be shown on ZTV. This led to repetition of the same shows or older Zimbabwean shows that people had seen many years ago. A frequent complaint about programming was that shows were broadcast again and again and again. DStv offered viewers a chance to see new shows and the foreign content that they missed.

There are legitimate arguments to be made for former colonies like Zimbabwe to wean themselves off of foreign content, particularly that produced by their former colonizers. Some discussion of these arguments will be explored later in the chapter. However, only a small portion of the population accepted this stated reason for the content change on ZTV.

In addition to production quality problems, Zimbabweans recognized that journalistic standards of professionalism were not being upheld on ZTV news and current affairs programming. In other words, the quality of journalism was questioned. For both economic and political reasons, many seasoned, well-trained broadcasters left ZTV and many also left the country. A purge of the more independent minded journalists at ZTV was framed as a restructuring exercise (Bratton, Chikwana and Sithole, 2005). The reporters, commentators and magazine show hosts who replaced them were often perceived as being less competent, less well-trained and less professional than their former colleagues.

Tom Rhodes of the Committee to Protect Journalists (2010) recently reported that about 48 Zimbabwean journalists have gone into exile in the last decade, with many, not all, having to abandon their journalism careers as they try to build new lives for themselves in other countries. For a country the size of Zimbabwe to lose so many journalists in such a short space of time was a blow to the profession. It would appear that the Zimbabwean media audience is correct in their observations that those who followed do not exhibit the same journalistic skills and values such as being objective and confirming stories from multiple sources.

Escapism

A third reason that DStv became popular has to do with the social environment that obtained in Zimbabwe. Living in Zimbabwe over the past decade, no matter what a person's socio-economic class, race, or political persuasion, has without a doubt been extremely stressful. A desire for entertainment in a culture experiencing extreme social, economic and political upheaval may be another reason that DStv became so popular. While all media messages have an ideology built in to them, the Zimbabwean audience sought entertainment that took them out of their lives for a while and allowed them to enjoy themselves.

Zimbabwe, as in many developing countries, sometimes relies on donors to fund production of television programs. Some of the programming on ZTV that was not overtly political in nature still smacked of "edutainment".

Last year, the United Nations poured vast sums of money into filming six pilot episodes of "Heart and Soul", a television drama set in Kenya. But the programme flopped, with writers and audiences complaining that the donors were pushing their educational messages too hard, at the expense of the entertainment. (Salmon, 2003, p. 1)

Similar to the Kenyan experience described above, some Zimbabwean TV productions are funded by aid donors who have a message or agenda that they wish to get across to the audience: messages about HIV/AIDS prevention and treatment or domestic violence and women's rights. While these messages are important, often the programs can be seen as "preachy," are really not that entertaining, and do not remove the viewers from their reality. If you have a family member dying of AIDS, and most Zimbabweans have experienced this tragedy in their lives, why would you want to constantly see shows with characters like that in your living room? DStv provides such a vast array of choice that people can find stories that remove them from their own realities for a while, transport them to places they can never hope to go to, and make them laugh.

The preceding discussion in the chapter has focused on the three elements hypothesized as drawing Zimbabweans to DStv. The following section of the chapter will examine two major concerns about satellite TV in Zimbabwe: exacerbation of social stratification and cultural influences.

Social Stratification

People's perceptions of their culture and what is going on in their country are affected to some degree by media exposure, with TV being one socializing agent in a society. ICTs can unite, but they can also divide a community (Odendaal, 2006). A concern that needs to be considered is the potential further social stratification of Zimbabwean society that access to satellite TV may present. Two major areas of social stratification that already exist are the rural-urban split and the rich-poor divide. Adding this digital divide may further exacerbate this stratification.

Those who live in urban areas and those who are affluent enough to afford the equipment and pay the subscription fees necessary for satellite TV have access to multiple perspectives on Zimbabwe, Zimbabwe's relationship with other countries and cultures, and how the rest of the world views what is happening in Zimbabwe. Those who live in rural areas without the ICT infrastructure to access satellite TV nor the money to pay for it do not have the chance to access those multiple perspectives. If that portion of the population consumes any TV, it would be ZTV's take on all those issues, giving the audience only ZTV's government directed perspective.

If local ownership of TV broadcast is limited to a government monopoly, in a society that is polarized and stratified and has gone through political, social and economic troubles, it is problematic for sections of the population to have dramatically different interpretations of similar situations.

Localism in media ownership is most critically important when it comes to providing news to the citizens of a region. It has been said that the media serve a surrogate role for the public. In no case is this more important than when it comes to delivering local information to citizens about their political leaders, the areas economy, the region's educational progress, social standards, and countless other matters that confront residents each day. Without this vital information, citizens cannot effectively monitor their civic leaders and effect change to assure that those leaders are serving the public will. (McCall, p.124)

At a time when Zimbabwe needs to resuscitate the economy, bring back tourism, re-build agriculture, re-build the health system, focus on reducing the spread of HIV/AIDS, provide a social security safety net for the most vulnerable in society, etc., it is detrimental to developmental efforts of both the private and public sector to have media-widening divisions in an already sharply divided population.

The Parasocial Contact Hypothesis (Schiappa, Gregg and Hewes, 2005) posits that people process messages from the media, from TV for example, in a way that is like an interpersonal interaction so media might be *one way* to reduce prejudice and influence attitudes towards certain groups or individuals. While their research made no specific claims about the specific quantity or quality of programming required to induce specific attitude changes, the larger message that can be applied to the Zimbabwean context is that *portrayals of individuals and groups matter*. The quality provided by ZTV without complex and thoughtful portrayals of the diversity of the population could reinforce negative and divisive attitudes, while programming that has that complexity and quality may help change attitudes toward those groups that have been misrepresented on ZTV. Zimbabweans who sought that diversity shifted to DStv.

Whether the quality programming comes from ZTV to provide competition for DStv or from another future Zimbabwean TV station remains to be seen. As reporter Dumisani Ndlela noted:

Zimbabwe's ZBC TV has overhauled its programming as it battles to lure viewers dejected by pathetic programming and a blackout on foreign content.

Since 1 July 2009 viewers on the state-owned broadcaster woke up to an unfamiliar line-up of programmes, largely dominated by foreign content. (Ndlela, 2009)

It appears that ZTV has noted the competition that satellite TV has given them and is taking steps to try and improve their service.

Cultural Concerns

A further reason to consider the rise of satellite TV in Zimbabwe has to do with a set of cultural value concerns. As mentioned earlier in the chapter, the government moved to a 75% local content on ZTV virtually overnight; one reason for a move like this it to "protect" your culture.

After apartheid fell and sanctions were lifted, South Africans began investing heavily in the rest of Africa. They are displacing former colonial powers like Britain and France as the top foreign investors in industries from mining to media, raising talk of a new imperialism. Prof. Guy Berger, head of journalism at Rhodes University in Grahamstown, says South African tolerance for TV sex and violence don't always "go down so well." (MacGregor, K, 2003, p. 39)

As indicated above, what is acceptable in terms of depictions of sex on TV, for example in South Africa, where DStv is headquartered, may be offensive to other more conservative African cultures. Homosexuality is another example. Sexual orientation is protected under the South African constitution; it is illegal in Zimbabwe. So a concern for some is that those electing to watch DStv will be exposed to messages about various elements of culture that run counter to what the majority of Zimbabweans believe.

Another example of a cultural concern surrounding satellite TV is music. Pfukwa (2008) suggests that popular music is a way for a culture to shape its identity and that this is especially important in former colonies like Zimbabwe. Pfukwa's own research is in Chimurenga music, now a genre of Zimbabwean music that is regularly heard on ZTV. However, a larger point about shaping cultural identity after a colonial period can be made about all music in Zimbabwe on TV in general. Is a young singer who raps in Shona "diluting" or "polluting" Zimbabwean culture because the genre of music is foreign, or is she expanding the definition of what Zimbabwean music is by co-opting a foreign musical genre? There are legitimate arguments to be made for taking each position. However, if you take the former position, you may consider exposing a Zimbabwean audience to Shona rap to be a negative cultural influence and not want it shown on local TV.

Musiyiwa (2008) suggests that music is important in nation building and achieving unity. In Zimbabwe music has been important for the nationalist movement, immediately after independence and in the decade this chapter focuses on. If local music, of all varieties and from all sectors of society are allowed to be shown on ZTV, then perhaps that would be one step to helping the nation develop and open a dialogue as to what Zimbabwean culture is in the 21st Century.

It would seem a logical reaction to what has been described as "a tremor of anxiety at the rising power of South Africans in the region" (MacGregor, K, 2003, p. 39) to make local TV fare more appreciated by opening up the TV broadcast space. If this happened, it could lead to Zimbabweans shaping their cultural identities instead of others doing it for them.

The hegemonic fears that some have about South African television influence are also compounded by the fears of Zimbabwean culture being "re-colonized" by European and American television programming. The solution ZTV came up with was to remove most of that programming from their broadcast. A portion of their audience moved to DStv and found that programming there. This fear of negative media influence on culture and dilution or homogenization of culture has been accelerated in the era of globalization. DStv brings a more global view to the audience than does ZTV. As Masuku (2009) points out:

Our youngest daughter, Phathi, always pleads with us to pay our monthly subscriptions for DSTV so that she can always be up-to-date with her school friends when it comes to latest movies, musicals and celebrity shows.

Do Zimbabweans want their children or their culture to follow the latest trends in media entertainment and be sucked into the world of celebrity worship that troubles many in the West? As African researcher Professor Egziabher suggests "In this age of globalization, local differences are seen as a nuisance. We are all being pushed towards congealing into an indistinguishable human mass, all of us becoming approximately American" (Quoted in Alemneh and Hastings, 2006, p. 14).

There is also a legitimate concern about people consuming news and information about Zimbabwe and Africa in general from mainly Western sources and perspectives that can be found on DStv's programming. As Zimbabwean James Makawa who started the The Africa channel for cable service in America and the UK points out:

The news has portrayed a certain image of Africa that's slanted toward war, famine, and abject poverty. We recognize that those things exist, but on the continent as a whole there are pockets of opportunity and excellence. (Quoted in Alexander, 2008, p. 70)

There is a perception on the continent of the "Western biased reporting on Africa by international news channels such as CNN and the BBC" (De Jager, 2007, p. 18). Is it harmful for Zimbabweans to consume news about Zimbabwe from BBC, CNN, and Sky News? Does it skew the reality on the ground with a kind of journalism that only focuses on crises? Many people believe this is a significant problem with images of Africa produced from outside the continent. Eko (2001) notes:

...Africans are probably the most caricatured and stereotyped people in history. To this day, the world is full of mass media images of Africa presented through the eyes of non-Africans to other Africans as well as to the rest of the world. (p. 366)

Eko (2001) further suggests that "African presentation of African images" (p. 366) is the solution to having African culture distorted and mischaracterized. A solution to this concern is to allow for Zimbabweans to present Zimbabweans and the rest of the continent and the world with images of themselves. ZTV is part of the DStv satellite programming now, but would most Zimbabweans want that to be the image that the continent and the world sees of them?

Kenyan government official Dr Bitange Ndemo makes the point that government generated information is "not enough and we need people to develop high quality, locally-produced content that will be of much benefit to our people" (Quoted in Ganthunji, 2010 p. 21). While Ndemo was making reference to content on the internet as a new ICT for some Kenyans, a similar argument can be made for content on television in Zimbabwe.

THE FUTURE

Government "made sure that ZBC continues to hold a de facto monopoly of the broadcasting airwaves despite a Supreme Court ruling in the year 2000 that declared ZBC's monopoly unconstitutional because it violated Zimbabweans' right to freedom of expression" (Media Monitoring Project Zimbabwe, 2007). In Zimbabwe, ZTV has been seen as a nation-building tool, as television is seen in many African countries (Harding, 2003). The new political dispensation in Zimbabwe that exists at the time this chapter is being written *might* open up the media space in the country. The transitional unity government is working to revise media and other laws that have restricted new free-to-air television stations from being opened and run locally.

If one or more new TV stations that are privately owned are allowed to broadcast free-to-air programming in Zimbabwe, the platform for showing local productions of television shows will expand. It is possible that this expanded TV space will allow many more Zimbabwean voices to be heard and to tell their stories. These stories might give voice to those that have been marginalized in the current news and current affairs agenda of ZTV. These stories might be dramas, comedies, TV talk shows and soap operas. The popularity of satellite television, most of which is not Zimbabwean or African in production or focus, does not mean that the audience for quality, diverse local content does not exist.

People do not necessarily decide to watch DStv because they do not want to watch local productions. Bourgault (1995) reports that there is some evidence that African audiences' like or even prefer local television programming. When a Zimbabwean production has high production value, people watch it and as has been the experience in other African countries, often will prefer to see and hear their own cultures stories over those from other countries. A ZTV soap opera called *Studio 263* was a hit for the station and dealt with contemporary Zimbabwean issues. It was generally well acted, scripted, directed and edited, so people enjoyed watching it. The show is coming back after being off the air for a number of months (Mbiriyamveka, 2009).

Passing through Harare International Airport when *Studio 263* was playing was an enlightening experience as airport workers and Zimbabwean passengers would be crowded around the TV monitors in the departure lounge enjoying the show; laughter, clapping, and loud commentary all demonstrating how the audience liked the program. In stark contrast, when the drums sounded to signal the start of the main local news broadcast following the conclusion of *Studio 263*, most of the audience rapidly left the TV monitors to return to work or find other things to do while awaiting their flights.

If the owner(s) of a new Zimbabwean TV station take time and effort to consider the media tastes and desires of the audience then it is more likely that the station will develop programming that will be successful in attracting viewers. As McCall (2007) aptly points out:

It doesn't take a logic expert to figure that the closer a media owner is to the community, both geographically and emotionally, the more conscientious and committed that owner will be to understanding and serving that community. (p. 118)

In order to compete with satellite TV, it is important that any new TV stations that are licensed and allowed to broadcast in Zimbabwe must be operated by people who are professionals, and who have the knowledge and skill to operate their businesses not only for profit motive, but for the Zimbabwean audience they will serve. McCall (2007) attests to this; "Owning a media property brings with it a larger responsibility because the products being peddled are news, entertainment, culture and the very identity of a society…" (p.118)

The institutional bodies that govern licensing procedures and oversight of any new TV stations needs to be transparent and run by those who are held in high regard in the industry. This is especially important given that independent or privately owned media may well be looked at with some suspicion by whatever political party is in power. Opening up TV broadcasting space that is local will be an experiment in a country that has long held a government monopoly of the local TV airways.

CONCLUSION

As Salmon (2003) suggests "Story-telling is one of the arts Africans have traditionally used to make sense of their lives and to help people respond to new events and challenges" (p. 1) It is quite possible that if Zimbabweans are given a choice between DStv and local free-to-air TV that features a range of voices, telling Zimbabwean stories with a high production value and an objective, professional stance on news broadcasting that the audience will select to watch the local TV station. It will be extremely interesting to monitor the television media space in Zimbabwe to see what new players in the market bring to audiences, how the satellite DStv responds to these potential new challenges to their market dominance and how audiences react.

Baran (2009) suggests that it is vital to realize that the media are places that tell our culture's stories and that the media act as a cultural forum: a place for discussion and debate. Whether the stories Zimbabweans see on TV are news and current affairs shows, songs, soap operas, comedies or dramas, they are stories. The audience has clearly rejected Zimbabwe Broadcast Holding's arrogant and often repeated claim that they are the country's "permanent choice" of television station (Media Monitoring Project Zimbabwe, 2007). Until Zimbabweans are satisfied with the stories that they are receiving from local TV, satellite television will have an important role to play in the lives of the Zimbabwean audience as they use this ICT to help meet their media needs and gratifications.

REFERENCES

Alemneh, D. G., & Hastings, S. K. (2006). Developing the ICT infrastructure for Africa: Overview of barriers to harnessing the full power of the internet. *Journal of Education for Library and Information Science*, *47*(1), 4–16. doi:10.2307/40324333

Alexander, G. (2008). Channeling better images. *Black Enterprise*, *39*(5), 70.

Baran, S. J. (2009). *Introduction to mass communication: Media literacy and culture*. New York, NY: McGraw-Hill.

Black, J. (2001). Semantic and ethics of propaganda. *Journal of Mass Media Ethics*, *16*(2-3), 127–137.

Bourgault, L. M. (1995). *Mass media in sub-Saharan Africa*. Bloomington, IN: Indiana University Press.

Bratton, M., Chikawana, A., & Sithole, T. (2005). Propaganda and public opinion in Zimbabwe. *Journal of Contemporary African Studies*, *23*(1). doi:10.1080/0258900042000329466

De Jager, C. (2007). SABC gives Africa a voice in the news. *Variety*, *407*(10), 18.

Eko, L. (2001). Steps toward Pan-African exchange: Translation and distribution of television programs across Africa's linguistic regions. *Journal of Black Studies*, *31*, 365–379. doi:10.1177/002193470103100308

Fontein, J. (2009). Anticipating the tsunami: Rumours, planning and the arbitrary state in Zimbabwe. *Africa: The Journal of the International African Institute*, *79*(3), 1–21. doi:10.3366/E0001972009000862

Gathanju, D. (2010). Kenya builds digital villages. *EContent*, *33*(4), 18–22.

Gustainis, J. J. (1990). Demagoguery and political rhetoric: A review of the literature. *Rhetoric Quarterly*, *20*(2), 155–161.

Harding, F. (2003). Africa and the moving image: Television, film and video. *Journal of African Cultural Studies*, *16*(1), 69–84. doi:10.1080/1369681032000169276

MacGregor, K. (2003, August 18). Africa's satellite effect: They are creating the first Pan-African networks—and hits. *Newsweek (International Edition), 39*.

Masuku, J. (2009, October 20). Letter from Zimbabwe: Digital tv taking over. *Radio Netherlands Worldwide*. Retrieved from http://www.rnw.nl/english

Mbiriyamveka, J. (2009, September 19). Studio 263 Back! *The Herald*. Retreived from http://www.allAfrica.com/stories

McCall, J. (2007). *Viewer discretion advised: Taking control of mass media influences*. Lanham, MD: Rowman and Littlefield.

Media Monitoring Project Zimbabwe. (2007). *Free the airwaves-your right to the freedom of choice*. Retrieved from http://www.mmpz.org.zw

Merrill, J. C., & Lowenstein, R. L. (1971). *Media, messages, and men: New perspectives in communication*. New York, NY: McKay.

Mokwetsi, J. (2005, July 24). Satellite tv proves a hit. *The Standard*. Retrieved from http://thestandard.co.zw/entertainment/14296.html

Musiyiwa, M. (2008). The mobilization of popular music in the promotion of national unity in Zimbabwe. *Journal of Music Research in Africa*, *5*(1), 11–29.

Mutula, S. M. (2004). Making Botswana an information society: Current developments. *The Electronic Library*, *22*(2), 144–153. doi:10.1108/02640470410533407

Ndlela, D. (2009, July 15). ZBC tv spruces up to win viewers, adverts. *Bizcommunity*. Retrieved from http://www.bizcommunity.com/article/238/66/37933.html

Nevin, T. (2002). Africa's new challenge: Crossing the digital divide. *African Business*, *277*, 24–26.

Odendaal, N. (2006). Towards the digital city in South Africa: Issues and constraints. *Journal of Urban Technology*, *13*(3), 29–48. doi:10.1080/10630730601145997

Pfuwka, C. (2008). Black September et al: Chimurenga songs as historical narratives in the Zimbabwean liberation war. *Journal of Music Research in Africa,5*(1), 30-61.

Polikanov, D., & Abramova, I. (2003). Africa and ICT: A chance for breakthrough? *Information Communication and Society*, *6*(1), 42–56. doi:10.1080/1369118032000068778

Ranger, T. (2004). Historiography, patriotic history and the history of nation. *Journal of Southern African Studies*, *30*(2), 215–234. doi:10.1080/0305707042000215338

Rhodes, T. (2010). *In African hot spots, journalists forced into exile*. Committee to Protect Journalists. Retrieved from: http://cpj.org/2010/02/in-african-hot-spots.php

Salmon, K. (2003, June 12). Drama-Africa: Persuading a man without hitting him on the head. *Global Information Network, 1*.

Schiappa, E., Gregg, P. B., & Hewes, D. E. (2005). The parasocial contact hypothesis. *Communication Monographs*, *1*, 92–115. doi:10.1080/0363775052000342544

Thram, D. (2006). Patriotic history and the politicization of memory: Manipulation of popular music to re-invent the liberation struggle in Zimbabwe. *Critical Arts: A South-North Journal of Cultural and Media Studies, 20*(2), 75-88.

Wright, B. (2008). Is triple play the answer? *African Business*, *346*, 44.

Zheng, Y., & Walsham, G. (2008). Inequality of what? Social exclusion in the e-society as capability deprivation. *Information Technology & People*, *21*(3), 222–243. doi:10.1108/09593840810896000

ADDITIONAL READING

Allen, R. C., & Hill, A. (Eds.). (2004). *The television studies reader*. London: Routledge.

Black, J. (2001). Hardening of the articles: An ethicist looks at propaganda in today's news. *Ethics in Journalism*, *4*, 15–36.

Darnolf, S., & Laakso, L. (2003). *Twenty years of independence in Zimbabwe: From liberation to authoritarianism*. London: Macmillan. doi:10.1057/9781403948120

Gillespie, M. (1995). *Television, ethnicity and cultural change*. London: Routledge.

Hartley, J. (1999). *The uses of television*. London: Routledge.

Hartley, J. (2007). *Television truths: Forms of knowledge in popular culture*. Malden, MA: Wiley-Blackwell.

Jacobs, S. (2007). Big Brother, Africa is watching. *Media Culture & Society*, *29*(6), 851–868. doi:10.1177/0163443707081691

Kamalipour, Y., & Kuldip, R. R. (Eds.). (2001). *Media, sex, violence and drugs in the global village*. New York: Rowman and Littelfield.

Mano, W. and Willems, W. (2008). Emerging communities, emerging media: The case of a Zimbabwean nurse in the British Big Brother show. *Critical Arts: A South-North Journal of Cultural and Media Studies*, 22, 1, 101-128.

Potter, W. J. (2005). *Media Literacy* (3rd ed.). Thousand Oaks, CA: Sage Publications.

Windrich, E. (1981). *The mass media in the struggle for Zimbabwe: Censorship and propaganda under the Rhodesian Front rule.* Gweru: Mambo Press.

Zilberg, J. (1995). Yes, it's true: Zimbabweans love Dolly Parton. *Journal of Popular Culture*, *29*(1), 111–125. doi:10.1111/j.0022-3840.1995.2901_111.x

KEY TERMS AND DEFINITIONS

Agenda Setting: Deciding what issues are important to consider.

Chimurenga: War of liberation.

Homogenization: The process of blending to make different things similar.

Propaganda: Messages purposefully designed to serve and support a specific agenda.

Shona: Most widely spoken language indigenous to Zimbabwe.

Uses and Gratifications: Refers to using the media to meet specific needs and wants.

Chapter 8
'Wiring' African Newsrooms:
The Internet and Mainstream Print Journalism Practice in Zimbabwe

Hayes Mawindi Mabweazara
University College Falmouth, UK

ABSTRACT

This chapter uses a qualitative case study approach to critically examine the appropriations of the Internet by Zimbabwean mainstream print media journalists. It draws on social constructivist approaches to technology and the sociology of journalism to highlight the impact of the Internet on practices and professionalism. The chapter argues that the deployment of the Internet by Zimbabwean journalists (and indeed in Africa at large) is relative and contingent upon the 'internal' newsroom context(s) and the wider socio-political and economic circumstances in which the journalists operate. Among other functions, the technology shapes the mainstream press' news agenda as well as avails information often censored by government. However, as the chapter shall demonstrate, the use of the Internet is replete with ethical and professional implications.

INTRODUCTION

The advent of new technologies (ICTs) in Africa in the 1990s "sparked celebratory, almost utopian bliss" (Banda et al. 2009, p. 1) among its proponents. It was accompanied by the "hype about the continent's possibility of 'leapfrogging' some stages of development" (ibid.). From the outset, the adoption of ICTs in various sectors was largely motivated by ideas of bridging the divide between the rich and the poor and promoting socio-economic progress. In the context of journalism practice, new technologies were also seen as having the 'potential' to increase

DOI: 10.4018/978-1-60960-591-9.ch008

journalists' work efficiency and thus overcoming the barriers associated with 'traditional' means of journalism practice. However, the realisation of such over-hyped mono-causal explanations of the perceived potential of new technologies that disregard contextual realities which shape the processes of ICT adoption and appropriation in Africa has always remained elusive.

This study takes a critical view of what Slack and Fejes (1987) describe as the "optimistic naïve" (p. 202) perspective, which sees the economically developing countries as having the most to gain from new ICTs. It is premised on the view that to understand the impact of new technologies on journalism practice in Africa, we must put journalists into a critical and analytical context and begin to question the immediate and wider social context in which they deploy the technologies. This approach finds root in the collective strengths of two broad theoretical concerns: the *sociology of journalism* and *social constructivist* approaches to technology. Although these theoretical bodies were conceptualised before the 'new media age' – in the 1970s and 80s – together they provide a basis for conceptualising the interplay between journalists, their everyday practice and the wider social factors that coalesce to structure and constrain the deployment of new technologies.

The early newsroom studies whose work crystallised in the *sociology of journalism* offer enduring insights into the working practices of journalists and thus provide a default setting against which most news production studies have been rooted. As Paterson (2008) contends, without these "early ethnographic investigations of news production, our understandings of journalism would be limited to what little we are able to glean from the observation of news content, or from what journalists say they do" (p. 2). Similarly, *social constructivist* approaches to technology, whose work emphasise the 'social shaping' and 'interpretive flexibility' of technology offer a key guiding frame in terms of which technologies can be understood as not necessarily replacing existing social realities and dynamics in which they are appropriated, but rather as continuous with and embedded in them (Bijker, 1995, p. 6; Miller & Slater, 2000, p.5). In deploying these influential theoretical frameworks therefore the chapter acknowledges that journalism (and research about that work) is not performed in a vacuum; independent of the 'shaping impact' of contextual influences.

Against this backdrop, this chapter departs from 'deterministic' approaches to the social function of technologies and reinvigorates traditional sociological approaches to journalism and technology to closely examine how Zimbabwean mainstream print journalists across the divide of state-controlled and private press deploy the Internet in their everyday professional practices. It specifically seeks to highlight patterns and trends in the appropriation of the Internet and its impact on the professional normative ideals of journalism as a social practice.

BACKGROUND AND METHODOLOGICAL APPROACH

Although Zimbabwe, like most sub-Saharan African countries faces major challenges of socio-economic development that manifest in poverty, disease and low level access to social services, including political turbulence (Kupe, 2004, p. 354), it nonetheless provides a good case for a close examination of the deployment of new technologies by mainstream journalists in sub-Saharan Africa for two key reasons. First, while the Zimbabwean media scene can not be generalised to scenarios in individual African countries, its mainstream press remains a central and vibrant platform for the struggle for control of public discourse between the opposition and the incumbent elite.[1] Second, relative to other sub-Saharan African countries, new technologies have proliferated and permeated key facets of social life in the country (see Moyo, L. 2009).

They constitute an important factor in shaping communications, not least, the mainstream press' newsmaking practices.

The study examines newsrooms drawn from the dominant state-controlled Zimbabwe Newspapers Group (Zimpapers) and the small but vibrant private press, comprising of two dailies (*The Herald* and the *Chronicle*) and four weeklies (the *Sunday Mail*; the *Sunday News*; *The Zimbabwe Independent* and *The Standard*), all located in Zimbabwe's two major cities: Harare, the capital and Bulawayo, the second largest city.[2] In the late 1990s the state-controlled press through its mother company Zimpapers committed itself to raising editorial standards through the use of new technologies to produce the best quality of newspapers and thus sustain and increase the company's competitive edge. With the setting up of Information Technology (IT) support units within its newsrooms, Zimpapers took the lead in harnessing the perceived power of the Internet in its newsmaking operations. It provided journalists with 24-hour access to the Internet.

As with newspapers under Zimpapers, the privately-owned press (in particular, two sister weeklies owned by Alpha Media Holdings, *The Zimbabwe Independent* and *The Standard*) also embraced new technologies as a 'backbone' to their operations. Although all the newspapers studied assumed an online presence in the late 1990s, in 2007 *The Zimbabwe Independent* and *The Standard* refurbished and modernised their websites with interactive features that enable readers to contribute content to the newspapers. The weeklies sought to use their websites to get leads and insights from their readers in the wake of the intensifying political impasse at the time (Moyo, D. 2009, p.11). A focus on these two weeklies in this study is therefore not without justification.

To gather empirical data the study adopted a qualitative case study approach that employed in-depth focus group interviews in conjunction with individual interviews and observations, to offer 'thick' descriptions and 'insider' perspectives of practices and cultures of Internet use in news production. The flexibility of a qualitative approach enabled the construction of meanings out of the ongoing interactions between journalists, their immediate context of practice and the wider social milieu. The approach revealed the contingent nature of cultural production and facilitated a close understanding of the dynamics of newsmaking practices in which new technologies are deployed.

Overall, a minimum of 96 journalists were involved in the study, either through interviews, observations or a combination of all the methods. The figure represents the minimum number of journalists involved in the study since it was highly probable that some reporters were involved in observations or other interactions but were not counted due to their limited presence or lack of striking contribution to the focus of the study. However, it is also important to highlight that the research's aim was primarily to understand particulars rather than generalising to universals, in keeping with the key assumptions of qualitative research. The study was 'concerned with *explanation* in a wider sense than measurement or causation' (Mason, 2006, p.16). Interviewees were purposively sampled using convenience/theoretical sampling rather than random sampling. Of significance in the sampling procedure, however, was the need to select journalists from different desks in order to cover all key 'beats' in the newsrooms.

Due to the political unpredictability of the research context and the sensitivity of some of the responses provided by journalists, the chapter uses pseudonyms in place of the actual names of journalists. It also refers only to generic titles in order to ensure confidentiality and anonymity. As Mason (1996) notes "given the personal nature of data generated from qualitative interviews. Such data can be recognised by the interviewee...and so may be recognisable to other people" (p.56).

THE APPROPRIATION OF THE INTERNET IN NEWSROOM PRACTICES

This section discusses the salient uses of the Internet across the newsrooms studied. These range from basic and general uses that transcend the confines of individual newsrooms to the more intricate uses resulting from the 'internal' as well as 'external' factors that characterise the context in which the journalists operate. These include: the sourcing of news and story ideas; research (for background information; researching specialised topics; learning new writing styles and skills); social networking and source cultivation; interacting with readers and the mobilisation of user-generated content. I discuss these in turn below.

THE INTERNET AS AN ALTERNATIVE SOURCE OF NEWS AND STORY IDEAS

The use of the Internet as an alternative source of local and foreign news was one of its key functions across all the newsrooms studied. This function was closely tied to the general challenges that the news media in Zimbabwe face in news sourcing because of the polarised political context in which they operate. While journalists in the state-controlled press generally find it easy to source information from government departments, sourcing information from opposition political parties has over the years been a challenge. Similarly, the private press has challenges in sourcing information from government-related sections of society. It is in this context that the Internet has become an alternative source of news and story ideas. As the news editor of the sate-controlled *Sunday News* explained:

Gilbert: There is one thing that I want to make very clear... because of our continued lambasting of the MDC and supporting ZANU-PF's policies we have lost our sources among opposition politicians... They choose to maintain contact with our fellow reporters who clandestinely write for foreign-based online newspapers. So, these online newspapers carry a better Zimbabwean story than us...

If you look at it closely there is an interesting irony there, we now have to rely more on the Internet to get an understanding of what is happening in our own country. Yes, we have other sources who give us news but they do not tell us everything, they leave out other details which end up on online publications... If you have noticed in most cases we make follow-ups on stories that we would have seen on the Internet...

These circumstances have resulted in foreign-based online newspapers run by exiled Zimbabwean journalists always being the first to break news on many topical issues in the country ahead of the traditional mainstream press which is on the ground. Moyo, D. (2007) observes that often the online newspapers "set the agenda for the mainstream media in Zimbabwe by coming first with the stories partly because of the speed and instantaneous publishing enabled by new communications technologies" (p.91). This scenario further finds explanation in the fact that "online publications...enjoy 'unfettered liberty' to publish what they want" (Chari, 2009, p.23) hence the pervasiveness of online newspapers that focus on Zimbabwe. The Internet therefore offers mainstream journalists an unfettered alternative space for exploring and accessing local news from diverse sources and perspectives including stories they cannot pursue because of the constrained nature of the context in which they operate.

A senior political reporter at *The Zimbabwe Independent*, Luthando, highlighted that the pervasiveness of Zimbabwean news on the Internet spurs mainstream journalists to keep track of online news. Particular focus is given to news websites run by exiled Zimbabwean journalists such as *Newzimbabwe.com*, *Zimonline.co.za*, *Zimbabwesituation.com*, not only as a way of keeping abreast of general news and current affairs, but also as a way to avoid publishing 'stale' news. The Internet has thus exerted profound influence on news coverage in Zimbabwe's mainstream press (see Chari, 2009, p.13; Moyo, D. 2009, p.2).

The Internet is also central as a source of foreign news. Unlike in the past where the newsrooms accessed international stories mainly through the Zimbabwe Inter-Africa News Agency (ZIANA), a government owned news agency that supplied stories to the newsrooms via telex, the Internet provides newsrooms with a wide selection of international stories. This entails that international news is now published more timeously than previously.

The selection of news websites was patterned and shaped by a broad variety of factors which that included established professional imperatives such as: news values; readership considerations; and personal journalistic judgements. Following these criteria, a number news websites were branded as 'lacking credibility' or completely 'untrustworthy' by most reporters. Resultantly, journalists used a number of criteria and ratings to assess the credibility of online news sources. Thus, in testing for credibility (or lack of it) in online news sources, journalists tend to lean on traditional journalistic culture and normative values.

However, the appropriation of the content sourced from the Internet was shaped by the newsroom's editorial policies. These set the parameters, not only in terms of how the journalists navigated the Internet in search for news, but also in terms of how they made use of the news content in their news construction.

INTERNET AS A RESEARCH TOOL

The use of the Internet as a research tool emerged as one of its key functions in all the newsrooms studied. For most journalists, the Internet provided access to a new frontier and tool for research and data collection. It helped to overcome some of the barriers imposed by traditional research approaches. Newsroom observations revealed that online research was widely used for background information as well as to seek clarifications on various subject areas.

Emphasising the importance of the Internet as a research tool for journalists, Machill and Beiler (2009) highlight that with the Internet, "even editorial offices with poor research capacities now have at their disposal a relatively good means of conducting research" (p. 201). A number of journalists highlighted that the Internet provided them with a platform to conduct research on how to tackle specific assignments as well as learn from other writers thus improving their writing skills. As one reporter at the *Sunday Mail* explained:

Mehluli: When I want to "featurise" a certain subject, I usually Google my proposed headline... for instance: "woman scalds daughter with hot water" and see how various international reporters have tackled that headline... I then borrow one or two writing styles...

Of significance in the extract above is the journalists' reliance on Western journalistic forms as a standard measure of their own newsmaking practices. This emphasises the fact that newsmaking practices in Zimbabwean newsrooms continue to be influenced, at least at a general level, by Western patterns of newsmaking (see Louw, 2001, p.159; Nyamnjoh, 2005, p.39).

However, a number of journalists had difficulties or limited skills in exploiting various online search options which included using: topic-specific search engines; evaluating the quality of online content; and properly attributing material abstracted from the Internet. These challenges related to individual levels of Internet use proficiency and pointed to "a moderate level of search-engine competence on the part of journalists" (Machill & Beiler, 2009, p. 201).

Figure 1. Popularly visited websites across newsrooms

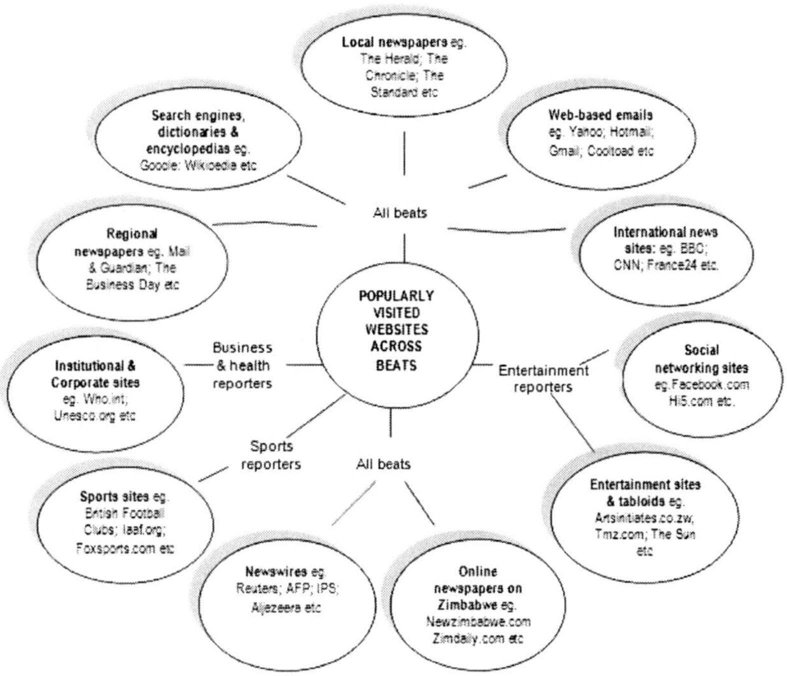

BEAT INFLUENCES ON THE USE OF THE INTERNET

Although the discussion above points to a general trend in the use of the Internet that appeared to cut across different news desks, the beat system also shaped and influenced how journalists deployed the Internet in the newsrooms. Observations revealed that journalists were less inclined to stray to Internet functions outside the sphere of their beats. Each beat required "very different coverage routines, different work rounds individually tailored to the particular activities within the beat's territory" (Fishman, 1980, p.36). Time constraints and the pressures of deadlines also restrained journalists from exhaustively exploring aspects of the Internet outside the sphere of their beats. A close examination of the journalists' Internet browsing history revealed a particular pattern in terms of websites frequently visited by specific beat reporters, as well as across beats as illustrated in Figure 1 below.

This illustration indicates a deliberate and conscious attempt by journalists to select particular websites based on their day-to-day beat routines. In the sections that follow I closely discuss the influence of specific beats on the use of the Internet among generic beats in the newsrooms.

Political Reporting

The polarised political environment in Zimbabwe had marked implications on how political reporters deployed the Internet in their daily news construction and sourcing routines. The interview extract below sums up how the political editor of the state-controlled *Sunday Mail* conceived of the role of the Internet in his beat.

Mutepfa: I can't imagine my job without the Internet because as a political editor I'm responsible for monitoring the political environment within and outside the country. I do that through contact-

ing the political players themselves, but I also get useful tips from the Internet. I read stories from online publications on Zimbabwe. I also read news from outside Zimbabwe, looking at how other media houses are reporting on Zimbabwe. In brief, I would say the Internet is playing a very big role because it keeps me informed on the latest political developments...locally and globally.

Interviewer: Are there specific news websites that you visit?

Mutepfa: Because of our strained relationship with Britain, it is a must to read what the BBC is saying about Zimbabwe...I take the BBC to be the purveyor of British foreign policy.[3] *Once you get what the BBC is saying you understand what the British government is thinking at any particular moment. So, I make it a point that I visit the BBC website everyday... I also follow the CNN website.*

The extract above highlights how online news sources play a key role in shaping the political content of Zimbabwean mainstream newspapers.

For political reporters in the private press, the Internet provides unfettered space to access and explore issues that are not easily accessible because of the government's tight control on access to information. The Internet provides access to stories that counter the perceived 'propaganda' churned out in the state-controlled press. According to Chari:

The new technology has rendered stringent media laws such as the Access to Information and Protection of Privacy Act (2001) and the Broadcasting Services Act (2001) null and void resulting in journalists having free reign. Online publications now enjoy "unfettered liberty" to publish what they want and how they want. (2009, p.23)

This observation was also articulated by a senior political reporter at *The Standard*, Busani, who saw the Internet as an alternative source of information on issues censored by the government:

...we also rely on the Internet for most of the information that we cannot access freely because of the political situation, but we always try and localize the material so that it becomes relevant to our readers and editorial policy.

The Internet has also facilitated the emergence of what Kasoma refers to as a "journalism of hatred, revenge, and dislike" (1996, p.99) among political reporters, especially in the state-controlled press. This journalistic culture in which political reporters browse an assortment of news websites with a view to respond to stories that attack government policy can best be described as "activist journalism" (Chari, 2009, p.29) or "vendetta journalism" (Kasoma, 1996, p. 99). The practice assumed the character of a qualitatively different "patriotic journalism" (Ranger, 2005, p.8) concerned, as the former Minister of Information and Publicity in President's office, Jonathan Moyo, put it: "to protect, preserve and project...regime values" (Moyo, J. 2004). This practice was a direct result of political developments in the country and the general feeling by government that the nation was facing unwarranted vilification from foreign media. Chari (2009) argues that "vendetta journalism violates...the core tenets of [journalism] such as objectivity, balance, fairness and impartiality" (p. 23). Journalists "cast caution to the wind in order to *settle scores with perceived or real enemies*" (Chari, 2009, p.26) and moral and ethical standards are sacrificed in pursuit of 'self defence' and political expediency.

Entertainment Reporting

Entertainment reporters followed local, regional and international entertainment and celebrity gossip websites such as *Artsinitiates.co.zw*; *Zimvibes.com*; *Tmz.com* and tabloid newspapers such as the *Sun.co.uk* and the *Newsoftheworld.co.uk*. The issues pursued in the beat and the nature of the sources used tended to differentiate the reporters from their counterparts in other beats (see figure 1 above). As one entertainment reporter at the *Sunday Mail*, Mthabisi, explained:

The moment I sit on my desk I just find myself visiting those websites that relate to my beat – Artsinitiates.co.zw; Itsbho.com; Zimvibes.com; Tmz.com, you name them – it's instinctive, I don't even need to think about that because I have to update myself on what is happening in the arts and entertainment arena, otherwise there will be nothing to write about!

This response reinforces the fact that new technologies have become part of entertainment reporters' taken-for-granted daily newswork.

A regular observation of web-browsers in the newsrooms also indicated that entertainment reporters invested more time on social networking sites such as *Hi5.com* and *Facebook.com* much more than their colleagues in other beats. Facebook in particular, was seen as a valuable platform for connecting with potential sources scattered across the globe and was integrated into some reporters' newsmaking routines. The social networks' integration of email and 'instant messaging' allowed the reporters to connect with their sources in 'real-time'.

Users of Facebook also highlighted that the social platform was an invaluable reporting tool that instantly connected them to their sources. As the entertainment editor of the *Sunday News* explained in the interview extract below:

Interviewer: You talk about Facebook, how important is it in your daily business?

Lawson: I think Facebook is effective for instant communication unlike the general email. For instance, if I am conducting an interview on Facebook I am assured of instant and direct answers because I can immediately follow up for clarifications. But with email, if you send questions to your source that's that... if you want to follow up on some of the responses, you have to write another email and wait maybe for another day or two for a response. Whereas if you are chatting on Facebook as soon as something interesting is brought up you pursue it. Its like a conversation or one-on-one communication.

Interviewer: But how do you balance between the social side of the site and its professional functions?

Lawson: Well, you are right! We use Facebook usually for social purposes – just socialising with friends, but then you realise that most of our sources are also on Facebook. You will be surprised to know that a lot of public personalities who provide news are active users of Facebook. This makes them more accessible to us [journalist] because the truth is that you cannot know the contact details of every public personality, but it's easy to connect with them on Facebook... So as journalists we also benefit from sites – just discussing general issues sometimes gives us story ideas which feed onto our day-to-day activities in the newsroom.

From the interview extract above, it is clear that social networking sites are not without journalistic value for entertainment reporters. The example shows that the functions of social networking sites are not 'fixed' as technological determinists would suggest, rather they are a site of interpretive work that is entrenched in the intricacies of social interactions. This reinforces Bijker's (1995, p.6) view that: "[O]ne should never take the meaning of a technical artefact or technological system as residing in the technology itself instead one must study how technologies are shaped and acquire their meaning in the heterogeneity of social interactions".

Of significance is the fact that the social networking platform connects the reporter to 'society' and fosters trust and norms of reciprocity that constitute key antecedents to effective entertainment reporting. The short messages left by friends on the reporter's 'wall' (the space on each user's profile page that allows friends to post messages for the user and is visible to anyone among the user's network of friends on the site) reflects sentiments and call attention to what is happening on the social scene hence offering a conduit for maintaining relationships and engaging with readers.

The utility of the social networking sites as journalistic tools was, however, not without challenges. In reflecting on the effectiveness of online social networking as a journalistic tool, some entertainment reporters underlined the challenge of balancing between the 'personal' and the 'professional'. Referring to the activities on *Facebook.com* and their link with journalistic practice, reporters spoke about how they exercised restraint and professed the need for caution in terms of expressing their personal interests, emotions and opinions about issues raised by their network of friends on the platform. Doing so, in their opinion, would put them on the spotlight and subject them to criticism given the general understanding that journalists should maintain a public deportment that does not put into question their journalistic 'independence' in the eyes of the public.

This critical approach to the use of Facebook echoes Berger's (2005) observation that Southern African journalists are far from "lacking when it comes to critical perspectives with ICTs and global information networks" (p.1). Similarly, journalists also pointed to challenges embedded in independently verifying information sourced from Facebook by highlighting that some people tend make things up or exaggerate issues, especially given the very nature of social networks as "social settings".

Sports Reporting

The wide presence of sports news on the Internet also offered 'new' opportunities for sports reporting in the newsrooms studied. Like entertainment reporters, sports reporters across all the newsrooms have adapted their work routines to the varying opportunities offered by the Internet. Acknowledging the sport beat's influence on the deployment of the Internet, one sports reporter at *The Zimbabwe Independent*, Mthulusi, noted: "I rarely open sites that don't have anything to do with sport during the course of the day, especially football. So, my beat really does influence the way I surf the Internet…"

In general, sports reporters searched for the best selling stories with a potential for popularity among readers. To this end, they made an attempt to follow and understand the sporting interests of their readers as well as keep abreast with developments in different sporting disciplines. The notion of 'proximity' in the selection of sports websites and specific stories and was also evident in terms of the regional locations of the newspapers in the country.

For sports reporters, the Internet has thus been adapted to fit into local news interests. Most senior sports reporters also believed that they are now widely read on international sports and better able to cover a wide base of sporting events and disciplines than was possible before the Internet age.

Business, Health and Court/ Crime Reporting

Although Internet-based research was common across all beats, some beats like business, health, court and crime reporting used the Internet more intensively for research than their counterparts in other beats. These beats constantly required specialised knowledge and understanding of technical issues that reporters had to routinely interpret in their assignments.

In doing their research business reporters also regularly visited corporate and institutional websites (see figure 1 above) in pursuit of specialised information. However, the key challenge in researching for local stories on the Internet for business reporters was that most Zimbabwean corporate organisations and key government institutions took time to update their websites with current information. Consequently, a number of business reporters highlighted that the Internet was less helpful when it comes to local news assignments.

As with business reporters, health and crime/court reporters also used the Internet to search for clarifications on technical issues they routinely encountered in their beats. As one health reporter at *The Sunday Mail*, Paida, further explained: "As a health reporter, some of the issues I have to write about are too technical for my understanding, and medical practitioners may not always be available to explain…So, I always resort to the Internet for clarifications…". In the same way, court and crime reporters also extensively used the Internet to search for clarifications on complex legal issues. A senior court reporter at the *Chronicle*, Felicity, explained thus: "…at times we fail to find lawyers to explain difficult legal issues that we come across. In such instances, the Internet becomes our fallback".

The discussion above shows that the beat system has a defining impact on the deployment and appropriations of the Internet in the newsrooms studied. This is mainly because each beat has its own news coverage routines hence reporters adapt their work routines to aspects of the Internet that are closer to their "beat's territory" (Fishman, 1980, p. 36). Similarly, because of time constraints and the pressures of deadlines journalists across beats spend less time on aspects of the Internet that do not relate to their beat interests.

The section that follows discusses the adoption of interactivity and user-generated content in the private weeklies and the reasons for the state-controlled press's ambivalence towards opening up to the interactive aspects of the Internet.

The Adoption of Interactivity and User Generated Content

Although *The Standard* and *The Zimbabwe Independent*'s websites' still largely reproduce content from their print editions, the interactivity of the websites is a major shift from the common trend in Zimbabwe. The weeklies took the lead in innovatively blending traditional mainstream journalism with online journalism. As Mudhai (2004) contends, although "[o]ne may argue that the duplication of print content on the online editions does not add much value to the news [it, however,] is an important step" (p. 328) in the direction of maximising the use of new technologies.

The critical role of content posted by the readers on the newspaper's websites also emerged in editorial conferences where it often contributed to the agenda of the meetings. In this sense, readers inadvertently contributed to the issues covered in the newspapers. In some cases, the interactive features enabled readers to correct misnomers and factual inaccuracies. As one business reporter at *The Zimbabwe Independent*, Paulus, explained: "Some readers will tell you: 'you omitted something in your story about corruption in this company…can phone so and so, this will give you more details'…".

It is important to examine the adoption of interactivity and user-generated content by *The Standard* and *The Zimbabwe Independent* in light of the socio-political context in which they were introduced – a period during which Zimbabwe went through political turmoil following disputed presidential elections – in a context in which journalists generally faced challenges in accessing information owing to the polarised political environment governed by stringent media laws. In this crisis period as Moyo (2009) observes "news websites, including those that belong to mainstream newspapers such as *The Zimbabwe Independent*...sought to get leads and insights from their readers to help them cover this complex political impasse"(p. 561). In this light, the adoption of interactivity and user generated content by *The Standard* and *The Zimbabwe Independent* is not a mere shift towards global journalistic trends, but also an innovative attempt to access information in a context where the free flow of information is tightly controlled.

However, there are challenges related to "gatekeeping" and the traditional media's ambivalence towards fully embracing the voices of readers. The comments posted on the newspapers' websites are subjected to the traditional gatekeeping processes in which the editors select comments along the newspapers' editorial policies. One can thus surmise, as Hermida and Thurman (2007) observe in the British context, that "[w]hile there has been a change of attitude towards user-generated content, there remains a reluctance to relinquish the traditional gatekeeping role of journalists". For them, "[e]ditors tend to view moderation in terms of the traditional role of journalists, gathering and filtering information for the public". They conclude that "this approach may perhaps offer a model for the integration of user-generated content in professional news structures" (Hermida & Thurman, 2007).

On the contrary, the state-controlled newspapers have been reluctant to adopt interactivity and user-generated content. Until late 2010, their web-editions did not provide readers with space to comment or contribute content on stories. The restrictive newsmaking cultures in the state-controlled newspapers made it difficult for the newspapers to embrace the voices of readers. Resultantly, the newspapers interactions with the readers are largely limited to the traditional letters to the editor and the fixed phone, which are heavily censored. The reluctance by the newspapers to embrace interactivity and user-generated content found explanation in two reasons. First, the editors were clearly uneasy with the prospect of abusive comments from readers because of the newspapers' partisan reportage. Second, the newspapers' 'innate conservatism' (Hermida & Thurman, 2007) associated with their goal to maintain a unidirectional flow of information also made it difficult for them to embrace readers' voices. The response from the news editor of the *Sunday News* was particularly informing:

Gilbert: Let us be frank about this, there is nothing wrong with opening up to readers' input... but sometimes readers can be abusive. The experience that we had in this newsroom after the March 29 [2008] general elections was shattering to say the least... People would phone the newsroom just to shout at you or they would send very abusive emails...we know that from the way we cover our news, we were going to be insulted everyday...

From the above one notes that the state-controlled newspapers are sceptical and cautious of the idea of opening up to reader's scrutiny as a deliberate way to avoid abusive contributions from a readership frustrated by the newspapers' partisanship editorial content. This scenario echoes Deuze's (2003) view that news organisations tend to expand their operations to the Internet based on their existing journalistic culture, including the way they relate to the public. Similarly, Pavlik (2001) contends that "...newsroom traditions... ultimately determine whether journalists fully utilize online capabilities" (p.17).

Although there is no doubt that the Internet offers a number of advantages to journalism practice in Zimbabwean mainstream newsrooms it has, however, magnified traditional professional and ethical challenges. As Pavlik (2005, p.245) puts it: "new ethical problems are arising' and old ethical concerns have taken on 'new meaning". The sections that follow discuss these challenges.

THE IMPACT OF THE INTERNET ON PRACTICE AND PROFESSIONALISM

The Internet and 'Armchair' Journalism

The impact of the Internet on journalism practice was apparent in the generational tensions between senior conservative journalists and junior reporters. The former lamented the dearth of traditional 'shoe leather' journalism as a result of overreliance on the Internet by younger journalists – they saw the Internet as promoting a culture of laziness. As one long serving senior journalist at *The Zimbabwe Independent*, Paulus, put it: "Journalism is now devoid of the 'organic feel of things'. Journalists are now reduced to *armchair* practitioners who shun going out there to get the story…as a result their stories are dry" (emphasis added). These findings echo Pavlik's (2000) observations in the American context that with the development of the Internet "more and more journalists spend increasingly less of their time out in the field" (p.229).

The more conservative senior journalists further argued that journalists are now spending too much time in the newsrooms wedged on their armchairs surfing the Internet for story ideas instead of being "out in the field observing directly the events and processes on which they report" (Pavlik, 2000, p. 229). For the senior journalists, therefore, the Internet is stifling creativity and promoting laziness among young journalists.

Junior reporters, on the other hand, articulated contrasting views. They saw senior reporters' emphasis on "shoe leather" reporting as counterproductive. One junior reporter at *The Standard*, Nqaba, expressed his frustration with his seniors' refusal to "move with the times" by cynically suggesting the need to "digitise their minds". Some reporters argued that the Internet was "saving the situation" as newsrooms, like most institutions in the country, were operating under constrained budgets due to economic challenges facing the country. These journalists saw the Internet as enabling newsrooms to cut costs while at the same time continuing to cover a wide range of stories.

The tension between senior and junior journalists was reflective of the impact of age differences and academic backgrounds in shaping attitudes towards the use of the Internet as a journalistic tool. However, the inclination towards 'armchair journalism' in newsrooms was also as a result of the economic circumstances in which the newsrooms operated which limited their operational budgets.

Plagiarism and the Challenge of Verifying Internet Content

Related to the challenge of "armchair journalism" were ethical challenges of plagiarism and the failure to verify Internet content. These problems raised questions of "news accuracy and credibility" (Mudhai & Nyabuga, 2001) that newsrooms had to contend with. Although plagiarism has always been a problem in Zimbabwean newsrooms, newsroom observations confirmed that the Internet has made plagiarism "increasingly simple and tempting" (Pavlik, 2001, p.47). While there was awareness among editors of the increase in plagiarism from the Internet, it is important to highlight that the anonymity of the technology made it difficult to gauge the magnitude of the problem because of the wide usage of pseudonyms on the Internet particularly on news websites that focused on Zimbabwe.

The problem of plagiarism was closely related to the challenge of verifying or judging the reliability of information sourced from the Internet. Although there were attempts at being selective and cautious in using Internet content, with journalists deploying a number of criteria to ascertain the credibility of content from news websites, there were a number of sustained challenges in verifying Internet content due to low barriers to access which make it open to manipulation and therefore problematic in terms of "credibility and reliability" (Machill & Beiler, 2009, p.201). It is in this light that Adamolekun (1996, p.34-35) warns that:

If we must know... [the information superhighway] is full of several booby traps for unwary and unprepared information wayfarers. The sheer great immediacy and drama we now experience and the speed at which we can view what is happening world wide... makes it extremely difficult to easily process the data we have on our hands at any given time.

Other countervailing factors such as limited budgets and the pressure to deliver stories within a limited time frame were also key limiting factors in terms of verifying Internet content. With respect to deadline pressures, Pavlik (2001, p.94) writes: "[f]requently, deadline pressures have serious negative consequences for the truthfulness of the news. Journalists under intense deadline pressure... have little time for fact checking...".

Although no excuse can be proffered for plagiarism and the use of the Internet without verifying content, from my newsroom observations, the problem was partly rooted in the absence of formal policy frameworks and training guiding journalists on how to effectively use the Internet as a journalistic tool. Commenting on training in the newsrooms, the news editor of *The Zimbabwe Independent*, Costa, stated: "I have been in this game for the past 14 years, I don't remember receiving any training on Internet as a journalistic tool, you learn as you go". These findings support Berger's (2005, p.9-10) observations that Southern African newsrooms have no formal policies on the training of staff in the use of the Internet, thus resulting in a large proportion of "peer-to-peer learning and self-teaching in the newsrooms".

The Impact of the Internet on News Access and Sourcing Routines

The increase in the reliance on the Internet as a source of news and story ideas by journalists as discussed earlier has had implications on news access and sourcing patterns. From observations and interviews, it seemed clear that the Internet promoted the sourcing of stories from elite sections of society with access and the means to contribute content to the Internet. This scenario coupled with growing impact of online newspapers in 'setting the agenda' of news coverage promotes an elite news culture, particularly in terms of the political news.

The reliance on the Internet cements established relations between elite forces and newsmakers which partly influence who gets on the news. As the Glasgow University Media Group (1980, p.114) argues "[a]ccess is structured and hierarchical to the extent that powerful groups and individuals have privileged and routine entry into news itself and to the manner and means of its production".

The appropriation of social networking sites by entertainment reporters as discussed earlier provides a good example of the impact of the Internet on news access and sourcing. As the entertainment editor of the *Sunday News*, Patson highlighted:

...the problem with sourcing stories from social networking sites is that it slants your stories towards the more affluent and educated... In fact, the whole process of interviewing people "digitally" leads to journalists missing a lot of good ideas outside these networks...

The response above highlights that reliance on the Internet for story ideas and connecting with sources tends to annihilate potential story ideas from many other sources that are not digitally connected. This reinforces the social impact of the "digital divide", in particular, the question of "access" to technology which also includes "(digital) skills or competencies and media or technology use and applications" (van Dijk, 2006, p.224). It also mirrors extant problems of widespread illiteracy, poverty, infrastructure, management, policy 'regimes' and other socio-economic and political issues (Adam & Wood, 1990, p.313-314). Thus, those who actively participate on social networking sites become the "primary definers" (Hall et al., 1978, p.61) of entertainment news. They command the discursive field and set the agenda for issues covered by entertainment reporters much to the annihilation of the voices of those without the means or the knowhow to effectively deploy the Internet.

The reliance on online research, in particular if traditional 'shoe leather' approaches are dispensed with also presents 'risks of a distorted reality' (Machill & Beiler, 2009, p.179) in terms of news coverage. For Machill and Beiler (2009), "[t]his situation requires careful reflection since the danger of a *distortion of reality* or of *one-sided* reporting exists if trust is placed unhesitatingly in the logic of a single search engine" (p.201, emphasis added). Journalist also highlighted that the practical pressures of constantly working against the clock and the need to meet required story quotas combined to structure "over-accessing to the media" (Hall et al., 1978, p.58) for those in positions of privilege with sustained online presence and visibility.

While it is indeed reductionistic to attribute elite news accessing to ICTs alone, it seems clear from the discussion above that the Internet is promoting routine elite access to the news. It places the elite at a vantage point in terms of defining and pontificating on newsworthy events. In the same way, the sociology of journalism has also proved that reporters tend to privilege sources who wield "social hegemony" mainly because they have the appropriate *infrastructure* to guarantee a reliable and steady supply of raw materials of news production (Schudson, 2000, p.184).

The Internet and 'Moonlighting': Some Professional Implications

The Internet has also facilitated the bourgeoning of 'underground' extra paid work for journalists in the mainstream press. Most journalists clandestinely correspond for foreign news organisations as a way of supplementing their poor salaries. Although moonlighting is not a new phenomenon in Zimbabwean journalism, it has taken on new meanings with the exposure of local journalists to international news organisations through the Internet and the prohibition of foreign media houses from practising in Zimbabwe.[4] Making money took precedence over professional and ethical standards. This development was clearly articulated by the political editor of *The Zimbabwe Independent* in a lengthy interview set at the Harare Press Club:

Jabulani: The Internet has economically empowered a lot of journalists in Zimbabwe where the majority of top journalists were forced to flee the country. Those that have remained have managed to use the opportunities offered by the Internet to break into global media organisations that were traditionally very difficult to break into...

There are people right here in this very club who are doing jobs for Reuters; BBC; The Washington Post; and The New York Times: some of the biggest news organisations in the world and numerous other media organisations that are banned from practising here. They are being serviced by local journalists, and it has all been made possible by the Internet...we're talking here of journalists in Zimbabwean newsrooms. These are journalists that can now be classified as some of the richest guys in Zimbabwe...some of these guys drink here [in the press club] everyday and drive very good cars...and all these are benefits accrued through the Internet...

Interviewer: But doesn't that whole scenario have negative professional implications?

Jabulani: It does to a certain extent, because we have had instances where we discuss rumours in this the press club and ten minutes later you find the rumour on the Internet as a story. A case in point was during the March 29 [2008] parliamentary and presidential elections, rumours were circulated here in the press club that Mugabe was losing and that his sister, Sabina, had collapsed and died in shock. In no time at all, there was a story on the Internet that Mugabe's sister had died. Someone had picked it up and written a story under a pseudonym and the editor of that particular news website published it without crosschecking.

From this interview, it is clear that the Internet has provided Zimbabwean mainstream journalists with opportunities for economic survival. However, these opportunities are fraught with negative professional and ethical implications.

Journalists' use of pseudonyms as a way of disguising their identities from their regular employers and protecting themselves from the perceived threat to personal security by the authorities further compromised ethical standards as accountability is buried in anonymity (Moyo, D 2007, p.91). Moreover, the speed with which the Internet allows for the publication of stories also fuels the publication of "unsubstantiated and often highly opinionated stories" (ibid.).

The professional and ethical challenges posed by the increase in moonlighting among mainstream journalists were also a result of competition for breaking news among foreign news agencies and online newspapers with an interest in Zimbabwe's unfolding socio-political. As one reporter at *The Standard*, Busani, explained the competition among journalists to be the first to break a story exposed the profession to ethical challenges:

...because of the competition among news agencies and online publications for the Zimbabwean story...each journalist wants to be the first to break the story and make more money. This has damaged the profession as people don't even wait to adhere to the cardinal rules of journalism of verifying and crosschecking issues.

CONCLUSION AND SCOPE FOR FUTURE RESEARCH

This chapter has discussed wide-ranging uses of the Internet in Zimbabwe's mainstream press. It has highlighted a number of opportunities which the technology has brought to the newsrooms. The importance of the Internet is shown in its shaping of the mainstream press' news agenda and in its availing of information often censored by government, among other functions. Although all the newspapers' online editions predominantly duplicate their print versions' content. The private weeklies have, however, taken significant strides towards creatively deploying interactivity and user-generated content to enrich their newsmaking practices. This development is, however, perceived with ambivalence and resistance by the state-controlled newspapers. As the study has revealed, the resistance is not just a matter of professional conservatism, but should be understood in the broader context of the state-controlled newspapers' newsmaking culture.

While the Internet and email have undoubtedly added value to the newsrooms' newsmaking practices, they have also magnified the traditional ethical and professional challenges surrounding news production (Pavlik, 2001, p.87) and the very nature of journalism itself. Myriad professional and ethical problems associated with the use of the technologies in the newsrooms emerged in this study. Plagiarism has increasingly become simple and tempting and journalists do not adhere to the ethical cannons of the profession. In the same way, the technologies are contributing to the development of 'new' forms of journalistic practices such as "armchair journalism", where journalists are increasingly spending more time in the newsrooms than in the field. "Activist journalism" or "vendetta journalism" among political reporters has also become more defined in the state-controlled newsrooms.

The Internet has also seen a surge in a "mercenary" approach to journalism as journalists strive to augment their poor salaries through moonlighting. Reliance on the technology is also promoting a shrinking diversity of voices in terms of who contributes to news content given significant inequalities in access to the Internet in the country. The absence of formal policy frameworks supporting training in the use of ICTs in the newsrooms has also exacerbated these challenges.

The 'internal' conditions of practice, professional ideologies of timeliness and the demands of specific beats also foster particular patterns of Internet use. In the same way, the wider socio-political and economic conditions in which the journalists operate also heavily shape and constrain the deployment of the technologies. This counters "technological determinism" and makes a stronger case for a "constructivist" approach that looks at new technologies as deeply embedded in the social context in which they are deployed.

Given the qualitative nature of data generated by the research, there is a need to test them in other countries. Indeed, there is a case for cross-cultural comparative studies that assess the impact of new technologies on mainstream journalism, especially between economically developing African countries like Zimbabwe and the economically developed countries of the North, which offer markedly different socio-economic, cultural and political conditions of practice for journalists to those in the global South. The value of cross-cultural comparative studies is not that universal truths about mainstream journalism in the age of the new media will emerge but to seek wide-ranging empirical evidence that might enable us to critically assess a number of utopian predictions that have been made about the impact of new technologies on journalism practice. For example, are the epistemological, ontological and ideological imperatives of journalism necessarily standardising in the face of the diffusion and permeation of globalising new technologies across various aspects of news production. Research from wider and varied contexts would provide a backdrop for testing and supplementing the predominantly Western examples and theoretical frameworks on which African researchers continue to rely. Such studies will draw us closer to (re) defining the ideological foundations of journalism in the context of the new media. There is much here that is yet to be mapped and understood.

REFERENCES

Adam, L., & Wood, F. (1999). An investigation of the impact of Information and Communication Technologies in sub-Saharan Africa. *Journal of Information Science, 25*(4), 307–318. doi:10.1177/016555159902500407

Adamolekun, W. (1996). The information superhighway and traditional communication: Where we stand. *Africa Media Review, 10*(2), 22–36.

Banda, F., Mudhai, O. F., & Tettey, W. J. (2009). Introduction: New media and democracy in Africa-a critical interjection. In Mudhai, O. F., Tettey, W. J., & Banda, F. (Eds.), *African media and the digital public sphere* (pp. 1–20). New York, NY: Palgrave Macmillan.

Berger, G. (2005). Powering African newsrooms: Theorising how Southern African journalists make use of ICTs for newsgathering. In Berger, G. (Ed.), *Doing digital journalism: How Southern African newsgatherers are using ICTs* (pp. 1–14). Grahamstown, South Africa: High Way Africa.

Bijker, W. E. (1995). *Of bicycles, Bakelite and bulbs: Towards a theory of socio-technical change*. Cambridge, MA: MIT Press.

Chari, T. (2009). Ethical challenges facing Zimbabwean media in the context of the Internet. *Global Media Journal: African Edition, 3*(1), 1-34.

Deuze, M. (2003). The Web and its journalisms: Considering the consequences of different types of news media online. *New Media & Society, 5*(2), 203–230. doi:10.1177/1461444803005002004

Fishman, M. (1980). *Manufacturing the news*. Austin, TX: University of Texas Press.

Glasgow University Media Group. (1980). *More bad news*. London, UK: Routledge.

Hall, S., Critcher, C., Jefferson, T., Clarke, J., & Roberts, B. (1978). *Policing the crisis: Mugging, the state, and law and order*. London, UK: Routledge & Kegan Paul.

Hermida, A., & Thurman, N. (2007). *Comments please: How the British news media is struggling with user-generated content*. Paper presented at the 8th International Symposium on Online Journalism, University of Texas, Austin, USA.

Kasoma, F. P. (1996). The foundations of African ethics (Afriethics) and the professional practice of journalism: The case of society-centred media morality. *Africa Media Review*, *10*(3), 93–116.

Kupe, T. (2004). An agenda for researching African media and communication contexts. *Ecquid Novi: South African Journal for Journalism Research*, *25*(2), 353–356. doi:10.3368/ajs.25.2.353

Louw, E. (2001). *The media and cultural production*. London, UK: Sage.

Machill, M., & Beiler, M. (2009). The importance of the Internet for journalistic research. *Journalism Studies*, *10*(2), 178–203. doi:10.1080/14616700802337768

Mason, J. (1996). *Qualitative researching*. London, UK: Sage.

Mason, J. (2006). Mixing methods in a qualitatively driven way. *Qualitative Research*, *6*(1), 9–25. doi:10.1177/1468794106058866

Miller, D., & Slater, D. (2000). *The Internet: An ethnographic approach*. Oxford, UK: Berg.

Moyo, D. (2007). Alternative media, diasporas and the mediation of the Zimbabwe crisis. *Ecquid Novi: African Journalism Studies*, *28*(1&2), 81–105. doi:10.3368/ajs.28.1-2.81

Moyo, D. (2009). Citizen journalism and the parallel market of information in Zimbabwe's 2008 elections. *Journalism Studies*, *10*(4), 551–567. doi:10.1080/14616700902797291

Moyo, J. (2004, November). *The role of the media in politics*. Paper presented at the Department of Journalism and Media Studies, National University of Science and Technology, Bulawayo, Zimbabwe.

Moyo, L. (2009). Repression, propaganda, and digital resistance: New media and democracy in Zimbabwe. In Mudhai, O. F., Tettey, W. J., & Banda, F. (Eds.), *African media and the digital public sphere* (pp. 57–71). New York, NY: Palgrave Macmillan.

Mudhai, F. (2004). Researching the impact of ICTs as change catalysts in Africa. *Ecquid Novi: South African Journal for Journalism Research*, *25*(2), 313–335. doi:10.3368/ajs.25.2.313

Mudhai, F. O., & Nyabuga, G. (2001, September). *The Internet: Triumphs and trials for journalism in Kenya*. Paper presented at the Annual Highway Africa Conference, Rhodes University. Retrieved June 24, 2004 from http://www.highwayafrica.org.za/presentations/55.doc

Nyamnjoh, F. (2005). *Africa's media: Democracy and the politics of belonging*. London, UK: Zed Books.

Paterson, C. (2008). Introduction: Why ethnography? In C. Paterson & D. Domingo. *Making online news: An ethnography of new media production* (pp. 1-11). New York, NY: Peter Lang.

Pavlik, J. (2000). The impact of technology on journalism. *Journalism Studies*, *1*(2), 229–237. doi:10.1080/14616700050028226

Pavlik, J. V. (2001). *Journalism and new media*. New York, NY: Columbia University Press.

Pavlik, J. V. (2005). Running the technological gauntlet: Journalism and new media. In De Burgh, H. (Ed.), *Making journalists* (pp. 245–263). London, UK: Routledge.

Ranger, T. (2005). The rise of patriotic journalism in Zimbabwe and its possible implications. *Westminister Papers in Communication and Culture*, 8-17.

Schudson, M. (2000). The sociology of news revisited. In Curran, J., & Gurevitch, M. (Eds.), *Mass media and society* (pp. 175–200). London, UK: Arnold.

Slack, J., & Fejes, F. (1987). *The ideology of the Information Age*. New Jersey: Alex Publishing Corporation.

van Dijk, J. G. M. (2006). Digital divide research, achievements and shortcomings. *Poetics*, *34*(4-5), 221–235. doi:10.1016/j.poetic.2006.05.004

ADDITIONAL READING

Berger, G. (2000). Grave New World? Democratic Journalism enters the Global Twenty-First Century. *Journalism Studies*, *1*(1), 81–99. doi:10.1080/146167000361186

Gyamfi, A. (2005). Closing the Digital Divide in Sub-Saharan Africa: Meeting the Challenges of the Information Age. *Information Development*, 21(1), 22-30.

Mabweazara, H. M. (2010) 'New' technologies and journalism practice in Africa: Towards a critical sociological approach. In Hyde-Clarke, N (Ed.) *The Citizen in Communication: Re-visiting Traditional, New and Community Media Practices in South Africa* (pp.11-30). Capetown: Juta & Co.

Mabweazara, H. M. (2010). Researching the use of New Technologies (ICTs) in Zimbabwean newsrooms: a qualitative ethnographic approach. *Qualitative Research*, *10*(6), 659–677. doi:10.1177/1468794110380516

Mudhai, O. F., Tettley, W., & Banda, F. (Eds.). (2009). *African Media and the Digital Public Sphere*. New York: Pelgrave Mcmillan. doi:10.1057/9780230621756

Nyamnjoh, F. (1996). Africa and the Information Superhighway: Silent Majorities in Search of a Footpath. *Africa Media Review*, *10*(2), 1–21.

Obijiofor, L. (1996). Future Impact of New Communication Technologies: A Bibliographic Analysis. *Africa Media Review*, *10*(2), 53–71.

Obijiofor, L. (2009). Journalism in the Digital Age: The Nigerian Press Framing of the Niger Delta Conflict. *Ecquid Novi: African Journalism Studies*, *30*(2), 175–203. doi:10.3368/ajs.30.2.175

Okigbo, C. (1995). National Images in the Age of the Information Superhighway: Afrcan Perspectives. *Africa Media Review*, *9*(2), 105–121.

Zuckerman, E. (2009). Citizen Media and The Kenyan Electoral Crisis. In Allan, S., & Thorsen, E. (Eds.), *Citizen Journalism: Global Perspectives* (pp. 187–196). New York: Peter Lang.

KEY TERMS AND DEFINITIONS

'Activist Journalism'/'Vendetta Journalism': Refers to a journalism of 'self-defence' in which political reporters browse an assortment of news websites with a view to respond to stories that attack government policy

Ethical Challenges: Refers to elements of journalism practice that compromise good practice in journalism.

Interactivity: In the narrow sense used in this study, it refers to the 'dialogue' between Internet users and a particular website.

Mainstream Journalism: As used in this study, the term refers to the most popular and dominant press as opposed to the 'alternative press'.

News Access: Is used in the limited sense of the voices and viewpoints that structure and inform news discourses.

News Beats: Refers to a general reporting subject that a reporter sticks to for a lengthy period of time.

Social Networking Sites: These are websites that enable users to connect with friends in cyberspace to chat, socialise, debate, and network.

ENDNOTES

1 Since 1997 Zimbabwe experienced unprecedented socio-political and economic problems which signify the inextricable relationship between society, politics and economics. The socio-political and economic turbulence has had marked implications on the operations of the press and the general practice of journalism in the country. The political divisions between the main opposition, the Movement for Democratic Chang (MDC) and the Zimbabwe African National Union Patriotic Front (ZANU-PF) which has been the main ruling party since independence in 1980 have led to marked polarisation between the state-controlled public press and the private press (two sectors of the press that offer a vantage point from which to closely examine the social practice of mainstream journalism in Zimbabwe). While the state-controlled press functions as a means to spread government policy and quash dissenting voices, the private press gives space to dissenting voices and subjects government policy to heavy scrutiny and criticism. As Chari (2009, p. 10) puts it, "[t]he state media is unapologetic in its support for the ruling ZANU-PF government while the private press seems to have signed a pact with the opposition to 'hear no evil', 'speak no evil' and 'see no evil' regarding its affairs". The conflicting editorial thrusts characterising these two broad sectors of the press mirror the political and ideological power struggles pervading public discourse in Zimbabwe.

2 Although these newspapers are a key part of the mainstream press in Zimbabwe, it is important to highlight that the shrinkage of the private press and the generally tight control of the media environment triggered an unprecedented mushrooming of alternative online newspapers by exiled Zimbabwean journalists. These journalists resorted to the Web for organising themselves and publishing fearless political content which the government cannot stop. Their mottoes and stated objectives, such as: "to tell the real Zimbabwean story" (Zimonline.co.za); "to provide you with the most independent uncensored news from Zimbabwe" (Zimdaily.com); and "to tell the Zimbabwean story like it is without fear or favour" (Zimbabwejournalists.com) sum up their desire to provide alternative viewpoints on the Zimbabwe situation (Moyo, D 2007; Chari, 2009).

3 This understanding was clearly based on the journalists' imprecise understanding of the role of the BBC as a public broadcaster whose editorial content is influenced by government as is the case with Zimbabwe's public media.

4 At the time of doing fieldwork for the present study, many foreign news organisations deemed hostile to government policies in their reportage had been banned from practising journalism in Zimbabwe.

Section 2

Emerging Media, Language, Pop Culture, and Health Communication

Chapter 9
Textualizing the HIV/AIDS Motif in Theater-Against-AIDS Performances in Kenya

Mahiri Mwita
Princeton University, USA

ABSTRACT

This chapter studies the theatrical and cultural texts that are performed through a Theater for Development (TFD) rubric known as Magnet Theater, which uses theater-based outreaches to mobilize people in low–income communities into forums that discuss HIV/AIDS and how its problems manifest in their localities. Using examples from performances of four theater groups that operate in Nakuru and Mombasa towns of Kenya, the chapter examines how the performances textualize, thematize, and theatise the main issues in HIV/AIDS as seen through the perspective of the performers and how the targeted audience reacts to these "AIDS performances." Beyond studying the theatrical outreaches, the research for this chapter surveyed communities in which these performances have taken place to further appreciate how the motifs discerned from the theatrical outreaches compare to realizations of the AIDS problem in the communities.

BACKGROUND

Taking account of the proliferation[1] of theater-for-AIDS-education programs that have happened especially in East Africa in the last decade, it is deserving to problematize this performance rubric as not only a major emerging genre of the TFD[2] field, but also as an illustration of new media that textualizes new issues which are confronting the changing socio-cultural scene in East Africa. Recent studies in Kenya[3] show general evidence that despite reported high levels of HIV/

DOI: 10.4018/978-1-60960-591-9.ch009

AIDS awareness and increasing availability of AIDS prevention services, far fewer people are changing risky behaviours, coming forward to utilize counselling and testing services or seek anti-retroviral (ARV) treatment. The main reason behind this AIDS intransigence is cultural taboos that sanction silence about HIV/AIDS, because it is inappropriate to talk about sex, AIDS, and its related dynamics such as death.

To address this intransigence, many prevention programs have adopted theater-based interventions to mobilize members of the community to combat behaviours that perpetuate AIDS stigma, and motivate them to use prevention services. These programs have been gradually mainstreamed into a critical media of targeting cultural stereotypes that minimize the people's confidence to confront the HIV/AIDS problem. The theater-based outreaches that form the base of these programs capitalize on the cultural cognizance that performance uses the ludic licence of theatricality to create non-judgemental forums in which people can speak out about sensitive issues without the inhibition of being stigmatized by their neighbours and friends.

Working from a premise that the theater-against-AIDS-interventions are an effective tool of conscietization against HIV/AIDS stigma, this chapter examines how the theater programs conceive, perform, and discourse on the HIV/AIDS themes in their performance texts. I examine the themes presented by the dilemma-skits, issues that are raised from the theatization of the AIDS motifs during the outreaches, the text, and context of the discussions that ensue and how these relate to what the society outside the outreach space think is the face of the AIDS problem in their communities.

The texts and contextual data for the performances analysed in this chapter were collected through participant observation and memmoing of outreach performances which I participated in three consecutive summers (June-August) in 2005, 2006, and 2007. During these months, two research assistants and I embedded with four theater groups that performed a theater-against-AIDS TFD rubric which they call "Magnet Theater", recorded their performances and interviewed participants while taking part in the outreaches as members of the theater groups. Two groups, REPACTED and TEARS work in low income areas in Nakuru municipality of Rift Valley province. The other two, KWACHA Afrika and The House of Courage perform in the Kisauni/Bombolulu/Bamburi and other low-income neighbourhoods in and around Mombasa city, coast province. We attended as many of their preparation meetings as possible, rehearsed with the performers, acted in the dramas, and took part in the post-performance discussions. After three years of researching the interventions, I conducted a survey in the communities where these performances happened to establish whether the community cited the same issues that featured in the performances as the face of their AIDS problem and how the performances may have impacted their attitudes towards seeking HIV/AIDS testing and treatment. The analysis in this chapter relates the motifs in the performances to the responses from the survey. Findings on the impact of the programs in these communities are presented elsewhere[4].

WHY THEATER?

The reasons as to why theater renders itself useful to the anti-AIDS movement lies in both the role of performance in East African cultural traditions as well as the aesthetic theory behind the value of theater as a 'socio-conscietization' medium, to paraphrase Augusto Boal's term. Firstly, popular theater is traditionally an integral part of the expressive cultures of most Kenyan, and indeed African communities. Africans are not only predominantly oral in their cultural practices, but also espouse theatrical performances as a spontaneous, public activity whose value includes entertainment and education. For example, Ngugi wa Thiong'o explains the phenomenon that there is nothing like a private wedding or funeral in most "traditional" communities in Kenya, because as the Swahili proverb of '*Mtu ni Watu'* goes, every person belongs to their community (in Byam 1999). Occasions such as circumcision, wedding, and funeral ceremonies that provide the 'formal stage-space' for community theater are unique, not only because they are open to all neighbours and passers-by, but also because they experience theatrical performances as part of the festivities and are sanctioned enactments of counselling the initiates, the brides and grooms or the bereaved, as the occasion may be. In the coming pages, I quote a performance of the Luhyia[5] traditional lyric of "*Mama mbohere mara mu nda*" ("Mother, tie for me my intestines") which is usually performed during wedding ceremonies to counsel the bride to be ready to "tie their intestines" (loosely interpreted as 'tighten their belts') as they go to their new homes, if and when things become tough. In weddings among the Kuria community[6], the festivities usually include an overnight performance of music and dance in which professional performers are hired to enact praise poetry specifically meant to tell the bride and groom of the ugly realities of married life. During these performances, the audience is allowed, with or without the permission of the lead performer, to jump onto stage and perform praise-poetry or chant specific personalised messages which they may deem necessary for the bride/groom.

These enactments draw from the ludic licence that the society associate with a theatrical platform — such enactments would be inappropriate in a "regular" non-theatrical performance such as a church wedding. It is important to emphasize that such performances are not a recapitulation of the oral past of African communities. Theater, dance, and song may today be part of a proscenium theater in Nairobi, Nakuru, Mombasa, and other cities in Africa, but in the streets and villages beyond the city centre, people still celebrate formal and informal occasions through these and other myriad forms of performance. In this context, therefore, theater renders itself as a suitable medium of performing cultural issues that are taboo for open discussion. These societies sanction theatrical performances as part of the spatial platforms where people can get into ludic roles and allow themselves to be provoked to debate openly the sensitive issues with regard to HIV/AIDS that are otherwise problematic for public discussion.

The second value for a theater-based intervention is informed by theatrical aesthetic theory. Theater for Development (TFD), as espoused and practised by Augusto Boal and other TFD practitioners[7] is a theory and process of "critical conscietization" that seeks to use a "theater of the people and by the people" (Boal 1992) as a tool of mobilizing "oppressed" communities to congregate in theater spaces to act, ask and answer questions about their situations. Rubrics in the TFD genre bring in the larger concept of theatricality as representing the holistic totality of theatrical qualities that effects a reciprocal actor-audience communication in a performance space that creates a non-judgemental space to exhale and confront issues that are otherwise inhibited by customary etiquette and therefore absurd to talk about in public dispositions[8]. Relating this conceptualization to theater-against-AIDS outreaches in the communities, I posit theatricality as the platform on which TFD outreaches are performed to effect social change. While TFD provides a critical method of social conscietization based on theater as a tool for communication, theatricality provides the poetic license and the technical platform on which the community mobilization process is played. The theatrical process, in this sense, acts as a frame within which the actors and the community negotiate the cultural meanings and realities of HIV/AIDS in their social milieu in an environment secured by the un-inhibitive forum where people speak out taboos without moral repression. Looked at from a theatrical semiotic point of view, this conceptualization positions the performances as signifying meaning in a way that transcend the cultural stigmatization of HIV/AIDS that is rampant in the society.

It is also significant that the actors involved in these performances are youth. This resonates with the target audience, most of whom are young men and women, to identify with the performance, because they are more at home discussing sex with their peers than with their parents, or older members of the community, as a CARE International's report advised:

"The problems (that) parents have talking to their kids (about sex and HIV/AIDS) don't exist between the kids themselves. That is why youth make such effective educators". (Driscoll, 2002:3)

In the following sections, I analyse the texts of the theater-against-AIDS performances following the chronology of the performance outreaches -- beginning from the songs, dances, and theater games that characterize the mobilization phase, to the content of the main skit and the facilitator's intervention to provoke participation and the post-performance discourse that ensue.

'MAKING FOOLS OF OURSELVES TO ATTRACT ATTENTION': MOBILIZATION PERFORMANCES AND THE TEXTS FOR NEGOTIATING PERFORMANCE SPACE

The beginning of any outreach performance – also referred to as "the gathering phase" (see Kershaw 1992:24) - is not only spontaneous, but also unpredictable, wild and hilarious. There is no formula or any set rules of how the performance will begin. The guiding practice is that performers arrive at a selected empty space and they use all theatrical techniques at their disposal in order to attract residents and passers-by to stop and take interest in their performance. In this context, the beginning of any outreach thrives on spontaneous theatricality which at times demands for the actors to be outrageously funny, to the extent that I call "making fools of themselves to make sense and attract an audience". To achieve this, the following examples observed in performances in Nakuru (a town located about 150 kilometres west of Nairobi) and the coastal city of Mombasa, illustrate the content of the performance process during this challenging phase of mobilization.

Table 1 summarizes sample texts and contents of songs, dances, and games that were enacted for mobilization purposes in Nakuru performances, trying to retain their sequence as much as possible.

The first lyric can be viewed as a disclaimer – right from the point of entering the community's stage, the actors are declaring that they are there to make noise, and making noise need not be of a lesser value, compared to say, working in an office, a statement that easily conjures a contestation against Plato's assigning of a lesser value to literature because it is thrice removed from the "truth". This disclaimer could also be interpreted as deconstructing the notion that comic artists should not be taken seriously, simply because they just "entertain"[9]. The fact that the artists choose such music for their gathering phase performances speaks to their recognition of the need to challenge their potential audience to take theatrical performance, especially this particular performance, seriously. The re-creation of the Paul and Silas lyric (no. 2 (a) and 2 (b)) draws from the biblical story in Acts 16, 26-40 (The Holy Bible), in which Paul and Silas were Christian evangelicals who were arrested, flogged, and jailed by authorities in the city of Philippe, Ancient Greece for the crime of preaching Christianity. According to the biblical story, the evangelists prayed to God until their chains were broken and the prison gates burst open to gain them freedom. Beyond the comical value that the "corrupted' lyrics may provoke to an audience that is already familiar with the Christian narrative and its hymn, the recreated version maintains the theme of suffering and endurance, but changes the context of this motif to reflect local material conditions that speak to the life as experienced in the local community. The new lyrics, "soap and water-basin are avail-

Table 1.

Lyric sequence	Original lyrics	English translation	Context and role in the outreach
1) *Usidharau kazi ya kelele* (Never underestimate the value of noise-making)	*Usidharau kazi ya kelele x2 / Kelele ni kama kazi nyingine x2*	Never underestimate the 'work' of noise-making x2/Noise-making is a job like any other x2	Provoke/challenge passers-by to think twice before dismissing the performers as noisemakers who may not have anything to offer
2) (a) *Paulo na Sila* (Paul and Silas)	*Paulo na Sila waliombax2/Milango ya gereza ikafunguka...*	Paul and Silas prayed x2/(Until) prison doors opened x2	A Church hymn based on the biblical story of Paul and Silas
2. (b) *Paulo na Sila* (Paulo and Silas – A re-created versión)	*Taulo ndio sina ningeoga x2/ Sabuni na karaya ziko kwa bafu x2*	It's the towel that I do not have x2/Soap and basin (of water) is (already) in the bathroom	The church hymn above is recreated and "corrupted" for comic effect as well as a deeper metaphorical meaning
3. *Mama mbee mbohere* (Mother, tie for me my intestines)	***Muimbaji:****Mama mbee mbohere...* ***Kiitikio:****Mbeeeee...* ***Muimbaji:****Mama mbee mbohere* ***Kiitikio:****Mbee mukoye, mbohere mara mu'inda*	**Soloist:** Mother tie for me... **Chorus:** Give me... **Soloist:** Mother tie for me... **Chorus:** Give me a rope (so that) I can tie my intestines...	A traditional Luhyia lyric used for comical effect. Could be interpreted for deeper HIV/AIDS related message.
4. *Tujikinge jamani* (Let's protect ourselves people)	*Tujikinge jamani tujikinge x2/ Tujikinge jamani, tujikinge/ Tujikinge jamani tujikinge/Daktari wa Ukimwi amekufa*	Let's protect ourselves people, let's protect ourselves/Let's protect ourselves ladies, let's protect ourselves/ Let's protect ourselves people, let's protect ourselves/ (Because)The AIDS doctor is dead	An HIV/AIDS conscietization lyric sang in a popular local tune. It is used performed to transit the performance from mobilization/warm-up comics to the main theme.

able in the bathroom/but I cannot take a shower because I do not have a towel", conjures the image of a life of hardship in low-income residences in Kenyan towns. They use the example that in this condition, people wash themselves from a bucket of (cold) water and without the luxury of a towel. This recreated version transcend the suffering/torture motif of the biblical script and ascribe these meanings to the conditions of hardship due to poverty and deprivation in the communities that they target. The connection between the biblical and the applied context do not make sense in terms of how they relate to HIV/AIDS. However, viewed as a mobilization text, it makes sense in the context that it adapts the religious theme to reflect the problem of deprivation and poverty in the community, which may resonate with the audience. Lyric no. 3 is a popular Luhyia traditional song usually performed during wedding ceremonies. The literal translation of this song conforms to its traditional meaning. It signifies a traditional view in the Luhyia community that marriage is a difficult institution, because oftentimes, young girls were married off to husbands and homes that they hardly knew anything about, because the marriages were arranged between elders in the two families. The song uses the concept of "tying ones intestines with a rope" to suppress feelings of hunger when there is nothing to eat. This symbolizes perseverance in times of hardship. In the original Luhyia version, the bride is asking for a rope to tie her intestines, because she should be prepared to survive in her new home whatever the conditions she finds there. In the Magnet Theater version both the literal and figurative meanings go deeper than the original, in that the subject is urged to get a rope and tie their stomachs, or better, take control of their sexual organs because these are dangerous times of HIV/AIDS. These lyrics use metaphors from traditional customs to textualize the scary perception of AIDS, as opposed to slogans such as "be careful, AIDS kills", which are common in a lot of anti-AIDS literature seen in many public places in Kenya. This rendition also functions to connect with members of the Luhyia community who are familiar with the culture that is invoked in the song. The church and Luhyia performances are therefore interpretable as appealing to both the contemporary (modern religion) and customary (Luhyia dance) repertoire in order to appeal to an audience that reside in modern towns but may still have strong connections to their customary cultures.

Lyric no. 4 is a deliberate composition that targets the HIV/AIDS message of "let's be careful because AIDS has no cure". It is performed during the transition from the gathering phase to the main drama, ostensibly to enhance the focus of the performance to the HIV/AIDS message. Indeed, the song serves as the cue for the troupe to switch gears from mobilization to assuming their roles in the HIV/AIDS intervention outreach. As soon as the soloist (usually the facilitator of the day) commences this song, the members of the troupe will intensify whatever act they are doing and at the same time break from the centre of the stage and mingle with the audience. They shake hands with the audience and in the process encourage them to sing with them while urging them to come close and approach the space designated for the stage. When the song ends, members of the troupe will have withdrawn to become part of the audience while the performers take their position in the centre-stage, ready to start the main part of the day's performance. In this context, the text of the song is used as the medium that concludes the transformation of this forum from a group of villagers watching some drama, to participants in an anti-HIV/AIDS outreach.

In the outreaches in Mombasa, a larger coastal Kenyan city which merges a long tradition of Swahili poetry with modern lifestyles enriched by a large presence of local and foreign tourists, the mobilization exercise blends the theatrical media with both the traditional Swahili poetic tradition as well as modern discotheque techniques. Rather than depend on songs and dances only as in Nakuru, the fast-paced performance begins with a DJ-led disco with huge speakers and a public system that serves to draw attention to the venue. While the artists perform dances and mimic the music, the DJ organises a dance competition for the many children who are easily attracted to these forums, in which the winning dancers are rewarded with a soda or candy. As soon as there is a sizeable audience, the DJ transforms into a facilitator and ushers in an *ngojera*[10] poetic performance which may thematize the outreach. I noted that these *ngojera* performances are quite popular with participants who would often recite the verses along with the performers signifying that this is a medium that resonates with them. Table 2 is an excerpt of one of the texts performed as part of the disco-comics-*ngojera*-comics-main skit transition in most of the Mombasa outreaches.

In this performance, the poetry presented a gender debate: between women and men, who are to blame for the risky behaviours that perpetuate the spread of HIV/AIDS? The content of this poetic dialogue debates cultural stereotypes such as "men are more reckless with casual sex than women", "women are more reckless because they go to men for money, not love", and so on. The facilitator/DJ may interject during the performance to theatricalise important messages for emphasis or to involve the audience.

That this multi-media performance renders itself to the theater-against-AIDS outreach in contemporary Mombasa is instructive of how the Magnet Theater performances are both a product of spontaneity as well as blending local performance traditions into their contemporary theatrical media. Compared to Nakuru where the artists use any contemporary performance texts for their outreaches, in Mombasa the artists used disco, children performances, as well as serious poetic performances to achieve the same mobilization function. Also, while poetry is appealing in Mombasa because of the rich tradition of Swahili

Table 2. Ni nani? (Who is responsible?) – by House of Courage, Mombasa

Wanaume: *Ni nani? Ni nani lakini, aliyezusha taabu hizi* *Dhiki nyingi na mashaka, wanadamu wahangaika,* *Kote ulimwenguni* *Ni nani, ni nani lakini?* *Enhe!* *Tumewakumbuka! kizazi cha mashetani* *Ni hawa, binti sultani, kwa kujisasha na kujishaua* *Ni nani awashindaye?*	**Men:** Who is it? Who is it that created all these troubles? Problems and tribulations, humans are suffering All over the world Who is it? Who is it that causes this? Yes! We have discovered them! This bedevilled gender They are the ones, our flowering and "cosmetized" princesses Who else beats them?
Wanawake: *Ni nyie, nambari wani,* *Kwa kila mahali mwaponda mali kwa kifua* *Huku mwajitia waume, mkileta ngazitume.* *Kwamba: "Oh wake wafaa tohara..."* *Je huu ni ungwana?*	**Women:** Who else but you, you are number one! Everywhere you indulge with your money Posing as the conquering bulls You say, "Oh women need to be circumcised..." Is that your civility?
Wanaume: *Ni ungwaana, tena ulio sawia!* ... *A-a-a-h nyamazeni...* *Mbona msemeni, wake watueelezeni?* *Kila mahali wambwiii!.... wan'jaaa* *Si mabaa, si vilabuni, kote wan'jaa* *Ukimwi kuenea, shauri hawa wanawake.*	**Men:** It is civil, and just too! ... A-a-a-h! May you (women) be silent! How dare you say, what can women tell us? You are all over, everywhere you go From bars to clubs The spread of AIDS is your responsibility!
Wanawake: *Yoo jamani, hujioni, wanaolala walala* *Heri mimi vilabuni,* *Je mume wewe kiondo kipya wangojani?* *Ukiulizwa: "Eti nangoja* spring chicken*"* *Ni watoto bado, jituze!*	**Women:** Ooh my god! Don't you see yourselves? (Truly) let the sleeping sleep. Better me and the clubs, What about you men hunting for the young You say: "I seek the *spring chicken*" Shame on you! You hunters of our children!
Wanaume: *Hata hata hata!* *Simfuati mwanamtu, wanifuata wenyewe...*	**Men:** No! No! No! I never follow anybody's daughter, They follow me on their own...
Wanawake: *Tuliupata wapi? Kama si kwa nyie ma-*sugar daddy*?* *Kwanza... "money talks"... pili una gari,* *Tatu...* "Babe, everything is under control*".* *Ndivyo mwaeneza ukimwi!*	**Women:** How else did we get AIDS? If not from you sugar daddies? First (you say)... "Money talks"... Secondly... your cars, Third... "Babe, everything is under control". That's how you spread AIDS!
Wanaume: *A-a-a-a! Mwatukubaliani, na mwajua tuna ukimwi?*	**Men:** A-a-a-a! Why then do you take us, knowing we have AIDS? .....
Wote: *Wake kwa waume tubadilisheni tabia* *Tubadilisheni tabia kwani tumepotoka.* *Tubadilisheni tabia ili tupate okoka...*	**All:** Women and men, we should all change behaviour We should all change our behaviours, because we are lost We should all change if we want to survive...

(Recorded at House of Courage outreach at Kwa John, Mombasa, August 2007)

classical poetry in the culture, the use of disco and public address system may be a response to the need to compete with numerous powerful media that happens in the city. At almost every street corner, there were Islamic or Christian preaching sessions most of which used complex sound systems to attract audiences.

Generally, therefore, two narratives emerge from the texts for the gathering phase in both towns. The first is adapting traditional lyrical performances from a variety of ethnic and aesthetic backgrounds, especially Luo, Kikuyu, Kalenjin, and Luhyia in Nakuru and Swahili in Mombasa, to their theatrical purpose. Second, is to adapt and recreate both traditional and popular cultural repertoire of each of the communities to give them a new functional meaning both for mobilization purposes as well as to serve a role within the behaviour-change communication module. The remaking of the traditional Luhyia songs and popular gospel hymns into an AIDS performance illustrates this skill in the Nakuru performances. In Mombasa, the theatrical re-engagement of the *ngojera* performances into a theatrical mobilization technique for an HIV/AIDS outreach is an example of this creativity.

Transiting from the gathering phase to the central theme of the outreach, the main performances usually present a dilemma-skit that focuses on a particular HIV/AIDS theme for the outreach. An analysis of these themes in all the skits in performances in both towns brought out several motifs that recur in most outreaches. I examine these motifs in four main categories: the role of the family as a cultural institution that perpetuate stigma and how gendered interactions affected family members with regard to HIV/AIDS stigmatization and individuals ability for positive change; the variables of condom use as a preventive measure, stigmatization sanctioned by religion, and the advent of the culture of HIV/AIDS-related fatalism in undermining individual and social change in the communities.

THEATIZING THE ROLE OF THE FAMILY IN HIV/AIDS INTRANSIGENCE

The family is the central cultural unit through which an individual first experiences acceptance or rejection when they first disclose their HIV/AIDS status. Responses to the community surveys in which I asked respondents to name examples of incidences of HIV/AIDS stigmatization that they had witnessed in their communities supported what other studies on stigma have reported with regard to the role of immediate family. For instance Gwyn (2002) confirmed that most people who have been diagnosed with HIV/AIDS are first rejected by their family members (who are often the first they disclose to), even if the same relatives later learn to forgive and accept them back. Table 3 shows sample incidences that relate to family stigmatization as reported by the survey in Nakuru and Mombasa.

These incidences point to a phobia that would discourage any person from disclosing their HIV status to their families and friends for fear of rejection. The stigma equally deters people from testing to know their status, especially if this knowledge would most probably lead to rejection. This initial rejection by one's immediate family is therefore a fundamental problem not only because it is the most painful that people associated with HIV/AIDS encounter, but also for the role it plays in hindering disclosure, which perpetuates silence on the disease and also influences the way people who are living with HIV/AIDS (PLWA) continue relating with the society.

The Magnet Theater outreaches that I researched featured family-related stigma in a variety of contexts in both Nakuru and Mombasa. As I cannot analyse here all the skits that thematized the HIV/AIDS problem in the family context, I quote below a substantial excerpt of a Mombasa performance that seems to dramatize most of the issues involving the family institution. In this performance titled *Sitaki Virusi katika Nyumba Yangu*(I don't want viruses in my house) by House of Courage at Kwa John in Mombasa in August 2006, Mzee Chamos, a strict traditional man has just arrived at home with his daughter after the doctor diagnosed her as not only to be pregnant but also HIV-positive (Table 4).

Although pre-marital pregnancy for a young woman like Azile in this performance is a serious and shameful "offense" in this community, in the

Table 3.

Area/Resp.[11] *Code*	*What incidence of HIV/AIDS problem such as stigmatization or discrimination that you have witnessed in your community in the last one year?*
NAKURU	
5.3	"I have cousin who was rejected, his parent refused to take him to the University after he was diagnosed with HIV, saying that it was a waste of money, they would make him wash and do exercises saying that he was lazy"
7.34	"A guy who was positive - so his family would hire a taxi to take him to the hospital when they had a car, and would be discriminated at any family event due to his status"
7.46	"The brothers and some male relatives could not share a bed, towels, and clothes with him."
7.47	"... She was unmarried, lived with parents and was pregnant... and was not allowed to attend any family function or even visit anyone and always remained indoors"
MOMBASA	
1.13	"My brother's wife is in her second year of AIDS sickness and she (together with my brother) have been kicked out of our home, they have been isolated like (people suffering from) leprosy"
1.38	"My uncle was HIV-positive and his mum was so scared of him that whenever he used a house hold item it was thrown away"
5.8	"My sister had the virus and she got problems until her death, my dad was angry because my sister had been sent away from high school, she was so sick until she was discovered to having AIDS, my dad refused to take care of her and he sent her away from home because she had embarrassed him and brought bad omen to our family"
6.6	"My neighbour who was infected with AIDS and was kept alone in his/her room and was not allowed to share utensils with others"
7.12	"When people die from AIDS they are not washed by their families"

performance, the mother is ready to defend her daughter from her father's wrath as long as pregnancy is the only issue. However, the mother literary freezes, and pregnancy is no longer a taboo, when the father reveals that the crime is not merely being pregnant, but an HIV/AIDS infection! The advent of HIV/AIDS on this scene instantly creates a bond between the mother and the father, who were hitherto fighting along gender lines on the issue of pregnancy. They declare that '*sisi hatutaki kuishi na virusi katika hii nyumba*" (we do not want to live with viruses in this house). They decide that Azile has to move out, immediately. In this case, the crime of being HIV/AIDS-positive is stronger than any family or customary taboo like pre-marital pregnancy. The anti-AIDS stigma is therefore stronger than both gendered and general social stigma against girls, who 'shame' their family and society with pre-marital pregnancy, and the mother is the first to chase their daughter out of their home.

In typical traditional society -- say such as the Kuria[12] which I am well acquainted with -- the punishment to a girl like Azile who "shames" her family with pre-marital pregnancy could entail forcing the responsible young man to marry her at a punitive bride price. The boy's family would also be forced to appease elders with some form of punitive measure before the couple were cleansed and admitted back as regular members of the community. Even though it is safe to assume that this custom may not be in practise in present-day Mombasa, the text of this performance follows the same narrative. The purpose is to show that HIV/AIDS infection is such a serious issue that would not even be accommodated by any punitive cleansing in this community. When Mzee Chamos takes the daughter to look for the young man who is responsible for this misfortune, the mother orders them never to come back again, and the boy's parents immediately disown the affected youth. Mwalago, from a neighbouring home of Mzee Chamos' friend is the boy responsible for

Table 4.

***Baba**: (Akimwambia Mama) Mwanao namuona nje ya nyumba hapo nyuma... Atapika kabisa... Nikamwambia mwanangu: "una nini"? "oh... mi ni mgonjwa, mi ni... oh mi..." Nikafikiri ana homa. Nikamchukua, bwana nikaenda naye hapo ehe, bwana kufika naye... Naambia daktari ampime huyu mwanangu atapika kweli kweli. Aah, daktari akatoka akanieleza.... daktari akaniambia bwana mwanao hana homa, huyu mwanao ana mimba!*	**Father:** (Speaking to Mother) I saw your child outside, behind the house... Yes, she was vomiting a lot... I asked her, "My child, what's with you?" "Oh... I am sick, I am... oh... I...," I thought she had a cold; I took her to the doctor and asked the doctor to test her because she vomits a lot. Aah, the doctor came out and told me that my daughter has no cold... he said, "your child is pregnant!"
***Mama**: Basi mimba ninaona, waniita nini mwanaume unapenda kutia mtu pressure, wewe mimba tu, sasa kwani mimba ni nini?*	**Mother**: Okay, pregnant I see, why have you called me you man, you just want to subject me to unnecessary pressure! So what if she is just pregnant?
***Baba**: Mimba tu?*	**Father**: Just pregnant?
***Mama**: Mbona mimi ukanitia mimba nikamzaa..., nikamzaa yeye. Muache pia azae aah..*	**Mother**: Why! Didn't you impregnate me and I gave birth to her? Let her give birth too!
***Baba**: Basi sikiza nikuambie nyumba hii hatujaagana mambo ya mimba hapa!*	**Father**: Okay then listen, we have not had any agreements in this house about pregnancy issues.
***Mama**: Mi nakwambia ana mimba na atakaa.*	**Mother**: I am telling you that she is pregnant and she will stay!
***Baba**:Nakwambia kama simba mwenye hii nyumba... na si mimba tu, mwanayo kisha akapimwa tena. Daktari akaniambia mwanayo, mbwana ana virusi!*	**Father**: I am telling you as the lion of this house...and it is not only pregnancy, your child was then tested again. The doctor told me that your child has the virus.
***Mama**: Ati?*	**Mother**: What?
***Baba**: Oh sasa ndio wakuja chini... Mwanayo ana virusi na nyumba hii leo nataka yuaniambie yeye ndio simba ama simba ni mimi?*	**Father**: Oh, now you are coming down...your child has the virus and in this house today, I want her to tell me if she is the lion of this house or the lion is me?
***Mama**: Ati?*	**Mama**: What?
***Baba**: Mwanayo ana virusi, na virusi sivitaki hapa. Hatuwezi kuishi na mtu ana virusi kabisa. Mi nakataa hiyo...*	**Father**: Your child has the virus, and I do not want viruses here. There is no way we will be living with someone with viruses at all. I refuse!
***Mama**: Yaani ana...*	**Mother**: You mean she has....
***Baba**: Aaah*	**Father**: Aaaah
***Mama**: We Azile, hata sijui nimfanye nini huyu mtoto Azile...*	**Mother**: You Azile! I do not know what to do with this child Azile...

the pregnancy. Although Mwalago's father, Baba Mwalago, is aware that "*mwenye virusi hatengwi*" (one with the HIV virus should not be discriminated against), he cannot admit his son's responsibility in the matter, so he rejects "contaminated Azile" to inhabit his home (Table 5).

During this performance, the facilitator froze the action and sought the audience's intervention; they all sympathized with the young girl. In the post-performance discussion that followed, a very intensive discourse ensued on what exactly these youth couple need to do. A summary of the issues that the outreach discussed included: Will the young couple continue having sexual intercourse if they are going to live together, or will they abstain? If they do, will intercourse put them at worse risk of re-infection and AIDS-related sickness or not? Can using condoms help them when they are already infected? What about the baby, will it automatically be infected, or can it be protected? Should they smoke or drink? What kind of food should they eat? Table 6 contains the scene on how these issues manifested in the actual performance.

As the underlined responses from both the audience and the participants suggest, the participants stepped forward to offer their opinions as to why the couple should not be abandoned by their families. Three hours later, the participants were still discussing these suggestions, and ulti-

Table 5.

***Baba Mwalago**: Ngoja nikuulize swali, ni mimba peke yake, tosha?*	**Father Mwalago**: Let me ask you a question is it the pregnancy only or is there something else?
***Mzee Chamos**: Mimba pekee heheheee* ***Azile:**Na virusi!*	**Mzee Chamos**: Pregnancy only heeei! **Azile**: HIV too!
***Baba Mwalago**: Usinimalize, usinimalize kabisa wewe mama wa-zungumza... kuhusu virusi, wewe ndio kumbe wewe ndio wampatia mwanangu virusi sio?*	**Father Mwalago**: Don't kill me, don't kill me totally, you my daughter, you are talking about... about HIV. So...so, you are the one who infected my child with the virus?
***Azile**: Yeye ndio amenipa*	**Azile**: He is the one who infected me!
***Baba Mwalago**: Heeei*	**Father Mwalago**: Heeei!
***Mzee Chamos**: Mwanao amempa mwanangu virusi*	**Mzee Chamos**: Your child infected my child with the virus.
***Baba Mwalago**:* (Anavua shati) *sasa ngoja nikuonyeshe vile watu... hufanya. Yaani mwanao, Mwalago, ngoja atakuja. Mwalago...*	**Father Mwalago**: (*taking off his shirt*) Now let me show you what people do....you mean your child? Mwalago, just wait, he will come... Mwalago!
***Mzee Chamos**:* (Anavua kanzu) *wewe vua na mimi nivue nguo...*	**Mzee Chamos**: (Taking off his *kanzu*). You just take off your shirt, and I will too...
***Baba Mwalago**:* No! *ngoja, mama... hii mimba unasema ni ya Mwalago mwanangu?*	**Father Mwalago**: No wait, my daughter, you say that my son Mwalago is responsible for this pregnancy?
***Azile**: Ni ya Mwalago.*	**Azile**: Yes, it is Mwalago's.
***Baba Mwalago**: Halafu unasema ati ndio amekuja tena na virusi sijui nini?*	**Father Mwalago**: Then you say that he's come up again with viruses?
***Mzee Chamos**: Mwanao ememtia mimba mwanangu, mwanao amempa virusi mwanangu.*	**Mzee Chamos**: You son has impregnated my daughter. And your son has infected her.
***Baba Mwalago**:Sasa huyu aletwe hapa ili atueleze kabisa.*	**Father Mwalago**: Now bring him here to explain to us properly.
***Mzee Chamos**: Hapa ndio wakae pamoja, na virusi...*	**Mzee Chamos**: Yes! So that they can stay here together, with viruses...
***Baba Mwalago**: Hapana! hapana bwana!*	**Father Mwalago**: No! No! No way!
***Mzee Chamos**: Mimi kwangu hakukai virusi*	**Mzee Chamos**: Viruses do not stay in my house!
***Baba Mwalago**: Mwanangu hana tatizo nilisikia habari kwamba mwenye virusi huwa hatengwi, lakini hatuwatengi. Ni kwamba mwanao sitaki akae kwangu nyumbani, TOSHA!*	**Father Mwalago**: My child has no problem, I have heard that people with HIV should not be discriminated against, and we are discriminating against them. It is just that I don't want your child to stay in my house, THAT'S ALL!
***Mzee Chamos**: Na mimi atakaa HAPA HAPA!*	**Mzee Chamos**: As for me, HE WILL STAY HERE!
***Baba Mwalago**: HAKAI HAPA!*	**Father Mwalago**: HE WILL NOT STAY HERE!
***Mzee Chamos**: ATAKAA HAPA HAPA!*	**Mzee Chamos**: HE WILL STAY HERE!
***Baba Mwalago**: Hatuwezi kuwashikanisha hivi!*	**Father Mwalago**: We cannot join them like this!
***Mzee Chamos**: ATAKAA HAPA HAPA!*	**Mzee Chamos**: HE WILL STAY HERE!

mately requested the parents to take back their youth because they were the only ones who should understand their predicament.

There is no gainsaying the importance of this dialogue on family rejection in the HIV/AIDS problematic, as this ostracization may in reality lead to further socio-economic consequences for the affected individuals. These include disinheritance from the family estate, which was especially reported to be rampant where the affected were female members of the family. A respondent in Nakuru (quoted above as Nakuru 5.3) shows that family rejection may include denying the affected member basic needs such as paying school fees for young members attending school. This means that people who have experienced rejection by their

Table 6.

Chamos: *Atakaa hapa, nakuambia atakaa hapa hapa!*	**Chamos**: She will stay here; I tell you she will stay here!
Mshirikishaji: (anawakatisha) *Basi sasa huyu mtoto afukuzwe ama abaki? Afukuzwe ama abaki huyu mtoto?*	**Facilitator**: *(cuts in)* Ok … now… Should this girl be chased away from home or not? Should she be chased away or not?
Hadhira:*Abaki!*	**Audience:**She should stay!
Mshirikishaji: *Mama unasema nini?….* Okay… *haya ndugu yangu njoo hapa kidogo njoo hapa. Habari ya leo bwana mimi naitwa Siswa sijui wewe waitwa nani?*	**Facilitator**: What do you say ma'm? …. Okay … here, my brother, come closer… how are you today. My name is Siswa, what about you?
Oigo:*Oigo….*	**Oigo:** Oigo…
Mshirikishaji: *Oigo, sasa wewe umesema mtoto abaki?*	**Facilitator**: Oigo, so you say that the girl should stay?
Oigo:*Afadhali abaki nyumbani ni apate chakula ya kumfaa kwa sababu amekua na virusi, sasa akienda mahali popote si atafariki mapema?…*	**Oigo:**It's better if she stays at home and be able to eat appropriate food since she has HIV, if she is to go anywhere, she may end up dying soon…

families are likely to be intransigent in their coping and dealing with behaviour changes because their sense of security and belonging is destroyed and therefore, have no interest for positive progress including use of HIV/AIDS treatment services. The research also reported many cases where family members are taken to hospital/health care institutions and abandoned there with no money to pay for their health care. In a society without any public health care program, and where poverty is the other face of the AIDS problematic, this is among the worst stigmatization that an individual can face, because it literally means that they are abandoned to die. This is especially a critical dynamic, because traditionally, caring for the sick, especially those who suffer from illnesses that may expose them to stigmatization by the wider society, is done by family members, at the family level. With the advent of HIV/AIDS, Respondent 7.44 of Mombasa asserted that *"watu wakifa na ukimwi hawaoshwi na familia zao"* (those who die of HIV/AIDS cannot even be washed by their families). This adds another level to the severity of stigmatization because in most African cultures, cleaning the dead and burying them honourably is one of the most important symbols of "final respects" that a dead person is entitled to from their families. In this context therefore, this research suggests that HIV/AIDS stigma compounds with the existence of unequipped health care institutions in helping the family institution to abdicate a traditional role -- by replacing "home" with the health care facility as the place that takes over the responsibility of caring for the sick and abandoned.

THE GENDER MOTIF IN PERFORMING INTRANSIGENCE IN THE FAMILY THEME

Because the gender factor and the dynamics of wife-husband relationship came into play in significant ways in the family and other themes that emerged in this study, the motif calls for a deeper examination of its variables beyond the 'family' theme. The *ngojera* poem which I quoted as used for mobilization purposes in the Mombasa performance shows how the gender theme easily emerges in these debates. The *ngojera* poetry dramatized both sides of the debate of whether women are to blame more than men in spreading the HIV/AIDS disease. In the poem, the two sides conveniently came up with a resolution - they agreed that both genders are responsible for the AIDS problem and need not blame each other but change risky behaviours from both sides. However, other texts in these performances did not present such agreeable resolutions. For instance, in the performance on the infected youthful couples whose parents wanted to disown, a participant came up with a question that illustrates the complex cultural dynamics that inform the gender-based problem in the HIV/AIDS debate (see Table 7).

From this exchange, it is clear that Participant 1 is unwilling to change his 'belief' that some women are immune to HIV and can infect and kill men without themselves dying. I should mention that part of this notion may be attributed to newspaper reports in Kenya on on-going research

Table 7.

***Mshiriki 1**: Swali langu.. nafikiri kama si swali ni maon...*	**Participant 1**: My question, maybe not really a question, but a comment...
***Mshirikishaji**: Haya ni sawasawa...*	**Facilitator:** It is okay
***Mshiriki 1**:* Okay *nafikiri kuna taswishi ambayo watu wengi wako nayo hata mimi pia inaniuma sana kwa kichwa yangu, swali lenyewe ama maoni ni kwamba nafikiri kuna mke na mume... pengine mwanamke niseme ana virusi... Sasa kila akilala na mwanaume... - na hapa tumehojiana ama katika mjadala tumezungumza ya kwamba unapogawa zaidi ndipo unapata zaidi... -sasa yeye anagawa lakini kufa hafi, ushanielewa? Sasa sijui kama ako na mienendo gani ndani yake ama ako na nini zaidi kulingana na wanawake na wanaume kwa sababu tukiangalia idadi pengine huyo mwanamke anaigawa lakini kufa hafi yeye ndiye anapeana tu wanaume wanakwenda wakienda kwani huwa ako na nini ndani yake?*	**Participant 1**: I think many people have a problem with some issue; it is also disturbing to me too. The question or comment rather is that in the case of men and women... there may be a woman, who, let me say has viruses. Now, each time that she sleeps with a man - as we have discussed here that the more one has intercourse while infected is the more viruses they infect and get infected with HIV - now this woman infects men, the men die but she doesn't die, do you know what I am saying? Now I wonder what mechanisms does the woman possess in gender terms because if we look at the numbers it seems that the men that she infects die but she doesn't... what does she have as woman that protects her? What does she have in her?
***Mshirikishaji**: Sote tumepata swali na sasa majibu... (kwa mshiriki mmoja wa kike) Mama...*	**Facilitator**: We all understand the question and now it is time for answers *(to a female participant)* ma'm...
***Mshiriki 2**:Ningependa kumjibu huyu*	**Participant 2**: I would like to answer this man
***Mshirikishaji**: Haya mjibu*	**Facilitator**: Okay, answer him
***Mshiriki 2**:Umeuliza jinsi mtu anaweza kuwa anasambaza ule ukimwi sana kwa mda mrefu na hafi. Wacha nikuambie, nilikuelezee vizuri ... vile unazidi kugawa ile virusi ndio* the more *unagain ile virusi tena. Kwa hivyo huwezi kaa* more than three years, *unaweza kaa kwa mda mchache pengine uwe unatumia dawa ile ya kuzuia kidogo tu, lakini kama ni bure hukai more than* three *years unaweza kaa kama mwaka mmoja halafu wa pili wafa na wale umezidi kugawa tayari watakuwa wamegain na wakizidi kugawa vile vile pia hawakai kwa mda mrefu. Basi hivyo kuna tofauti lakini ile ya kusema kuna mtu aweza kuwa anasambaza na hafi hio ni uongo.*	**Participant 2**: You are asking how a person may be spreading HIV/AIDS to many for a long period yet she does not die. Let me explain it to you well, the more one continues spreading the virus to others, the more viruses she acquires in her body. You cannot live for more than three years. You may live for only a short period, unless you are using preventive medicine, but if you are not on medicine you cannot live for more than three years... same with those infected So there may be some difference but it is not true to say that that there is a person who can continue to infect others without dying.
***Mshiriki 1**: Dada kuna kisa ama mkasa fulani nafikiri kama mimi nimeona mkasa ama nimeelezewa mkasa kwamba fulani anao na anagawanya, sasa wale wanagawiwa wanakufa na idadi inaongezeka, sasa mimi nataka kujua ana damu gani yeye kuwa anayo na hafi yeye ndio anaua wenzake?*	**Participant 1**: Sister, there is a certain incidence that I know of or probably may have heard of one who was infected, continued infecting many others she still remained alive as her victims died, I want to know what type of blood she has that she does not die but kills others.
***Mshirikishaji**: Sasa ndugu yangu ana....*	**Facilitator**: Now my brother...
***Mshiriki 1**: Pengine kuna damu fulani yaani mtu akipata ingawa na yeye hafi labda kuna hiyo* blood group...?	**Participant 1**: Maybe there is some type of blood that if one has she becomes immune to HIV, maybe there may be such kind of a blood group?...
***Mshirikishaji**: Sawasawa...*	**Facilitator**: Okay. Okay....
***Mshiriki 1**: Tunataka tujue maanake watu wengi wanajua yule anayo anauwa wenzake lakini yeye hafi, huwa ako na siri gani...?*	**Participant 1**: We want to know because many people know of someone who kills others yet does not die, what secret does she have?

in Nairobi with female prostitutes who are said to be resistant to infection even when they are exposed to HIV multiple times[13]. Regardless of these reports however, this exchange points to a notion in this community that "it is the women who kill men with AIDS" (note that there may be opposite notions that could feature in such a discussion). This notion is an indication that although people associated with HIV/AIDS are generally stigmatized, females with HIV/AIDS may be perceived to be evil-minded and can decide to infect and kill men without themselves dying. This discussion took a long time, and indeed ended without a resolution. This context also raises the question of the extent to which the performers can be professionally competent on the technical issues of the HIV/AIDS problem, despite their training in basic HIV/AIDS education. Acknowledging this dynamic, the facilitator promised to bring a health professional to address the unresolved issues in the next outreach.

On the marriage motif, the community survey reported the incidences shown in Table 8 as illustrating the AIDS problem in the male-female marriage contexts.

According to the incidences cited above, women in all these marriages were punished by not only the HIV disease but also their husbands, the extended family and the social systems of their community. When any of the parties were discovered to be HIV-positive, the women bore the guilt of their husband's dying of HIV/AIDS whose consequences included being vilified, chased away from the marital home and ultimately losing claim to any property in the marriage. On the other hand, only one incident in over 813 responses reported that an infected man was kicked out of his home because of his HIV status, an affirmation of the rarity of such ostracization happening to the male gender in the communities. This scenario becomes critical when we consider that in communities around Nakuru and Mombasa, the institution of "marriage" is often defined to include polygamous families especially in Islamic as well as traditional African customary marriages[14]. It implies that the women belong to those homes for as long as they are presumed

Table 8.

Area/Resp. Code	*What incidence of HIV/AIDS problem such as stigmatization or discrimination that you have witnessed in your community in the last one year?*
NAKURU	
1.23	"The husband died of HIV/AIDS, then the family came and took everything from the house and left the wife with children alone because the wife refused to go home to bury the husband"
2.10	"Last year our neighbour's house help was sent out of the house at two in the morning after a quarrel by the couple of the house. It was after the wife told the husband that she came from the prenatal clinic and that she tested positive. The husband disclosed to her that he had only had an affair with her and the house help. It was then that the wife got furious and sent the house help out of the house in the middle of the night. The husband has since looked for the house help and is now living together with her while the wife has no one to turn to. It is perceived that the wife must be unfaithful since the house help was a virgin when the man took to having an affair with her"
7.49	"The in-laws took the inheritance and the husband took custody of the children and couldn't let them or her visit"
MOMBASA	
4.29	"I came to Mombasa because my husband died from AIDS and I was sent away from the family compound by his family and am not welcome there anymore"
6.45	"Up to now, I haven't gotten a husband to marry me after my fiancé died suddenly from a startling illness that they thought was AIDS"
7.53	"My neighbour's husband was sick and died, the husband's family send her away saying that she is the one who killed the husband"

"innocent" by the husband's extended family and for as long as the husband is alive. And this is despite evidence from research that has shown that these women have no say over their husband's sexual conduct. Dube and Kanyoro explain:

"In most cases women may well know that their spouses are unfaithful, but they still are not in a position of insisting on faithfulness, abstinence or protected sex in fear of losing their economic support.... With HIV/AIDS, heterosexual marriage has turned out to be one of the most deadly institutions due to its patriarchal distribution of power" (in Dube and Kanyoro, 2004:9)

Supporting this argument for the vulnerability of women as compared to men, another research by Baylies and Bujra found out that:

"... (although) both men and women are affected by HIV but women particularly so, given how gender relations configure with sexual behaviour and economic security. Gender relations not only underlie women's particular vulnerability; they also inhibit women's attempts to protect themselves and their families. If interventions around AIDS are to be effective, they must address ... such factors (that) are deep-seated and intransigent, embedded in the very power relations which define male and female roles and positions, both in intimate relations or the wider society" (Baylies and Bujra, 2000:1)

This factor is not only limited to gender relations in a marital context; the current study found similar trends amongst the youth. In a focus group with youth in both towns on this theme, it was revealed that the male youth hold equally chauvinistic perceptions. In one particular session with youth in Nakuru, I asked the young men whose responsibility it was to ensure safety in a sexual encounter. All reiterated that it is usually the responsibility of the girl to: 1) resist any advances for reckless sex from men, 2) make sure that they have a condom, because the men may not care to carry any, and, 3) make sure they "are clean in order not to scare the men who may suspect them of being "carriers" of HIV/AIDS or sexually transmitted infections". This is the challenge that women and girls face in the societies like the one quoted above where these performances take place. In such cultures, men are customarily sanctioned to engage in risky multiple sexual relationships, yet they are the ones who wield the deciding power on whether and how couples can practice protective sex, whether in marriage or casual/informal relationships. In the context of these studies and in view of the results from the current study, I discern that dealing with HIV/AIDS intransigence in contexts that involve the gender phenomenon requires a deeper understanding of the role of female vulnerability in any intervention, which I unfortunately did not witness in many of the Magnet Theater outreaches in Nakuru or Mombasa.

"Avae Ama Asivae": Thematizing the Condom-Use Debate

The third most prominent motif in these outreaches is the campaign to use condoms, as the primary preventative measure of HIV infection through sexual intercourse. Although the community surveys did not report responses that mentioned intransigence to condom-use as their problem with AIDS, the response *"siku hizi watu wengi wanatumia kondom"* (nowadays many people use condoms), was the most common expression of HIV/AIDS behaviour-change that I received in the study. On the other hand, promoting correct use of condoms was a major theme which dominated not only the Magnet Theater outreaches, but also most of behaviour-change communication programs in Kenya[15]. As such, by the time of this research, condom-use campaign was one of the strategic objectives of the IMPACT[16] project that supported the theater programs. Condoms were also part of the materials that were distributed during the outreaches. The theme therefore dominated the performances in the outreaches, the post-performance discussions, the interviews, and the focus group discussions with demonstrations on the correct use of the male and female condom becoming one of the most applauded sessions of the outreaches.

Perhaps, owing to this prominence of the condom theme, the phrase: *"Avae au asivae?"* (Should he put it on (use a condom[17]) or not?), was a very popular slogan in youth jargon amongst the performers and participants, including the youth that I met outside the outreaches. In most of these contexts, the slogan was used as a metaphor that prodded the targeted individual to remind him or her of the dilemma they were likely to face if they were considering sexual intercourse. The slogan emanated from the performances as the question that the facilitator asked at the moment of dilemma in scenes that dramatized the why/ whether characters should use condoms or not. The phrase was so popular that audiences in frequented venues knew exactly when the facilitator was going to freeze the drama and ask the magic question. At that moment, the audience would chorus the question and different opinions quickly shot up as different sections of the audience answered either for or against the debate.

In the following scene from a performance on condom-use titled *"Avae au Asivae?"*(Should he put on a condom or not?) at Bondeni in Nakuru, the main character is Katapue, a long-distance truck driver who is being persuaded by his friends against using condoms, despite the fact that he has more than one sexual partner that he visits with in the various stations that his job takes him. It should be mentioned that the selection of Katapue's character as a truck driver is contextualized

Table 9.

Katapue: *Yaa… Leo iko mi naenda kusheherekea… Ndio mimi anaenda huko. Lakini iko kitu kimoja. Hiyo miezi 'tatu ni mingi sana. 'pana tumia, hapana enda kutumia…*	**Katapue:** Yaah..., today I go to celebrate… Yes, I am going to her… but there is one thing. Those three months (that I have not seen her) are too long a period. (so) I will not use 'em, not me...
Karapuna: By the way *Katapue, mimi nakubaliana na wewe.*	**Karapuna:** By the way Katapue, I agree with you.
Katwanga: *'Situmii hiyo!*	**Katwanga:** I cannot use that!
Karapuna: *Nakubaliana na wewe… usitumie hiyo kitu…*	**Karapuna**: I agree with you…don't use that thing…
Katapue: *Hiyo kitu ni mbaya sana… kwanza, anapunguza nini… nini… oroto… hehehe, anapunguza… Hapana kuwa tamu... he he he.*	**Katapue:** That thing is very bad…first, it reduces what…what… umh…*oroto*… ha ha ha….it reduces…the sweetness….ha ha ha.
Karapuna: *Hiyo ni makosa ya kwanza…*	**Karapuna:** That is mistake number one…
Katwanga: *Na ya pili… inapunguza* speed…	**Katwanga:** And the second…it reduces speed…
Karapuna: Speed *ni makosa ya pili… na ya tatu… ukizoea hiyo kitu saana… Unaona mtu akizeeka mpaka anatembea hivi.* (Anaashiria mtu mkongwe anavyotembea kwa shida)	**Karapuna:** Speed is the second mistake…and the third…if you get used to that thing so much…you see how some people walk like this when they get old. *(Demonstrates an old person walking with difficulty)*
Karapuna/Katwanga*: Utaanza kutembea hivyo.*	**Karapuna/Katwanga:** You will start walking like that.
Katapue: *'aki a'mungu subuhana wa taala sitatumia hiyo, leo… leo hatumii! Nimekasirika, nimekasirika hatumii leo!*	**Katapue:** Truth to God, *subuhana wa taala,* I won't use it today… today, I won't use it! I am angry, I am angry, I won't use it today!
Katwanga: *Kitu kingine, kitu kingine… unaona hawa watoto wa majirani, watoto wazuri-wazuri, wewe hapana taka watoto kama hao? Hiyo kitu wanaweka kwa hayo makaratasi si nzuri* ***Katapue:*** *Haki ndio maana mi napenda nyinyi… We ulitumia siku moja?*	**Katwanga:** Another thing, another thing… Do you see these neighbour's kids, cute, beautiful kids. Don't you want such kids of your own? The things they put in those plastics is not good at all… **Katapue:** Truly, this is the reason I like you guys… By the way, did you use it one day?
Karapuna*: Siku moja nilitumia nikajikuna mpaka ikatoboka!*	**Karapuna**: I used it one day; I scratched and scratched myself until it was torn!

in the general stereotyping of long distance truck drivers as a group that are vulnerable to reckless sex because of the nature of their work. In the following scene, his friends, Karapuna and Katwanga, cite numerous negative stereotypes that stigmatize condoms (see Table 9).

This setting assumes that these partners have been using condoms, but today Katapue is determined to follow his friends' advice and stop the habit. Katapue's resolve not to use a condom in the above excerpt is what sets the stage for a showdown with his girlfriend. He has to convince her that condoms are bad and the role of the performance is to seek the community's views on condom-perception as represented by Katapue and friends. The following scene enacts the anticipated meeting between Katapue and his girlfriend Mshenee. After the initial niceties of greetings between two lovers who have not been together for three months, the events roll on to the main moment of dilemma (see Table 10).

This is the point at which the facilitator seizes the moment. The intense argument between the couple quickly climaxes into the dilemma of the performance and the hilarious response from the audience create the perfect moment for capturing the audience and thrusting them into the desired discourse. The facilitator therefore quickly seizes the moment (see Table 11).

From the above text, the major cause of intransigence to condom-use as represented by Katapue and friends is the perception that condoms are uncomfortable, uncouth, and one cannot enjoy sex with them. The youth joked about this perception as "... it takes away the sweetness/reduces speed/it's like eating a sweet without removing the wrapping paper/... like eating a banana without peeling it". These texts also portray elements of a macho notion that condoms hasten aging, and may also cause inability to walk upright in old age. Also in the underlined texts above, Katwanga and Participant 2 alluded to the role of condoms in the family planning campaign in Kenya to give them the negative attribute of causing infertility:

Katwanga: *Another thing, another thing... Do you see these neighbours' kids, cute, beautiful kids? Don't you want such kids of your own? The things they put in those plastics is not good at all...*

Table 10.

Mshenee:*Jitayarishe wewe... vaa*	**Mshenee:** Get ready…put it on…put it on
Katapue:*Sivai hiyo!*	**Katapue:** I will not put that (thing) on!
Mshenee:*Nini mbaya sasa... aah haukubeba...)* (Anatoa pakti ya kondomu mfukoni na kuanza kuchana kuifungua)	**Mshenee:** What's the problem now? Aah! Didn't you carry one? *(She takes out a condom pack from her pocket and tries to open it)*
Katapue: (anajaribu kumnyanganya) *Ngoja kidogo...*	**Katapue:** (*tries to grab it*) Wait a minute…
Mshenee:*Acha nikuvalishe... unajua nachemka...*	**Mshenee:** Let me help you put it on …you know I am boiling up!
Katapue:*A-a mimi sivai hiyo kitu!*	**Katapue:** No! I won't put on that thing!
Mshenee:*Kwa nini sasa? Nini mbaya na wewe?*	**Mshenee:** What's wrong with you now?
Katapue:*Hiyo kitu iko na madudu ndani!*	**Katapue:** That thing has viruses in it!
Mshenee:*Sasa unataka tufanyeje? Lazima tuvae...*	**Mshenee:** Now what do you want us to do? We must put it on…
Katapue:*Sivai mimi hiyo kitu...*	**Katapue:** I won't put on that thing…
Mshenee:*Aaa, acha bwana... vaa...*	**Mshenee:** Aah, stop it…put it on…
Katapue:*Hata, sivai...* (Vicheko na shangilio kutoka kwa hadhira, na mshirikishi anaingia jukwaani kukatisha mchezo)	**Katapue:** Not at all, I won't put it on… *(Hilarious applause from audience... and the facilitator steps in to interrupt the action)*

Table 11.

Facilitator:*Avae ama asivae?*	**Facilitator:** Should he put it on, or not?
Hadhira:*Avae!*	**Audience:** Put it on!
Facilitator:*Avae au asivae!*	**Facilitator:** Should he put it on, or not?
Hadhira:*Avae!*	**Audience:** Put it on!
Facilitator: *Wangapi wanasema avae?* *... Wangapi wanasema asivae?*	**Facilitator:** How many say he SHOULD put it on? .... How many say he should NOT put it on?
Facilitator:*Asivae?*	**Facilitator:** He shouldn't put it on?
Mshiriki 1:*Asivae-- twende mambo kwa mambo*	**Participant 1:** Don't put it on - let's "go for real".
Facilitator:*Haya nani anasema yule asivae – wewe unasema avae... Haya kuja, kuja tu usiogope. Mimi naitwa Facilitator wewe unaitwa nani*	**Facilitator:** Okay, who says that he should not put it on? ... You? Ok, you say he should... Okay come, just come don't be afraid... My name is Facilitator what is yours?
Mshiriki 1:*Mato*	**Participant 2**: Mato
Facilitator:*Mato,... wewe unasemaje Avae?...* (Kwa Mshiriki mwingine) *Na wewe unasema nini? Asivae?* (Mshiriki 1 anaingia jukwaani)	**Facilitator:** Mato... what do you say? He puts it on? *(To another participant)* And what do you say? He doesn't use it *(Participant 1 comes on to stage)*
Facilitator: *Wewe unasema asivae unaitwa nani*	**Facilitator:** You who says he shouldn't use it, what's your name.
Mshiriki 2:*Julias*	**Participant 1:** Julius
Facilitator:*Julias. Sawa. Unasema asivae? Sawa... Nani mwingine anasema asivae?* (Mama anayebeba mtoto mdogo anaingia jukwaani)	**Facilitator:** Julius, okay, you don't want him to use it... Okay... Who else says he shouldn't use it? *(A woman holding a baby comes forward)*
Facilitator: *Unaitwa nani Ma'am*	**Facilitator:** What is your name ma'am
Mshiriki 3:*Ninaitwa* Mum	**Participant 3**: My name is Mum
Facilitator:*...*Mum *anasema asivae...*	**Facilitator:** ....Mum says he shouldn't use it...
Facilitator:*Sawa... sawa...Hapa kunao (watu) wawili wanasema avae. Na hawa wawili wanasema asivae.* (Anamleta Katapue) *Sawa, mnaona huyu... huyu jamaa ni beste wenu. Na mnasema asivae sasa nyinyi wawili ninataka m'm-*convince *kwamba asi-vae...*	**Facilitator:** Okay...okay...here we have two people who wants him to use it, and these two don't want him to use it (Positions Katapue in their midst) Okay, do you see this guy... this guy is your friend, and you don't want him to use the condom... now I want you two to convince him not to put it on?
Mshiriki 2:*Haya, ndugu... Katapue... unajua wewe ukivaa hau-tapata watoto...*	**Participant 2:**Okay brother...Katapue...you know if you put it on you will not be able to get children...

Participant 2: *Okay brother...Katapue...you know if you put it on you will not be able to get children...*

These perspectives seem to be emanating from local stereotypes of condoms as unreliable and a western symbol of sexual immorality. This perception seems to allude to the 'condoms for family planning', campaigns of the 1970's and 80's which preceded the HIV/AIDS campaign. These campaigns marketed condoms as symbols of contraception -- against unwanted pregnancies; as well as for those who were engaging in unsafe sex. This campaign fuelled a backlash from African traditionalists who quickly linked with religious values that opposed condoms characterising them as promoting sexual immorality, and Western disrespect of the African traditional moral values behind sexuality. Therefore, from these perceptions, condoms came to be associated with attitudes that people who use condoms are most likely to be immoral – prostitutes, the defiant, unfaithful, and promiscuous individuals. Indeed, there were several occasions where participants came up with strong opinions that they did not trust condoms because condoms and AIDS are Western vices that America has brought to Africa to control African sexuality. Of great interest in such cases, is the fact that such opinions were applauded by the audience, an indicator that the views resonated with the participants.

Another dimension in this motif is the extent to which this association of condoms with immorality hinder behaviour-change especially because of the phobia that anybody buying condoms is likely to be engaged in immoral sex. A social worker with CARE international in Kisumu summed up this phobia in one workshop:

"I went into a shop to buy a condom once, just to see what would happen. All business stopped. Suddenly, I was the only one being served. I was quite open about it, because I wanted to see what the reaction would be, and everyone in that shop assumed I was a prostitute" (Driscoll, 2002:2).

Dramatizing the effect of this phobia on people's resistance to use condoms, a KWACHA Afrika performance in Mombasa shows how older people are really traumatised by the idea of just asking for a condom in a local shop. In the following performance titled *Nataka Kununua Kondom* (I want to Buy a condom), Kokoro has a date with a married woman whom he has been seducing for a while. He has borrowed 2,000 shillings for lodging and entertainment. Because he also needs to look stylish, he goes to his friend Habib, to borrow his new pair of clothing to impress his date. To his chagrin, Habib insists that he will only lend him the clothing if Kokoro promises to use a condom. Kokoro has never bought a condom, believes it is not meant for 'responsible adults' like himself and does not imagine how he can even utter such offensive words in a public place like a shop (see Table 12).

Kokoro's abomination of condoms is typical of the role of customary values in stigmatizing condoms. Despite his failings such as being involved in an adulterous relationship with another person's wife, having to borrow money and clothes to impress and entertain his date, these are trivial failing compared to him, "... a grown up adult... being noticed asking about such stupidity as condoms in a shop". His culture treats promiscuity as less repulsive than buying a condom, or else that is what men of his generation think. Apart from the strong urging from his more exposed friends, the drama shows that Kokoro has to witness another (young) couple buying condoms for him to take the initial and most important step of buying one. He has yet to go through the trauma of using it, which other texts in the performances also showed to be another challenge. All in all, the performance shows that knowledge about condoms was not sufficient for individual behaviour-change. People need social change-- they want to feel part of, and witness incidences of changed behaviour from others in their community, which in turn becomes an incentive and support mechanism for individual change.

Focus group discussions on this theme affirmed resistance to condoms, and men were blamed to be the most averse to the practice, compared to women. In one group discussion in Nakuru with women who were formerly in commercial sex, all participants had a story to tell on encountering men who were willing to pay more money for sex without condoms. In the same context, outreach participants and female interviewees cautioned that women should always insist on having lights on during sexual intercourse because men who did not like condoms took advantage of the darkness to lie about having a condom on, or some could put one on and then tear it during intercourse. The condom-use demonstrations therefore urged women to keep lights on during intercourse and to ascertain that the condoms were actually worn and worn appropriately for effective protection.

Table 12.

Kokoro:*Basi wacha niende mara moja, ndio hivo-*	**Kokoro:** Ok, let me go at once, ok--
Habib:*Sawa bwana. Aah Kokoro, Kokoro hebu ngoja kwanza kidogo, hebu njoo kidogo*	**Habib:** Ok *bwana*[18]. Aah Kokoro, Kokoro wait a minute.
Kokoro*: Aha sema*	**Kokoro**: Ok?
Habib:*Nilikuwa nakuambiaje ndugu yangu, tafadhali ndugu yangu…*	**Habib:** I wanted to tell you, my brother—please, brother…
Kokoro*: Unipe wasia ndugu yangu…*	**Kokoro**: Yes? Advise me, your brother…
Habib:*Nilikua nakuambiaaje, fanya yote mpaka ufanye yoote lakini usisahau moja… Sasa sikiza bwana kabla hujafanya kitendo… fanya lakini usisahau kutumia kondom sawa? Usisahau kabisa.*	**Habib:** I wanted to tell you— go do everything you want to, but do not forget one thing… Now listen to me, before you do the thing… Do it but don't forget to use a condom, ok? Don't forget at all!
Kokoro*: Unasema? Unasema nini wewe sasa unataka kunitusi?*	**Kokoro**: What did you say? You now want to insult me?
Habib:*Ahaa sijakutusi*	**Habib:** Ahaa I have not insulted you.
Kokoro*: Ngoja umezungumza yote namesikia lakini sasa wataka kunitusi, mpumbavu wewe*	**Kokoro**: Wait a minute. I have listened to all you have to say, but now you're insulting me, you stupid fellow.
Habib:*Sikiza Kokoro, bwana sikiza--*	**Habib:** Listen Kokoro, please listen--
Kokoro*: Sisikizi, sisikizi sisikliizi hata kidogo sisikilizi kabisa huwezi kuniambia mimi nitumie mambo fulani ambayo mimi nime gharamika pesa nyingi sana.*	**Kokoro**: I wont listen, wont listen to you any more. You cannot tell me to do such nonsense in an issue that has expensed me to this extent.
Habib:*Sasa ndugu yangu mimi nakulilia maisha yako… ndugu yangu...*	**Habib:** Now my brother, i am just concerned about your life...
Kokoro*: Maisha yangu yaache hivi!*	**Kokoro**: Leave my life the way it is!
Habib:*Yaani hujui huyu mama anatembea na watu gani*	**Habib:** I mean, you do not not who else this woman relates with.
Kokoro*: Ahaa! Najua ni watu wawili tu, najua bwana yake alikuwa airport akafutwa kazi*	**Kokoro**: Ahaa! I know that she has only two people, her husband who used to work at the airport, and me.
Habib:*Sasa sikiza Kokoro. Ndugu yangu mi najali maisha yako. Kabla hujafanya--*	**Habib:** Now listen Kokoro. You are my brother and I care about your life. Before doing anything—
Kokoro*: Mimi huwezi kuniambia lolote ndugu yangu. Kaa maisha vile unataka we mimi si mtoto mdogo... Situmii bwana, situmii, ala situmii, utafanya nini?*	**Kokoro**: You cannot tell me anything my brother. You live the way you want, I am not a small child... I can't use it, I cannot. What will you do?
Habib:*Yaani waniambia nitafanya nini Kokoro. Sawa hakuna shida nyingi (grabbing the clothes from him) Ahaa leta leta bwana.*	**Habib:** Now you are daring me with what will I do Kokoro. Ok, no problem *(grabbing the clothes from him*) Ahaa give me back my clothes.
Kokoro*: Ngoja, ngoja, ngoja!* *... (Mwalago anaingia)*	**Kokoro**: wait, wait, wait! *... (enters Mwalago)*
Mwalago:*Aah! Habib, namna gani bwana, mie nakungojea tangu saa ile kule bwana.*	**Mwalago:** Aah! Habib, what is happenning *bwana*, i have been waiting for you since that time....
Habib*: Ni huyu huyu Mgiriama sijui atoka wapi huyu...*	**Habib**: It is this Mgiriama (Giriam person), I don't even know where he comes from....
Mwalago:*Bwana kuna nini hapa?*	**Mwalago:** What is happening here *Bwana*?
Kokoro*: Ngoja ndugu yangu huyu bwana ananikosea heshima nimekuja hapa kihekima. Nilikuwa na deni naye bwana nikachukua.... Naenda zangu bwana sasa huyu bwana ananiambia we ukienda kule oh-oh!*	**Kokoro**: Wait a minute my brother. This guy is disrespecting me on my honourable visit. I came here because I was indebted to him... Now on my way out he begins to tell me that "when you go there oh-oh"
Mwalago:*Si umalizane naye basi twende zetu.*	**Mwalago:** Then clear with him and we get going
Habib*: Wacha nikuambie kuna mambo yuakudanganya huyu….Amekuja hapa akaniazima nguo, nikampa nguo zangu, nikamwambia: "Nje ndugu yangu ukienda kufanya mapenzi tumia condom!"*	**Habib**: Let me tell you, there are things he is lying about. He came to borrow my new clothes. I consented and gave him the clothes, with a brotherly caution that, if you have an affair, please use a condom.
Kokoro*: Situmii bwana! Hayo ndio nazungumza naye bwana.*	**Kokoro**: I will not use it *bwana*! That's the conversation we having here.
Mwalago:*Sasa hayo ndio yameleta kelele hii?*	**Mwalago:** Now that is the source of all these arguments?
Habib*: Hataki, hataki.*	**Habib**: He does not want to, at all!
Mwalago:*Angalia Kokoro, huyu anao ukweli bwana-*	**Mwalago:** Look here Kokoro, He is right--
Kokoro*: Mimi ni mtu mzima, mimi ni mtu mzima!* *Na upumbavu huo siwezi nikautumia! Ushaona, siwezi nikapoteza wakati wangu--*	**Kokoro**: I am a grown up, I am a grown up man! And cannot engage in that nonsense. You understand? I cannot waste my time—

continued on following page

Table 12. continued

***Habib**: Sikiza bwana, tutapelekana nawe mbio kama huwezi tumia (ana-fanya kumnyangaya nguo)*	**Habib**: Listen *bwana*, you and I cannot agree on this if you don't-- *(tries to gram the clothes from Kokoro)*
***Kokoro**: Haya, haya, haya!*	**Kokoro**: Ok, ok, ok!
***Mwalago:**Bwana ni hivi. Huyu yuakwambia ukweli mtupu, siku hizi kume-haribika bwana huyu unayekwenda naye hujui ameenda na wangapi bwana!*	**Mwalago:***Bwana*, it is this way. Habib is telling you the bare truth, this days it is really dangerous, you don't know how many other people she has slept with.
***Habib**: Si hiyo ndiyo namwambia lakini sasa hanielewi.*	**Habib**: That is what I am telling him and he doesn't seem to understand.
***Mwalago:**Sasa wewe wamwamini vipi huyu Kokoro?*	**Mwalago:** Ok Kokoro, how do you trust this woman?
***Kokoro**: Namwamini kwa sababu ana masikio kama ya huyu bwana... Haya tuache hayo. Ananiambia ati mimi nitumie. Nitaipataje mahali kama hapa. Na yenyewe ni dharura.*	**Kokoro**: I trust her because she her ears resembles Habib's... Ok, let's leave that alone. He tells me to use a condom. How do I get a condom in a place like this? And as you understand this is a very urgent matter.
***Mwalago:**Huo ndio ugumu? Huo sasa ndio wampigiza kelele huyu? ... Ndio ugumu? Utazipata-- Si dukani zimo tele... shilingi kumi.*	**Mwalago:** Is that the problem? Is that why you are being stubborn? You can get condoms here. There are plenty in all these shops, for just ten shillings.
***Kokoro**: Unaniambia kwenye maduka haya yote-- upambavu huo upo?*	**Kokoro**: You tell me that all these shops sell such nonsense?
***Mwalago:**Sio upumbavu huo, si upumbavu kabisa!*	**Mwalago:** Condoms are not nonsense at all!
***Kokoro**: Ni nini? Dukani kunauzwa vitu kama sukari, mahindi mambo mengi lakini vitu kama hivo?*	**Kokoro**: What is it then? Shops sell things like sugar, corn – many things, but not such things.
***Mwalago:**Kondom ziko basi, kondom ziko hapo!*	**Mwalago:** They sell condoms too.
***Kokoro**: Wacha mchezo!*	**Kokoro**: You are joking!
***Habib**: Kondom zauzwa dukani shilingi kumi tu!*	**Habib**: We are not joking. Condoms are sold in the shops at only 10 shil-lings.
***Kokoro**: Haya basi mtu kama mimi, nitaanzaaje hapo dukani? Niende dukani--*	**Kokoro**: Ok, a person like me, how do I start at the shop?
***Mwalago:**Kwani we ukienda dukani ukitaka mkate wasemaje?*	**Mwalago:** What do you do when you go to buy bread at the shop?
***Kokoro**: Afadhali mkate nitamwaambia mwenye duka: "Nisaidie na super-loaf". Na loaf nitapewa. Sawa bwana?*	**Kokoro**: better bread. I just tell the shop keeper: "Please let me have *super-loaf*"'. And *super-loaf* I will be given.
***Habib&Mwalago:**(Kwa pamoja). NA UKIENDA DUKANI UNAMWAAM-BIA MWENYE DUKA: "NATAKA KONDOM"!!*	**Habib & Mwalago:***(In a chorus)*. AND WHEN YOU GO TO THE SHOP, TELL THE SHOPKEEPER: "I WANT A CONDOM"!
***Kokoro**: A-a-a-ah!*	**Kokoro**: A-a-a-ah!
***Habib&Mwalago:**Sasa?*	**Habib & Mwalago:** Ok?
***Kokoro**: Hayo! Hayo ni makubwa-- na hata hizo pesa ndio balaa kupata.*	**Kokoro**: That? That is tough—and where do even get that money to buy the...
***Mwalago:**We huna shilingi kumi?*	**Mwalago:** You! You don't have ten shillings?
***Kokoro**: Hapa nina shilingi elfu mbili tu za kukopa kwa ajili ya ile shughuli.*	**Kokoro**: Here I have only 2,000 shillings which I borrowed for that busi-ness.
***Mwalago:**Sawa mpatie (Habib anapmoatia Kokoro shilingi kumi)*	**Mwalago:** Ok, give him *(Habib gives Kokoro ten shillings)*
***Kokoro**: Sasa hii shilingi kumi yatosha hii?*	**Kokoro**: Now, is these ten shillings really enough?
***Habib**: Yatosha hiyo!*	**Habib**: It's enough!
***Kokoro**: Kwa heshima zenu nitajaribu*	**Kokoro**: For your sake, I will try--
***Mwalago:**Mi sitaki ujaribu bwana Kokoro, namna gani wewe?... (Igizo linabadilika na inakuwa kwenye duka. Kokoro anakaribia duka kwa uoga na wasiwasii)*	**Mwalago:** I don't want you to try Kokoro, what is wrong with you?... *(The scene changes to be outside a shop. Kokoro approaches hesitantly)*
***Mwenye Duka:**Mambo?*	**Shopkeeper:** Hi!
***Kokoro**: Salama, hili duka lako?*	**Kokoro**: Ok... is this your shop?
***Mwenye Duka:**Eeh!*	**Shopkeeper:** Eeh!
***Kokoro**: Nzuri sana mzee! ...*	**Kokoro**: Great, great....

continued on following page

Table 12. continued

Mwenye Duka:*Basi sema nikusaidie na nini bwana.*	**Shopkeeper:** Ok, how can I help you?
Kokoro*: Unauza aje ile? How much?*	**Kokoro**: How do you sell—that—that? *How much?*
Mwenye Duka:*Nini*	**Shopkeeper:** What?
Kokoro*: Ile yaaani-- ni shilingi ngapi unauza ile ya packet?*	**Kokoro**: That, I mean—how much is it—the one in a packet?
Mwenye Duka:*Hapo huwa tuauza shilingi mia*	**Shopkeeper:** We sell at one hundred shillings--
Kokoro*: A-a-a-a, ile ya nini-hii-- ile ya packet?*	**Kokoro**: No-no, A-a-ah, that, that one in a packet?
Mwenye Duka:*Sasa ngoja kijana tusifuatane maneno, bado ni asubuhi.*	**Shopkeeper:** Ok, now don't confuse me this early in the morning--
Kokoro*: A-a-a! I'm a customer, I'm a customer*	**Kokoro**: No-no-no! *I'm a customer---*
Mwenye Duka:*Sasa ukinuliza hiyo ya nini-hii, hiyo inaitwaje wewe?*	**Shopkeeper:** Now, if you ask me – you don't seem to know what you are asking.
Kokoro*: Sasa sikiza bwana nilikuwa nataka ile yaani—ni-ni, aaah, kama uko na ile ya nini, inaitwa co- co- co- co…* *(Baada ya kusumbuana kwa mda wanaingia wateja wengine – mvulana na msichana. Mwenye Duka anawakabili kuwahudumia)*	**Kokoro**: Listen *bwana*, I just wanted that—I mean aaah; I mean if you have the one that... the one called co-co-co... *(a young male and female couple enter the shop and the shop-keeper turns to attend them)*
Mwenye Duka:*Habari zenu.*	**Shopkeeper:** How are you?
Vijana:*Nzuri mzee.*	**Young man & woman:** Fine sir.
Msichana:*Tupe kondom tukachemshe roho.*	**Young woman:** Give us condom...
Mwenye Duka:*Haya, leteni pesa basi.*	**Shopkeeper:** Ok, bring the money.
Mvulana:*Shika, pesa hizi.*	**Young man:** Here, here is our money.
Mwenye Duka:*(Anampa mvulana) Shika basi.*	**Shopkeeper:***(Gives them the condom)* Ok, here you go.
Mvulana:*Aaah, sawasawa, kwaheri.*	**Young man:** Ok, great, bye.
Mwenye Duka:*Nafikiri wajua matumuzi bwana.* *Mvulana: Matumizi nayajua bwana, sawa...* ...	**Shopkeeper:** bye, I think you know how to use it appropriately.... ...
Kokoro*: (Anamkabili Mwenye Duka kwa hamasa) Wewe, nipe kama ile, lete kama hiyo, ee, hiyohiyo!*	**Kokoro**: *(Confronts the Shopkeeper in a confused hurry)* You, give me like that one, the same one you gave them!
Mwenye Duka:*Pesa?*	**Shopkeeper:** Ok, give me your money.
Kokoro*: Pesa, pesa. Mp--... lete haraka kumbe unayo na huniambii we--.*	**Kokoro**: Money, money! Bring the thing quickly – so you knew what I wanted....

PERFORMING INTRANSIGENCE IN THE RELIGIOUS INSTITUTIONS

Another important component that plays into the dynamics of HIV/AIDS and immoral sexual behaviour is religion. The texts from this study indicated that association of HIV/AIDS with defiant sexual behaviour was one of the worst tragedies of the HIV/AIDS narrative. While HIV/AIDS was initially associated with, "homosexual men" in the United States and other western cultures, in Africa HIV/AIDS was generally narrated as spreading through reckless and immoral sexual behaviour. Indeed, it was variously known as a disease of the prostitutes (meant to be women) and people who were regarded to be defiant and indecent.

Table 13.

Pasta Mdogo*: Pastor anasema kanisa hii haibagui. Inapenda watu wote ambao ni wazima na wale ambao ni wagonjwa... na anasema* (Pasta Mdogo anasitasita kwa wasiwasi) *watu wa Ukimwi ni watu wa Mungu... wanahitaji kuombewa.... enhe enhe... hayo mengine si mnajua....?*	**Pasta Mdogo**: Pastor says that this church does not discriminate. (This church) loves all the people who are not sick and those who are sick... he says (*hestation*) people who have AIDS are god's people... they need prayers.... enhe enhe... the rest you people know what he means...?

Table 14.

***Mshirikishaji**: Je ni sawa kuwataka watu wenye ukimwi kuji-tokeza, kujitokeza kuombewa?(*Kimya...) *Kwa nini mtu mwenye ukimwi alazimishwe kusema hali yake hadharani?*	**Facilitator:** I ask, is it right for the pastor to ask for those who HIV-positive to stand up in church...to stand up to be prayed for? *(Silence)*Why should a PLWA be forced to declare their status in public gathering?
***Mshiriki 1**:Itawasaidia hao wengine...*	**Participant 1**: It will help the other worshippers...
***Mshiriki 2**:Pastor ni mtu wa Mungu kwa hivyo hawezi kufukuza mtu wa Mungu kwa kanisa.*	**Participant** 2: Pastor is a person of God so he cannot chase chase away God's people from the church.
***Mshirikishaji**: Mtu akiwa na virusi vya ukimwi ni lazima ajitokeze mbele ya watu aseme ana ukimwi au si lazima?*	**Facilitator**: If one is HIV-Positive, must they and declare to others their status, or not?
***Mshiriki 3**:Si lazima!*	**Participant 3:** Not a must!
***Mshirikishaji**: Si lazima?... Kama si lazima yule mtu ambaye tu-meona ana virusi vya ukimwi katika hii kanisa afanyeje?... Kanisa inapaswa kumsaidiaje mtu mwenye uukimwi?*	**Facilitator**: Not a must.... if it is not, what then should the person who is infected do in this church?... How does the church need to help one who is infected?
***Mshiriki 4**:(Pastor) aombee huyu mtu, ukimwi iende katika jina la yesu! (Wanashangilia)*	**Participant 4:** (Pastor) should pray for that person (so that) the AIDS is banished in Jesus' name! *(Applause)*
***Mshirikishaji**:Ati kama Pastor ameokoka aombee huyo mtu menye virusi, ili itoke katika jina la yesu.*	**Facilitator**: So if Pastor is saved, he should pray for the person....
***Mshiriki 5**:Hata kama pastor akiomba na mwenye ako na ugonjwa kama hana imani haiwezi ikatoka.*	**Participant** 5: Even if Pastor prays, if the person has no faith, (the AIDS) cannot go away.
***Mshirikishaji**: Ati pastor akiombea mwenye virusi na yeye hana imani virusi havitatoka. Lakini, mtu mwenye virusi inatakikana aishi vizuri, si ndio? Nini tutafanaya huyu, mwenye ana virusi vya ukimwi.... tufanyeje ili aishi vizuri?* (kwa mshiriki mmoja*) Unaitwa nani?*	**Facilitator**: So even if Pastor prays, if the person infected has no faith, the viruses cannot be banished..... But don't a PLWA need to live well? What do we do so that that the person can live well? *(Addressing one participant*) What is your name**?**
***Mshiriki 6**:Wanjiru*	**Participant 6**: Wanjiru
***Mshirikishaji**: Haya....*	**Facilitator**: Ok....
***Mshiriki 6**:Tumpende, na tumwonyeshe mapenzi, tusimwonyeshe kwamba hapaswi kuwa karibu na sisi.*	**Participant 6**: We love him... show him love we should not show him that he is not suppposed to be close to us.
***Mshirikishaji**: Nini ingine?*	**Facilitator**: What else?
***Mshiriki 7**:Apelekwe counselling, ili asiwe na uoga.*	**Participant** 7: S/He should be taken to counselling so that s/he does not live in fear.
***Mshiriki 8**:Huyu mgonjwa wa ukimwi awe msitari wa mbele katika* praise and worship team, *katika* prayers *ili hiyo dhambi aliyofanya iweze kusamehewa....*	**Participant** 8: The PLWA should be in the forefront in the praise and worship team, in prayers so that the sin they have committed can be forgiven....
***Mshirikishaji**: "Dhambi aliyofanya"... na swali yangu inatokea hapo. Ukiwa na ugonjwa wa ukimwi, ni dhambi umefanya ama si dhambi umefanya? Ehe, mshiriki, niambie jina lako kwanza.... Sophie, ehee*	**Facilitator**: "The sin they committed"... and that is what my question is about. If you are infected with HIV/AIDS, have you committed a sin or not? (to a participant) Yes, please tell me your name first.... Sophie, ok....
***Mshiriki 8**:Ni dhambi amefanya...*	**Participant** 8: They have committed a sin..
***Mshikishaji**:Kwa nini?*	**Facilitator**: Why?
***Mshiriki 6**:Ukilala na msichana ambaye anaye....*	**Participant 6:**If one sleeps with a girl who has it....
***Mshirikishaji**: Kama ukilala na msichana ambaye anayo, umetenda dhambi? Ehe, na je ukilala na bibi yako ambaye anavyo,umetenda dhambi?*	**Facilitator**: If you sleep with a girl who is infected ypu have com-mitted a sin? Ehe, what about if you sleeep with your wife who is infected, have you sinned?
***Audience:**(kwa pamoja) Ndiyo!*	**Audience**: *(in chorus)*Yes!...
***Mshirikishaji**: Je tunakubaliana na yeye?... Yaani mtu yeyote anayelala na bibi yake ambaye ana ukimwi anatenda dhambi?*	**Facilitator**: Do we we agree with him? That anybody who sleeps with with his wife who is infected is sinning?
***Mshiriki 9**: Hapana, kwa sababu ni bibi yake.*	**Participant 9**: No. Because it is his wife.

continued on following page

Table 14. continued

Mshirikishaji: *Kwa sababu ni bibi yake kisheria, wameunganishwa pamoja katika ndoa, yeye ni bibi yake, si ndiyo...*	**Facilitator**: because she is his wife by law, they have been joined together by law, she is his wife, is not so...
Mshiriki: 10:*Swala hilo ni gumu kwa sababu kuna njia nyingi za kupata ukimwi. Si lazima ufanye mapenzi ili kupata ukimwi. Unaweza..upata hospitali, ikiwa umekwaruzwa-kwaruzwa mahali, unawezapata ukimwi...*	**Participant 10**: That question is difficult because there are so may ways of contracting HIV/AIDS. You dont have just to get by having sex.. you can get it in hospital, if you are injured....
Mshirikishaji: *Anatwambia si lazima upate ukimwi kwa kupitia njia ya mapenzi... unaweza upata kwa kutumia sindano, kupata damu iliyo na virusi n.k. Sawa hapo bado kuna swali: Huo ukimwi, ni dhambi, au siyo dhambi?*	**Facilitator**: She is telling us that you don't have to get AIDS through sex only, you can get it through injection, if you get blood transfusion and the blood is infected etc... Ok, but the question still stands: Is AIDS a sin or not?
Mshiriki9: *Inategemea umepataje..*	**Participant 11**: It depends on the way you get it.
Mshiriki 10:*Ni bahati mbaya....*	**Participant 12**:: It is bad luck....
Mshirikishaji: *Tunakubalina na yeye?* Mshirikishaji mwingine anaingia...	**Facilitator**: Do we all agree with that? *Another facilitator takes over...*
Mshirikishaji: *Huyu ni mwenye ukimwi, huyu ni rafiki yake ambaye pia ni Pasta mdogo. Na huyu ni Pasta mkubwa. Pastor mkubwa amesema kwamba watu wenye ukimwi hawahitajiki katika kanisa lake. Je baada ya Pasta kusema hivyo, huyu (mwenye ukimwi) atakuwa na marafiki? Na urafiki wa Pasta mdogo na mwenye ukimwi utaendelea?*	**Facilitator**: This is the person with AIDS, and this is his friend, who is also the assistant Pastor. And this is the Principal Pastor. The Principal pastor has said that people infected with AIDS are not needed in his church. Now after pastor says that, will the infected person continue being friends with the other worshipers?
Hadhira:*(Sauti chache) Utaisha!*	**Audience**: *(A few voices)* It will be over!
Hadhira:*(Sauti nyingine) Utaendelea!*	**Audience**: *(Other voices)* It will continue!

This narrative quickly led to an alliance between customary and religious values to condemn HIV/AIDS and anybody associated with it as a product of immorality. Musa Dube and Musimbi Kanyoro's work is among the emerging studies that examine how the conceptualization of HIV/AIDS as God's punishment against sin, and immorality has contributed to the double tragedy of AIDS suffering through disease and stigma. They write:

"..... Biblical texts unfortunately, contributed more to the suffering of the disease. Many texts were found that linked the illness and plagues with God's punishment for disobedience and immorality... the disease was thus held to be sent by God as a punishment to the immoral and disobedient. Sufferers were accused of immorality, sexual sins and regarded as those punished by God... These interpretations contributed to the rise of another vicious epidemic; namely, the HIV/AIDS stigma..." (Dube and Kanyoro 2004:6).

In this perspective, people infected by HIV/AIDS are sinners, immoral and deserve to be punished by both God and society. In their study in South Africa, Kalichman et al (2005) found that even the PLWA themselves have not only accepted the blame that they are sinners and should be punished, but also thought they did not deserve to work in public service. For example when they asked them on whether people who are HIV-positive should be allowed to "teach our children in schools", the majority of the PLWA responded in the negative (139-140).

In the current study, the challenge of religion as an institution that sanctions intransigence against de-stigmatizing HIV/AIDS is evident in the responses of the participants who speak during the outreaches that thematized the issue. In the following performance titled "*Muumini Mwenye Ukimwi* (The worshiper who is HIV-positive), the performance captures the irony in which the church is caught preaching "love your neighbour as you love thyself", yet condemning HIV/AIDS infection as a manifestation of sinful and immoral behaviour. In the following scene, Pasta Morani is the senior (foreign) pastor of the church and is assisted by Pasta Mdogo, a local who also acts as his translator. Pasta Mdogo is also a caregiver and friends to Nuhu, a member of the church who is HIV-positive but healthy because he is on ARV treatment and observes principles of healthy and positive living. In the sermon in the scene, Pasta Moran asserts the message of "love thy neighbour" and says they do not discriminate against any people including people living with HIV/AIDS.

"This is a church of love, and the Bible says, you should love thy neighbour as you love thyself... My church does not discriminate, we love everybody... including those who have HIV/AIDS. We however would like to protect our congregation from AIDS. So let those who are infected stand up and be known, we will pray for them... *hallelujah!"*

Yet he asks those who may be HIV-positive to stand up and be known in the name of "protecting the congregation". Pasta Mdogo does not want to translate this statement to the congregation because he fears its effect on his friend Nuhu, so he feigns a misunderstanding of that particular statement. Table 13 shows how Pasta Mdogo struggles to translate the above statement to the Swahili speaking congregation.

Table 14 shows the context that the facilitator seizes to explore the texts and meaning of this stance. I underline the texts that manifest the notions of religious stigma.

From this excerpt, one can appreciate how much religious-related stigma plays out not only in the church context, but also in the community's perceptions, judging from the participants' views. In this particular outreach, it is credible to conclude that most of the participants held the perceptions that: a) HIV/AIDS is largely contracted through immoral sexual behaviour and is therefore a sinful disease; b) a partner/spouse who is HIV-positive is likely to have sinned so s/he should not have sex with her/him; and c) people with HIV/AIDS have sinned, they need to pray hard and seek forgiveness from God, the church and society. In this context protective sex in terms of condom-use is not in the picture as that is not encouraged in religious repertoire. It is also clear that though the church should love everybody and not discriminate, people with HIV/AIDS are an exception; they should be sympathized with, be prayed for, but be known and be kept on check so that they do not infect others.

Figure 1. Poster in Nakuru. English translation: The origin of AIDS is sin. Jesus is the Medicine. (Mathew 8:17)

Relating this to the community surveys, I encountered the poster shown in Figure 1 in a neighbourhood in Nakuru which incidentally was echoed in the survey, as we see in the tabulated responses below.

Just as it is inscribed in this poster, the incidences reported as happening in the religious contexts emanate from the religious values that equate HIV/AIDS and related variables to be sinful and immoral behaviour. Table 15 contains the illustrative responses from the community surveys in both towns.

THE ROLE OF A CULTURE OF AIDS-RELATED FATALISM IN INHIBITING CHANGE

The final and most critical motif that I would like to discuss with regard to the whole of the theater-against-AIDS interventions is the role of what I will call a 'culture of AIDS-related fatalism' in inhibiting positive change in these communities. The fact that elevation from poverty is the antidote to the problems facing poor African communities, including HIV/AIDS, has been underscored by many studies (see for example, Poku 2002, Campbell 2003). An analysis of the face of poverty as it emerged in various texts and contexts of this study suggests that there is a new aspect to the cultural mix of poverty and HIV/AIDS, a state of HIV/AIDS-related fatalism. A recent research in South Africa cite the characteristics of an HIV/AIDS-related culture of fatalism as manifesting in a general sense of hopelessness and despair resulting from low levels of self-efficacy, low living standards, feelings of hopelessness to effect positive change in one's life, and unclear life-goals (Meyer-Weitz, 2005). In Nakuru, Nairobi, Mombasa and other towns of Kenya, there appears to emerge an interplay between the effects of HIV/AIDS, poverty and other socio-cultural problems like unemployment, low health literacy, lower levels of 'household resilience', and cultural

Table 15.

Area/Resp. Code	*What incidence of HIV/AIDS problem such as stigmatization or discrimination that you have witnessed in your community in the last one year?*
NAKURU	
2.23	"A man was giving a testimony of living positively in a church sermon and after he sat down the worshippers, who were sitting on the same bench with him immediately left their seats. After the event no one wanted to talk or associate with him"
3.21	"Pastor in church does not contract marriages to people who are HIV positive"
6.15	"When my husband disappeared when he knew that he was positive, I confessed in a church crusade where I said I was tested positive for AIDS, little did I know that I was destroying my life, that's when people started fearing most of them didn't want to buy the maize which I sold".
7.4	"A parent overheard children talking about HIV/AIDS and scolded them saying that people infected with the disease were dangerous and evil and in a way cursed by God for their sins"
7.45	"She was demoted from being an elder in the church after it was known that she was HIV-positive"
MOMBASA	
6.21	"One person at my place of work has the HIV virus and we serve her separately in everything. As Muslims we don't like a woman who despise others and also have more than one husband"
8.12	"Girls are sent away from church because of the way they are dressed, i.e. trousers or short skirts, saying that their aim is to provoke men sexually not prayers. It is not allowed for African women, especially Mijikenda women, to wear male clothing or clothes that reveal one's body shape. Those women lack respect, especially by going to places of worship with such strange clothing"
8.24	"A boy is ostracised by his family after they learnt that he was HIV-Positive. The family claimed that the boy has embarrasses them in the (eyes of the) community people because his family and parents are very religious"

disintegration that conspire to create a culture of despair and fatalism which undermine individual and social change. Mildred Ngesa, a reporter with a prominent Kenyan newspaper *The Daily Nation that*:

"... The drugs given at no cost to AIDS patients across the country, are being sold by desperate families to raise money to feed the children... some Aids patients, especially family members, share the drugs for fear of the stigma associated with being seen lining-up to obtain them... Early this week, a woman from Mathare slums in Nairobi, who has been on the drugs for two years, disclosed that she sold her medication to buy food for her four children... "I know I should not do it because it interferes with my treatment. But when you have no food and your neighbour is willing to buy the medicine from you, what choice do you have but to sell the medication to them because they fear being seen going to Blue House?" ... "Blue House" is an HIV and AIDS clinic run by Medicines San Frontiers (Doctors without Borders)" (Ngesa, 2007)

Ironically, poverty has contrived a unique relationship with stigma to undermine the effect of available free HIV/AIDS services to the poor who cannot afford to buy them. Consequently, this context shows that even free HIV/AIDS services does not solve the problem of AIDS without addressing the fundamental problem of poverty. In the texts of outreaches, interviews, focus group sessions, and the community survey for this study, examples of this fatalism abound, albeit in more subtle nuances. In the focus groups in Nakuru, youth reported that hopelessness from unemployment; lack of meaningful life-goals as well as the general culture of poverty, often make them feel that they did not have the power to control their behaviours, including those of making safe choices in relation to HIV/AIDS protective behaviour. For instance, they will readily join an alcohol drinking party, disco dance or any merry-making occasion, especially if it was free. Indulgence in "free" drinking sprees lead to unplanned exposure to alcohol, drugs, and risky sexual behaviour without the financial means of excising one's prudence in the circumstances.

In a performance titled *Kodi ya Nyumba* (House Rent) by TEARS theater group in Kanyoni site in Nakuru, a female tenant is vulnerable to risky sexual relationships with both her landlord and neighbour because she is unable to raise money for her rent. In repeat performances of this skit, the character is variously improvised to play the role of an unemployed student in a local middle level college, thus elevating her desperation for money to include school fees. The more desperate she becomes the more vulnerable she will be to get into a compromising sexual relationship. In another Nakuru play titled *Kwenda Kazi* (Joining the job market) by REPACTED theater group at Pembe-mbili site, Baba Nyangi and Nyaroka, the main characters, are a couple faced with a sick child, lack of food, and numerous debts. The only option available, as presented by a female family friend, is for the wife to join her girlfriend in the "market place" of prostitution. In the Mombasa play titled *Nataka Kununua Kondomu* (I want to buy a condom) by House of Courage at Mlaleo, Mzee Kokoro borrows 2,000 shillings to entertain his female date, but wonders how "a poor man like me will afford" a 10-shilling packet of condoms. In another Mombasa performance titled *Nikiacha Nitafanya Kazi Gani?* (What job can I change to?) by KWACHA Afrika, the audience fails to "convince" the main characters who are playing the roles of prostitutes who have managed to upturn their economic fortunes through the trade to leave their occupations. In a very heated discussion, the performance could not suggest a convincing alternative form of employment that will afford them any income close to what they get in prostitution.

In these performances, the performers were not able to inspire "acceptable" alternative solutions that can substitute the income that was anticipated from some of the risky behaviours. For example, alternative economic activities to prostitution include vegetable vending, hawking, going back to the rural villages, and practising peasant farming. The girl in the rent dilemma is advised to go back to her parents and wait until she can raise enough rent and fees before she comes back to the city to continue her education. These solutions proved impractical because some of the individuals involved may already have tried out the options and given them up. The conclusion therefore in these scenarios is that the issue of poverty is complex and the theater-against-AIDS programs can only deal with it superficially because they do not have the technical capacity to engage it.

CONCLUSION

The texts and contexts discussed above provide a picture of not only what the theatre-against-AIDS interventions are, but also a basis of conceptualizing their effectiveness in providing a culture-focused forum of addressing the various issues at the centre of HIV/AIDS prevention, care and cure. Among the issues worth emphasising in this presentation is the way they replicate cultural behavior and regurgitate it back to the community for their analysis and reactions. The issues tackled by the performances are real in these societies, the way they are presented also resonates with their cultural perspectives and medium allows them to laugh at taboos issues that nobody dares joke about in regular non-theatrical community spaces. And this is the basis of the role of theatricality as a medium that provides a free, non-judgemental space where members of the same community can dare break their own taboos and be able to laugh about their discourses without social repercussion of stigma to individuals. That these are rare and unique forums that have no parallels in these communities need not be overemphasized.

These renditions also explain the complex nature of the poverty factor that affect positive change, or lack of it, and how these factors have permeated the cultural space in society. The overall effect of the AIDS presence in the social psyche is a strong phobia against HIV/AIDS and anything related to it, which in turn leads not only to stigma, but also to a general sense of hopelessness that affects individuals' resolve to change and a weakening of a social network within which this change can be sustained.

Furthermore, the attention that the AIDS problem enjoys in society produces a fatalistic repertoire in the air that everybody breathes, such as: "This is the era of HIV/AIDS, be careful, or be ready to die". This may be one of the reasons why there is a near universal awareness about HIV/AIDS in Kenya which does not translate into an equally positive behaviour-change trend in the population. The theatre-based intervention is one media that is experimenting with engaging communities using content and contexts of their value systems in order to appeal to their common cultural psyche that impacts their individual and communal perceptions about HIV/AIDS and its challenges in behavior-change.

REFERENCES

Andanda, P., & Cook, L. J. (2007). *Majengo HIV/AIDS research case: A report for GenBenefit.* Retrieved on 19th June, 2007, from www.uclan.ac.uk/genbenefit

Baylies, C., & Bujra, J. (2000). *AIDS, sexuality and gender in Africa: The struggle continue.* London, UK: Routledge.

Boal, A. (1979). *Theater of the oppressed.* New York, NY: Theater Communication Group.

Boal, A. (1992). *Games for actors and non-actors.* New York, NY: Routledge.

Boal, A. (1998). *Legislative theater: Using performance to make politics.* London, UK: Routledge.

Byam, L. D. (1999). *Community in motion: Theatre for development in Africa*. Westport, CT: Bergin & Gorvey.

Campbell, C. (2003). *Letting them die: Why HIV/AIDS prevention programs fail*. Bloomington, IN: Indiana University Press.

Campbell, C., Nair, Y., Maimane, S., & Nicholson, J. (2007). Dying twice: A multi-level model of the roots of AIDS stigma in two South African communities. *Journal of Health Psychology, 12*, 403. doi:10.1177/1359105307076229

Davis, T. C., & Postlewait, T. (2003). *Theatricality: Theater and performance theory*. Cambridge, UK: Cambridge University Press.

Driscoll, W. (2002). Fire for our children. CARE International/Kenya special report. Retrieved on January 1, 2006, from http://www.careusa.org/newsroom/special reports/aids/fire.asp

Dube, M. W., & Kanyoro, M. (Eds.). (2004). *Grant me justice: HIV and gender readings of the Bible*. Maryknoll, NY: Orbis Books.

Frank, M. (1995). *AIDS-education through theatre*. Bayreuth African Studies, Bayreuth.

Gwyn, R. (2002). *Communicating health and illness*. Thousand Oaks, CA: Sage Publications.

Hatar, A. (2002). Theatising AIDS in rural communities: Lessons learnt. *Utafiti*, 4.

Hatar, A. (2005). Are we answering the right questions? Challenges in participatory development communication among the Barbaig in Northern Tanzania. *Tanzania Journal of Population Studies and Development, 1*.

Kalichman, S. E., Simbayi, L. C., Jooste, S., Toefy, Y., Cain, D., Cherry, C., & Kagee, A. (2005). Development of a brief scale to measure AIDS-related stigma in South Africa. *AIDS and Behavior*, *9*(2). doi:10.1007/s10461-005-3895-x

Kershaw, B. (1992). *The politics of performance: Radical theater as cultural intervention*. New York, NY: Routledge. doi:10.4324/9780203412282

Levert, L., & Mumma, O. J. (Eds.). (1995). *Drama and theatre: Communication in development*. Nairobi, Kenya: KDEA.

Mda, Z. (1993). *When people play people: Development communication through theatre*. London, UK: Zed Books.

Messer, D. E. (2002). *Breaking the conspiracy of silence: Christian churches and the global AIDS crisis*. Minneapolis, MN: Fortress Press.

Meyer-Weitz, A. (2005). Understanding fatalism in HIV/AIDS protection: The individual in dialogue with contextual factors. *African Journal of AIDS Research*, *4*(2), 75–82. doi:10.2989/16085900509490345

Mlama, P. M. (1991). *Culture and development: The popular theatre approach in Africa*. Uppsala, Sweden: Nordiska Afrikainstitute.

Mumma, O. J., Mwangi, E., & Odhiambo, C. (Eds.). (1998). *Orientation of drama, theatre, and culture*. Nairobi, Kenya: KDEA.

Mwarogo, P. (2007). *From behavior change communication to strategic behavioral communication on HIV in Kenya, 1999–2006: Impact, project experience, and lessons learned*. Nairobi, Kenya: USAID/FHI.

Mwita, M. (2010). *Breaking cultural barriers of HIV/AIDS intransigence: An examination of theatre-against-AIDS interventions in Kenya*. Unpublished PhD Thesis, University of Dar es Salaam.

Mwita, M. (2011). 'From a cathartic function to confronting stigma: Theorizing the conscietization efficacy of theater-against-AIDS interventions in Kenya.' in *East African literature: Essays in written and oral traditions*. Eds. J. K. S. Makokha, Egara Kabaji & Dominica Dipio. Berlin. Logos Verlag. pp. 319-344

Ndigirigi, G. (2007). *Ngugi wa Thiong'o's drama and the Kamiriithu popular theater experiment*. Trenton, NJ: African World Press.

Ngesa, M. (2007, March 17). Setback for anti-AIDS fight as patients sell drugs to buy food. *Nation Newspapers*.

Obanyi, G., & Pyne-Mercier, L. (2000). Realizing the HIV prevention-to-care continuum in Kenya. *Impact on HIV: Making a Difference*, *2*(2), 11–15.

Odhiambo, C. J. (2004). Theatre for development in Kenya: In search of an effective procedure and methodology. Unpublished PhD Thesis, University of Stellenbosch.

PATH. (2005). *Magnet Theatre: A facilitators training manual*. Nairobi, Kenya: PATH/UNAIDS.

Poku, N. K. (2005). *Aids in Africa: How the poor are dying*. Cambridge, UK: Polity Press.

Republic of Kenya. (2001). *AIDS in Kenya: Background, projections, impacts, interventions, policy* (6th ed.). Nairobi: Ministry of Health.

Riccio, T. (2004). Kenya's community health awareness puppeteers. *A Journal of Performance and Art, 26*(1), 1-12.

Taylor, P. (2002). Afterthought: Evaluating applied theatre. *Applied Theatre Researcher, 6*(3). Retrieved on February 3, 2007, from http://www.gu.edu.au/centre/cpci/atr/journal/article6_number3.htm

Taylor, P. (2003). *Applied theater: Creating transformative encounters in the community*. Portsmouth, NH: Heinemann.

The Holy Bible, King James Version, Acts 16, 26-40.

wa Thiongó, N. (1986). *Decolonising the mind: The politics of language in African literature.* Nairobi, Kenya: Heinemann.

LIST OF LIVE PERFORMANCES CITED[19]

1*Avae ama Asivae* (Should he put on a condom or not?), by REPACTED, Nakuru.

4*Kodi ya nyumba* (House Rent), REPACTED, Nakuru

2*Kwenda sokoni* (Going to the job Market), TEARS, Nakuru.

3*Muumini mwenye Ukimwi* (The worshiper who HIV-positive), REPACTED/TEARS combined performance, Nakuru.

8*Nataka kununua kondomu!* (I want to buy a condom!), Kwacha Afrika, Mombasa.

6*Ni nani (Ngojera poetry)* (Whose Responsibility is it?), House of Courage, Mombasa.

7*Nikiacha Nitafanya kazi gani?* (What job can I change to?), Kwacha Afrika, Mombasa.

5*Sitaki virusi katika nyumba yangu* (I do not want viruses in my house), House of Courage, Mombasa.

KEY TERMS AND DEFINITIONS

BCC: *Behaviour Change Communication:* A term used by community health educators to refer to a multi-media and multi-level communication approach that employs different strategies of community mobilization to design, and execute tailored messages to individuals, families and communities aimed at influencing positive change in their attitudes and action with regard to HIV/AIDS-related risky behaviours.

Conscietization: A term used in Theater for Development (TFD) literature to refer to the process of using theater education to create critical consciousness about local problems and local solutions to address them in impoverished populations.

Ludic Role: The role that a theater audience assumes when they are participating in a theatrical or ritual performance. The role is supposed to free an individual from their ordinary roles in their culture so that they act as any characterization may demand upon them in the 'life/reality' of stage performance.

Outreach site: (Magnet Theater outreach/performance site): The specific space within the community or village where the theater-against-AIDS performances happened on a recurring basis.

Performance: Anthropology scholars have defined "performance" as a theory of the nature of all human activities – artistic and otherwise (Turner, 1986; Schechner, 1988; Bauman, 1992). However, the term is used in this research to represent the act and occasion of doing a theatrical presentation to an audience.

Popular Theater: This research uses the term to denote the various community-based theatrical models in use by agencies and theater groups in Kenya. They are "popular" because they attempt to use local theatrical genres, and do the theater with the intention of provoking the people to participate in the process of discussing their situation and how to improve it. It is therefore "theater by the people and for the people".

Research Site: The particular 'open' space in a market, church or any other community point where the research assistant and I conducted interviews during the outreaches and/or the community surveys.

Text: This research adopts Daniel Chandler's definition of "text" as "a message which has been recorded in some way (e.g. writing, audio- and video-recording) so that it is physically independent of its sender or receiver. A text is an assemblage of signs (such as words, images, sounds and/or gestures) constructed (and interpreted) with reference to the conventions associated with a genre and in a particular medium of communication." (Chandler, 2002, p.2-3)

Textualization: This term is operationalized in this research to mean the study of the holistic totality of verbal, non-verbal, written, and contextual components of a theatrical communication that includes and transcends the written text.

Theatre (Theater) for Development (TFD): Deriving from Augusto Boal's theory and practice of Theater of the Oppressed in the 70's, Theater for Development is the term that has been employed to define participatory programs that involve local populations in brainstorming and "trying out" solutions to their problems through a theatrical forum. Augusto Boal employed this theory in various techniques such Forum, Invisible, Legislative, or Theater of the Oppressed. In Europe, it has sometimes been referred to as Street Theater. In North America and Australia scholars have also christened practice of theater-for-social-change as Applied Theater, while in Africa it has been mostly referred to as Community Theater or Popular Theater. This study regards all these concepts as rubrics under the umbrella of TFD.

Theater/Drama: Theater and drama are used synonymously in this research to include the multiple artistic techniques of organizing, rehearsing, mobilizing and performing a theatrical/dramatic presentation to an audience. Drama includes the act of performing as well as the written text. Theater and Drama merge when a text, oral or written, is enacted on stage.

Theater-Against-AIDS Outreach: A gathering in a community place that is created and sustained by theatrical performances, that targets to inspire the audience to be involved in discussing the HIV/AIDS related issues in the community.

Theatricality: The term Theatricality is used in contemporary theater studies to define the conceptual framework of drama and theatrical performances as activities that play a role in representing real life in a way that positively impact human progress. Theoretically, the term is viewed against the Aristotelian mimetic presentation of theater as a poor imitation of reality. In this study, I use theatricality as both a theory and technique that understands theater as a medium whose enactment is a rendition of life in a unique form that gives us a deeper and critical understanding of real life.

Voluntary Counselling and Testing (VCT) centre.: These centres are localized agencies specifically equipped for HIV-AIDS counselling and testing. They account for a large percentage of all the AIDS-related counselling, testing, and post-test education activities in Kenya. The term has been revised recently to include and emphasize "confidentially" as one of the important characteristics that these centres espouse in their practice – thus Voluntary Counselling and Confidentiality (VCCT) centre.

ENDNOTES

1. Studies that have documented TFD-for-AIDS-education programs include: Frank (1995), Hatar (2002, 2005), Mutembei (2001) Levert and Mumma, (1995), Mumma, Mwangi and Odhiambo (1998), Riccio (2004), and Odhiambo (2004), Mwita (2010).
2. TFD is used throughout this chapter as the acronym for Theater for Development, a practice of theater for social conscietization popularized by Augusto Boal in his many writings including 1979, 1992, 1998.
3. See Obanyi and Pyne-Mercier (2000), Republic of Kenya (2001).
4. An evaluation of the theatrical and conscientization efficacy of these programs is presented in Mwita (2010 and 2011).
5. Luhyia is an ethnic community that lives in Western province of Kenya.
6. The Kuria community lives in the Mara region of Tanzania, with a section in the South-Western part of Nyanza province in Kenya.
7. Augusto Boal (1931- 2009) is credited for the highest development of TFD, through his practice of its theory and method especially in Brazil, but also in Europe and North America. In Eastern Africa, the practice was applied and practiced to significant successes by practitioners such as Ngugi wa Thiong'o in collaboration with the Kamirithu community Project (see Ngugi wa Thiong'o 1986, Ndigirigi 2007, Zakes Mda 1993, Penina Mlama 1991, among others). Although these practices run concurrently with Boal's movement in Latin America, the African TFD movement bears its own original innovations as well as influences from Boal's highly successful, consistent and lifelong practice.
8. See Davis and Postlewaite (1992), Baz Kershaw (1993), Zakes Mda (1993), Philip Taylor (2002) and Boughault (2003) for a detailed account on how the theatricality of a community performance transforms their audience from their ordinary roles into a "ludic" state that allows them to participate freely in a performance.
9. In East Africa, artists are generally not held with much esteem, as compared to other professionals such as lawyers or economists because artists hardly make any wealth for them to be respected. Remmy Ongala, a prominent Tanzanian rhumba musician has decried this notion in one of his popular musical album entitled "*Muziki ni wa Nani*?" (To whom does music belong to?). The music urges society to take music and

art seriously because it serves an important role in the society's education.

10 "*Ngojera*" is the generic term for various forms of Kiswahili dialogic poetry. A dramatized (often choral) verse presents two or more sides of a controversial issue through dialogic poetry. The opposing sides would "argue" out their points and eventually come to a compromise, or have a third "wise" party help resolve the issue.

11 Abbreviated throughout this chapter for "Respondent".

12 The Kuria community lives in South Western Kenya/North Eastern Tanzania in Kuria County in Kenya and Mara Region of Tanzania.

13 For more on this research see Pamela Andanda and Cook, L. J (2007).

14 Although 'Westernised' families are shying away from polygamy in present times, the practice is legitimate and widely accepted/practised in the Islamic cultures in Mombasa. In up-country communities, there are many families that accept polygamous marriages and families. In this context, it is therefore appropriate to expand the definition of the "marriage" institution to be consisting one man with one or more wives.

15 For more on this, see PATH (2005) and Mwarogo (2007).

16 IMPACT stands for 'Implementing AIDS Prevention and Care' Project (1999 – 2006), a USAID/FHI supported program on AIDS prevention that supported the theater-based interventions in this study.

17 "Condom" in the context of these outreaches always refers to male condom, unless otherwise stated.

18 "Bwana" translates literally to "mister". It is an informal way of addressing a friend or colleague, a near antonym to "sir".

19 Each of these skits was performed in many repeat outreaches in multiple locations, sometimes by different theater groups, thus making it invalid to assign specific dates and places. The theater groups reflected against each skit above refers to the groups whose performance I video-taped.

Chapter 10
New Media and Health Communication:
Communication Strategies in Malaria Control in Nigeria

Nwachukwu Andrew Egbunike
University of Ibadan, Nigeria

ABSTRACT

Malaria is endemic in the tropics and is responsible for a very high infant mortality, killing more than 3,000 children in Africa daily. The Nigerian government's control measures are targeted at nursing mothers and children. However, a significant portion of the population–the youth–are also very vulnerable. The new media are gradually gaining ground as a dependable channel of meeting the communication needs of young Nigerians. This chapter discusses how the potential associated with the Internet and social networks can be incorporated in the campaign for the Insecticide Treated Nets (ITNs) by Nigerian youth. It also proffers solutions and recommendations based on the concept of participatory development communications.

INTRODUCTION

Malaria is endemic in Nigeria (97%) and is responsible for one out of every five childhood deaths (United Nation Population Division: 2002). Nigeria also ranks first among the thirty-five countries that are responsible for 98% of the total malaria deaths world-wide and contributes 96% to the total number of malaria cases (Roll Back Malaria: 2008). Malaria is responsible for the morbidity and mortality from illnesses such as respiratory infections, diarrhea, iron-deficiency

DOI: 10.4018/978-1-60960-591-9.ch010

diseases, anemia and malnutrition (Opiyo *et al*Opiyo *et al*: 2007). Onwujekwe *et al* (2005) stated that malaria is the number one public health problem in Nigeria.

Malaria control is now a major global development priority and is critical in the attainment of the Millennium Development Goals. It is undeniable that malaria imposes a huge economic burden on the nation's health development. Ukwoma and Muanya (2009), summarises the cost implication of the endemic disease to individual and government thus:

Cost to individual and their families include: purchase of drugs for treating malaria at home; expense for travel to and treatment at dispensaries and clinics; lost days at work; absence from school; expenses for preventive measures; expenses for burial in case of deaths. Cost to government include: maintenance of health facilities; purchase of drugs and supplies; public health interventions against malaria, such as insecticide spraying or distribution of treated bed nets; lost days of work with resulting loss of income; and lost opportunities for joint economic ventures and tourism.

BACKGROUND

Current malaria elimination strategies rely more on individuals and communities to protect and treat themselves. Such strategies will only succeed if there is effective communication and genuine understanding around the causes, symptoms and means of preventing and treating malaria to protect the most vulnerable people. A current challenge facing Nigeria is how to achieve widespread distribution and use of insecticide-treated nets (ITNs) for the control of malaria (Onwujekwe *et al:* 2005).

The Abuja Declaration of 2000 reaffirmed international commitment to roll back malaria and called upon member states in Africa to undertake health system reforms, including promoting community participation in joint ownership and control of the Roll Back Malaria alliance (RBM). The Abuja goals include ensuring that 60% of those with malaria have access to treatment within 24 hours of the onset of symptoms, at least 60% of at-risk pregnant women receive preventive drugs and at least 60% of at-risk groups sleep under bed nets (African Summit on Roll Back Malaria: 2005).

Vertical malaria control programmes of the 1970s focused on indoor residual spraying, uniformly applied across continents. Shifting from 'eradication' to 'control', today's approaches are more horizontal, with malaria control integrated into the primary health care system and relying on the participation of local people. The National Malaria Control Programme (NMCP) has developed a strategic plan which aims to halve the burden of malaria by 2010 and 75% by 2015. Nigeria's former Health Minister, Professor Babatunde Osotimehein explains that NMCP planned to have "at least 80% of households with two or more Insecticide Treated Net (ITN) by 2010 and sustained at this level until 2013."

COMMUNICATION AND MALARIA CONTROL

The principles of participatory development communication have great potential to be applied to the malaria control strategy. Development communication is a specialist field that has emerged around the mid-1960s with the emphasis on using news reporting for constructive and developmental ends. According to Folarin (1998), its focus is on long-term development. It is characterised by people's rights to information and to a voice, freedom of all communication channels, participation, ownership of knowledge, accountability of governments and societies, and people's improved ability to put informed choices into practice (Dunn: 2005).

Communication needs to be included in the process of malaria control right from the start, not as an afterthought or a discretionary appendage. This means thinking strategically about communication from the planning stages through to implementation and evaluation.

New Media

The so-called new media, based on the internet and mobile phone, is overtaking the market of existing media swiftly (Wang: 2009). The internet typifies the new media. With its convergence of radio, film and television, and distributes them through 'push' technology (McQuail: 2007). The contrast between the new and old media (newspapers, radio and television) can be lumped up in the formers digitalisation and interactivity. This can be aptly appreciated in the internet which is a multimedia per excellence due to its capability to make text, graphics, audio, animation, video, data, etc., available on one medium. The internet also provides the opportunity to selectively choose who one converses with using the immediacy characteristic of face-to-face communication, while maintaining an unparalleled degree of anonymity (McCabe: 2009).

The conversation about the future of old media (especially newspapers) continues when compared with the widespread acceptance of the new. Though this global controversy may not have similar validity in Nigeria, the fact remains the sales of some dailies have been plummeting. It seems more people read papers at no cost or for a token. A visit to a vendors stand always reveals a large amount of middle aged Nigerians who constitute the 'Free Readers Association'. Asides these, it appears like most Nigerians spend more money talking than reading (Egbunike: 2009a). This may have precipitated this angry outburst from Onome Osifo-Whiskey, Editor of Nigeria's *TELL* magazine:

Now we have the luxury of seeking all shades of an issue, we work harder, we report deeper, but even the deeper you go about it, the more you are losing your readers to the tribe that seeks that pleasure of a roaming GSM (mobile phone). People, I mean, artisans, mechanics, drivers, can spend fortune on GSM, but they cannot spend 1 kobo on a newspaper or a magazine.

Figure 1. Arial view of the old Ibadan City from Mapo Hall, Ibadan. Photo credit: Bola Famuyiwa (2009)

Social Networks

Social networking sites primarily comprise of online (internet based) communities of people with similar interests and/or activities. Most of these social networks provide a variety of ways for users to interact, communicate and share information. These networks provide e-mail and instant messaging services, like other regular sites. The list of social networks includes: Facebook, MySpace, Twitter, LinkedIn, Bebo, Hi5, do12day, Orkut, Wretch, Xiaonei and Cyworld. Asides Facebook (which is the most popular in Africa) Nigerian youths make use of NaijaPals, NaijaFriends, Nigerian Village Square, Nairaland, etc.

The surge in social networking prompted the *Economist* (2010) to declare that Facebook is now the second most popular site on the internet after Google. Facebook after just four years of existence has 350 million users with 55 million updates per day and shares more than 3.5 billion pieces of content with other members of the site every week. Facebook certainly has a wider outreach than any old (mass) media can ever dream of.

At the same time most educated Nigerians especially the young ones are becoming more passionate about the internet. Thus, the propensity to hook up to new technologies as source of information is gradually on the increase. Many young Nigerians are signed up to Facebook, Twitter and other social networking sites. Since the addition of "the new electronic media can be viewed initially as an addition to the existing spectrum rather than a replacement" (McQuail: 2007), how potent is the new media in the control of malaria in Nigeria, vis-à-vis the old media?

THE YOUTH: VULNERABLE BUT FORGOTTEN?

Government focus on malaria prevention is tilted towards gravid women, post-natal mothers and children. The five year strategic plan for RBM (2001 to 2005) had the following objectives: to reduce morbidity and mortality by malaria by 25% by the end of the year 2005 and to reduce mortality due to malaria among pregnant women by 25% by the end of the year 2005 (Edugie *et al*Edugie *et al*: 2004).

Unfortunately Nigerian youths (specifically between 15-35 years) constitute a vulnerable proportion of the population that suffer from malaria. This group ironically have more access to information (especially the new media) as shown in Figure 2. It is therefore strange that while so much have been done to immerse the youths with information on HIV/AIDS, a deafening silence envelopes malaria campaigns. As this chapter will show, most of these people have been burdened with malaria more often than with AIDS.

Though most of the blame can be correctly laid on the doorsteps of the Nigerian government, due to its focus on children and mothers, it is also trite to encourage a new generation who will adopt the use of ITNs. Since this has to do with behavioural change, the earlier young Nigerians key in into using treated nets, the better and easier it will be to control the wildfire of malaria in the country.

This chapter will attempt to answer the following questions: which medium do Nigerian youths get most of their information on malaria from? Is the new media (internet) being of help

Figure 2. Level of respondents' exposure to the internet

in communicating strategies for malaria control? Have communication strategies (old or new media) focusing on ITNs been able to convince the youth to adopt it? How aware are the youths about government policy or programme for eradicating or controlling malaria? How do people perceive the use of ITN? Do they see it as an effective intervention in the control of malaria? It will also test the validity of social change and development – new communications, mass information and persuasions – in the context of ITN's adoption amongst Nigerian youths

This chapter will aim to:

1. Examine the level of awareness among Nigerians youths on malaria prevention particularly on the use of ITN as a tool for malaria prevention.
2. The influence of new media technologies in comparison with the old media in health communication.
3. To establish if the affinity of youths towards the new media, necessarily means adoption of health information, in the context of malaria control.

Findings from this research will have tremendous impact in the public health situation of malaria in Nigeria. It will also access the effectiveness of the campaign to use and adopt ITN in the country.

MALARIA CONTROL AND NEW MEDIA

Ibadan

This small scale study was designed to ascertain the aims of the chapter as enumerated above. Since the study was directed at youths, the natural choice was undergraduate students of the University of Ibadan (UI), Ibadan and Advanced Level students of Educational Advancement Centre (EAC), Ibadan.

Ibadan is a cosmopolitan city in South West Nigeria, formerly capital of the Old Western Region and currently hosts the seat of government of Oyo State. It is located about 110 km northeast of Lagos. Aside being made famous by J. P. Clark's poem, *Ibadan,* the city of 'running splash of rust and gold – flung and scattered among seven hills like broken china in the sun' (Moore and Beier: 1999) has an estimated population of 2 to 3 million (Fourchard: 2002).

The University of Ibadan, formerly University College, Ibadan, was founded in 1948 (UI Calendar, 2008-2009: 2008) by the colonial British government. As the first university in Nigeria, it also boasts of renowned alumni like Chinua Achebe (1953) author of *Things Fall Apart* and Wole Soyinka, 1986 Nobel laureate, and the first African to win the Nobel Prize for Literature (1954) [Osofisan: 2008]. Although situated in a predominantly Yoruba town, the university and the A level school attract students from all parts of the country.

The University of Ibadan has thirteen faculties and the total undergraduate population in the 2009/2010 session is 12,232, consisting of 6,834 male students and 5,398 female students (University of Ibadan-Management Information Systems, 2010). UI is mainly residential and operates twelve halls though a proportion of the students reside off campus by choice. Nine of the halls of residence are for undergraduate students; two for postgraduate students, while the last one (Awolowo Hall) is for both undergraduate and post graduate students. Eleven of the halls are located on the main campus while the Alexander Brown Hall, the clinical students' hostel is on the grounds of the University Teaching Hospital about five kilometres away. The total population of undergraduates resident on campus in the 2009/2010 session is 2, 887 (UI-MIS, 2005).

Methodology

Combining convenience with stratified sampling, 200 people responded to our questionnaire. In addition, four people (three students of UI and a consultant to GFATM) were interviewed. The respondents were volunteers (willing and available). They were not forced, coerced or unduly induced to participate in this study (Bordens and Abbott: 2008).

Non-random sampling was employed due to the unavailability of a sampling frame. UI for instance, lacks an up-to-date database of the entire student population. EAC on the other hand is practically in a fluid state. They have many programmes – A' level May/June, A' level Nov/Dec, UME (University Matriculation Examination) class, etc., – running concurrently throughout the year. In addition, as each examination is completed, new students are being admitted for another set of exams. Consequently, we had no better option than the non-random sampling.

Out of the 250 questionnaires administered, only 200 were properly filled, the remaining 50 had defects which disqualified the respondents from the study. The fact that 83% responded accurately to our questioner is particularly impressive. Marczyk *et al* (2005) justified the statistical validity of a large sample size as that with 40 participants and above.

Of these, 77% were undergraduate students of UI while the other 23% were from EAC. Also the gender showed that 76.5% of males, 23% were females and 0.5% gave no response. The age distribution of our respondents is shown in Table 1.

Most of the questionnaire items were close-ended in order to maintain anonymity and confidentiality of the respondents. This is because they focused on the attitudes and behaviour of the respondents. To ensure representiveness, the sample population included students from diverse disciplines – economics, law, veterinary medicine, electrical and electronic engineering, industrial and production engineering, civil engineering, food technology, medicine and surgery, and dentistry. Others were from computer science, physiology, biochemistry, psychology, sociology, political science, theater arts, zoology, physics, geography, agricultural economics, mathematics, microbiology, geography/ urban and regional planning, statistics, pharmacy, accounting and, human nutrition. A breakdown of the respondents – into content categories using their faculties – is shown in Table 2.

Table 1. Age range of respondents

Age Range	Frequency	Per cent
15-20	107	53.5
21-26	65	32.5
27-30	13	6.5
31 and above	4	2.0
No response	11	5.5
Total	**200**	**100**

Source: From the data generated from the study

Note: The above applies to all figures (except Figure 1and 7) and tables used in this chapter

Interpretation of Data

The breakdown of the level of respondents' exposure to the internet was shown in Figure 2. It is

Table 2. A stratum of respondents by their faculties

S/ No	Faculty	Frequency	Per cent
1.	Agriculture	6	3
2.	A' Level	36	18
3.	Arts and Law	15	7.5
4.	College of Medicine	46	23
5.	Technology	24	12
6.	Science	20	10
7.	Social Sciences	11	5.5
8.	Veterinary Medicine	42	21
Total		**200**	**100**

obvious that most young Nigerians have affinity for the new media. However, the regularity of internet use exposes the infrastructural limitation associated with information technology in these parts as we have shown in the preceding paragraph. Since power is almost unavailable in Nigeria, it is not surprising that only 36.5% of the respondents could use the internet daily. While 41% had access to the web weekly, 14% did same monthly and 1.5% used it once a year.

The popularity of the sites visited shows that Goggle has the largest traffic (33.5%), followed closely by Facebook and Yahoo with 31.5% and 31.1% respectively; while Hotmail tagged along with 3.9%. Majority of the respondents belong to a social network (76.7%) with only 23.3% not subscribing to any. The fame of Facebook as the network of choice was reinforced as shown in Table 3.

The respondents displayed a high affinity for health news (Figure 3), though most went out of their way to look for it (44.3%), very few stumbled upon them (16.1) and 39.6% had it both ways. Our claim on a deafening silence in the media on malaria when compared with the surfeit of information on HIV/AIDS was validated by the respondents reply to the health problems that they are well familiar with (Figure 4). Majority (46.3%) had particular interest in HIV/AIDS, 20.3% (malaria), 11.9% (diabetes), heart disorders (11.5%), hypertension (10.1%).

Table 3. Social networks of choice amongst Nigerian youths

Social Networks	Percentages
Facebook	42.5
Hi5	19.2
My Space	12.7
Twitter	7.1
NaijaPals	4.7
Bebo	4.1
NaijaFriends	4.1
Badoo	1.5
Nigeria Village Square	0.9
Orkut	0.9
LinkedIn	0.9
Cyworld	0.6
Dp12day	0.3
Wretch	0.3
Xiaonie	0.3
Total	**100**

Since the Federal Ministry of Heath's (FMOH) focus on malaria prevention is biased against youths, it logical that only 41.1% are aware of the establishments' campaign, while a greater majority (54.5%) are entirely clueless. Does that mean that the young people are immune to malaria bouts, of course not? Table 4 reveals that a higher proportion has been incapacitated with the illness than those who have not. This necessarily translates to unquantifiable time, human and financial capacity lost to a preventable disease.

Most respondents skewed towards curative rather than preventive response to malaria. As Table 5 illustrates, 30.1% used anti-malaria medicine and 6.4% took herbs for treatment. The preventive measures showed that many were conscious of their hygiene (25.2%), while very few slept under ITN (21.7%) or used malaria prophylactic commonly called, Sunday-Sunday medicine (1.7%).

Figure 3. Frequency of finding health news on the internet

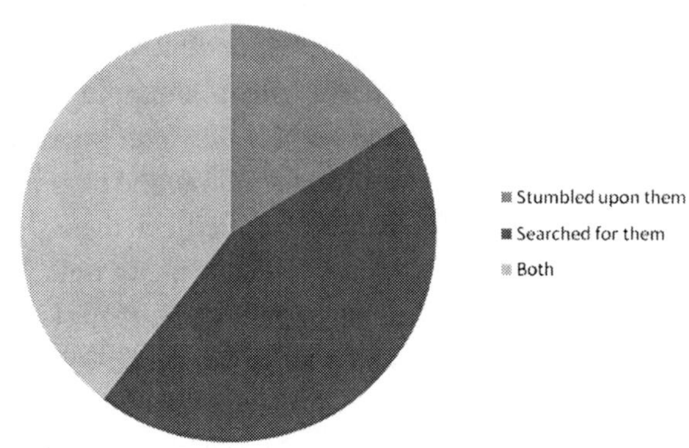

Figure 4. Health issues of interest

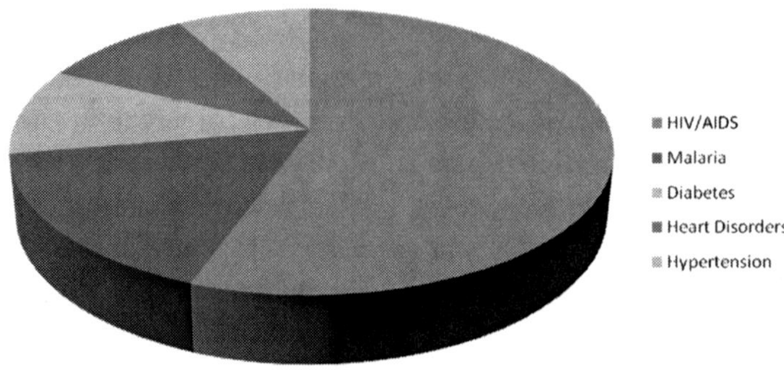

Table 4. Incidence of Malaria attack

Incidence of malaria attack	Frequency (%)
Every 2 Weeks	1.0
Monthly	4.2
Every 3 Months	16.1
Every 6 Months	15.1
Once a Year	24.5
Can't Recall	32.8
Never	6.3
Total	**100**

Table 5. Malaria therapy of choice adopted by most Nigerian youths

Malaria Therapies	Frequency of Use (%)
Anti-Malaria Medicines	30.1
Insecticide Treated Nets (ITNs)	14.9
Sunday-Sunday Medicine	1.7
Good Hygiene	25.2
Insecticides	21.7
Taking Herbs	6.4
Total	**100**

One of the respondents whom we interviewed, admitted using ITN but 'I always have to bear the brunt of friends' crude jokes. I am not particularly bothered since am the one who suffers bouts of malaria not them.' He however admitted that it took a lot of will to adopt and continually use the nets. Another claimed that using nets were clumsy, "I have to fix the net before sleeping, that's too much work. The ritual is too much *wahala* (trouble). Another respondent stated that "with the heat, I can hardly sleep, imagine what will happen when I add nets. If they can provide treated nets for the doors and window, and provide the chemicals that can be applied on them by spraying or painting, then I'll start using nets."

Respondent's source of news about malaria is reproduced in Table 6. Most satisfied their information needs from non-media (old and new) channels. The percentage of those who got malaria information from the internet (47.5%) was higher than those who did not (43.5%) – Table 7. And many stumbled upon them (25.0%), searched for it (15.5%), both ways (7%) and others (52.5) as shown in Figure 5.

This buttresses the fact that little information is available on the internet. Although one may be superficially imply that since 47.5% depended on the internet for malaria news (Table 7), yet Table 6 reveals that non-mass media sources (interpersonal and group avenues) like parents, schools, books, friends, hospital accounted for the greater presence of news items on malaria in Nigeria.

If there was to be any doubt that FMOH's malaria campaign is exclusively focused on gravid and nursing mothers, then Table 7 will lay that to rest. Majority of the respondents (54.5%) were ignorant, 38.5% were aware of FMOH's work while 7% declined to respond.

Table 6. Source of information about Malaria

Source	Frequency (%)
Parents	10.4
School	9.5
Books	10.3
Friends	7.4
Hospital	11.6
Radio	11.0
Internet	7.8
Posters (IEC Materials)	5.9
Newspapers	10.5
Television	13.2
Others	2.6
Total	**100**

Table 7. Frequency of finding information about Malaria on the internet

	Frequency	Per cent
Yes	95	47.5
No	87	43.5
Missing	18	9.0
Total	**200**	**100**

When prodded further, for those who were claimed to be aware of government's malaria policies, to list the programmes, we got this melange of answers. From the blunt: "I don't remember", "forgotten", "I only heard of a policy, I can't remember any", "No, it is not publicised, but I think there's a subsidy or something on mosquito treated nets" or "my uncle is in one but I don't recall the name. He works for WHO"; to the enlightened: "the Kick Malaria Out/Say No to Malaria", "Use of Insecticide Treated Nets", "Building more hospitals where patients can go for treatment", "Operation Roll-Back Malaria, prevalent in Cross Rivers State where I reside". And the totally oblivious ripostes were: "NACA (National Agency for the Control of AIDS) and "Society for Family Health". These two Nigerian

Figure 5. How respondents found news about Malaria on the internet

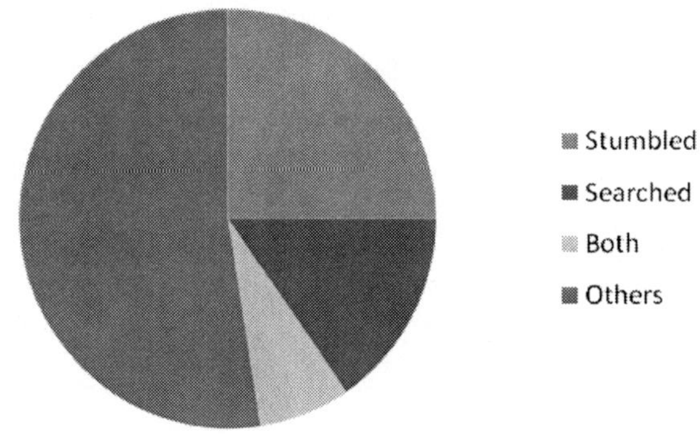

agencies are primarily dedicated to reproductive health.

Other interesting reactions are: "UNICEF: for pregnant women and children; NHIS (National Health Insurance Scheme)", "Roll-Out Malaria Programme", "keeping the environment clean", "MAMA", "use of combination therapy for treating malaria", "AMPCH (Amokoko Partners for Community Health), "Lagos State Malaria Control Programme. I get free drugs from there", "Fight against Malaria in Uganda", "Subsidised anti-malaria drugs", "Youths for Positive Health Control", etc. Nonetheless, the most publicised anti-malaria agenda was the Roll Back Malaria Programme.

Little wonder that only a frugal (2%) of the respondents belonged to a malaria group on the various social network sites in the internet. Majority (91%) did not subscribe to any group while 7% of the respondents declined to give an answer (Figure 6).

Table 8. Awareness level about any government programme on Malaria control

	Frequency	Per cent
Yes	77	38.5
No	109	54.5
Missing	14	7.0
Total	**200**	**100**

Figure 6. How respondents found news about Malaria on the internet

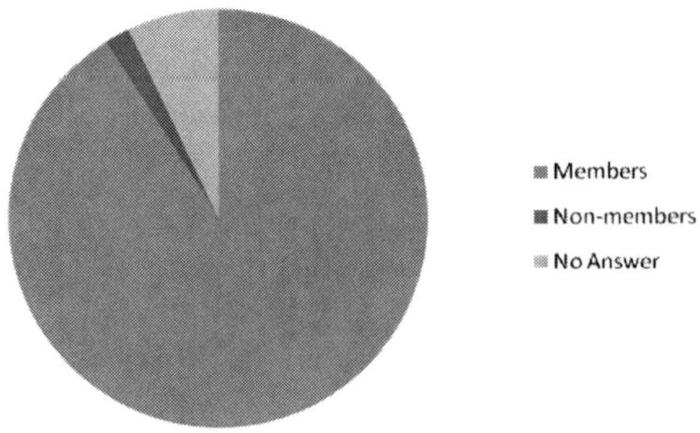

With the scenario in Figure 6, one can be tempted to conclude that apart from the lack of awareness; perhaps the social network sites are deficient of malaria groups. On the contrary, Facebook alone has the following number of malaria groups as listed in Table 9.

INFRASTRUCTURAL CHALLENGES AND NEW MEDIA IN NIGERIA

The barriers against using the internet in Nigeria are legion but only the major ones are discussed below.

Table 9. Malaria groups on Facebook

S/No.	Name of Group	Category	Membership	Accessibility
1.	Africa Fighting Malaria	Organisations	173	Open and Public
2.	Burma Empowerment Action Team: BEAT Malaria!	Organisations	152	Open and Public
3.	Buzz and Bite Malaria Prevention Campaign	Organisations	55	Open and Public
4.	Bring Back DDT	Common Interests	35	Open and Public
5.	Faith Acts - Malaria: A Devastating but Easily Preventable Disease	Common Interests	191	Open and Public
6.	Great Malaria Facebook Race	Organisations	151	Open and Public
7.	Help Cure Malaria	Organisations	60	Open and Public
8.	Help the fight against Malaria, TB and HIV	Organisations	62	Open and Public
9.	Join One Million faces Against Malaria for Positive Change	Organisations	47	Open and Public
10.	Malaria Action Now	Student Groups	9	Open and Public
11.	Mosquitoes should FOAD	Common Interests	146	Open and Public
12.	Net-a Sucker!	Organisations	31	Open and Public
13.	Nothing But Nets: Broughton	Organisations	40	Open and Public
14.	Nothing But Nets: Campaign	Organisations	21	Open and Public
15.	Nothing But Nets Elon	Organisations	39	Open and Public
16.	Nothing But Nets	Organisations	92	Open and Public
17.	One Million Faces Against Malaria - Project Blackout Task Force	Organisations	17	Open and Public
18.	Stop Malaria in Africa!	Common Interests	12	Open and Public
19.	The Malaria Foundation International	Organisations	96	Open and Public
20.	UMaine Fights Malaria	Common Interests	5	Open and Public
21.	Upper Valley Nothing But Nets (Student-to-Student)	Organisations	4	Open and Public

Source: Facebook.com (accessed through authors account on Facebook on May 14, 2010)

Electric Power

References to Africa as the "dark continent" are rather irritating. There's so much sunlight, so many vibrant people, so much daily excitement. However, the equatorial sun rises in Lagos at about 6.30 am and sets at about 6.30 pm. Then it is dark, really dark. Street lights are rare and power blackouts are common. So common, in fact, that selling generators in Nigeria will be a good business for quite some time (African Review of Business and Technology: 2005, Egbunike: 2009b).

It is sad that although Nigeria has "a total grid capacity of 5, 924.7MW, only 4, 586MW is available" (Okoro and Chikuni: 2007). Matters are more complicated with the huge funds that have been lost in the bid to generate power for the country. Since 1999, government has spent some $10 billion on revamping the electricity sector (Smith and Fagbule: 2010). Nigerians have long realised that it makes no sense connecting to the national grid; with *I better pass my neighbour* (Nigerian pidgin for a 2 KVA generator), power is assured.

Therefore it is no news that with this abysmal situation, that power ranks high as the major restrain for a new media culture to flourish. Since Nigerian youths are not secluded from the prevailing circumstances, despite living in better-off university campuses and hostels, their potency to use the internet will be necessarily curtailed.

In addition, Emenari (2003) observes that the technological backwardness of Nigeria has restricted internet use to the cities and shutting out over 70 percent of the country's population which reside in the rural areas. Other information technology limitations in Nigeria, (Oni *et al*Oni *et al*: 2009) include poor services, poverty and socio-cultural factor among others.

Poor Internet Services

Although phone and internet services have flourished since 1999 when Nigeria was restored to democratic rule, yet the tales abound about the poor service of some Internet Service Providers (ISP). Most cybercafés can hardly function when it rains. Mark Nwagwu, a retired university professor, narrates his disgust in an op-ed article in the *Guardian* (2010) thus:

Recently, I was in Owerri, at one of the mobile phone companies trying to get them to attend to my internet phone modem that I had purchased at great sacrifice. Nobody could tell me what was going on and they parried me, giving me one reason after another why they could not yet solve my problem. This went on for over one hour with their repeats of, 'please come back in fifteen minutes, sir.'

Poverty

Poverty and computer illiteracy are Siamese twins as regards the inability to embrace information technology. The minimum wage of civil servants was recently increased from N7,500 (US$50) to N18,000 (US$120). As such few can afford personal computers and amongst those who can, it serves other purposes like video and music players.

MALARIA AND YOUTHS: THE NIGERIA STORY

It will be no exaggeration to affirm that most young Nigerians are internet aficionados, and their resilience in overcoming the harsh IT climate is inspiring. Nonetheless it will be an illusion of grandeur to conclude that the new media alone has been able to satisfy their information needs about malaria. While most have not been spared the ordeal associated with malaria, it seems that much more is still to be done.

From the study above, most youths satisfy their information needs about malaria from non-media sources (parents, school, books, friends and from the hospital). Although Table 7 shows that the old media still fares better than the new media, nonetheless the potentials of the internet should be explored. This validates the position of some scholars that the new media would only complement the old ones and not totally annihilate them. Besides it only shows a lack of depth in planning; as if communication is seen as an appendage to the anti-malaria campaign. What is particularly pitiful is that this illness which has more dreadful manifestations is not getting as much media attention/mileage as HIV/AIDS.

Going by the high frequency of the respondents who stumbled upon information about malaria in the internet, we dare to assert that the internet has been helpful in *communicating* strategies for malaria control. However, this has not translated into *adoption* of control measures for the malaise. One the contrary, we noticed that most young people are more prone to treatment than control of malaria. This attitude can be appreciated since anti-malaria therapies are more convenient and cheaper. However there is wisdom in the old saying that 'prevention is better than cure'. Preventive measures, nonetheless, require the cultivation of certain habits, and this takes time.

One of such healthy and life saving habit lies in the adoption of Insecticide Treated Nets (ITNs). This chapter has shown a dismal record of affinity for ITNs by the youths. It has also validated the concept of social change and development – new communications, mass information and persuasions – in the context of ITN's adoption amongst Nigerian youths. In summary it shows that knowledge does not necessary translate into acceptance. This does not mean that all hope is lost, because behavioral change can be achieved with a gradual but constant appeal to people. And the youths are no exception. If the millennium development goal on health is ever to be realised, then now is the time to start.

As grand as the FMOH's focus on making ITNs available to nursing mothers and children are, much more still needs to be done. The reality, as this chapter has demonstrated, is that the youths of Nigeria are totally at sea as regards FMOH's policy on malaria control. This shows a certain attitude of policy makers and the media who set public agenda by talking *for* the people rather than talking *with* the people. As Ojebode (2008) rightly noted, this is one more manifestation that the last vestiges of long military rule in the country has not been erased. "The result is a government or an agency which decides what the citizens want and goes ahead to do just that, the way it understands it, using the means it is pleased with." And by so doing, thinks that their wishes will become reality by divine or magical fiat.

It is high time we leap from rhetoric to action. The future of Nigeria lies on the development of her youth. This holistic progress cannot come about with a generation of sickly young people. Besides, in a decade of two, these youths will become parents, policy makers and media practitioners. If they do not form the habit of using ITN now, then it's dooms day awaiting Nigeria.

While we insist that the government should change focus, NGO's and donor agencies should also follow suit. It makes no sense sinking billions of dollars into the provision of ITNs only for it to get stacked in a warehouse, accumulating dust. It is therefore worthwhile to commend the Global Fund against AIDS, TB and Malaria for "providing close to US$250 million worth of ACT in our Round 8 of malaria intervention to Nigeria." Dr Nze Sylva Ifedigbo – a consultant to GFATM – also informed us that the intervention is pursuing a two modes approach: private and public sector. "We also increased acceptance of the Long Lasting Nets (which is the ITNs and the chemical for treatment) in northern Nigeria by distributing same on the National Immunisation Days."

RECOMMENDATIONS

1. The future is bright to those unafraid of change, those who know how to absorb the best of globalisation and yet retain the best of the past. Nigerian youths are high users

of new media and this should be explored for malaria control.

2. The new media should also be embraced as a *necessary* channel for ITN adoption in malaria control. This entails a reversal of approach and the digitalisation of ideas in order to log into this popular medium amongst youths. More local malaria campaign should be present in social network sites, especially those that are frequently used by the youths.
3. Since infrastructure is necessary for ITN adoption and the flourishing of new media, government should urgently revamp the nation's public utilities. More regular electric power for instance, will make it easier for not only young people but also for other citizens to sleep under a mosquito net in a tropical region like Nigeria.
4. There has to be a major policy shift by the FMOH in malaria control. As noble as the present emphasis on lactating mothers and children is, the other vulnerable and active population – the youths – should also be addressed. This is because they constitute a major part of the labour force of the country and if they key into the habit now, then they will likely pass same to their offspring.
5. Since adoption of ITN involves a behavioural change, then the message of this advocacy should be attractive to youths. These young people should be made to see that it is also trendy for them to sleep under mosquito nets. This is where personality ambassadors like music stars, comedians and Nollywood actors will make a great impact.
6. There should either be a shift from ACT's to ITN's or a combination of both. Young Nigerians should be convinced that it is easier to prevent than to cure. Nothing stops NGO's (like GFATM or Kick-Out Malaria) and the FMOH to also hand out free ITN's together with ACT's in medical centres in tertiary institutions in Nigeria.
7. More advocacies about malaria in both old and new media should be encouraged. E.g. having live musical shows, capacity building seminars, competitions like debates, talent hunts, etc., in the tertiary institutions on the malaria malaise. However, this should be done *with* the youths. Thus a good approach is to work with student groups – and there many of these associations on the various campuses in the country.

ACKNOWLEDGMENT

We are particularly grateful to the following for making this chapter a reality. Dr Ikechukwu Obiaya (School of Media and Communication, Pan African University, Lagos) like Virgil leading Dante, guided us through the treacherous valleys that marked the path of this work. Dr Ayobami Ojebode inspired the flame of research through his lectures in the post-graduate class of the department of Communication and Language Arts of the University of Ibadan. We are also indebted to Dr Anthony Obeameta, department of Sociology, University of Ibadan. Also a million thanks to the anonymous reviewer whose assessment turned this work into a decent chapter.

REFERENCES

African Review of Business and Technology. (2005). Genset sales continue to soar: The world wants more diesel generator sets, and manufacturers are hard pressed to supply. *Entrepreneur*. Retrieved on May 13, 2010, from http://www.entrepreneur.com/tradejournals/pub/gen sale.html

Bordens, K. S., & Abbot, B. B. (2008). *Research design and methods: A process approach* (p. 165). Boston, MA: McGraw-Hill.

Clark, J. P. (1999). Ibadan. In Gerald, M., & Beier, U. (Eds.), *Penguin book of modern African poetry* (p. 256).

Dunn, A. (2005). Participatory communication in malaria control: Why does it matter? *Findings, 4*, October.

Economist. (2010, January 30). *A world of connections: A special report on social networking.* (pp. 3-20).

Edugie, A., Mosanya M. E., Amajoh C., Otsemobor O., Ezedinachi E. N. U., Afolabi B. M.,... Aaron, K. O. (2004). *Nigeria roll back malaria consultative mission: Essential actions to support the attainment of the Abuja targets.* Nigeria RBM Country Consultative Mission, Final Report, 07 March.

Egbunike, N. A. (2009a). *Peering into the future: The fate of newspapers in a digital world.* Unpublished Seminar, Department of Communication and Language Arts, University of Ibadan.

Egbunike, N. A. (2009b). Power play. *Interpress Newsletter, 34*(3-4).

Emenari, O. G. (2003). The Internet as a challenge to broadcasting and allied media. In Onyeka, N. (Ed.), *Communication and national development.* Onitsha, Nigeria: African Links Books.

Folarin, B. (1998). *Theories of mass communication: An introductory text.* Ibadan, Nigeria: Stirling-Horden Publishers Ltd.

Fourchard, L. (2002). *The case of Ibadan, Nigeria.* Ibadan, Nigeria: Institut Francais de Recherche en Afrique (IFRA), University of Ibadan. Retrieved May 13, 2010 from http://www.ucl.ac.uk/dpuprojects/Global_Report/pdfs/Ibadan.pdf

Marczyk, G., DeMatteo, D., & Festinger, D. (2005). *Essentials of research design and methodology.* Hoboken, NJ: John Wiley & Sons, Inc.

McCabe, J. (2009). Resisting alienation: The social construction of Internet communities supporting eating disorders. *Communication Studies, 60*(1), 1–16.

McQuail, D. (2007). *McQuail's mass communication theory.* London, UK: Sage Publications Ltd.

Nwagwu, M. (2010, April 28). Where is Nigeria? *The Guardian*, p 67.

Ojebode, A. O. (2008). Low patronage of development radio programmes in rural Nigeria: How to get beyond the rhetoric of participation. *Journal of Multicultural Discourses, 3*(2), 135–145. doi:10.1080/17447140802240019

Okoro, O. I., & Chikuni, E. (2007). Power sector reforms in Nigeria: Opportunities and challenges. *Journal of Energy in Southern Africa, 18*(3), 52–57.

Oni, A. A., Oni, O., & Omidiji, O. G. (2009). *The World Wide Web.* Seminar Paper presented at the Department of Communication and Language Arts, University of Ibadan, Ibadan.

Onwujekwe, O., Uzochukwu, B., Ezumah, N., & Shu, E. (2005). Increasing coverage of insecticide-treated nets in rural Nigeria: Implications of consumer knowledge, preferences and expenditures for malaria prevention. *Malaria Journal, 4*, 29. doi:10.1186/1475-2875-4-29

Opiyo, P., Mukabana, W. R., Kiche, I., Mathenge, E., Killeen, G. F., & Fillinger, U. (2007). An exploratory study of community factors relevant for participatory malaria control on Rusinga Island, Western Kenya. *Malaria Journal, 6*, 48. doi:10.1186/1475-2875-6-48

Osifo-Whiskey, O. (in press). *Witness to justice: An insider account of Nigeria's Truth Commission.*

Osofisan, F. (Ed.) and School of Post Graduate Studies UI. (2008). *Sixty iears of the muse at UI: A commemorative directory.* Lagos, Nigeria: Concept Publications.

Osotimehein, B. (2009, April 24). In Nigeria, others race against time in the fight against malaria. *The Guardian*, pp. 24-25.

Roll Back Malaria. (2008). *The global malaria action plan for a malaria free world.* Retrieved on November 19, 2009, from http://rollbackmalaria.org

Smith, P., & Fagbule, T. (2010). Reform stutters amid calls for power shift. *Africa Report, 22*, 45–48.

UI. (2008). *60th anniversary brochure*. Ibadan, Nigeria: Publications Unit of the Registrar's office, University of Ibadan.

United Nation Population Division. (2000). *World population prospects: Population database*, 2002 revision. Retrieved on November 19, 2009, from http://esa.un.org/unpp

Wang, S. (2009). Is convergence enough to make existing media a brighter future? *China Media Report Overseas, 5*(1).

World Health Organization. (2005). *African Summit on Roll Back Malaria*, Abuja, Nigeria, April 25 2000. Geneva, Switzerland. (Document WHO/CDS/RBM/2000.17).

ADDITIONAL READING

Ajuwon, G. A., and Rhine, L., (2008). The level of Internet access and ICT training for health information professionals in sub-Saharan Africa. *Health Information & Libraries Journal*, Sep, Vol. 25 Issue 3, p175-185.

Boyle, M. P., Schmierbach, M., (2009). Media Use and Protest: The Role of Mainstream and Alternative Media Use in Predicting Traditional and Protest Participation. *Communication Quarterly*; Jan, Vol. 57 Issue 1, p1-17.

Carpini, M. X. D., (2000). Gen.com: Youth, Civic Engagement, and the New Information Environment. *Political Communication*, Oct-Dec, Vol. 17 Issue 4, p341-349.

Coleman, R., Lieber, P., Mendelson, A. L., and Kurpius, D. D., (2008). Public life and the internet: if you build a better website, will citizens become engaged? *New Media & Society*, Apr, Vol. 10 Issue 2, p179-201.

Department for International Development. (2008). *Malaria Factsheet*. December, Crown.

Diop, W., (2000). From Government Policy to Community-Based Communication Strategies in Africa: Lessons from Senegal and Uganda. *Journal of Health Communication,* June, Vol. 5 Issue 3, p113-117.

Evans, W. D., Blitstein, J., Hersey, J. C., Renaud, J., and Yaroch, A. L., (2008). Systematic Review of Public Health Branding. *Journal of Health Communication*, Dec2, Vol. 13 Issue 8, p721-741.

Fork-Kintz, L. (2007). *Current Media Practices and Trends in West Africa: A Case Study Analysis of the Media of Ghana and Nigeria*. Conference Papers - National Communication Association, p1.

Hesse, B. W., (2009). Cancer Communication: Status and Future Directions. *Journal of Health Communication*, January Supplement 1, Vol. 14, p109-127.

Hoffman, L. H., Appiah, O., (2008). Assessing Cultural and Contextual Components of Social Capital: Is Civic Engagement in Peril? *Howard Journal of Communications*, October, Vol. 19, Issue 4, p334-354.

Langlois, G., Elmer, G., McKelvey, F., & Devereaux, Z. (2009). Networked Publics: The Double Articulation of Code and Politics on Facebook. *Canadian Journal of Communication, 34*(Issue 3), 415–434.

Lines, J., Whitty, C. J. M., & Hanson, K. (2007). *Prospects for Eradication and Elimination of Malaria: A Technical Briefing for DFID*. December, London School of Hygiene & Tropical Medicine.

Locke, G., (2005). Just another day in Africa? Hardly. *Media*, spring, Vol. 11 Issue 2, p20-21.

Munthali, K. (2007). Africa: What Went Wrong? *Nieman Reports*, spring, Vol. 61 Issue 1, p116-116, 1/2p.

Pasek, J., Kenski, K., Romer, D., and Jamieson, K.H., (2006). America's Youth and Community Engagement: How Use of Mass Media Is Related to Civic Activity and Political Awareness in 14- to 22-Year-Olds. *Communication Research*, June, Vol. 33 Issue 3, p115-135.

Pratt, C. B., Ha, L., and Pratt, C. A., (2002). Setting the Public Health Agenda on Major Diseases in Sub-Saharan Africa: African Popular Magazines and Medical Journals, 1981-1997. *Journal of Communication*, Dec, Vol. 52 Issue 4, p889.

Shah, D. V., Kwak, N., & Holbert, R. L. (2001). "Connecting" and "Disconnecting" With Civic Life: Patterns of Internet Use and the Production of Social Capital. *Political Communication. Apr-June*, *18*(Issue 2), 141–162.

Smith, R., Hull, S., & Catona, D. (2009). *Mixing It Up: Investigating Engagement in Multiple Vector-Control Behaviours to Prevent Malaria in Mozambique*. Conference Papers -International Communication Association; 2009 Annual Meeting, p1-24.

Urban, M., (2002). Studying Health-care systems in Africa. *News Photographer*, Feb, Vol. 57 Issue 2, p36.

World Bank. (2007). *The World Bank Booster Program for Malaria Control in Africa: Scaling-Up for Impact (SUFI), A Two-Year Progress Report.* Retrieved on January 15, 2010 from www.worldbank.org/afr/malaria.

KEY TERMS AND DEFINITIONS

EAC: Education Advancement Centre
ITN: Insecticide Treated Net
GFATM: Global Fund against AIDS, Tuberculosis and Malaria
IEC: Information Education and Communication
ISP: Internet Service Providers
NGO: Non Governmental Organisation
NMCC: National Malaria Control Committee
NMCP: National Malaria Control Programme
LLITN: Long Lasting Insecticide Treated Nets
UI: University of Ibadan
GSM: Global System of Mobile Communication

Chapter 11

Translanguaging and Negotiation of Ethnicity: Reproduction of Hegemonic Structures in Communication Media

Peter Githinji
Ohio University, USA

ABSTRACT

Looking at the way the new communication technology has dominated Kenyan's daily lives, it is hard to imagine how they functioned without it. In spite of these developments, the traditional institutions that mediate social relations of power remain unchanged. Essentially, the new communication technologies have only succeeded in providing a new arena where the traditional discourse is reproduced. Standing at the intersection between these new technologies and communication is language—the medium that facilitates interpersonal interaction both in traditional modes of interaction and in digital interactions.

We shall be examining how the linguistic practices interact with these new communication technologies. By emphasizing on the centrality of language in mediating social interactions, we argue that post-traditional discourse fails to extricate itself from cultural hegemony due to global trends that favor the hegemonic languages as well as the ingrained habitus[1] that predisposes Kenyans to reproduce linguistics structures inspired by the discourse of power of hierarchy of languages in the linguistic marketplace. Our discussion opens with a brief overview of the place of language in a technological world.

INTRODUCTION: LANGUAGE AND COMMUNICATION TECHNOLOGY

In the past decade, communication technology in Kenya has experienced a dramatic growth. The CIA World Factbook data shows that in 2008, there were 252,300 phone lines, 16 million mobile phone subscribers and 3.36 million internet users in Kenya. This unprecedented development in new communication technologies has completely altered the interaction habits and information access among Kenyans. The government control on the flow of information is diminishing as the industry unleashes innovative products that are

DOI: 10.4018/978-1-60960-591-9.ch011

redefining the nature and forms of interaction, information gathering and consumption as well as the way people conduct their daily business. Institutions and services that were previously controlled by the powerful elite are slowly being demystified and sometimes dismantled. Kenyans no longer have to resign themselves to censored information, thanks to increasing access to new communication technology. In the economic sector, the poor farmers do not have to put up with middlemen and brokers to come and determine the price of the produce as they can easily monitor the market trends without making a single trip.

The marginalized rural poor who for a long time have been shut out by the banking industry can now send and receive money on their mobile phones. Looking at the way the new communication technology has dominated Kenyan's daily lives, it is hard to imagine how they functioned without it. In spite of these developments, the traditional institutions that mediate social relations of power remain unchanged. Essentially, the new communication technologies have only succeeded in providing new arena where the traditional discourse is reproduced. Standing at the intersection between these new technologies and communication is language—the medium that facilitates interpersonal interaction both in traditional modes of interaction and in digital interactions.

We shall be examining how the linguistic practices interact with these new communication technologies. By emphasizing on the centrality of language in mediating social interactions, we argue that post-traditional discourse fails to extricate itself from cultural hegemony due to global trends that favor the hegemonic languages as well as the ingrained habitus[2] that predisposes Kenyans to reproduce linguistics structures inspired by the discourse of power of hierarchy of languages in the linguistic marketplace. Our discussion opens with a brief overview of the place of language in a technological world.

With respect to communication technology, Walther (2004) offers three rationales for studying language and social psychology. First is to interrogate how technology affects communication, use of language and resultant processes in order to see if it imposes, alters or changes the nature of conversation, relationships, attitudes or outcomes that are shaped by communication transactions. The second reason is that it illuminates the users' triumph in the control of technology during the transformation of one symbolic system to another while the third rationale is that studying language and social psychology enables us to adapt technology to make communication more effective. These rationales scaffold our discussion of the dialectics of ideological hegemony and communication technology. In essence, language users in a multilingual setting exploit linguistic resources to challenge the hegemony of mainstream interactive norms while at the same time using communicative technology to engage in linguistic practices that reconstitute hegemonic structures. This chapter focuses on translanguaging and the consumption of ethnic discourse by Kenyans at home and in the Diaspora. An understanding of the intersection between language and ethnocultural identity allows us to understand the resilience of ethnic languages as they fight for survival in a linguistic environment dominated by transnational languages of wider communication. Do these new technologies adapt themselves to operate in a multilingual environment or do people adapt their everyday communicative norms in order to operate in a technological world?

It is perhaps important to point out that even before the coming of technology, human interaction as far as languages are concerned has been characterized by transformations as speakers modify their languages to reflect their realities. As new phenomena prompt linguistic modification, the symbolic representation of phenomena that do not feature in everyday discourse drift away from the speakers' semiotic system instigating linguistic changes with far reaching ramification. Migration of people to novel environments or conquest by a power that imposes its cultural system can result in linguistic changes that can alter the surface representation of the language such that it becomes almost unrecognizable from its former self.

One need only compare modern English with old English to see how a series of historical events dating way back to the medieval period have had a lasting effect on English. Unlike English, many languages are not able to withstand the outside pressures and influences and gradually die out, while those that were lucky to be documented have been preserved as artifacts. On the other hand, some languages and dialects have been created as a result of human contact, though sometimes brutal. African American Vernacular English (AAVE) in the US, Haitian Creole in Haiti and Tok Pisin in Papua New Guinea are just some few examples of the outcomes of linguistic convergence. In Africa, Cape Verdean Creole, Krio in Sierra Leone, Nigerian Pidgin and emergent urban linguistic codes like Sheng in Kenya, Urban Wolof in Senegal and Town Bemba in Zambia illustrate how dynamic languages are as they adapt to new realities. In our examination of the emergent trends occasioned by new communication technologies, we do not regard technology as the root of all evil with regard to the marginalization of African languages in post traditional interactions. We shall however emphasize that the interactive practices that have emanated from the integration of technology into our daily lives not only mirror familiar trajectories of historical events and upheavals that have left lasting impacts on languages, but also that new practices reflect the adaptive and regenerative aspect of languages as they struggle for relevance in hostile environments that threaten their very existence.

Languages and especially ethnic languages have usually been associated with communities of speakers. The relationship between language and identity has been constantly cited in previous literature. In Fishman's (1991) explanation of Xish/Xmen[3] connection for instance, he emphasizes the importance of language as a key component of ethno-cultural vitality and identity even while appreciating that there are other elements which are equally important in defining the essence of someone's ethnicity. This fact is emphasized by Myhill (2003) who points out that while such a perspective can be valuable in certain contexts, it risks excluding some categories of people who prefer not to use their languages but like to retain their ethnic identity[4]. This is especially so in cyberspace where participants build communities modeled along the lines of their ethnic communities but does not necessarily revolve around the use of an ethnic language.

We shall therefore argue that while ethnic languages may be important components in ethnic categorization, they do not need to be the sole defining features of a community, especially an imagined online community. Before looking at language and identity politics, we shall briefly look at triglossia and translanguaging as characteristic practices among Kenyans. Understanding the allocation of functions to social and official domains illuminates our appreciation of how linguistic practices propagate ideological hegemony in the era of new communication technologies.

TRIGLOSSIA AND TRANSLANGUAGING AMONG KENYANS

Translanguaging involves linguistic practices in which more than one language is used in social interactions. According to Garcia (2008: 45), translanguagings are "*multiple discursive practices* in which bilinguals engage in in order to *make sense of their bilingual worlds.*" This definition includes all kinds of bilingual behavior like codeswitching, linguistic hybridization and other forms of linguistic compromises or accommodations that characterize bilingual or multilingual speech. In Kenya, the 'triglossic' language policy (Mkilifi

1978) assigns English the official and high status functions while Swahili serves the national and cross ethnic interactional functions. On the other hand, ethnic languages are confined to local interaction and cultural functions. This hierarchical order accords English the highest symbolic value while consigning lower symbolic value to national and vernacular languages. Similar to many post colonial societies, the English hegemony and its preference by the Kenyan elite can be viewed within Bourdieu's (1991) metaphor of linguistic marketplace where languages are regarded as commodities with a market value. In social transactions, speakers opt for the language with the highest symbolic capital in different interaction types. Although English might function in official and some non-official domains, it might be totally irrelevant in minimal group interactions where mastery of a local speech variety is held in high esteem. Nevertheless, languages of wider communication are still a preserve of a few among the nation's elite[5] which limits their application across domains. This allows Kenyans to spend their remaining symbolic linguistic capital in Swahili, the national language and other vernacular languages.

In many multilingual societies in sub-Saharan Africa, translangauging represents areas of convergence, divergence and overlap of discursive practices as speakers spend their symbolic capital. Codeswitching is one area where speakers are able spend almost their entire symbolic capital. Use of codes such as Sheng (Githinji 2006, Githiora, 2002, Ogechi, 2005) Town Bemba (Spitulnik 1999) or Camfraglais (Kuoega, 2003) for instance, indicate areas of linguistic convergence occasioned by language contact and linguistic and social divergence as urbanites negotiate a cosmopolitan identity. Other translanguaging practices involve cases of language crossing (Rampton 1985)[6] where speakers of one ethnic language attempt to accrue symbolic capital by using the code of a different ethnic group. The linguistics choices made by a person who has learnt some few words in Gĩkũyũ or a Kikuyu who can speak some sentences in Dholuo are borne out of the desire to accumulate as much symbolic capital as possible in cross-ethnic encounters where either of the two languages dominate. It is this complexity of interactions in a multilingual setting that makes it inaccurate to characterize many ethnic conversations as purely driven by ethnic loyalty.

Translanguaging may be regarded as an outcome as well as manifestation of the pervasive hegemony of languages of wider communication. The fact that speakers mix languages in order to gain status attributed to languages of prestige without losing the solidarity rewards associated with local languages illustrates that they are willing to sacrifice linguistic purity for the sake of expediency. When the prestigious languages provide the grammatical template during codeswitching, the prominence of ethnic languages is diminished since the same language is likely to provide a higher frequency of linguistic material. Translanguaging amongst educated Kenyan has a high frequency of English which sustains the linguistic status quo as far as language and power are concerned. On the other hand, it is possible that language mixing subverts linguistic hegemony by lexicalizing the matrix language with ethnic words, creating hybrid lexemes, displacing the prestigious languages as the matrix languages as is the case with Sheng in Kenya.

Linguistics practices such as translanguaging involve making choices which can either be marked or unmarked. Myers Scotton's (1993) defines marked choices as the linguistic norms that are consciously made by speakers after evaluating that costs and rewards of their utterances. They are therefore conscious and normally deliberate. Unmarked choices on the other hand are conventional interactive norms that are made subconsciously. According to Myers-Scotton, unmarked choices are made because interlocutors want to sustain the status quo while marked are made when they want to challenge the status quo. In face-to-face interactions, codeswitching employs marked choices and thus challenges the linguistic prescriptivism by rejecting linguistic purism. Breaching the purity of the standard languages demystifies the elite closure (Myers-Scotton 1990) by making a prestigious code accessible to non-elite. One does not need high mastery of English or French to insert some French or English words in their utterances and in absence of someone who can reveal the speaker's linguistic background no one can fault the speaker's competence in the language of status. Codeswitching also challenges the interactive dynamics of ethnic languages. While ethnic language purists lament the mutilation of their languages, switchers continue to extract what they need for their interaction.

Another reason for the prevalence of codeswitching among Kenyans is the breakdown of the traditional family under the pressures of modernity which has adversely affected acquisition of standard language by the youth. The choice of foreign languages as media of instruction in the educational system has challenged the traditional modes of language socialization. Children no longer have the support of the social institutions in learning structural and sociolinguistic rules of their ethnic languages. Semilingualism which is an outcome of disruption in channels for language acquisition leaves some speakers with no option but to codeswitch. Once this becomes the norm, codeswitching is no longer a marked choice but transforms into an unmarked choice. Since codeswitching to a large majority of Kenyans, takes place at a subconscious level, it has become an unmarked choice such that interlocutors find it as the most natural and only logical form of interaction especially in face-to-face interactions. If we assume that interlocutors strive to ensure the success of an interaction (see Grice, 1975) then we can reason within the Cooperative Principle, that they internalize certain interactive rules that orient their linguistic choices towards the unmarked. Why then do online discussion forums where internet anonymity reduces the threat to face (Goffman, 1967) still bound by the rules that operate in face to face discourse? The three possible explanations are 1) comprehensibility 2) the notion of imagined community and 3) the pervasiveness of hegemonic ethnic discourse.

The pervasiveness of ethnic discourse in online discussions is a reflection of the way ethnic languages are deployed in face-to-face interactions. It is important to note that internet participants do not live in isolation. The dialectics of internet interactions are that it feeds and is fed by face-to-face discourse. Internet does not operate in a vacuum; although it can create a type of unique discourse, its primary function is to facilitate the consumption of discourse. To a large extent, the language of the internet is an echo of the language of face-to-face interactions. Those contributors who codeswitch in their face-to-face interaction are likely to post codeswitching material in their online interactions. Hence, the hegemony of ethnic discourse in face-to-face interaction inevitably finds its way into the internet.

LANGUAGE AND ETHNIC IDENTITY

Language is a vital component in the affirmation and contestation of ethnic identities, especially in face-to-face interactions. Whether intended or not, speakers' speech behaviors reveal certain information about their identity. Indeed, the relationship between language and social identity has been a dominant topic in social psychology, sociolinguistics, linguistic anthropology and language acquisition studies among others. Le Page and Tabouret-Keller (1985) regard all linguistic utterances 'acts of identity' in appreciation of the centrality of language in social transactions. Krauss and Chiu (1998) inform us that languages not only function as markers of identity, but they also maintain social identities in addition to providing listeners with information that allow them to categorize speakers of a language or a speech variety. Studies in language variation such as Blackledge and Pavlenko (2001), Meyerhoff

(1998), and Hazen (2002) among others have stressed the pivotal role that language plays in negotiating the speaker's identity. On their part, Myers-Scotton and Bolonyai (2001) have argued that speakers make rational choices in order to gain rewards in interpersonal communication. In social psychology, insights from communicative accommodation theory (e.g., Giles, Taylor & Bourhis 1973; Giles, Copeland & Copeland 1991; Giles & Powesland 1975; Williams 1999) have shown how intergroup interactions are characterized by linguistic modifications relative to contexts and interlocutors. In acts of linguistic convergence, speakers minimize social distance by engaging in linguistic practices that align with those of desirable others while in linguistic divergence, they increase or overplay social differences by engaging in practices that intensify distance with undesirable others. This is more evident in multilingual setting such as urban areas where neutral codes are chosen for cross-ethnic interaction. In Sub Saharan Africa the most logical codes are international languages like English, French and Portuguese, as well as national or regional lingua francae such as Swahili in East and Central Africa, Somali in the horn of Africa and Lingala in the two Congos. Although ethnic languages may feature in cross ethnic interaction as in the case of language crossing (see Rampton 1985), they serve minimal interactive roles in cross cultural interaction, and are mostly confined to intra-ethnic interactions. The use of the so-called neutral languages does not eliminate the possibility of ethnic identification. Thanks to transfer issues or the so called 'cross linguistic influence' (Dopke, 2000Sinka, 2000) in bilingualism literature, linguistics structures from ethnic languages such as deviant phonology or insertion of lexical materials are always evident in the speech of bilinguals. A good illustration is the following sentences from *kisii.com*;

- okay pati tonti tu tati again
  Okay, but don't do that again
- i goti iti furomu mae girandimata
  I got it from my grandmother

The first sentence is taken from a dialogue between a mother and her son as observed by a participant in the discussion forum while the second example comes from a different participant who was advancing the dialogue. Although these sentences are in English, the cross linguistic structures at the phonological level completely alters the sentence such that it is unrecognizable to native speakers. In Kenya for instance, the Kikuyus are known for their leveling of phonemes [r] and [l], the Luos are famed to have a problem in their production of phoneme [š], while the Kambas are known for their dropping or addition of word initial [h]. Such cross linguistic structures in English are fodder to the ethnic stereotypes associated with speakers of those ethnic groups even when they are not speaking their ethnic languages. Other examples of cross-linguistic structures come from the following two sentences from Myers-Scotton's work on codeswitching in Africa;

- bene o-nwongo gi-using Swahili ka-terrorizing lu-civilians
  One finds they are using Swahili to terrorize civilians
- *Sasa unaweza nieleza juu ya dini, tuko confused*
  Now you can tell me about religion, we are confused

(Myers-Scotton 2002)

Although Acholi-English and Swahili English codeswitching in these two examples may be regarded as cross linguistic influence, Myers-Scotton (1992 & 1993) warns against regarding all instances of codeswitching as evidence of linguistic deficiency. Speakers sometimes codeswitch because they want to exploit the symbolic benefits associated with the codes involved (Myers-Scotton 1993). While speakers use

Swahili, the national language to identify with the national culture, they switch to English in order to identify with the status and power attributed to English. In a different context, the same speakers may use languages of their ethnic affiliation to cultivate ethnic solidarity. Therefore, during the negotiation of ethnic identities, language not only categorize a speaker as a member of a certain ethnic affiliation (see Krauss & Chiu 1998) but they also mediate social transaction where ethnicity is affirmed. Myers-Scotton (1989) cites a situation where two Kikuyu speakers were engaging in a conversation in Gĩkũyũ in the company of Kisii and Kalenjin speakers. The Kisii speaker protested by codeswitching between Swahili and English while the Kalenjin speaker opted to speak in Swahili;

- "Ubaya wenu Wakikuyu ni ku-assume kila mtu anaelewa Kikuyu"
  The problem with you Kikuyus is to assume that everyone understands Kikuyu,
- Si mtumie lugha ambayo kila mtu hapa atasikia?"
  Why don't you use the language that everyone here can understand?

By engaging in what Bucholtz (1999) calls negative identity practices, the Kikuyus engage in linguistic divergence to create social distance with their companions. The Kisii speaker uses a non-ethnic language to affirm the identity of the two Kikuyu speakers by saying 'Ubaya wenu wakikuyu…*the problem with* ***you Kikuyus***… The Kikuyus themselves are aware of these charges of linguistic arrogance by non-Kikuyus. In the jokes section at *kikuyu.com* for instance, a contributor posted a common joke that Kikuyu men think all light skinned women are from their tribe. Therefore, if they see one passing by, they talk to them in Kikuyu and if the girl happens not to speak Kikuyu, they accuse her of pretending.

Just as Krauss and Chiu (1998) had claimed, language not only categorizes others, but as conscious acts, linguistic practices play a critical role in the negotiation of identity. Members of criminal gangs use coded languages to isolate themselves from non members while elites choose to use the high status languages in order to mark status boundaries. People opt to use a form of speech identified with other groups because they want to identify with those groups. We however need to mention that although the primary function of language is to facilitate communication, it can sometimes loose the communicative function and only retain the symbolic one (Edwards, 1985) such as identity. In this context we may view the quest for national language as aimed at solidifying the perception of an imagined national community indexed by a shared common language. In summing this section we take into account the fact that the role of language in identity is not just limited to ethnic and national languages. When a monolingual speaker chooses one variety of the language over others or when bilinguals opt for one language over others in their repertoire, the main rationale is to optimize communication outcomes. A multilingual speaker who wishes to identify with members of his or her ethnic group will always communicate using the shared ethnic language. A non-ethnic language may however be used by speakers who share the same ethnic language if both are familiar with the non-ethnic language.

LANGUAGE, POWER AND HEGEMONY

We take Woolard's (1985:739) definition of hegemony as referring to "the legitimation of the cultural authority of the dominant group, an authority that plays a significant role in social reproduction…" This definition is informed by the Gramsci's cultural hegemony where a social class exerts power and control over others. Contrary to classical perception of hegemony that rely on force and coercion, the Gramscian perspective capitalizes on the notion of ideology where the ruling class cultivates a common sense way of doing things which makes the dominated implicitly work against their own self interest. Let's take a multilingual country as an example. The high value placed on national languages which are overwhelmingly foreign, maintain the status quo as far as power relations are concerned. The elite who control political power also control economic power and they maintain the status quo by erecting social structures that limit accessibility of the languages of status by the dominated. In Kenya for instance, as the school performance in under resourced public schools continues to plummet; private institutions continue to celebrate their students' impressive performances. Since these private institutions are only accessible to those students from high socioeconomic backgrounds, students from poor background remain deprived of the very means they need for their socioeconomic advancement. This reality calls for a reexamination of the fairness of pedagogy based on a hegemonic language. Inadequate funding and endemic corruption compound the problems in under-resourced public schools.

The resource gap between public and private institutions means that lack of qualified language teachers in public schools compromise students' education success. At the same time, students in private schools are guaranteed better education which gives them an advantage in the job market. This also has implications for the use of new communication technology; until recently, internet domain names were all in English, a situation that limited internet participation by many non-English speakers. As a result, e-commerce and vital information for self actualization was inaccessible to those without good English skills.

At another level, studies that follow the Sapir-Whorf tradition (e.g. Fantini 1985) emphasize the connection between language and worldview. We are interested here in Whorf's recognition that a language contains cryptotypes that are encrypted into its structural categories that help in accessing the meaning of phenotypes (Whorf 1956). While phenotypes are words or morphological categories identified in grammar, cryptotypes are the "submerged, subtle, and elusive meaning, corresponding to no actual word, yet shown by linguistic analysis to be functionally important in the grammar" (p 71). These cryptotypes in the languages of status make the language opaque which increases its mystique and its reverence. Although one may argue that such meanings might not apply to Kenya since they are not the native speakers of English, we cannot underestimate the fact that a majority of the speakers, having adapted English as an instrument for their power are happy to adopt the worldview that that language espouses.

We mentioned above that the elite that wield socioeconomic authority possess an advantage when it comes to participation in global modes of consumption, thanks to their advantage at different levels. First, their children participate in good schools where computer instructions are offered as opposed to public schools where even the administrators do not have computer access. Second, the parents have the economic means to buy their personal computers which introduce their children to new communication technology at a young age and finally, the children, thanks to quality education, have the language that allows them to navigate in global discourse. All these factors act in tandem to propagate English hegemony in addition to sustaining the status quo as far as power and control are concerned.

Coming to the national language Swahili and the ethnic languages, we note that the elite is keen to provide a space for their use in order to avoid the backlash by the powerless. By providing a place for these languages in an alternative linguistic marketplace (Woolard 1985) the elite is co-opting them into the hegemony so that they do not act as opposing linguistic marketplaces that challenges the status quo. Declaring Swahili as a national language is a token appeasement that sooths Swahili loyalists who come to support the status quo as the only logical option. On the other hand, ethnic fragmentation does not pose a real danger to the elite. However, popular countercultural codes like Sheng that draw from the language of status and local languages are a threat to established order because they cannot be controlled. Introducing a hybrid discourse not only acts as an antidote to ethnic fragmentation, but it also empowers the powerless and challenges the status quo by deliberately breaching the 'conventional' norms (see Mbugua, 2003). The constant denunciation of Sheng by language purist as unstable and corrupt should be seen in this light.

If hegemony relied on force and coercion alone, it would not withstand the current onslaught from the new communication technology. However, hegemony works through persuasion; it manipulates the consciousness and negotiates its place into the new dispensation so that the technology sustains it without the conscious knowledge of the consumers. The power class uses its financial advantage in different areas to ensure that it does not yield its control on vital institution that it needs to sustain itself by 1) direct engagement, 2) controlling ownership, 3) advertisement and collaborative venture 4) selling their services to the new communicative technology and 5) controlling the services among others. Driven by market forces and desire for profit, the new technology owners rationalize that it serves their entrepreneurial purposes to work with hegemony rather than engage it. In direct engagement, the hegemony comes with its alternatives to the new technology. We see this in Kenya by the entry of telkom into the mobile phone industry.

A state controlled corporation that enjoyed monopoly since colonial times and had only managed to connect less that 300000 people is suddenly caught by the entrepreneurial bug and wants to offer consumers cheaper alternatives! As an interested party, the state's mandate to legislate on media ownership means that decisions on who controls the new communication technology are not based on the operator's innovative ideas or financial ability to roll out and run the enterprise, but on who the authorities believe can work with them. As the elite with economic power, they have the rights to invest in enterprises that work in their favor through advertisements or funding of new projects. The hegemonic control of economic power thanks to the old order also enables it to sell their services to the communication technology. A good example is the current efforts to charge for online news by major news organization. Since communication technology only mediates rather than create the information, it has no alternatives but to buy these services from the news organization. On the part of control, it would be erroneous to think that censorship is over. As recent attempts to introduce punitive media laws in Kenya have shown, the power elite is not ready to give up on its privilege to control the flow of information and its content.

LANGUAGE USE IN THE ERA OF NEW COMMUNICATION TECHNOLOGIES

The rise of online communication in Sub-Saharan Africa especially at the beginning of the last decade can be attributed to the phenomenal expansion of mobile telephony and computer accessibility. In Kenya, increased internet access that was first enabled by the mushrooming of cyber cafés in urban areas and launching of mobile phone services such as Blackberry has enabled Kenyans to participate in national and global discourse unconstrained by time and location. Social interaction in terms of frequency, linguistic medium and the content has also changed. Mobile phone users can give misleading information about their location or the activity they are engaged in without fear of being audited. Since airtime is only charged to the sender and not the receiver, senders have become resourceful and innovative when coding their message in order to optimize their airtime, or limit the number of characters in texting messages. A sentence like "I feel lyk holdn 2 it a lil longer. I wud ratha the devil i knu than the angel on the other syd" *I feel like holding to it a little longer. I would rather have the devil I knew than an angel on the other side* economizes on space without compromising the message's intelligibility.

The fact that many Kenyans are accessing the internet via their mobile phones has given net lingo a prominent role in Kenyans' discursive practices. Although some of the net lingo used in Kenyan digital interactions may not have originated in Kenya, the globalized nature of digital interactions allows internet users to download the current net lingo and share it among friends. In spite of the fact that a great part of the net lingo is borrowed, it would be erroneous to assume that this 'borrowing' waters down the authenticity and innovativeness that pervades Kenyans' discourse. Thanks to translanguaging, the borrowed or imported net lingo is often modified or recreated to give it a Kenyan flavor. The now widespread *LOL* (laughing out loud) has been given a superlative form *LOLlest* in order to add extra information. The non-verbal gesture of a clicking of the mouth to indicate anger which is uncommon in western verbal behavior but very common among Kenyans is represented by NKP.

Coming back to the issue of translanguaging in Kenyan interactions, we note that all languages do not participate in the same capacity. The languages of power and status dominate in terms of frequency of use. The dominance of English in face-to-face interactions amongst the educated people who dominate internet transactions hinders innovation of the net lingo in other Kenyan languages. Translangauging reduces the need for coinage of net lingo in African languages because it is already available in English, one of the participating languages. That said, we note that innovation of net lingo in African languages is not a barren area. Kenyans are quickly adapting their languages to make them compatible with the peculiar needs of the internet and mobile phone communications. In Swahili for instance, traditional shorthand way of phrasal representation has always followed initialization and acronyming. Three common examples are UM for "Umoja wa Mataifa" *United Nations*, k.m. for "Kwa mfano" *for example*, and n.k. "na kadhalika" *etcetera*. Current trends in Swahili shorthand coinage differ from the traditional trend in that it involves elimination of vowels and other phonological adaptations as we can see in Table 1.

Table 1.

standard Swahili	derivation	gloss
sisi	cc	us
katika	ktk	in, at, on
tofauti	tft	difference
kitabu	ktb	book
mbalimbali	mbali[2]	various

We must bear in mind that innovations are able to spread because people take them up and use them in interactions. When coding the message, senders intend their messages to be understood. In internet and mobile phone interaction, senders are careful not to obscure their message in a language that cannot be decoded. However, issues of language choice, acceptability and adoption of linguistic innovations are determined by intergroup relations. When sending a text message, a mobile phone user employs the lingo that is comprehensible amongst members of his/her interactional network. A mobile phone user of any of the Swahili derivations above assumes that his interactional network is familiar with the lingo. The norms of interaction that are favored by a group of speakers shed light on the socio-psychological consequences of linguistic choices in digital interaction.

In the internet, the use of language in Kenyan web forums might lead one to assume that there is a clear cut line between hegemonic and counter-hegemonic languages. In the newspaper blogs such as *nationmedia.com* and *standardmedia.net*, English is used throughout, reflecting the formal norms of language use.

We must note that as the most dominant media houses in Kenya, they serve as mouthpieces of official ideology. Strict adherence to norms or interaction and the supremacy of the official language signals the hegemony's rule of discourse on its terms. Discussion boards on the other hand mirror the non-formal discourse where language mixing is permitted or tolerated. Although freedom in language choice may be regarded as subversion of mainstream discourse championed by the hegemony, entrenched linguistic ideology makes it impossible to make a dent on the values and linguistic norms of the mainstream. Labov's (1972a) definition of a speech community could help us understand this scenario. According to Labov, a speech community does not just include people who speak in the same way; it also includes people who converge on evaluative behavior of speech norms. Although discussion boards and personal blogs might not have a strict 'language policy' which creates room for translanguaging, the official language is still most frequently used because Kenyans have internalized idealized patterns of language use that favor English. A look at some of these discussion forums is in order.

We may divide these discussion boards into two categories; 1) those with a national or international orientation like *mambogani.com*, jukwaa.proboards.com, *rcbowen.com*, *mashada.com* etc, and 2) those with an ethnic orientation led by *kikuyu.com* and *kisii.com*. Swahili forums are conspicuously absent in spite of the fact that it is the national language. Swahili titled forums like jua *kali.com, kenyaniyetu.net* and *kelele.com* are no longer operational. Absence of Swahili titled forums does not mean complete exclusion of Swahili in Kenyan forums. Apart from the newspaper blogs, the choice of language in Kenyan web forums is very unpredictable, in most cases reflecting the country's triglossic and multilingual practices mentioned earlier. In the topic "the awful Kiswahili language" at *mashada.com* for instance, the languages in use vary from English, Swahili, and Swahili-English mixing as we can see in the following sentences;

1. Unity of a people may not be based solely on language, but it has a *heck of a lot to do with language*. These complex and intricate patterns of behavior you're referring to include language.
2. .... Sheng' ni lugha ya Nairobi, sheng' haina stable backbone, sheng' haina institutional infrastructure wala trained teachers wala a longer than 200 year history.
   Sheng' is a Nairobi language, sheng' has no stable backbone, sheng' has no institutional infrastructure, nor trained teachers nor a longer than 200 year history.
3. Kamwe Kiswahili hakitayafikia malengo kama tutendelea (sic) kukiponda kama mfanyavyo hapa badala ya kukienzi!
   Swahili will never reach its target if we continue trampling on it they way you are doing here instead of honoring it!

From these examples, we can see that translanguaging takes place at both individual and interpersonal level. We see in (ii) that a participant can switch from one language to another without fear of being misunderstood. This is accomplished because the contributor possesses the background knowledge of interactive habits of Kenyan that allow them to exploit all the linguistic resources in their repertoire. Such types of language mixing are not peculiar to *mashada.com*. A statement like "Nasikia Ongeri is playing stubborn na anapuuza hii kitu" *I hear Ongeri is playing stubborn and is disregarding this thing* taken from *Jukwaaforums.com* or "...kama mnataka picha yangu ulizeni tu and i will put it up for a day" *If you want my picture, just ask and I will put it up for a day* from *kisii.com* are testament that language mixing is a characteristic feature of almost all Kenyan web forums or blogs. Thanks to multilingualism, availability of many languages and Kenyan's poorly enforced language policy results in limitless possibilities of linguistic output in verbal interactions which have found their way into cyberspace. The internet and mobile telephony have minimized the power of institutional organs like the state or corporate news media to legislate over language of interaction or influence the national debate, however, ideological hegemony goes on unabated even in the websites with an ethnic focus like *kikuyu.com* and *kisii.com* as we shall see below.

LOW USE OF ETHNIC LANGUAGES IN KIKUYU.COM AND KISII.COM

The 1990s and early 2000 witnessed the proliferation of ethnic based websites that were mostly set up by Kenyans in the Diaspora. The Kenyans who were emigrating to America and Europe in search of economic opportunities in droves were seeking social forums that would connect them with their kinfolk and friends back home. These forums became the social platforms or discourse sites in which Diaspora Kenyans could reconcile with their alienated selves. *Kikuyu.com* was one of the pioneering ethnic website that originally did not require registration. This policy made it popular even among the non-Kikuyus who could participate in non-Gĩkũyũ forums. The success of *kikuyu.com* inspired other ethnic forums like *jaluo.com*, *kalenjin.net* and *kisii.com*. Currently only *kikuyu.com* and *kisii.com* are operational but participation in these websites has waned due to the increasing popularity and convenience of other social networking sites like Facebook.

Contrary to what their names suggest, very little interaction in these forums takes place in ethnic languages. The organization of the discussion forums in these websites makes it possible for people who do not belong to the ethnic affiliation of the website to participate. It is noticeable that in both websites, forums are organized into categories such as general discussion, politics, religion, relationship, entertainment, economy etc, as well as Gĩkũyũ or Ekegusii forum for *kikuyu.com* and *kisii.com* respectively. Assigning of an ethnic language forum seems to send a message to the participants that the ethnic language should only be used in that forum only. At the same time, it indicates that posts in other forums should be in English. In *kisii.com* for instance, the administrator seems to give explicit message for participants to post in Ekegusii by naming the forum "Ekegusii gioka" accompanied by the requirement that that postings should be "…in Ekegusii language ONLY". However, this advice is ignored in topic headings, majority of which are in English. From a total of 37 topics under this forum, 24 are in English while only 13 are in Ekegusii. The marginalization of Ekegusii is also evidence in Ekegusii headed topics where some participants prefer to post in English, Swahili or mixing between the three languages in spite of the requirement prominently displayed under the forum header "Postings in Ekegusii language ONLY". In the thread "Ninki ominetie", a frustrated contributor complains "I thought this is a kisii chat only." The irony is that the protest is in English which breaks the very rule s/he is accusing the other person from breaking.

The story is not different at *kikuyu.com* where language mixing and English dominance prevails. The Swahili forum fares the worst compared to English forum and two Gĩkũyũ forums— "Learning Kikuyu" and "Ndai (Riddles)." The distribution of languages according to topic titles shows that English dominate the topics as illustrated in Table 2. The three most active forums are main board (427), election 2007 (203) and politics (124) while the least active are pregnancy/parenting (17), Swahili (21) and religion (33). My familiarity with the languages involved in this website coupled with my active participation accords me the benefit of insider knowledge to comment about the nature of interactions in the forums. Low participation in the pregnancy/parenting and religion forums can be attributed to the fact that these forums were started after members asked for them. In addition, some of the topics that fall under these

Table 2. Forum languages in kikuyu.com

Forum	No. of topics	English	Swahili	kikuyu	mixed
Main board	427	269	7	119	22
Election 2007	203	192	4	4	3
English speakers	104	99	1	3	1
Swahili	21	10	9	2	0
Business/finance	50	50	0	0	0
Politics	124	111	3	7	3
Learning Kikuyu	88	43	0	40	5
Pregnancy/parenting	17	14	1	2	0
Religion	33	29	1	3	0
Ndai (riddles)	82	56	0	24	2
Totals	1149	873	26	182	36

two topics are also covered in the main board or English forums. However, there is not a good explanation for the low participation in Kiswahili forum apart from the fact that Swahili competence is very low amongst the participants. The discomfort with speaking Swahili is evidence from, self evaluative comments like 'kiswahili changu, sio sanifu/sahihi, kimethofiika." *My Swahili, is not standard/correct. It is weak,* "Ningeoma muniwiye radhi ikiwa kiswahili changu si sanifu" *I pray you excuse me if my Swahili is not standard* and "Kiswahili ni kigumu" *Swahili is difficult.* To a large extent, insecurity about low skills in Swahili by members may account for the low participation in the Swahili Forum.

Logically, the topic header should cue the kind of language used throughout the thread but this is not what always happens. More often than not, the topic header does not always predict the type of language that follows. A topic header in Gikuyu might only have the first and second posts in Gikuyu before giving in to English, Swahili, Sheng or a mixture of the four. In the topic "atumia aingi" *many wives* that appears on the main board forum, the heated debate on polygyny that quickly turns to a discussion on homosexuality, there are 58 posts. The first three posts in Gikuyu are followed by posts that mix Gĩkũyũ and English or English and Swahili. After post 10, the rest are all in English. The distribution of languages was even worse at the *jaluo.com* and *kalenjin.com* which are no longer operational. In these two websites, there was very little interaction in Dholuo or Kalenjin and Kiswahili was virtually non-existent. Still, translanguaging between English and those languages was rife.

Earlier we mentioned that triglossic linguistic practices assign different functions to domains of languages, translanguaging blurs the boundaries demarcating these domains. It is common to hear people using their mother tongues in offices while politicians address political gatherings in the rural areas in English or Swahili, even though very few people speak English and their command of Swahili is not so good. In the internet, language domains may be conceptualized in terms of institutional blogs like statehousekenya.go.ke and nationa.co.ke that do not mix languages or forum categories such as English forums, Swahili forums or ethnic language forums. Similar face-to-face interactions, translanguaging can be random with languages being used where they are not expected. In *rcbowen.com* for instance, we would expect that since it does not lean on any ethnic group, vernacular material would not appear; however, the Dholuo phrase "Ase mako dhogi" (I have cornered you) with its translation provide challenges such assumptions. One question that begs answers is this; if ethnic websites were launched with the sole purpose of "assisting kikuyus in the diaspora to ""touch base" and for people like me who live far from home, to get connected to the "everyday"" as one contributor at *kikuyu.com* puts it, why then is English so pervasive? A good explanation may be found from a combination of factors that emanate from English hegemony.

Overemphasis on English has generally marginalized literacy in African languages. In education for instance, the ministry guidelines to teach the first three years of primary school in mother tongue or language of the catchments area[7] is not strictly enforced and schools take the liberty to implement the medium of instruction that suits their purpose. Three years of instruction in mother tongue are also insufficient to build a strong foundation in mother tongue literacy. Scarcity of reading materials to consolidate mother tongue literacy results in deterioration of literacy as soon as mother instruction ceases. The situation is worse in urban areas where children are raised by housemaids from different ethnic groups. At school, these children speak English with their teachers and peers, but at home, they speak to the housemaids, or neighborhood peers in Swahili. The little time that they spend with their parents is not enough to master their parents L1. The situation is worse if the parents are educated and come from middle class families since these parents are known to speak to their children in English. A good anecdote comes from Ngũgĩ wa Thiong'o (1981) who recounts how he visited a middle class family in Nakuru and the parent was proud of his child's inability to speak Gĩkũyũ. Such attitudes result from the general perception of English as the language of upward mobility while ethnic languages are perceived as symbols of backwardness, and traditionalism which go against the dash for modernity. A generation of children who have grown up under such circumstances cannot be expected to display vernacular literacy on the internet

Another factor has to do with the documentation and orthographic nature of African languages. Diacritics and conventions governing representation of tones and accent marks in many African languages make writing a strenuous exercise and many people get frustrated and give up. In Gĩkũyũ and other languages of the Thagichu family, graphemes ĩ and ũ have been innovated to represent phonemes [e] and [o] respectively. The clicks in Nguni and Khoisan languages in Southern Africa have also posed orthography problems. In addition, phonetic opacity in non-phonemic languages like Akan in Ghana have posed a great challenge to literacy. In Gĩkũyũ for instance, it is very hard to represent the diacritics in ĩ and ũ, yet, they are very important in distinguishing meaning in writing since one does not have the benefit of intonation or context. Without diacritic marks, the following words would not constitute minimal pairs;

matu/matũ clouds/ears
ruma/rũma insult/bite
tuma/tũma sew/send
rira/rĩra fend off/cry
kira/kĩra be quiet/exceed, surpass, pass

Many internet users are unwilling to spend extra time to represent these diacritic marks and resort to English. These orthographic challenges are compounded by the unavailability of updated dictionaries. I have spoken to many Kikuyus who are not even aware that there is a Kikuyu dictionary. This is a common trend in may African languages. Since many of these dictionaries were written by missionaries with the sole intention of assisting missionaries learn local languages to ease evangelization, most of them are bilingual. Today, very few African languages have monolingual dictionaries. This lack of monolingual dictionaries is a linguistic handicap which hinders L1 speakers from developing or improving their descriptive ability in their respective native languages. Although linguists have spent a good deal of time describing African languages, there has been very little development of instructional material in those languages. While writing of reference grammars is a very encouraging venture, their scope is very limited and not intended for learning purposes. The descriptive nature of reference grammars thus aids in linguistic analysis but not language learning. Lack of coordinated work on lexicography or standardization has especially made the language insufficient in mediating contemporary discourse that is replete with alien concepts. To supplement this lexical handicap, Kenyan languages have resorted to nonce borrowing[8] (Poplack, Sankoff & Miller 1988), adaptations of borrowed lexical items as wells as codeswitching. The low skills in native languages heighten linguistic insecurity which manifest itself in mother tongue avoidance in the form of translanguaging as a 'face saving' strategy (Goffman 1967). As a result, the new communication technology rather than challenging the linguistic hierarchy, only replicates the existing relations of power between these languages which weakens them further. Although new technologies may boast of providing new forums for the conduct of ethnic discourse, decades and sometimes centuries of marginalization has solidified ideologies of inferiority among users of ethnic languages.

CULTIVATING ETHNICITY IN DIGITAL COMMUNICATION

Whether engaged in social interaction or intellectual activities such as language learning or research, the internet and mobile phones gives their users a great deal of autonomy. But as rightly pointed out by Park, Crocker & Mickelson (2004), human beings are "inherently social creatures" and the desire for emotional and psychological attachments which overrides the desire for autonomy.

Since the structural set up of post traditional reality makes traditional communities impractical, consumers of new communication technology find themselves confronted with a problem of reconciling their post traditional selves with their innate yearnings. This dilemma is resolved by exploiting the new innovations in software that enables them to compensate for unavailability of time for face-to-face interactions. Some of the online relationship originates from existing social networks and relationships. A mobile phone user for instance sends messages to people s/he is familiar with. Text messaging and phone conversations allow the sender to maintain social networks that could otherwise have been lost as s/he tries to deal with the pressures and demands of technology driven global world. In this way, technology facilitates the maintenance of traditional identities albeit in an altered environment. Computer mediated communications on the other hand not only have the ability to sustain old relationships, but it is also able to build new ones with others that one has never met before. Rather than participants mapping into inherited social categories, they choose who they want to interact with and construct their identities based on a set of shared characteristics. The difference between these new online communities and the traditional ones is well captured by Diamandaki's (2003) remarks that;

" ...community in late modernity is "adopted" rather than "handed down," it is a community of choice and free will, and something we search for rather than inherit. And it is, to a large extent, this voluntary quality of post-traditional communities that allows us to define ourselves and face the trends of abstraction, globalization, homogenization, and atomization." (p 5)

While this conceptualization of social construction of community provides us with a locus for discussing agency in self identification in cyberspace, the notion of social constructionism is not new. Berger and Luckman's (1966) influential *Social construction of Reality: A Treatise in the Sociology of Knowledge* outlines how everyday subjective reality becomes objective reality. According to Berger and Luckman, all reality is socially constructed. The activities and experiences of everyday life which falls within the realm of subjective reality can transform into objective reality through frequent performances. Iterative enactment creates patterns which are habitualized and reproduced in similar activities in future. When these habitualized patterns are reciprocated by others, they become institutionalized and are regarded as the conventional practice that every member of a community subscribes to. In other words, practice builds habitus (Bourdieu 1990:53) that predisposed members to act in a regular and ordered manner without conscious attempts to do so. Bourdieu defines habitus thus:

systems of durable, transposable dispositions, structured structures predisposed to function as structuring structures, that is as principles which generate and organize practices and representations that can be objectively adapted to their outcomes without presupposing a conscious aiming at ends or an express mastery of the operations necessary in order to attain them. Objectively "regulated" and "regular" without being in any way the product of obedience of rules, they can be collectively orchestrated without being the product of the organizing action of a conductor. (p 53)

The translanguaging and replay of familiar ethnic discourse in cyberspace flow from the engrained habitus amongst Kenyans. The negotiation of virtual ethnicity is not so much about the inherited traits, as the imagining of a communion. With the diminished role of ethnic languages as shared common trait, participants in online ethnic forums look for other areas of convergence that can define their identity. Take for instance the sharing of jokes and stereotypes about members of the ethnic group. A good example comes from *kisii.com* discussion on the Abagusii people and witchcraft. This stereotype is overplayed when a contributor poses; "Does anyone know of any omugusii [Kisii person] that has died of any causes other than oborogi [witchcraft]?". Posed by a non-Kisii with no background knowledge of social limits of Kisii discourse, this could have been regarded as crossing the line by engaging in face threatening act (Goffman 1967), but the fact that it is posed by a fellow member who share the same background erodes the threat and qualifies it as a joke. At *kikuyu.com*, the jokes about the exploitative nature of kikuyu women, the Kikuyu men's love for money, poor romantic tastes, and their 'addiction' to sevco jeans and godfather hats are transformed from a disparaging discourse to laughter. These kinds of verbal discourse can be compared to Labov's (1972b) analysis of African American ritual insults. According to Labov, statement in verbal art are not denied or mitigated because they are known to be false. To treat them as if they were true statements signals that what is being implied is indeed true. Ignoring such a statement on the other hand, indicates that the participant understands the discourse rules and know that they are not real.

The consumption of ethnic discourse in ethnic internet forums parallels offline interactions in that the participants want to bond with members of an imagined community. Our working definition of imagined community draws from Anderson's (1983) discussion of nations and nationalism. Anderson argued that all communities are imagined communities because members of even the smallest nations will never know most of their fellow members, meet them or even hear of them, yet in the mind of each lives the image of their communion" (p.15). This conceptualization regards communities as abstract constructions in the imaginations of members. During production and reproduction of social structures, these images of communion, though some may have originated from subjective reality become part of members objective reality. Although the internet is considered to give participants anonymity and autonomy, the participants are still bound by their perceived sense of communion that flows from habitus. Participants are attracted to these ethnic websites because they provide forums for reconnecting with their imagined ethnic communities. This is especially so for the Diaspora participants[9] separated by time and space from other members of their ethnic community. These sites are also education sites where participants pose questions ranging from meaning of words to cultural issues like folk dance, cultural rites and traditional marriage procedures. However, the most active threads in both *kikuyu.com* and *kisii.com* are those where participants share traditional literature like proverbs, riddles, songs, stories and common jokes. Even in the defunct forums like *luo.com* and *kalenjin.net*, popular stories derived from the community's oral literature, and jokes attracted many comments and hits. While some of these stories are retold due to participants' nostalgia, they nevertheless solidify the social bond in these websites which highlight their role as cultural bonding sites. Revitalization of ethnicity in these forums can be tapped to accomplish agendas that reflect mainstream norms especially in political arena. From the foregoing discussion, we can see that although the internet interactions are peculiar in their mode of delivery, the negotiation of ethnic identity is not radically different from the offline discourse. Diamandaki (2003) succinctly characterizes it thus;

...while cyberspace is a place where national or ethnic identity can be manipulated anew, it ends up not being totally different from its non-virtual manifestation. In most cases these digital corners of national/ethnic orientation function as spaces for the re-articulation of the same "social dialogues" that usually preexisted in the offline worlds and around which issues of national determination were produced and reproduced. The Internet becomes a new stage where old issues and conflicts, polarizations and discourses, claims and fantasies of the preexisting identity are played out anew. In the digital world, the memorable and historical past is blended with the challenging and contradicting present that individuals are faced with.

Rather than ushering in a post ethnic discourse, the digital world has intensified the conflicts of identity in post traditional discourse. The dilemma stems from the hypocrisy of the elite that advocates for a unified national discourse while at the same time manipulating social divisions to its advantage. The status quo is not ready to cede or share the privileges and perks that are bestowed by existing social hierarchies. New communication technologies only provide superficial empowerment to the marginalized and end up validating existing social relations that are masked in new consciousness and freedom. As modern communication extends tentacles to the once remote corners that were for a long time cut off from the national dialogue, ethnic polarization is likely to be reproduced on a mass scale. The (mis)use of mobile phone in the disputed election in Kenya in 2007 is a case in point. The political parties not only used chain text messages to mobilize supporters, but some of these messages promoted ethnic hatred that had serious consequence in the outcome of the election as well as the aftermath of the stolen election. However, ushahidi project[10] has also shown how the same communication technology was used to report incidents of human right abuses that could have gone unreported. Of interest here is that communication technology was used to transmit ethnically charged messages, some of which were malicious propaganda that revived Kenyan's ethnic mistrust. Whether the scale of ethnic violence could have been different if mobile phone technology did not play a role in mobilizing tribal militias cannot be ascertained. However, it is important to note that the antagonists were simply manipulating ethnicity for personal motives as post violence reconciliation has shown that the antagonists can work together and indeed protect each other since both are part of the elite.

WAY FORWARD: FUTURE AND EMERGING TRENDS

Up until 2009, internet forums have been an important site for the consumption of ethnic discourse. Recent trends however points to its decline as more people become drawn to social networking sites like Facebook and personal blogs. Unlike the discussion boards, Facebook allows one to choose members of their social network and exclude people who one no longer feels comfortable with. The Facebook also allows individuals to post pictures while at the discussion board, pictures had to be sanctioned by the administrator. In addition, at least as far as Kenyans are concerned, Facebook is also available on mobile phones which makes it convenient since internet connection is still poor. While these new communication technologies have allowed people, especially those in the Diaspora to reconnect with their imagined communities, they have also reconstructed ethnicities that are not bound by social rules as it were the case in the traditional set up. The veneration of elders in traditional societies regulated social practices among members of ethnic communities. There was also a mechanism for reconciliation and conflict resolution in cases of inter-ethnic conflict. In digital interactions however, everyone, even those without adequate knowledge of traditional norms claim authority to cast them any way they choose without responsibility or accountability. In the lead up and follow up to the disputed 2007 election, the vernacular radio stations especially Kass FM and Inooro FM became mouthpieces of ethnic chauvinism with devastating consequences. While we do not advocate for media censorship or infringement of individual freedoms, there is need for a well articulated media policy in which certain protocols have to be enforced. This policy is better negotiated by the stakeholders in the media industry rather than the government whose only interest will be to curb freedom of information. There should also be penalty for those who instigate and engage negative ethnicity. Registration of mobile phone users with the service provider can allow tracing the source of malicious chain-text messages.

As far as languages are concerned, there is need to revitalize and empower local languages so that they become relevant in today's technological world. Decentralization of language policy should be encouraged in the new dispensation of devolution. The ministry of culture should encourage the formation of language committees for each language in order to ensure that languages are not being regarded as static objects that do not change with the time. In order to encourage the current generation to value their languages, especially in the context of pressure from English hegemony, status and corpus planning must be taken seriously.

These language committees will enable these languages to come up with new vocabularies to describe contemporary phenomena as well as equip them for technology discourse. They will also recommend ethnic words to be tagged in order to monitor malicious discourse or offensive phonographic material. Although we do not see a time when African languages will compete evenly with hegemonic languages, there is a possibility of safeguarding them through translanguaging rather than exclude them altogether. Rather than lamenting the corruptive influence of translanguaging, it should be seen as a way of celebrating our diversity, as long as the languages concerned are not relics of the past millennium. More importantly, there should be efforts to devise computer keyboards that are compatible with the African language. One way to do this is the development of touchpad keyboards that can be easily adapted for any language through downloadable software. Unless new generations of keyboard that can be adapted for any language are developed, African languages are still going to be marginalized in communication technology. We may even argue that the development of software such that makes online lexicon in African languages available does more harm than good. It gives users a false sense of competence while that can hardly use it in a contextualized environment. Unless these languages are revitalized and empowered to operate in contemporary technological climate, technology will only hasten their demise rather than preserve them.

CONCLUSION

In postmodernist discourse where the technological advancement in news media shapes everyday opinion and perspectives, one would expect that the power of ethnic languages in influencing national discourse would have diminished. The onslaught on ethnic languages aided by the rapid urbanization has transformed many African cities into centers of linguistic convergence. This convergence has created the need for linguistic strategies that would facilitate interethnic communication. The search for a neutral language has come in the form of local lingua franca like Swahili in East Africa or languages of former colonial powers. As these powerful languages become associated with social prestige and economic clout, local languages are gradually being abandoned. Ironically, the pressure on African languages does not come from foreign languages alone; the aggressive promotion of Swahili in Tanzania has been to the detriment of local languages as captured by Batibo (1992) who compares Swahili to a Nile perch that swallows all the smaller fish.

Although some African languages have been as hegemonic as the foreign languages that they wish to replace, we note that when it comes to new communication technology, all of them participate on almost an equal footing. Most surprising was the marginalization of Swahili in ethnic forums which shows that simply promoting the national language at the expense of ethnic languages will not work. Empowering of these languages and promotion of literacy in mother tongue will make it possible to encourage participants to make boundaries in their communication. In this sense translanguaging will operate under well defined boundaries where each language will be associated with specific functions. Instead of having random mixing, speakers will be sensitized about the best practices in language mixing such that they learn to be sensitive to others who might not understand those languages. Since random mixing in ethnic forums was shown to illustrate the state of illiteracy due to English hegemony, empowerment of ethnic languages will enable Kenyans to feel secure about their languages and learn to appreciate the diversity that all the language bring. More importantly, ethnicity need not be viewed in negative terms, but can be seen as something to be celebrated.

REFERENCES

Batibo, H. M. (1992). The fate of ethnic languages in Tanzania. In Brenzinger, M. (Ed.), *Language death: Factual and theoretical explorations, with special reference to East Africa* (pp. 85–98). Berlin, Germany: Mouton de Gruyter.

Berger, P., & Luckman, T. (1966). *The social construction of reality: A treatise in the sociology of knowledge*. New York, NY: Doubleday.

Blackledge, A., & Pavlenko, A. (2001). Negotiating identities in multilingual contexts. *The International Journal of Bilingualism*, *5*(3), 243–257. doi:10.1177/13670069010050030101

Bourdieu, P. (1991). *Language and symbolic power*. Cambridge, MA: Harvard University Press.

Bucholtz, M. (1999). Why be normal? Language and identity practices in a community of nerd girls. *Language in Society*, *28*, 203–223. doi:10.1017/S0047404599002043

Diamandaki, K. (2003). Virtual ethnicity and digital diaporas: Identity construction in cyberspace. *Global Media Journal*, *2*(2).

Dopke, S. (2000). The interplay between language-specific developments and cross linguistic influence. In Dopke, S. (Ed.), *Cross-linguistic structures in simultaneous bilingualism* (pp. 79–104). Amsterdam, The Netherlands & Philadelphia, PA: John Benjamins.

Edwards, J. (1985). *Language, society, and identity*. Oxford, UK: Basil Blackwell.

Fantini, A. (1995). Introduction-language, culture and worldview: Exploring the nexus. *International Journal of Intercultural Religion*, *19*(2), 143–153. doi:10.1016/0147-1767(95)00025-7

Fishman, J. (1991). *Reversing language shift: Theoretical and empirical foundations of assistance to threatened languages*. Clevedon, UK: Multilingual Matters.

Garcia, O. (2009). *Bilingual education in the 21st century: A global perspective*. Wiley-Blackwell.

Giles, H., Copeland, N., & Copeland, J. (1991). Accommodation theory: Communication, context, and consequence. In Giles, H., Coupland, J., & Coupland, N. (Eds.), *Developments in applied sociolinguistics*. Cambridge, UK: Cambridge University Press. doi:10.1017/CBO9780511663673.001

Giles, H., & Powersland, P. (1975). *Speech style and social evaluation*. London, UK: Academic Press.

Giles, H., Taylor, D., & Bourhis, R. (1973). Towards a theory of interpersonal accommodation through speech: Some Canadian data. *Language in Society*, *2*, 177–192. doi:10.1017/S0047404500000701

Githinji, P. (2006). Bazes and their shibboleths: Lexical variation and Sheng speakers' identity in Nairobi. *Nordic Journal of African Studies*, *15*(4), 443–472.

Githiora, C. (2002). Sheng: Peer language, Swahili dialect or emerging Creole? *Journal of African Cultural Studies*, *15*(2), 159–181. doi:10.1080/1369681022000042637

Goffman, E. (1967). On face-work: An analysis of ritual elements in social interaction. In Jaworski, A., & Coupland, N. (Eds.), *The discourse reader*. London, UK: Routledge.

Grice, H. (1975). Logic and conversation. In Jaworski, A., & Coupland, N. (Eds.), *The discourse reader*. London, UK: Routledge.

Hazen, K. (2002). Identity and language variation in a rural community. *Language*, *78*(2), 240–257. doi:10.1353/lan.2002.0089

Krauss, R., & Chiu, C. (1998). Language and social behavior. In Gilbert, D., Fiske, S., & Lindzey, G. (Eds.), *Handbook of social psychology* (4th ed., *Vol. 2*, pp. 41–88). New York, NY: McGraw-Hill.

Kuoega, J. (2003). Camfranglais: A novel slang in Cameroon schools. An account of the intricate hybrid used by young people in West Africa. *English Today*, *19*(2), 23–29.

Labov, W. (1972). *Sociolinguistic patterns*. Philadelphia, PA: University of Pennsylvania Press.

Labov, W. (1972). *Language in the inner city: Studies in Black English vernacular*. Philadelphia, PA: University of Pennsylvania Press.

Le Page, R., & Tabouret-Keller, A. (1985). *Acts of identity: Creole-based approaches to language and ethnicity*. Cambridge, UK: Cambridge University Press.

Mbugua wa, M. (2003). *Identity politics in Nairobi Matatu folklore*. Unpublished doctoral dissertation. Jerusalem, Israel: Hebrew University.

Meyerhoff, M. (1998). Accommodating your data: the use and misuse of accommodation theory in sociolinguistics. *Language & Communication*, *18*, 205–225. doi:10.1016/S0271-5309(98)00006-8

Mkilifi, M. (1972). Triglossia and Swahili-English bilingualism in Tanzania. *Language in Society*, *1*(2), 197–213. doi:10.1017/S004740450000049X

Myers-Scotton, C. (1989). Codeswitching with English: Types of switching, types of communities. *World Englishes*, *8*, 333–346. doi:10.1111/j.1467-971X.1989.tb00673.x

Myers-Scotton, C. (1990). Elite closure as a powerful language strategy: The African case. *International Journal of the Sociology of Language*, *103*, 149–163.

Myers-Scotton, C. (1992). Codeswitching in Africa: A model of the social functions of code relations. In Herbert, K. R. (Ed.), *Language and society in Africa*. Witwatersland, South Africa: Witwatersland University Press.

Myers-Scotton, C. (1993). Social motivations for codeswitching: Evidence from Africa. In Romaine, S., & Mulhausler, P. (Eds.), *Oxford studies in language contact*. Oxford, UK: Claredon Press.

Myers-Scotton, C. (2002). *Contact linguistics: Bilingual encounters and grammatical outcomes*. Oxford, UK & New York, NY: Oxford University Press.

Myers-Scotton, C., & Bolonyai, A. (2001). Calculating speakers: Codeswitching in a rational choice model. *Language in Society*, *31*(1), 1–28. doi:10.1017/S0047404501001014

Myhill, J. (2003). The native speaker, identity and the authenticity hierarchy. *Language Sciences*, *2*, 77–97. doi:10.1016/S0388-0001(01)00023-7

Ngũgĩ wa, T. (1981) *Writers in politics: Essays*. London, UK: Heinemann.

Ogechi, N. (2005). On lexicalization in Sheng. *Nordic Journal of African Studies*, *14*(3), 334–355.

Park, L., Crocker, J., & Mickelson, K. (2004). Attachment styles and contingencies of self-worth. *Personality and Social Psychology Bulletin*, *30*(10), 243–254. doi:10.1177/0146167204264000

Poplack, S., Sankoff, D., & Miller, C. (1988). The social correlates and linguistic processes of lexical borrowing and assimilation. *Linguistics*, *26*, 47–104. doi:10.1515/ling.1988.26.1.47

Rampton, B. (1985). *Crossing: Language and ethnicity among adolescents*. London, UK: Longman.

Sinka, I. (2000). The search for cross-linguistic influences on the language of young Latvian-English bilinguals. In Dopke, S. (Ed.), *Cross-linguistic structures in simultaneous bilingualism* (pp. 149–174). Amsterdam, The Netherlands & Philadelphia, PA: John Benjamins.

Spitulnik, D. (1999). The language of the city: Town Bemba as urban hybridity. *Journal of Linguistic Anthropology*, *8*(1), 30–59. doi:10.1525/jlin.1998.8.1.30

Walther, J. (2004). Language and communication technology: Introduction to the special issue. *Journal of Language and Social Psychology*, *23*(4), 384–396. doi:10.1177/0261927X04269584

Whorf, B. L. (1956). *Language thought and reality: Selected writings* (Carrol, J. B., Ed.). Cambridge, MA: MIT Press.

William, A. (1999). Communication accommodation theory and miscommunication: Issues of awareness and communication dilemmas. *International Journal of Applied Linguistics*, *9*(2), 151–165. doi:10.1111/j.1473-4192.1999.tb00169.x

Woolard, K. (1985). Language variation and cultural hegemony: Toward an integration of sociolinguistic and social theory. *American Ethnologist*, *12*(4), 738–748. doi:10.1525/ae.1985.12.4.02a00090

ENDNOTES

1 Bourdieu (1990: 53) defines habitus in terms of a set of disposition that organize

social practice. These dispositions generate structures in an organized way even though no conscious rule is being followed. This happened because agents internalization of certain modes of actions drive conscious actions into the subconscious such that one is not aware that they are doing something that have been internalized.

2 Bourdieu (1990: 53) defines habitus in terms of a set of disposition that organize social practice. These dispositions generate structures in an organized way even though no conscious rule is being followed. This happened because agents internalization of certain modes of actions drive conscious actions into the subconscious such that one is not aware that they are doing something that have been internalized.

3 Fishman explains that, Xman stands for a person or citizen e.g., Englishman, American, Frenchman etc while Xish stand for the language spoken by a person such as English, Irish, Japanese, etc

4 When there is penalty for using an ethnic language, e.g., the Berber people in Libya, one might avoid using the ethnic language but it does not imply the lack of alternative ways of expressing ethnicity.

5 Myers-Scotton (1990) uses the term 'elite closure' to describe how the elite promote language policies that limits accessibility to the language of prestige. It serves the elite to limit accessibility to the languages of status so that the social hierarchies that favor them can be sustained.

6 Crossing is the use of a language that is identified with another ethnic community. Rampton (1995) describes the use of stylized Asian English, Panjabi and Creole by Panjabi, Afro-Caribbean and Anglo adolescents. Cutler (1999) has also described a White American youth's attempt to speak the African American English as crossing.

7 Following the recommendations of the Phelps Stokes Commission and in line with the UNESCO 1953 recommendations, the government advocates for the mother tongue to be the language of instruction, in the first three years of school. If no education materials are available in a students mother tongue, then they have to use those of the dominant language in the area or the so called 'Language of the catchment area'. This rule is not strictly enforced.

8 Refers to one-term borrowing of lexical items that are not assimilated into the borrowing languages lexicon.

9 In Kenya and other African country, the notion of Diaspora extends to member who have moved from their ancestral lands. In this case, a Luo in Nairobi or Mombasa is as much a Diaspora Luo as the one in California or Beijing.

10 Software developed by Kenyan bloggers after the 2007 disputed election. According to a United Nations Foundation report (2009) *New technologies in Emergencies and Conflict: The role of information and social Networks,* this software combines SMS, Twitter, and Google Maps to crowdsource crisis information and has since been used in Afghanistan, Colombia, the Democratic Republic of the Congo (DRC), Gaza, India, and Lebanon.

Chapter 12
Veganporn.com & "Sistah": Explorations of Whiteness through Textual Linguistic Cyberminstrelsy on the Internet

Amie Breeze Harper
University of California- Davis, USA

ABSTRACT

Cyberspace can be a central site for excavating the invisibility of covert whiteness (a tacit form of racialized consciousness), which does not manifest itself at the surface level in the same overt manner that extreme white cyber hate "imagined communities" do. Through the application of Critical Race Theory and Black Feminist methodology-based discursive analysis, this chapter investigates performances of whiteness in a vegan/animal-rights-oriented website called Veganporn.com. As a progressive forum associated with social justice, Veganporn.com provides a radically different environment in which to examine white supremacist ideologies; ideologies typically found in more overtly-racist, "extremist" online dialogues already examined by critical research. Discourse analysis of a specific Veganporn.com forum topic revealed three major themes in the computer-mediated discussion: (1) discursive patrolling of epistemic borders to "protect" Standard English and colorblind expressions (whiteness) of veganism/animal rights from non-Standard English and non-white racialized expressions; (2) the use of blackface cyber-minstrelsy to reinforce the "superiority" of Standard English (whiteness) over the "inferiority" of speakers of Black English and Ebonics; (3) the premise among several white-identified Veganporn.com participants that a vegan- and animal-rights ideology is "colorblind" thus making invisible the current socio-historical implications of power structures created around white skin color. Though this chapter focuses on one discussion within a forum, the analysis of this event offers insight relevant to understanding whiteness as a system, an ideology, and a structure. Specifically, by employing certain theoretical components of critical race studies (racialized consciousness, social ontology of whiteness, and racial mapping), my analysis reveals how the World Wide Web can be an effective site for cyber-ethnographers focusing on "decoding" whiteness within progressive social justice movements.

DOI: 10.4018/978-1-60960-591-9.ch012

INTRODUCTION

In their seminal book, Race in Cyberspace, cyberspace researchers Kolko et al (2000), and Nakamura (2002) argue that not only is race often difficult to talk about in the academy, such difficulty (and often refusal) to talk about it manifests frequently within the realm of cyberspaces that consciously try to avoid it. In the words of UCLA Law professor, Jerry Kang, "Can cyberspace change the very way that race structures our daily lives? Cyberspace enables new forms of social interaction. How might these new communicative forms affect racial mechanics?" (Kang 2000, p. 1133). Furthermore, how does discourse analysis of digital rhetoric help researchers of critical race theory or anti-racist praxis, understand systemic whiteness, racism, or utopian dreams of cyberspace being a "post-racial" space?

Tara McPherson argues in her study of Neo-Confederate websites, that even deliberate and conscious efforts to elide questions of race online can manage to create unmistakably racialized spaces...[her work shows that the] virtual reality that is cyberspace has often been construed as something that exists in binary opposition to "the real world", but when it comes to questions of power, politics, and structural relations, cyberspace is as real as it gets.(Kolko et. al 2000, p. 4)

Nakamura (2002) concurs that raceless or post-racial, as concepts in cyberspace, equate to a default racial identity of "white." Research focused on constructions of raceless or post-racial cyberspace utopias- particularly amongst white identified users involved in progressive social justice movements, is a realm in need of further exploration (Banks 2006 ; Kolko et al 2000; Nakamura 2002). Yancy (2004) notes that most research about whiteness and racial discrimination is focused on the limited binary of "good whites" versus "bad whites" (i.e. "bad whites" equals engagement in overtly White Supremacist practices, such as the KKK or neo-Nazi group activity). This chapter will explore how discourse analysis of digital rhetoric, in one specific computer-mediated progressive "post-racial" forum, a) deconstructs the simplicity of white supremacist ideologies as part of the good/bad binary and b) reveals the illusion of post-racial or raceless cyberspaces.

METHODOLOGY

It is through the analysis of performance through language, via a Critical Race Theory (CRT) and Black Feminist discursive analytical framework, that I will show how systemic whiteness/white supremacist ideologies are manifested on a site dedicated to the discussion of veganism and animal rights. I have selected this topic because culturally in the West, collectively whites that practice such activism think they are incapable of participating in the overt racism one can normally find on radical right extremist white cyber hate Internet sites (Poldervaart 2001; Clark 2004). Yancy (2004) explains that white supremacist groups, such as the Ku Klux Klan, are collectively viewed by liberal progressive whites as the "real racists." Deeper explorations of progressive whites' own potential racism is "non-applicable" because "acts of racism" are viewed and compared to these "extreme" hate groups (Yancy 2004):

A key feature of the social ontology of whiteness is that whites attempt to avoid discussing their own social, political, economic, and cultural investments in whiteness. Many whites fail to see their complicity with the systemic workings of white supremacy. By perpetuating the dualism between the "good white" and the "bad white," whites attempt to mute the claim that white racism is not limited to the KKK, neo-Nazis skinheads, White Aryan Resistance, and other white racists groups...(Yancy 2004, p. 4)

I will investigate how systemic whiteness on Veganporn.com, a forum dedicated to discussions about animal rights and veganism, was masked as "post-racial" and normative behavior, maintained and enforced through the expected use of Standard English and Cartesian logic, key components in sustaining whiteness (Smitherman 2000). This examination of language in Veganporn.com

Demands that one go beyond the immediacy of the linguistic text to consider matters of socio-political and economic subordination and language, the perpetuation of inequality through language, and the historical backdrop against which these linguistic power-plays are enacted. (Smitherman 2000, p. 7)

Because the particular Veganporn.com discussion analyzed in this paper is centralized around the use of "proper English" and the construction of speakers of "Black English" and Ebonics as "inferior," I will be using Smitherman's analysis and definitions of Ebonics, Standard English and non-Standard English language, as it pertains to power maintenance of the white middle and upper class).

Nonstandard American English refers to those language patterns and communication styles that are non-African in origin and which are used by the working class. These patterns serve as linguistic borders that separate those who "carry on the affairs of the English-speaking people" from those who do not. (Smitherman 2000, p. 20).

Thus, socio-linguistic analysis of the selected discussion on Veganporn.com is central to understanding performances of whiteness.

Methods for Investigating Covert Whiteness on Veganporn.com

In the fall of 2005 I posted a request for papers for an anthology book project aimed at finding voices of black- identified females who practice veganism. The project is called Sistah Vegan. The call for papers was submitted to cyberspace groups that would most likely be interested in this project, such as vegetarian and vegan groups, black studies groups, alternative healing communities, animal rights advocates, and food studies scholars. Five months later, in February of 2006, a person named Herman Thrust[1] posted the request for papers on his Veganporn.com website. Not related to porn, this is a site for dialogues about veganism, animal rights activism, and other related topics. Upon checking responses to the request for papers on Veganporn.com, I found that the use of the word "sistah," in the title of the call for papers, sparked a debate on the forum.

More than 40 printed pages of this particular Veganporn.com discussion was birthed from the use of the word "sistah" in the posting of the call for papers. Though I did not participate in the discussion, I observed how systemic whiteness was performed when certain white identified cyberspace users were confronted with African American rhetoric (i.e., Black English) in a cyberspace dedicated to a progressive social justice movement focused on food production and anti-speciesism.

Nearly half of the participants in the discussion expressed their disapproval of the use of the word "sistah" and people [of color] who write or speak non-Standard English as well as engage in ethnic/racial solidarity. As an African American ethnographer, my "lurking" was invisible; therefore, I was able to observe the real or unobstructed views of how certain Veganporn.com participants felt without them feeling restricted because of my presence. If this forum had taken place in a physical location where they and myself were not "safely" hidden behind a computer monitor, this raw honesty may have never manifested. Kang (2000) notes that this is what cyberspace affords: a) the ability for a person- whose position, job or credibility might be compromised-, to state exactly how they feel about other people and b) the ability for me to observe this primary data "safely."

Recent scholarship in critical race studies theorize that many liberal- minded and progressive whites deny the accoutrements of white racial privilege and/or often engage in white supremacist ideology, at the structural level, within their social justice movements (Appel 2003; Nagra 2003; Poldervaart 2001; Rodriguez 2000; Slocum 2006; Yancy 2004;). Yancy's (2004) research has found that, collectively, whites see themselves as "raceless" and prefer that everyone else (non-Whites) do the same with their own racial identity in order to create a post-racial world. However, in this world systemic whiteness is not at the root of racial inequality; extreme hate groups (KKK, neo-Nazis, etc.) and resistance to being "post-racial" are. Such epistemic racism comes from a space of safety.

Systemic whiteness, a concept developed by Dr. Chithra Karunakaran, CUNY professor in Sociology and former NWSA Anti-White Supremacist Task Force Co-Chair. Systemic whiteness is literally the ideology of the superiority of whiteness at the systemic and structural level. It is separate from, yet inextricably linked to concepts of white privilege, white supremacist ideologies, institutionalized racism, First World imperialism and globalization. Dr. Karunakaran defines systemic whiteness as:

1. a *structure* operating on several organizational levels across the a) microstructure (e.g. civil society organizations and service organizations such as schools, churches, hospitals, systems of higher education, the media) and b) the macrostructure(e.g. nation-state).
2. a *system* of (un)earned privileges adhering to white skin, developed and maintained through advocacy of and subscription to supremacist beliefs and practices.
3. an *ideology* of supremacy, a set of beliefs underscored by everyday practices, openly expressed only under extreme organizational arrangements (KKK, neo-Nazi)but usually held covertly that advances and supports exclusionary, discriminatory and concealed practices vs. collectivist and inclusionary practices. (Karunakaran 2006)

Being "post-racial" reinforces systemic whiteness (Banks 2006; Nakamura 2002; Wright 2005) and whiteness and white supremacy are inseparable entities that maintain systemic whiteness and institutionalized racism (Rodriguez 2000; Yancy 2004). This is why I have chosen to explore the implications of systemic whiteness and how it functions in "post-racial" cyberspaces dedicated to progressive social justice movements.

Veganporn as Site for Covert Whiteness

I posted a request for papers for the Sistah Vegan anthology in the fall of 2005 requesting the voices of Black female vegans:

I am searching for Black identified sistahs that practice veganism to share their stories in an anthology I would like to put together. I am looking for poetry, narrative, creative stories or critical essays that link veganism to ending human oppression and ecological devastation. Topics and questions to consider can include but are not limited to:

- Choosing veganism and organic fair trade consumption because of its links to ecological harmony.
- Realizing that veganism connected you to other holistic practices that enhanced your health and managed life-long ailments
- Connecting a memory of sexual harassment and linking it to theories of veganism.
- The pivotal moment you realized that your dietary habits potentially supported the very types of oppression you were struggling against.

The post elicited disapproval from specific white Veganporn.com members. Their disapproval of the word of "sistah," other forms of non-Standard English, and the concept of community building among historically marginalized racial groups manifested as empirical data that closely aligned to the research data Whiteness Studies and Critical Race Theory scholars have published, in regards to a) progressive social justice movements dominated largely by white identified people or "goodwill" whites (Jones 2004); and b) [systemic] whiteness or white supremacist philosophy at the structural level of many of these movements (Appel 2003; Jones 2004; Karunakaran 2006; Nagra 2003; Poldervaart 2001; Slocum 2006). The heart of the progressive social movement of veganism calls for practice of Ahimsa or "compassion, kindness, and justice" for all living beings (including those beings who speak Black English).

One specific aspect that cyberspace affords to cyber ethnographers is the capacity to view digital rhetoric of "everyday people" engaged in dialogues that are often archived and accessible to anyone who has access to the Internet. Physically and socially, it would be extremely challenging- if not impossible- for me to document and archive casual dialogues throughout the "offline" world. In this sense, [cyber]mapping systemic whiteness- particularly within progressive social justice movements- becomes an easier and "safer" feat. Though the particular Veganporn.com forum discussion under analysis is an extremely small sample of performances of whiteness, one can imagine how effective it would be to gather and analyze a plethora of these "little performances" in vegan/animal rights oriented cyberspaces (or other progressive social movements). In Appel's (2003) research on how white supremacist ideologies manifested among white progressives that are against the prison industrial complex, she notes:

Through various forms of paternalism, reformist politics, romanticizing struggle and violence, exotification and exceptionalism, and neocolonial divide-and-counter tactics, whites propagate their privilege and power in their organizing and activism. (Appel 2003, p. 81)

Though she is speaking collectively about white progressives, one cannot assume that one will encounter this with every white progressive or white identified person involved in a progressive movement (such as veganism/animal rights). She, however, emphasizes in her work, that she critiques:

White progressives whose actions, though well intentioned, are laden with manifestations of white supremacy and challenge them to embrace a more radical politics, one the makes anti-racism a central tenet in any analysis and subsequent action against a white-supremacist state. (Appel 2003, p. 81)

Appel's comment about understanding the complexities of domination is noteworthy because the vegan and animal rights social justice movement seeks to dismantle the exploitation of non-human animals (Marcus 2005; Spiegel 1996). How foes systemic whiteness manifest- particularly amongst certain whites involved in progressive social justice movements? Let's take a look at Veganporn responses to "sistah."

EXCAVATING ACTS OF COVERT WHITENESS ON VEGANPORN.COM

The first responses to "sistah" was from Sardonicus, a white identified animal rights vegan.

Posted by Sardonicus Profile has photo! @ 2006-02-08 20:38:23 EST

"Sistahs"?

Rollng eyes.

Posted by Triddle Profile has photo! @ 2006-02-08 21:21:04 EST

You're JEALOUS!

[Reply]

Re: haw haw
Posted by Sardonicus Profile has photo! @ 2006-02-08 23:08:08 EST

Jealous of people who have institutionalized a speech impediment? Not to mention the spelling.

Re: haw haw
Posted by Sardonicus Profile has photo! @ 2006-02-08 23:33:24 EST

Sorry, I just can't stand to see my language sullied. People who don't value clear communications aren't very good social animals. That and I hate group mentality which this newspeak is a part of.

In retaliation, Swollen Glands, a white identified woman living in Germany, responded with:

Re: haw haw
Posted by Swollen Glands Profile has photo! @ 2006-02-08 23:30:43 EST

Just breathe deeply ten times and the dominance disorder interference will make way for clear thoughts to emerge.

*Because instead of poorly masked racist reflexes, what is actually to be criticized is the complete and utter lack of ethics and animal rights - the *actual* core foundations of veganism.*

However, Swollen Glands "corrects" herself with the following:

Re: haw haw
Posted by Swollen Glands Profile has photo! @ 2006-02-08 23:36:31 EST

Ok, I have to withdraw the "complete and utter lack", since ethics is 'hidden' behind "same oppression you are fighting against". But still, it needs to come first. Sometimes it's like using the word ethics will result in instantly breaking out with a fatal disease.

Ethics / ethical - the "dirty" word.

Sardonicus response is forthright and honest. On this particular online forum, he is comfortable in speaking openly. However, where does this honesty come from? Is it as simple as being "safe" and in cyberspace? Or, does he assume that because Veganporn.com is not a site dedicated to specific racially marginalized groups, is there a default assumption that this means there are no people of color using Veganporn.com? Does he assume that there are no Black English speakers on Veganporn.com ? If Veganporn.com was a forum occurring in real physical space, marked by multi-racial and multi-lingual visible bodies, what are the chances that Sardonicus would be so open and honest about his perceptions of "other" people? How does someone's relationship with the "other" shift or change when they move from physically bodied spaces to cyberspace? How does the performance of systemic whiteness shift among people like Sardonicus, when they interact with a cybertype as opposed to an actual human being they see in a physical space?

As a way to navigate her research on language of racialism Nakamura (2002) developed the term cybertype to

describe the distinctive ways that the Internet propagates, disseminates, and commodifies images of race and racism. The study of racial cybertypes brings together the cultural layer and the computer layer; that is to say, cybertyping is the process by which computer/human interfaces, the dynamics and economics of access, and the means by which users are able to express themselves online interacts with the "cultural layer" or ideologies regarding race that they bring with them into cyberspace. (Nakamura 2002, p. 3)

Sardonicus' discomfort in the Sistah Vegan! request for papers may stem not only from the use of the word "sistah," but also because Herman Thrust posted a call for papers that emphasized community-building among a particular group of women who continue to experience racial and gender oppression by the very system of whiteness that Sardonicus does not [cyber]publicly admit he benefits from. What Sardonicus as well as another white identified member do next in the forum is engage in a unique type of minstrelsy.

Jerry Kang writes, "Cyberspace enables new forms of social interaction. How might these new communicative forms affect racial mechanics?" (Kang 2000, p. 1135) How do certain Veganporn.com participants maintain systemic whiteness, particularly through demanding the use of typewritten Standard English?

The digital rhetoric amongst many of the discussants showed that their ideologies surrounding the treatment of the human animal paralleled dominating and oppressive ideologies; while discussing "sistah" and speakers of non-Standard English, several people rooted their arguments in unacknowledged whiteness.

As we shift from real space into cyberspace, the physical stage of the theater and television vanish. However, the absence of this physical space on the web does not mean an elimination of racism by those in progressive social justice movements who wish to reinforce values they may not consciously know are structurally rooted in systemic whiteness.

For instance, just because race is not signaled in cyberspace does not mean that race ceases to matter in cyberspace or, certainly, in real space. Race will continue to influence the "content" of one's communications even if the audience is not aware of the nature of that influence... even text permits racial mapping... One's language, grammar, and diction may also suggest race. (Kang 2001, p. 1156)

As the Veganporn.com dialogue progressed, Swollen Glands questioned Sardonicus' comments. After Sardonicus' wrote, "... I hate group mentality which this [Sistah Vegan] newspeak is part of" Swollen Glands wrote:

Re: Haw Haw
Posted by Swollen Gland @ 2006-02-09 00:0039 EST

*...But * YOUR * clinically sterile, restrictive and pedantic approach of English which you reserve for your social group * IS * group mentality par excellance. Showing that you are completely floating. You do not have to use newspeak, you don't even have to like it.*

It is interesting to note that Sardonicus does not respond negatively to Swollen Gland's misspelling of 'par excellence.' In this case, when are misspelled words in digital rhetoric acceptable by cyberspace users like Sardonicus that expect everyone to write their way? Is it the intentional misspelling of a word (i.e. "sistah") that makes it unacceptable by cyberspace users like Sardonicus? Or, is it deeper? Does it matter who has typed it? Does it matter who the cyberspace user thinks has typed it. Swollen Glands has presented herself as a white German identified woman whose first language is not English. Though she may not be who she claims to be in the Veganporn.com forum, does the cybertype of a white German woman allow her to misspell 'par excellence' without being reprimanded by Sardonicus, while the cybertype of a black female does not?

In his analysis of race on the Internet, Kang speaks of how "the architecture that mediates an interaction can alter racial mechanics in different ways" (Kang 2001, 1134). He gives distinct examples in which unique modes of technological communication influence the outcome of how he was treated by human beings he had never met in a physical space.

Can cyberspace change the very way that race structures our daily lives? To see why this might be possible, consider the following:

Car purchase: I have bought my last two cars through a buying agent, who charges me a flat fee of three hundred dollars over dealer's invoice. I use this service mostly because I am too busy to negotiate with car dealers. But there is another reason: I worry that I may receive worse offers than a similarly situated White male. By using the buying agent, I skirt the aggravation of wondering, "Am I being discriminated against?"

Tennessee roommate: One college summer, I was hired at Oak Ridge National Laboratories. Unfamiliar with Tennessee, I found housing over the telephone. I arranged to live with a graduate student, who was kind enough to pick me up at the airport. I told him my height, what kind of jacket I had, and that I wore glasses. He told me that he had red hair, which would make him easy to spot. I later learned that neither he nor my immediate supervisor knew that I was Asian American until we met face to face. My phone voice, grammar, and accent did not prompt them to flip out of the default assumption: White. As for my name, they somehow heard "Jerry King." (Kang 2001, p. 1133)

How is an individual's propensity for racially mapping another human being influenced by telephone communication versus online textual communication? Furthermore, in Kang's example in "Fighting Words," what are the ways in which Internet users, uncomfortable with "the other," display their need to dominate and control "the other" in a text based online forum?; a cybertyped "other" that exists socio-cognitively through this person' own unique sense of racial mapping and categorization? Sardonicus and another Veganporn.com member respond through minstrelsy.

In response to Swollen Gland's post, Sardonicus implemented a popular tactic people (mostly white) have historically used to falsely prove the inferiority of black people: blackface minstrel performance. However, because he could not perform minstrelsy in the traditional sense (on a physical stage in a theater), he achieved this by appropriating Black English in a manner that was negatively stereotypical. Before we go into these Veganporn.com cyber-minstrel performances, I will explain the strategy behind this technique and use several examples from recent cases of white identified people mimicking what they perceived to be "being Black in the U.S.A." In 1999, Chi Gamma Epsilon fraternity at Dartmouth College held a

'ghetto' party, in which students invited were expected to dress in the manner of residents of the inner city...students dressed as gangsta rap artists, wore Afro wigs, and carried toy guns. (Journal of Blacks in Higher Education 1999, p. 40)

Several years later, in 2001, Beta Theta Pi fraternity of Auburn University, in Alabama, threw a Halloween party "where members performed mock lynchings, paraded in blackface and donned homemade Ku Klux Klan hoods" (Lords 2001, p. 11). One year later, in October of 2002, University of Virginia fraternities, Zeta Psi and Kappa Alpha Order organized a Halloween party. A white student dressed as "Uncle Sam with an Afro wig, while two others dressed as tennis champions Venus and Serena Williams" (Hamilton 2002, p. 10). In November of 2002, the pictures were circulated on the Web site www.partypics.com (Hamilton 2002).

Tuhkanen (2001) argues that the strategy behind whites mimicking negative stereotypes of blacks has the sole purpose of "proving" to themselves that white people are superior while everyone else (non-whites) are inferior:

The dynamics of blackface should not be considered a thing of the past, a period in the history of US race relations, but that the minstrel mask perhaps continues to be a figure in how 'racial' visibility functions in the United States. (Tuhkanen 2001, p. 12)

In Veganporn.com, Sardonicus continues with the American tradition of racialization and "racial visibility" in his own performance of blackface minstrelsy.

Eat Me Glands
Posted by Sardonicus @ 2006-02-09 01:02:02 EST

*[[English is not *your* language]]*

Sure it is. I learned Enlglish good. I read it in a booook. I suppose you think you're very clever to act as if I implied ownership. When I make a colloquial use of "my language" and you try to pretend I meant "my" as if I own it... that's just very 5th grade debate club. You should work for Fox News. Dishonest but distracting.

[[and neither is it fixed but subject to and expression of the time in which we exist.]]

Biotch-ass ho, if you think dat you be illin! But OK, you be da ho like dat. Fine. But don't you be all gettin up in my grill like dat. Shiiit.

[[English is the language of anyone participating in it,]]

And if they can't be understood or sound like guttersnipes it doesn't serve them well does it?

[[you ignore and deny the flexibilities of creative expression]]

No I don't. I called you a "biotch-ass ho." Obviously I can deploy it for irony or to make a point.

Listen cranky German lady. When you have to live here and put up with commercials that are barely recognizable as English, because it's so cool, so gosh darn urban to sound like an uneducated minority hobbled by a history of having a crack-addicted mother then you can talk. They sound as if they just got out of prison.

Veganporn.com is a public site that is accessible to anyone with an internet connection. Sardonicus' minstrel performance, through appropriation of Black English, used the stage of the cyberspace to re-inscribe the "superiority" of whiteness by linguistically depicting blacks as "inferior."

[R]ather than mirroring African-American culture, blackface functions as a reflecting surface in which the image of the white audiences is projected according to social, political, and psychological exigencies—and at a considerable expense to African Americans. (Tuhkanen 2001, p. 16)

The mere fact that he said "Biotch-ass ho" and likened people who speak Black English as having been in jail and born to a crack addicted mothers, created an emotionally "unsafe" space for black identified people who may wish to explore veganism using "post-racial" sites such as Veganporn.com. t f lee, whose profile showed him to be a white man in college, joined the conversation:

Re: Eat Me Glands
Posted by t f lee Profile @ 2006-02-09 01:47:26 EST

The American black dialect is used to signify solidarity between people who share the common bond of being black. This happens in thousands of communities across the globe. The only difference is that most groups who develop a dialect for those reasons also speak a more standard dialect to communicate with the mainstream. Most black people don't do that, and I don't know why. If someone is raised immersed in American culture yet can't or won't use a standard dialect (note that I don't mean accent, I mean the syntax), they have no business being offended if no one will hire them. That's the way the world works. [Intentionally emphasized]

I might say to my friend, "Holy shit dude, we hella needa bomb it up to the dorms, fuck'n jeopardy is on and I'm not even crunk yet."

t f lee performed a hybridization of blackface, similar to Sardonicus. t f lee attempted to not appear racist, however, his mimicry of Black English revealed that his white privilege is dependent on the use of Standard English as the norm, particularly when it comes to securing employment. Standard English, as the norm, ensures that his job security (and economic mobility) will not be taken away by black people who speak Black English. t f lee's criticism and simultaneous use of false Black English, along with Sardonicus' cyberminstrelsy, resonate with Creese and Kambere's (2003) research on how figurative borders permit certain people in Canada from fully engaging in what the "imagined community" of real Canadian nationals have access to. Brint (2001) defines "imagined communities" as

communities of belief in which members are not in face-to-face contact with one another. The term was originally used by Anderson (1983) in relation to communities of believers in the nation-state, but it is equally applicable to... political tendencies, or philosophical schools. (Brint 2001, p. 11)

Veganporn.com is an "imagined community" of people connected through their beliefs and practices of veganism and animal rights. In "What Colour is Your English?", Creese and Kambere (2003) examine the treatment of African immigrant women in Canada who audibly have an "African English" accent.

'Canadian English' constitutes a border allowing only partial and provisional crossing for those with an 'African English' accent. The accent border is material and figurative, affecting access to material benefits such as jobs or housing, as well as shaping perceptions of who belongs in Canada. (Creese and Kambere 2003, p. 565)

Furthermore, their research "raises questions about ways in which the 'imagined community' is discursively patrolled through accents" (Creese, and Kambere 2003, p. 565). Similarly, the responses from "sistah," by certain participants on Veganporn.com, parallel such patrolling. Because Veganporn.com is an "imagined community" maintained through the written word, the cyber-panoptic lens of "expected usage of written Standard English" by vegans and animal rights activists like Sardonicus, t f lee, and similar, protect Veganporn.com's borders- albeit figurative, philosophical and virtual- from the socio-linguistic vegan and animal rights "outsiders."

Referring back to Creese and Kambere (2003), their research shows the consequences of Canadian nationalism (an "imagined community") for African female immigrants in Canada who are fluent in English. A majority in their focus group had advanced post-secondary degrees earned from English-language universities. The women in this study conveyed frustration with being mistaken as having limited English speaking skills and intellectual potential because of their "African English" within Canada.

The most common responses to "African English" identified in our focus groups were being ignored when speaking, and being corrected rather than responding to the content of speech. Comments by Mapendo [...] illustrate this point:

It seems that somehow they put you in a spot where you become defensive. You have to defend how you talk. When they correct me, I just say well this is the way you pronounce it, but this is the way I pronounce it. You continue to pronounce it the way you do and I will continue my way I do. We are from different schools so why should I listen to what you have to tell me. It becomes ridiculous when people are correcting you and not listening to what you have to say (Mapendo). (Creese, and Kambere 2003, p. 2)

No matter how the women in focus group tried to communicate with their "African English" in Canada, these women's "African" accents are the borders that keep them separated from becoming part of the imagined Canadian nation. Similarly, Sardonicus, t f lee, and others, as we'll see further, engage in a similar tactic that Creese and Kambere's focus group experience: ignore the content and context of the speaker (Sistah Vegan project founder) and correct rather than engage in the content of who is speaking. They do not attempt to fully engage with the real person, but with their stereotypical- or in this case, cybertypical- image of the person or community behind the spelling of 'sistah' that could perhaps never truly become part of their "imagined community."

My case study of responses to "sistah" reveal strategies in how systemic whiteness is maintained by certain white identified people who participate in a socio-politically progressive "imagined community" that is not predicated solely upon member's beliefs in advocating veganism and animals rights, but on structures that maintain the power and privilege of the dominant group within the "imagined community." In his research work on community building, Brint (2001) writes:

Early on, community studies researchers discovered the inequitable effects of social stratification (see, e.g. Lynd and Lynd 1937; Hollingshead 1949) and emphasized the structure of privilege as the hidden truth underlying nominally cohesive communities. Subsequently, the comforting image of community-centered governance was replaced by discovery of a self-interested and self-reproducing power structure ruling from behind the scenes (see, e.g., Huner 1953). Even the much-vaunted spirit of community cohesiveness was reinterpreted through the lens of the sociology of power. (Brint 2001, p. 6)

In the case study of the Veganporn.com 'sistah' discussion, power is embedded in racial (white) and socio-linguistic (Standard English) privilege which could very well be at the root of t f lee, Sardonicus's and other's own perceptions of an "imagined" vegan and animal rights virtual community. This could potentially apply to perceptions of an "imagined community" in real space, since Nakamura (2002) and Kolko (el al. 2000) indicate that cyberspace is a reflection of how power- along the lines of systemic whiteness, race and racialization- functions in the off-line world. In revisiting t f lee's Veganporn.com posting, he writes:

If someone is raised immersed in American culture yet can't or won't use a standard dialect (note that I don't mean accent, I mean the syntax), they have no business being offended if no one will hire them. That's the way the world works.

Such a textual linguistic [cyber]public performance reveals that t f lee does not fully understand Black English; he perceives it as not having syntax and that it is only a dialect. Smitherman (2000) has written that these are two popularly held false beliefs by many people in the United States of America that are unfamiliar with the fact that Black English and Ebonics are bona fide language systems and not speech impediments. Likewise, Creese and Kambere (2000) reveal that an embodied "African accent" renders women in their focus group as an "outsider within." Despite her legal citizenship status, "she knows she will never be perceived, or perceive herself, as 'Canadian'" (Creese, and Kambere 2003, p. 570) because collectively, the "Canadian English" speaking people in Canada (at least those with the power to give this women jobs, housing, etc.,) believe that "African English" is not a bona fide English language system and think it is a symptom of a speech and intellectual impediment (Creese, and Kambere 2003).

In the case of Sardonicus, t f lee, and the "Canadian English speaking" people who disparage the language of the African women in Creese and Kambere's focus group, each is engaging in a "systemic process of marginalizing and 'Othering' African women [and black people]" (Creese, and Kambere 2003, p. 570). If we engage further in t f lee's comments, who is "no one" in this "imagined community" of the world of employment? One could imagine that those who have the power and privilege of giving "African English" or non-Standard English speaking people jobs are people like t f lee: white identified college educated people who speak normal (Standard English) and without an accent. Accent is a relative term, as everyone has one; however, the accent of the socio-linguistic and racial status quo of Anglophone nations is rendered invisible yet simultaneously normalized as the default.

Though the focus group study and t f lee's comments indicate that people's accents or syntax prevent them from finding employment, does t f lee's comments extend to the "imagined community" of virtual progressive spaces such as Veganporn.com? For example, should a Black English speaking black identified vegan "have no business being offended if no one will" talk to them or let them enter the Veganporn.com community because "that's the way the world works"? Should they "have no business being offended if no one will" welcome them into other normal and Standard English speaking vegan and animals rights oriented cyberspace communities not based on racial/ethnic solidarity ("post-racial")? The questions I have posed emphasize how the "imagined community" of Canada and Veganporn.com have members that protect their borders through discursive patrolling. Sardonicus's, t f lee, and the Canadian focus group study lead me to ask the question, "Can a Black English speaking black vegan find full membership in a 'post-racial' Standard English speaking vegan and animal rights oriented virtual community, such as Veganporn.com with members who patrol their writing and/or speaking?" Or, "Would they be the 'outsider within' like the African immigrant woman in Creese and Kambere's research who are always being 'corrected' while the content of their words are made invisible?"

As we read further through the Veganporn.com discussion forum, TheRealPamela add her view on the word "sistah":

Posted by TheRealPamela Profile has photo! @ 2006-02-09 11:29:34 EST

I think the author is certainly well spoken with the rest of the title being "Black Women on the Social Implications of Veganism". She should drop the H on "sistah" though...

TheRealPamela was also white identified. Though she did not participate in cyberminstrel performance, what does it mean when a white identified female speaker of Standard English elects to correct the Black English of a black identified female? What does it mean that the project and the creator of the Sistah Vegan(a black identified female) be defined (discursively patrolled) by TheRealPamela's interpretation of Black English?

The insistence on Black female self-definition reframes the entire dialogue from one of determining the technical accuracy of an image [or word] to one stressing the power dynamics underlying the very process of definition itself. Black feminists have questioned not only what has been said about Black women, but the credibility and the intentions of those possessing the power to define. When Black women define themselves, they clearly reject the taken-for-granted assumption that those in positions granting them the authority to describe and analyze reality are entitled to do so (Hill Collins 1991, p. 38).

What does the "imagined community" of vegans on Veganporn.com look like to "outsiders within" who may visit the site and read: that particular members are advocating Standard English only; SwollenGlands' white ontological definition of how "veganism" should be defined; blackface cyberminstrel performances by Sardonics and t f lee; or TheRealPamela's critique that "sistah" is not valid Black English unless the 'h' is "dropped." How safe are certain "post-racial" cyberspaces for black females whose political consciousness and emotional survival is based on Black English or other form of non-Standard English. Furthermore, I am interested in how emotionally safe white dominated "post-racial" organizations, involved in progressive movements, are for people of color who knowingly must negotiate these territories of whiteness that are invisible to whites (Appel 2003 ; Farr 2004; McClendon III 2004).

It can be argued that this Veganporn.com forum is only one small example of how whiteness functions and that it is not reflective of off-line realities. However, when we begin to see patterns of these performances- particularly within progressive white dominated movements- within the research results of other case studies and research (i.e. Appel 2003; Creese and Kambere 2003; Farr 2004; Slocum 2006) these small occurrences become empirical data that begin to collectively construct a cultural narrative that depicts symptoms of systemic whiteness.

Whether it's the cyber blackface performance donned by UVA's Zeta Psi and Kappa Alpha Order fraternities' "ghetto party" pictures published on www.partypics.com, or the Veganporn.com forum case study, these cyberspace performances of whiteness add a broader dimension to the field of whiteness studies and critical race theory studies. Even more, cyberspace can help critical race theory researchers such as myself, gain safer access to empirical data on whiteness. Though I did not have the ability to physically attend the UVA party, cyberspace research permits safe access to these racialized performances without being in physical danger. However, what types of psychological dangers and issues of safety surround Veganporn.com?

Potential Psychological Consequences of Covert Whiteness for Black Females in Cyberspace

If you were a Black English-speaking black female interested in practicing veganism as a tool to combat the Black health disparities crisis in the United States of America, but did not know much about veganism, how would you feel if you encountered this particular Veganporn.com discussion? What if it were your first exposure to cyberspace vegan culture? What would your initial impression be of people practicing veganism? Would you feel that Sardonicus, TheRealPamela, and t f lee are non-human animal liberationists, yet, paradoxically don't have compassion and solidarity for the human animal and/or vegan that doesn't fit their rules of English language usage? Would you feel invited? Banks (2006) explores this question:

Black participation on the Web also begins to show the ways cyberspace can serve as a cultural underground that counters the surveillance and censorship that always seem to accompany the presence of African Americans speaking, writing, and designing in more public space- spaces that seem to consistently say to them that no matter what traditions they might bring to the classroom, the workplace, or to technologies- these spaces (and the written English that accompanies them) are, and will continue to be White by definition (Banks: 2006, p. 69).

Though Banks is speaking primarily of black web spaces that counter systemic whiteness, my indirect participation on Veganporn.com, via

Herman Thrust's posting of my project, showed surveillance and censorship of non-Standard English, making this particular site "White by definition."

As the conversation deepened, two new people entered:

Re: haw haw
Posted by Roethke Profile has photo! @ 2006-02-09 00:51:32 EST

... shoot me if I ever start referring to total strangers as though they were blood relations.

Posted by rain @ 2006-02-12 16:56:27 EST

Although I have a desire for the world to be more open to people with darker skin, I can't philosophically support the concept of banding together with anyone due to the color of their skin or their class...

Again, I find myself not wanting that to be the case just due to the fact that I want to be supportive of groups that have been marginalized for the simple reason that they are a minority and people feel they can get away with that[...] However, there is a difference between a science club (interest-based, not exclusive generally) and a group that clings together and shuns outsiders based on 'race.' Because skin color is only that, skin color. It doesn't make people who they are, it only says a little about where their ancestors came from[...] Part of the ideal of not seeing skin color when you look at people is that it works both ways. If 'white pride' is wrong then 'black pride' has to be too.

rain's placement and positionality of skin color is simplistic, untouched by socio-historical meanings influenced by the social constructions and power given to white skin color by the white racial status quo. rain's interaction with and the meaning placed on the Sistah Vegan call for papers conveys that, for her, systemic whiteness is not at the root of prejudice and racism. Instead, people of color who do not abandon their identification with "skin tone" and do not embrace the concept of a "post-racial imagined community" are at the root; they are the real racists who shun people like her ("post-racial" progressive white looking people) because of skin tone. "As the presumed sovereign voice, treating itself as hypernormative and unmarked, whiteness conceals its status as raciated, located, and positioned... [it is] presumed [to have] ahistorical stability and ontological 'givenness'" (Yancy 2004, p. 109). rain tried to leave her comments unmarked by the historical and social significance of her white skin tone. However, in her cyber performance of "post-racial," rain has made [in]visible the power that her own white skin tone has on her ontological perspective of the cyberspace forum discussion on the meaning of Sistah Vegan!.

Sardonicus continues the discussion.

Re:
Posted by Sardonicus Profile has photo! @ 2006-02-12 19:46:42 EST

I can understand a vegan who want to contact other vegans. I can sort of understand to some small degree refining that further. But when they exclude for both female and black it becomes somewhat troubling. [...] In my mind the main issue here is legitimizing illiteracy with intentional bad spelling of "sisters." As Roethke pointed out, that word and notion itself has problems. But by spelling it like that, the student only reinforces the notion that blacks can't spell, or worse, prefer to dis(respect) the English language and or further separate themselves linguistically from mainstream America. So much for integration efforts...

I make my fair share of typos. I usually give other people a pass on it (certain people around here use typos for attacks when they can't defend things logically) but I don't think "sistahs" was a typo.

... Talking like a gangsta or a dropout or drug mule doesn't represent one's group very well.

Sardonicus disapproves of the intentional wording of "sistahs" on the Veganporn.com forum but, again, is lenient when it comes to typos within this cyberspace forum. It is notable that the members discussing "sistah" have typed words incorrectly (typos), yet no one has been reprimanded or "corrected" for this. We first saw this with Swollen Glands' misspelling of 'par excellence.' This is an interesting example of how the norm- within systemic whiteness- is performed. "Traditionally, the concept of social norm has been considered one, if not the most important concept for understanding the impact of social phenomena on individual behaviors and judgments" (Dubois 2002, p. 1). Within systemic whiteness, is it a social norm to expect that [white] cybertyped human beings will accidentally mistype Standard English words in cyberspace forums? However, because Black English is not part of the social norm in Anglophone nations, is it socio-cognitively perceived as "deviant"? Are these the prescriptions with which members of Standard English Anglophone "imagined communities"- in cyberspace and real space- are supposed to comply? Are Standard English typos the acceptable social norm, and non-Standard English (regardless of types) unacceptable? The concept of social norm

Insinuates that whatever is prevalent is somehow worthy (valued), is a reflection of social values, is what one should do as a member of the concerned social body. In other words, it seems that even when it is supposedly being used in a descriptive way, the concept of social norm always has a prescriptive facet. (Dubois 2002, p. 2)

Beliefs and practices that are statistically performed routinely within a group set the social norm (Hofstede 1980 in Dubois 2002). Are typos in Standard English statistically performed more often than Black English is used in certain Anglophone "imagined communities"? If so, perhaps this is the sociocognitive process that is taking place in the minds of Sardonicus, t f lee, Roethke, and rain within this online "imagined community." However, Dubois and Hofstede's theories can only analyze this forum to a certain point, as one is unable to truly know what is occurring in the minds of these participants; their textual performances can only be analyzed.

Through textual performances, Roethke, rain and Sardonicus have displayed their disapproval of the black female political solidarity that the word "sistah" implies. Roethke's first message conveys that the Sistah Vegan call for paper assumes that all black females fall into the "sistah" category through filial association. The connotation of Sardonicus, rain, t f lee and Roethke's own socio-political identification of animal rights vegans means "progressive whites" in the Western World. Mentioned earlier in this chapter, veganism/animal rights is often considered a part of a progressive social justice movement and philosophy. This, for Roethke, rain and Sardonicus race/color should no longer be a justifiable reason for solidarity. In fact, according to them society should move beyond race and gender solidarity among marginalized groups and come together around similar ideologies- not similar "skin color."

To argue that race is myth and that it is an ideological rather than a biological fact does not deny that ideology has real effects on people's lives. Race serves as a "global sign," a "metalanguage," since it speaks about and lends meaning to a host of terms and expressions, to myriad aspects of life that would otherwise fall outside the referential domain of race. By continually expressing overt and covert analogic relationships, race impregnates the simplest meanings we take for granted. It makes hair "good" or "bad," speech patterns "correct" or "incorrect." (Higginbotham 1992, p. 255)

Roethke, rain and Sardonicus's textual performances of community building around a common ideology indicate that they feel ideology and not materiality (i.e. ethnicity or race) should be at the foundations of offline and cyberspace "imagined communities."

Understanding White Racialized Consciousness

It is important to note that even though these three Veganporn.com users identify as white animal rights vegans, this does not mean that their actions speak for the entire white identified animal rights and vegan population. However, their performances do connect to the recent bodies of scholarly work that collectively convey underlying concerns and themes about how systemic whiteness/white supremacist ideologies are reproduced in certain progressive social justice movements or organizations dominated largely by white identified or "white looking" people. Such literature analyzes how these "imagined communities" prefer to be rooted in common ideologies but do not address the dialectical materialism of their white skin color and the white ontological lens that influences the construction and practice of their "imagined community's" core ideologies (Appel 2003; Farr 2004; McClendon III 2004; Yancy 2004).

Dialectical materialism, the philosophy of Marxism-Leninism, holds that there is a material reality that is independent of our social consciousness and also that material reality is the basis for consciousness. Although material reality, from the vantage point of dialectical materialism, is independent of consciousness, it is nonetheless open to our cognition; knowledge is a form of social consciousness that provides an objectively approximate cognition of material reality...Furthermore, race has an ontological status, and that whiteness possesses an ontological status as well. (McClendon III 2004, p. 212)

Furthermore, Roethke's, Sardonicus', rain's and t f lee's textual performances, which indicate that they feel everyone should write and speak in Standard English and not organize themselves in the name of "skin tone" (minority racial group), is an example of "racial mapping." This is also an example of people who do not want to reflect on how their [racialized] consciousness is impacted by racist social structures but wish to engage in a progressive socio-political "imagined community" with a "raceless" ideology (an unacknowledged white ontological lens by default). Farr (2004) writes:

My attempt to make whiteness visible will be developed through what I call racialized consciousness. This term... will replace racism as the traditional operative term in discourses on race. The concept of racialized consciousness will help us examine the ways in which consciousness is shaped in terms of racist social structures. I will avoid the term racism because it tends to imply that one has a conscious commitment to race-based discrimination and acts of hate. (Farr 2004, p. 144)

Such an observation of Roethke, Sardonicus, rain and t f Lee is not judgmental, as this analysis is not about trying to prove that they are overtly racist ("bad whites"). Their perceptions of the Sistah Vegan call for papers are deeply complex and resonate more with Arnold Farr's work on and experience with white racialized consciousness. He writes:

Several years ago, a colleague approached me and asked: "What do people mean when they say that there are two Americas, one white and the other black?" I tried to explain to my colleague that there is a color line separating whites from blacks in America. This separation is not merely physical or geographical, but is manifest in terms of the distribution of social and economic resources, cultural capital, respect, and opportunities for self-development. My colleague refused to believe that there is such a color line producing radically different experiences of American society...

After I explained to him in more detail what philosophy of race was, he proceeded to warn me that we should be careful because "this sounds like what the Nazis were doing."... This colleague was not a mean-spirited racist, but just a white male whose entire epistemic grid for deciphering social data was too white to empathize with and comprehend the African-American experience. By "too white" I mean that my colleague's experience of the world as a white male produced a barrier between himself and those who experience the world in black bodies. (Farr 2004, p. 144)

When whites involved in progressive social movements fail to deeply engage with what their racialized consciousness means, and when they do not understand how they potentially can and/or do embody white ontological beliefs even though they may not overtly embrace racism, prejudice or domination (especially domination that involves coercive control), can they fully recognize the ways their actions support and affirm the very structures of racist domination and oppression that they presumably want to see eradicated?

Assimilation is the strategy...deeply rooted in the ideology of white supremacy and its advocates urge black people to negate blackness, to imitate racist white people so as to better absorb their values, their way of life. Ironically, many changes in social policy and social attitudes that were once seen as ways to end racial domination have served to reinforce and perpetuate white supremacy (hooks 1989, p. 113).

Furthermore, "racial mapping" and cyber-typing is what initially triggered the responses displayed by these participants.

Racial mapping triggers racial meanings- cognitive beliefs about and affective reactions to- people in these racial categories... Racial meanings are triggered, often automatically, without self-awareness, and once triggered, they alter our behavior... In these ways, racial mechanics alter interpersonal interactions. These microlevel changes accumulate over time and people to produce enduring macrolevel affects on our culture, economics, and politics. (Kang 2001, p. 1142-1145)

The use of the word "sistah" is socio-politically intentional in order to attract a specific group of black identified female vegans as opposed to all black female vegans. The goal was to find out how the vegan experience, amongst black identified females, has been shaped by a black female ontology. However, when the Sistah Vegan call for papers was initiated, I was fully aware that not all black-identified people support the use of Black English or Ebonics. However, Smitherman's (2000) research notes that most blacks in America speak Black English, Ebonics or some form of non-Standard English. Hence, the use of "sistah" should not be interpreted as such that it is an essentialization of all black female vegans. Often, feminist ethnographers assume that that "all women share some authentic feminine selfhood" (Zavella 1996, p. 140). The strategy in using "sistah" is political and consciousness-raising; a way to encompass non-white ontology and bring forth how a black racialized consciousness impacts black female ontology within "post-racial" progressive spaces and movements; it is in dialogue with and against white racialized consciousness.

Interestingly, "sistah" triggered responses of disapproval from several white identified members on Veganporn.com, paralleling Kang's consequences of racial mapping. It could be argued that those such as Sardonicus, t f lee, rain and Roethke were not given this information (in regards to my full intentions with the Sistah Vegan! anthology) in the Call for Papers and were therefore responding to this limited amount of information. However, through critical race theory based discursive analysis it is revealed that responses from Veganporn.

com forum did not manifest as "simple mistakes from having received limited information about the intent of the Sistah Vegan! anthology." The meanings that human beings apply to the textual performances they encounter and interact with, in cyberspace, do not manifest from within a void, untouched by influences of racist, classist, sexist, etc., social structures (Nakamura 2002; Smitherman 2000).

The decision to use the phrase "vegan sistah" to define how I practice veganism, as a Black woman, was a way of defining myself through the language and voice of my choice. Patricia Hill Collins references Barbara Christian in Black Feminist Thought:

Why this theme of self-definition should preoccupy African-American women is not surprising. Black women's lives are a series of negotiations that aim to reconcile the contradictions separating our own internally defined images of self as African-American women with our objectification as the Other (Hill Collins 2000, p. 99).

It was my self-definition, through non-Standard English, that caused a particular group of white-identified Veganporn.com members to engage in textual performance that made it quite clear that people "like me," employing African American rhetorical traditions, would not be welcomed into their "imagined community" of animal rights activism and veganism unless they undergo socio-linguistic assimilation.

Many of the Standard English speakers on the Veganporn.com forum colonized Black English speakers through expected usage of a Standardized language system based on the ideology that systemic whiteness is normal.

For many racial minorities, their racial identity is an integral facet of their self-conception. To strong-arm them into repressing that self-conception, even if only in cyberspace, is to ask some Americans to silence a part of themselves. (Kang 2001, p. 1158)

Much like Arnold Farr's encounter with his white colleague's white racialized consciousness, the people on Veganporn.com who opposed the use of Black English and racial/ethnic solidarity of marginalized people did not understand that for some black identified people, Black English, Ebonics, and African American rhetoric are significant tools of empowerment and liberation among many black people who wish to resist forced assimilation into "post-racial imagined community"(default whiteness)(Banks 2006; Smitherman 2000) in which skin color "doesn't matter anymore."

Conclusion: Employing Cyberspace in Developing a Model for Mapping and Studying Systemic Whiteness

The Internet is an effective tool within the field of Critical Race Theory because the multitude of public forums, blogs, discussion boards, and other dialogue facilitating cyberspaces offer digital rhetoric that is often more raw and honest than rhetoric that may take place in physical locations in which a) not all are welcomed and/or b) the visibility of other human beings may cause individual participants discomfort in "being themselves" and/or c) the space is inaccessible because of distance.

The World Wide Web can be a central tool in uncovering the invisibility of covert whiteness (a tacit form of racialized consciousness) (Farr 2004) that does not manifest itself at the surface level in the same overt manner that extreme white cyber hate "imagined communities" do. Access to archived digital rhetoric from public websites-such as the Veganporn.com- also allow cyber-ethnographers focused on Critical Race Theory to explore the subtleties of whiteness that are often made invisible through the use of "code words for racism" (Levine-Rasky 2002, 322). In regards to the Veganporn.com forum discussion analysis, an example of a code word or phrases that sustain racial (as well as class and socio-linguistic) dominance is "they should speak Standard English" or "they do not speak proper English."

When employing cyberspace for Critical Race Studies, particularly manifestations of whiteness within "post-racial" and/or progressive movements, researchers have the opportunity to discursively analyze "benign" forms of whiteness. Performances such as these that occurred during the Veganporn.com discussion lend themselves especially well for this sort of analysis (i.e. Swollen Glands' white ontological definition of what veganism should be and TheRealPamela's insistence that 'sistah' should be spelled without an 'h').

Frances Rains (1998) describes five "benign" manifestations of racism induced by whiteness: a sense of entitlement (on the basis of meritocracy); "citation of exceptions" (of racialized persons who over-came the consequences of social inequality); the I-can't-speak-for response (distancing whites from the claims of racialized persons); the reactions of guilt (diverting knowledge of racism to its effects on white participants); and color-blindness or racial neutrality (dismissing racialization as a context for understanding the experiences of persons of color). (Levine-Rasky 2002, p. 322)

Even more precisely, not only do such responses serve the interests of those who desire to take advantage of their white privilege, but cyberspace also gives Critical Race Studies scholars the ability to track performances of whiteness, whether "benign" or directly overt, in a way that reveals how such performances objectify racialized groups and "reduce 'black cultural practices to performance. [Thus]...Meanings of race and culture cultivated by racialized groups in opposition to white hegemony are obscured" (Levine-Rasky 2002, p. 323). The Internet allows us to see this in Sardonicus' and t f lee's appropriation of Black English as a "mere ignorant performance" detached from a socio-historical context of power and resistance against colonialism and systemic whiteness. Likewise, though I was unable to attend the 2002 Halloween party at the University of Virginia, in which several white students engaged in blackface minstrelsy, partypics.com allowed public documentation of white hegemonic appropriation of "blackness." Cyberspace allows us to archive many of these seemingly unrelated events and start seeing a metanarrative of white hegemony- particularly progressive social justice spaces like Veganporn.com.

After the discursive analysis of the Veganporn.com discussion about Sistah Vegan!, I am reexamining Jerry Kang's question: "Cyberspace enables new forms of social interaction. How might these new communicative forms affect racial mechanics?" Instead of focusing only on racial mechanics, however, I ask: How does cyberspace allow scholars of anti-racist praxis, social justice studies, and other disciplines, to map and understand manifestations of "benign" whiteness, particularly when it occurs within largely white dominated progressive social justice movements or organizations- especially if scholars who are non-white do not have emotionally "safe" access to these spaces?

This chapter also reveals how effective cyberspace can be as a "safe" tool for [cyber]mapping the possibility of systemic whiteness within the digital rhetoric of "imagined communities" focused on progressive social justice philosophies and movements.

REFERENCES

Appel, L. (2003). White supremacy in the movement against the prison-industrial complex. *Social Justice (San Francisco, Calif.)*, *30*(2), 81–88.

Banks, A. J. (2006). *Race, rhetoric and technology: Searching for higher ground.* New Jersey: NCTE Press.

Brint, S. (2001). Gemeinschaft revisited: A critique and reconstruction of the community concept. *Sociological Theory*, *19*(1), 1–23.

Collins, P. H. (1991). Learning from the outsider within: The sociological significance of Black feminist thought. In *Beyond methodology: Feminist scholarship as lived research* (pp. 35–59). Indiana Press.

Creese, G., & Kambere, E. N. (2004). What colour is your English? *The Canadian Review of Sociology and Anthropology. La Revue Canadienne de Sociologie et d'Anthropologie*, *40*(5), 565–573.

Dubois, N. (2002). *A sociocognitive approach to social norms*. New York, NY: Routledge.

Dyson, M. E. (2006). *Is Bill Cosby right? Or has the black middle class lost its mind?* Perseus Books Group.

Farr, A. (2004). Whiteness visible: Enlightenment racism and the structure of racialized consciousness. In G. Yancy (Ed.), What White looks like: African-American philosophers on the whiteness question (pp. 143-157). New York, Y: Routledge.

Flis, H., Miller, N., Senker, P., & Wyatt, S. (2000). *Technology and in/equality: Questioning the information society*. New York, NY: Routledge.

Hamilton, K. (2002). UVA lateest in string of black face incidents. *Black Issues in Higher Education*, *19*(22), 10.

Higginbotham, E. B. (1992). African-American women's history and the metalanguage of race. *Signs*, *17*(2), 25–274.

Hill Collins, P. (2000). Black feminist thought knowledge, consciousness, and the politics of empowerment (rev. 10th anniversary ed.). New York, NY: Routledge.

Hooks, b. (1989). Talking back: Thinking feminist, thinking black. Boston, MA: South End Press.

Jones, J. (2004). The impairment of empathy in goodwill Whites for African Americans. In Yancy, G. (Ed.), *What White looks like: African-American philosophers on the Whiteness question* (pp. 65–86). New York, NY: Routledge.

Kang, J. (2000). Cyber-race. *Harvard Law Review*, *113*(5), 1131–1209.

Karunakaran, C. (2006). *Personal email describing systemic Whiteness* (Harper, B., Ed.). New York.

Kolko, B. E., Nakamura, L., & Rodman, G. B. (2000). *Race in cyberspace*. New York, NY: Routledge.

Levine-Rasky, C. (2002). Critical/relational/contextual: Toward a model for studying Whitness. In Levine-Rasky, C. (Ed.), *Working through Whiteness: International perspectives* (pp. 319–352). New York, NY: SUNY.

Lords, E. (2001). Free speech becomes primary issues in Auburn blackface case. *Black Issues in Higher Education*, *18*(22), 11.

Maddox, A. H. (2003). Blackface, friend chicken, watermelon and a lynching - it's OK. *New York Amsterdam News*, *10*, 13.

Marcus, E. (2005). *Meat market: Animals, ethics, and money*. New York, NY: Brio Press.

McClendon, J. H. (2004). On the nature of Whiteness and the ontology of race. In Yancy, G. (Ed.), *What White looks like: African-American philosophers on the Whiteness question* (pp. 211–225). New York, NY: Routledge.

Nagra, N. (2003). Whiteness in Seattle: Anti-globalization activists examine racism within the movement. Alternatives Journal(Winter 2003), 27-28.

Nakamura, L. (2002). *Cybertypes: Race, ethnicity, and identity on the Internet*. New York, NY: Routledge.

Poldervaart, S. (2001). Utopian aspects of social movements in postmodern times: Some examples of DIY politics in the Netherlands. *Utopian Studies*, *12*(2), 143–164.

Rodriguez, N., & Villaverde, L. E. (2000). *Dismantling White privilege: Pedagogy, politics, and Whiteness*. New York, NY: Peter Lang.

Slocum, R. (2006). Anti-racist practice and the work of community food organizations. *Antipode*, *38*(2), 327–349.

Smitherman, G. (2000). *Talkin that talk: Language, culture, and education in African America*. London, UK & New York, NY: Routledge.

Spiegel, M. (1996). The dreaded comparison: Human and animal slavery (rev. and expanded ed.). New York, NY: Mirror Books.

Tuhkanen, M. (2001). Of blackface and paranoid knowledge: Richard Wright, Jacques Lacan, and the ambivalence of black minstrelsy. *Diacritics*, *31*(2), 9–34.

Wright, M. (2005). Finding a place in cyberspace: Black women, technology, and identity. *Frontiers*, *26*(1), 48–59.

Yancy, G. (2004). *What white looks like: African-American philosophers on the whiteness question*. New York, NY: Routledge.

Zavella, P. (1996). Feminist insider dilemmas: Constructing ethnic identity with Chicana informant. In Wolfe, D. (Ed.), *Feminist dilemmas in fieldwork*. Boulder, CO: Westview Press.

ENDNOTE

1 This is his handle and not his legal name.

Chapter 13
Language and Performing Arts:
East African Hip Hop and Public Sensitization for Political Change

E. Sangai Mohochi
Maseno University, Kenya

D. Ndirangu Wachanga
University of Wisconsin-Whitewater, USA

ABSTRACT

One of the many consequences of globalization and the new world order is the increased cross border interaction among people, leading to more transfer and exchange of knowledge, technology, values and virtues, vices and viruses, and other traits among nations. One area that has been impacted heavily by this flow, largely aided by the Internet and other emerging media, is culture. To a large extent, though, the transfer of cultural practices appears to be more from the western and more developed world to the weaker, economically and politically less powerful nations. But what is borrowed is indigenized, sometimes entirely altered, to meet the needs of local communities. These changes are reflected in Africa's music scene, dances, and other genres of popular culture. Within that context, this chapter aims at meeting two goals. First, to analyze the extent to which musicians, especially the youth, have managed to maintain a balance between educating and entertaining society at the local level, while keeping abreast with emerging global trends and influences. Particularly, it will show how the young generation of East African musicians uses music to sensitize the public by serving as critics of the management of public affairs and how this has contributed to political change. Secondly, it will investigate the effects that these emerging practices have had on the use of African languages in the performing arts.

INTRODUCTION

This chapter discusses East African hip-hop by analyzing its content, language use, and the role in political change, and social and cultural transformation. Also, it broadly examines the intersection of hip-hop as a cultural form with the emerging communication technologies, partly leading to the globalization of hip-hop. The first part, the transnationalization of hip-hop, traces the origin of hip-hop and explains some of the factors that catapulted it to the realms of acquiring traits as-

DOI: 10.4018/978-1-60960-591-9.ch013

sociated with global cultures. The second part examines the emergence of East African hip-hop, before proceeding to take a closer look at the content of songs by Kenyan and Tanzanian hip-hop artists. Section three examines the language used by the artists in conveying their messages. Also, this section discusses the role that hip-hop music plays in East Africa, how it is perceived, and its place in society. We conclude by noting the potential of hip-hop as a social and political mobilizing agent as well as an educational tool, but hasten to point how that potential has neither been realized nor fully utilized.

THE TRANSNATIONALIZATION OF HIP-HOP

Expanding on Benedict Anderson's (1991) notion of imagined communities, Corona and Madrid (2008, p. 3) have pointed the folly of insisting on the nation-state as a unit of analysis, especially in matters of culture where most nation-states are made up of diverse groups. Still, members of such groups are not restricted in their trans-border interactions:

Such a view emphasizes the fact that culture and the people who produce it, consume it, and identify with it continuously move through the borders of the nation-state via a wide range of technologies. These people group together in a variety of "imagined communities" that might be greater or smaller than the nation-State, but transcend it as a unit of identification (Corona & Madrid, 2008: 3).

Likewise, cultural theorists have accepted this position, especially considering that there is an increase in the flow of capital, people, information, and culture under globalization. This is evidenced in the growth of special interest groups with membership in several nations across the globe. In his article "*Disjuncture and difference in the global cultural economy*," Arjun Appadurai (1990), argues against the Cartesian plane of oppositions – center vs. periphery, North vs. South, global vs. local, which has been the strong point of globalization studies until recently. Appadurai suggests the need to focus on global dynamics from a viewpoint that acknowledges the ceaseless interaction of global relations- what he calls scapes - which mutate constantly. The Appaduraian "scapes" dissect the dynamics of global imaginary by examining the fluid constitutional make-up of our globe, which constantly transmutes, and especially the electronic ability to penetrate previously impervious national borders. This contemporaneous trait of our 21st century digital millennium facilitates and enhances deterritorialization of its citizens who have become subjects in search of cosmopolitanism and/or globality. The media, in their copious manifestations, are at the nucleus of these changes. It is not surprising how the mediascape, as Appadurai illustrates, possesses the power to form scripts about "reality" informing how different parts of the globe can be formed (See also Wachanga, 2010). We argue that the youth are more likely to be influenced by and open to such transnational movements. We will use this background to examine the growth and spread of hip-hop culture in East Africa. But using text to discuss music – a genre that is appreciated through lyricism – is "like dancing about architecture" (Goddard, 2002, p. 9). Even more poignantly put:

Words can never fully capture the electrifying rush and emotional punch that a great pop song can deliver. While a common objection among fans, journalists and performers is that attempts to analyze music often fail to grasp what makes the sound so important in the first place. (Carrabine, 2007 p. 231

Corona and Madrid (2008, p 5) see music as being "in constant flux," that "music is the perennial undocumented immigrant; it has always moved beyond borders without the required paperwork." If indeed music is that undocumented immigrant, it is, as Josh Kun (2005) notes, always from somewhere else. While it is produced in a specific place, its consumption goes beyond that place and may in the process acquire new meaning(s) and new expressive and performance styles. Ntarangwi (2007, p. 273) has signaled the lack of clarity on whether hip hop, as a cultural form, "is indeed a local indigenous East African hip hop or a Western (American) hip-hop expressed and mediated through cultural and creative processes." Scholars of hip hop, either as a music genre per se or as a cultural form, defy the tendency to confine it in restrictive categories of location – as local, foreign or a form of global culture (see also Cook 2004; Condry, 2000; Mitchell, 2001). Although Mwangi (2004) states that hip hop can be traced in the synthesis of Caribbean expressions and western rhythms by a Jamaican DJ, Kool Herc, in the West Bronx region of New York, dominant voices in hip hop scholarship see its origin as "elsewhere" - even if symbolically - as a borrowed form of expression that has undergone various transformations and mutations, and which is eventually creatively domesticated to reflect and serve local needs. But it is how hip hop is globally mediated, through various means of communication and novel convergence of technological instruments that generates the need to explore the "processes of production of difference in a world of culturally, socially, and economically interdependent spaces" (Gupta & Ferguson, 1992, quoted in Cook, 2004, p. 108).

The global characteristic associated with new communication technologies (NCTs) need to be understood from a perspective that recognizes unprecedented information capacity and speed with which it is shared and transferred across borders, even previously unreceptive ones. As Ntarangwi (2007) asserts, the prominence of East African hip-hop articles is predominantly in the cyberspace. But there are three key points that Cook (2004, p. 108) makes in relation to hip hop and NCTs:

- Studying technologies with global reach and global impact does not require or justify foregoing attentiveness to situated practice and localized meaning
- Global hip hop is not the sum total of its products: Its songs, dances, and graffiti images
- One can infer from the studies of hip hop around the world that global flows of aesthetic forms, cultural movements, and political stances (especially those mediated in some way by new technologies) are centered around youth.

While it initially started in the US in the 1970s, and served as an identity marker along class and ethnic lines, hip-hop is no longer an American affair. If Herc's (the Jamaican, West Bronx-based DJ) New York rap had its origins in Africa's verbal and rhythm games, as Mwangi (2004) suggests, it is through the processes of globalization, with NCTs at the core of this process, that it has found its way back. And upon its (re)turn to Africa, hip hop culture is largely perceived as "foreign because of its association with American urbanity."(Mwangi, 2004, p.7)." That is no longer the case anymore.

Globalization has contributed to hip-hop's transnationalization to the extent that it is now an international culture. In societies that are ethnically stratified such as Kenya, it is important to note how hip-hop has problematized ethnic membership, especially among the youth. Hip-hop is more of a generational, as opposed to ethnic or class identifier. While some of the artists come from lower-class urban neighborhoods, its acceptance cuts across all economic classes.

Its popularity among the youth converges at multiple points including mode of dressing, dancing style, and discourse construction and enunciation. In his description of hip-hop, Dennis (2008, p.185) states that "...generally speaking, hip-hop is defined as the culture, styles and discourses, while rap refers to the music, considered to be one of the four pillars of hip-hop culture: break-dancing, DJ'ing, MC'ing (rap), and graffiti." For the youth, it has become a global unifier. As Dennis (2008, p.185) indicates "it has become a vehicle for global youth affiliations in parts of Europe, Africa, Asia, and Latin America." The same is the case in East Africa. As much as hip-hop is associated with the American urbanity, "it is popular with the youth without much knowledge of traditional culture but, as in traditional oral literature live performance, the effects of hip-hop vary with the performer, the audience and the occasion" (Mwangi, 2004, p. 6).

Corona and Madrid (2008) have lamented that the production and distribution of music bring to mind the colonial relationship between the west and the less developed world. They invoke dynamics associated with cultural imperialism, especially when one considers that most songs attain global recognition if rendered in English, and that certain pre-recorded songs are brought to the West (especially America) from third world countries' musicians for more sophisticated elaboration. Such concerns have been raised by Carrabine (2007, p. 231), who eloquently examines a running motif in the literature of popular music. According to Carrabine, this motif is "a tension between describing the music industry as a constraining, exploitative and manipulative machine while perceiving the audience as active, autonomous and creative in its use of the texts manufactured by capitalist enterprise." This tension has not been adequately addressed with regard to hip-hop in East Africa.

The content, themes, motifs and styles associated with East African hip-hop divorce their creativity from simplistic imitation of the American urban expression. Ntarangwi (2007, p. 275) notes "East African hip-hop engenders a very specific local identity that is reflective of a history of colonial domination and a Western-inspired cultural identity." Recent developments in technology have made it possible for many recording studios to set camp in all major cities of East Africa and musicians are often spoilt for choice. Globally, the rise of hip-hop has been made possible by a combination of factors. In a discussion of the development of hip-hop in Colombia, Dennis (2008) mentions the important role played by the entertainment industry. American cinema and television played a critical role in the process. One readily remembers movies like *Breakin'* (1984), *Breakin'* 2: *Electric Boogaloo* (1984) and *Beat Street* (1984). Development of mass communication systems also facilitated the process. With more production and distribution of CDs and music cassettes, American hip-hop was consumed in almost all corners of the world. Not forgetting the role of television through programs in MTV for example. These same factors led to its spread to other parts of the globe.

EAST AFRICAN HIP-HOP

In the circumstances described above, it is no surprising how East African region got engulfed in the hip hop movement in the 1990s. Since then, over time, the number of hip-hop artists in the region has kept on rising at a very high rate. It is undeniable that music from different parts of the world has kept on influencing musicians and music production in other parts of the globe. What many people don't seem to appreciate is the fact that the influence is neither linear nor unidirectional. This has, however, been adequately captured by Corona and Madrid (2008, p. 9):

Just like western musicians have employed and been influenced by global musics, so musicians and producers everywhere have been adapting, mixing, and using modern popular Western music in a complex scenario of hybridization to the point that some music sociologists…call it "a universal pop aesthetic."

We need to keep in mind the complex dynamics informing the heavy localization of musical rhythms borrowed from elsewhere. This transformation is through various ways including, but not limited to, the use of local languages as well as infusing local rhythms, beats, and instrumentation. If we can compartmentalize hip-hop according to regions, there is a challenge of distilling the local from the borrowed, which, with time and creativity, is tightly fused and synthesized into an inseparable whole. But, there are certain mannerisms that current hip-hop artists in other parts of the world, for instance East Africa, have continued to borrow from the West that exude the air of "elsewhere" – baggy sagged trousers, sports shoes, sports jerseys associated with NBA, NFL, and European soccer clubs, and even dancing styles of Western hip-hop stars. The emergence of local clothing lines such as *Jamhuri* is challenging this trend. How successful it will be in competing with multinationals remains to be seen.

There are many hip-hop artists in East Africa who do not get to be known widely because their songs are never recorded for wider distribution, what would be referred in the west as "parking lot artists." They thus remain stage performing artists, known only to people in their locale. In a way, as alleged by Kitakufe in an article, "*The Rise of East African Hip-Hop,*" hip-hop is a culture and the same is the case in East Africa. The initial apprehension has given way to acceptance, albeit a little slower among the elderly who still view it with some suspicion. But its significance, arguably, rests in its ability to provide "insights into East African identities because it is performed by young people grappling with their sense of self. Still unofficial in a culture that would express itself in a language which appears to be unserious, the underground music offers a more genuine expression of what the young people would like to be" (Mwangi, 2004, p. 7).

While hip-hop has grown in popularity in the entire East African region, it appears to be more developed in Tanzania than in her neighbors. There are more artists in Tanzania, and they seem to have acquired more recognition, and are most likely making more income from their music than is the case in Kenya where piracy is still a huge obstacle. Tanzanian hip-hop, is popularly known as *Bongo-Flava*. *Bongo* is Swahili for brain and *flava* is a corruption of flavor. *Bongo* is commonly used to refer to Dar es Salaam, and at times to the country at large. It is claimed that the word originated from the fact that one needed brains to survive in Tanzania owing to the difficulty of life at some point in the country's history.

In Kenya, where it is mainly referred to as Kenyan hip-hop with some of its variants called *genge*, hip-hop took longer to be fully established as a career and a source of livelihood. However, it has been growing by leaps and bounds lately. There is now a very big band of hip-hop artists competing for space in the Kenyan media as well as leisure joints across the region.. In fact, it has taken over from Lingala (Congolese) music which had dominated the East African entertainment scene for decades. While for a long time, many Congolese musicians visited and crisscrossed East Africa, staging shows, that is no longer the case. Most entertainment spots, public functions, and public transport vehicles have now become arenas for the fierce competition among local East African artists, resulting in a reduced presence of Congolese music. Let us focus on the content of some of these songs.

CONTENT IN EAST AFRICAN HIP-HOP

For a long time hip-hop was taken as a form of cultural expression associated with the discourse of social protest. Similar sentiments emerged in an interview that Dennis (2008) had with Tostao, the leader of Choc Quib Town, a popular hip-hop group in Colombia on the role of hip-hop in society. While acknowledging the potential of hip-hop as a medium of awakening social consciousness and influencing social action, he advocates for experimentation with new forms of music within the wider hip-hop tradition. As a result he expressed the desire of his group to "spread the messages of their music beyond the traditional hip-hop circles." He mentioned Peter Tosh and Bob Marley as role models saying they were musicians who produced very popular music with lyrics that "provoked thought and consciousness." To what extent can we say the same about hip-hop artists in East Africa?

Kirkegaard (2002:15) rightly points out that:

...it is hard to talk about music without taking into consideration how the dance is performed, how the lyrics relate to social, cultural or political events and happenings...

In this section, we will analyze messages in East African hip-hop with a view to establishing the extent to which its content relates to the current social, cultural, and political environment. We will not dwell on how the dance is performed. Our analysis will be restricted to Kenyan hip-hop and Tanzania's *Bongo Flava.*

TANZANIAN HIP-HOP

There are several artists who are using their talents to point out social ills in society. These include the duo of Joni Walker and Ras Lion. In their album,"Nani Alaumiwe?" (who is to blame?), they do so through a number of their songs. They start by saying that they are very committed to their duty of educating society. In *Nimepania* (I am committed), they indicate that they have to talk, and put sleeping leaders on notice since without telling them, development will be delayed. Joni Walker and Ras Lion then proceed to lament, in the same track, about the Tanzanian situation in which many public corporations have been handed over to foreign ownership, for example TANESCO (Tanzania Electricity Supply Company), who then retrench locals without paying them their benefits. They wonder whether colonialism is coming back to the country. They state over and over again that: "hatutaki viongozi wanaosinzia" (we don't want sleeping leaders).

In the same album the song titled *Ajira ya Watoto* (child labor), they decry the habit of subjecting our children to child labor instead of giving them an opportunity to pursue an education for their own good and that of their children. They wonder when the children will go to school if they keep on working as house helps in the homes of people who themselves went to school and are successful. Overall, they urge people not to involve kids in activities like hawking, car wash services among other this is tantamount to abusing their rights and spoiling their lives at an early age. The same message touching on children's rights is found in *taifa la kesho* (the future of the nation) by Professor Jay, another Tanzanian artist.

Joni Walker and Ras Lion have a song titled *Walimu* (teachers) in which they talk about the plight of teachers, whom they lament, are very poorly paid, and have a very hard life despite their important role in society. Society laughs at them, undisciplined students abuse and attack them, yet nobody cares. They are responsible for preparing leaders and other human resources in the country but their working environment is very bad. When they start small income generating projects to augment their meager pay they are told they should concentrate on their job. This is a message that both teachers and other sections of society would readily identify with. In Tanzania and Kenya, there have been endless wrangles between teachers' unions and governments over teachers pay. The artists remind politicians, professors, doctors, and other leaders in society that they would not have been what they are now without the input of teachers and calls for more respect for them.

They also talk about more general issues. For instance, in *Moyoni Mwangu* (In my heart), they talk about love, and warn that it is not something to joke with. The singer assures his lover that he values and loves her genuinely, and requests that they take good care of their love. The other one on love is *Sofia* in which they are complaining to Sofia who does not appear to be willing to reciprocate the love she is shown. The two songs are on love but have good messages and the language used is not vulgar at all. They are tracks that can be listened to by one anywhere, without looking around to see who else is listening. Other "clean" songs on love include *Neno Nakupenda* (the word I love you) in which they claim the love the singer has is so much that it can't be captured by the words I love you. He wants assistance to come up with a word that will describe well what he has for her in his heart.

The duo has a song, *Mtukumbuke* (Remember us), which captures very well the habit that is rampant in East Africa among political leaders. They are very friendly and accessible during the campaign period but once they get elected they forget the electorate. They make numerous promises like their desire to build roads, schools, hospitals etc. None of that is done; they change their phone numbers and only resurface during the next election period. The artists conclude that people are tired of being cheated and next time they will just listen to them, laugh and elect someone else. In the current political dispensation where most elected leaders show very little regard for the electorate, it is becoming very necessary for citizens to be sensitized so as to be able to question them, and vote out those who don't want to listen. Lastly, in *Umaskini Huu* (This poverty), they regret that the poverty they witnessed when they were kids is still widespread in society. They wonder what needs to be done to rectify the situation in order to enable people live more dignified lives.

Another *Bongo Flava* musician whose songs I would like to analyze is Professor Jay from his *Mapinduzi Halisi* (Real Revolution) and *Aluta Continua* (the struggle continues) albums. In *Nang'atuka* (I retire/relinguish power), a leader has realized that he has been a big let down to society. He accepts that he has messed up after ascending to power and failing in his duties. After getting power he changed *kura* (elections) into *kula* (eating) in his leadership. He became selfish, misused his position and even organized his opponents' murders. He took no steps to rectify the health sector although he knew things were horrible. For instance, the maternity ward was so bad that his girlfriend had to travel abroad to go give birth. He got very corrupt and did nothing to stop the vice. As a result the economy has suffered. Not only did he defend robbers and other criminals so long as they shared with him, he also watched people die of hunger, and allowed foreigners to loot his nation's resources. To further show his excesses, he admits that what he has done does not deserve any sympathy and compares himself to Pik Botha of apartheid South Africa. His conscience has tortured him so much that he now regrets what he did, asks God for forgiveness, and decides to relinquish his position.

He has another song, *Kikao cha dharura* (an emergency sitting), in which he revisits the concern on the services offered by political leaders. In this emergency sitting, a politician begins by thanking the electorate for choosing him and wonders why he has been called to an emergency sitting. He accepts to take questions and the participants take him to task over his pledges. His responses show that he does no take their concerns very seriously. The police want helicopters promised and he tells them they will not be able to fly them, and a number do not even know how to ride a bike. Farmers remind him about tractors and he says that the government's budget is overstretched. Besides, he thinks that the current practice of using hoes in the farms is appropriate. As for the nation's civil servants who want vehicles, he is of the opinion that they want an easy life and should instead use public transport and concentrate on working. Lastly, he has the audacity to tell the people that if they elect him another time, he will fulfill all promises. This is exactly how the political elite take advantage of the electorate in East Africa. Professor Jay gives us hope because in this emergency sitting, the people openly tell their leader that they are not in agreement with him. In fact, they keep shouting that they disapprove and ask instead to share what he has since their living conditions are so bad. He responds by saying that is not possible since he has a family to take care of. Besides, everybody should be able to carry his own cross. In order to make our leaders accountable, the electorate in East Africa needs to start telling their leaders what their feelings are more openly like Professor Jay's group does.

In *Msilie* (Don't cry), he appears to be trying to give hope to the hopeless in society and asks them to kneel down and pray. He talks about the poor conditions in prisons and the fact that some of the people in them are actually innocent; laments that the state is buying arms rather than tackle other issues like increasing teachers' salaries; complains that churches have been turned into business premises; the county is smelling of corruption; the country's resources have been stolen by the rich, and rivers are drying up because of unchecked deforestation. He also states that the nation is sleeping on wealth yet they are relying on subsidies, perhaps in recognition of the fact that Tanzania has a lot of mineral resources bust still waits for assistance from abroad.

Professor Jay also sings about other issues affecting society. In *Naahidi* (I promise), he reaffirms his commitment to his lover whom he promises to stay with until death. He warns her against people who are not genuine but are out to use her for a while. Another song on love is *Zali la Mentali.* In this song, a poor young man (a push-cart puller) meets a rich young woman who falls in love with him to his and his friends' surprise. They get a baby, and to his dismay the lady's parents are very receptive. They accept their wish to get married which they do and live happily. He now is surprised to note that many people, including those who initially derided him, are very eager to be associated with him. *Narudi* (I return) *is* another song by the same artist in which he talks about a prodigal son who realizes his mistakes and asks God to forgive him and accept him in his kingdom.

When Professor Jay teamed up with Qchilla also from Tanzania, and Nonini, a Kenyan artist, in the song *Nakuita* (I call you), the focus changes. In the song, Nonini is simply praising himself for his exploits with women and calling this particular lady that he is targeting to come to him. He then praises her by mentioning how good her body is, and says she inherited a lot from her mother since she has a wonderful face and breasts. There is very little to learn from such a song which even goes on to mention parts of the body that we don't otherwise talk about openly in public.

Wagosi wa Kaya, in their song *Ushauri wa Bure* (Free advice) proclaim that they have a duty as artists. In this particular case, they are fulfilling it by giving advice to different categories of people in society. They start off with members of parliament in Dodoma Tanzania whom they tell to listen to the people who voted them in and heed their advice since they may be wiser than they are. Secondly, young girls who live in hostels while attending college are being warned against running around with men as their parents toil to raise their fees. They then turn to academics and tell them not to keep believing that Tanzania is poor yet it has lots of minerals. The same advice is extended to people in the villages who are urged to ask how they will benefit whenever minerals are discovered and extracted from their villages. Women are asked not to continue with the habit of putting on mini skirts as they go against the norms of society. In *Mjenga Nchi* (hardworking citizens) they highlight the plight of public servants who drive the country's development but are not well remunerated. They also want local investors to be given an opportunity instead of allowing foreigners to benefit from local resources. They also make a direct appeal to state house to look at these issues. They follow this up with *pole* (sorry) in which they sympathize with teachers, farmers, civil servants and doctors, among others, for the problems they face in life. They have spared time to express their sympathies to the president because development seems to be too slow in the country.

On a philosophical note, Mr. Ebbo, in his song *Njaa Inauma* (Hunger bites) wonders why hunger keeps stoking him and others all over the world since their birth. It knows no time, and attacks both the young and the elderly. Another vocalist suggests to him that he too should try to bite it, which is obviously not a solution. So fed up is Mr.

Ebbo with hunger that he is advising it to take a break and go on vacation for a while. He further advises that when it gets back, it should spare the young, the elderly and the disabled members of society. Lastly, he wonders why, if hunger is so much of a close friend, it abandons people immediately they die and comes back to continue torturing members of their family left behind and the rest of humanity. He would rather it accepted to remain buried with those who die. The song may sound funny but it talks about an important topic and has the potential to ignite some thought in peoples' minds on hunger and ways of looking at it and possibly dealing with it too.

On education, Mr. Ebbo, in *Kasome* (Go study), is appealing to his child to go back to school. Although the child is complaining that the environment in school is harsh and the teachers seem to be picking on him, Ebbo tells him that he has paid all the fees, bought him his requirements and given him enough money. He, therefore, pleads with him to go back because education will be very helpful to his life. This is in recognition of the importance of education, a message that would definitely apply to many young men who don't seem to be very keen on going to school.

This does not mean that all Tanzanian hip-hop artists tackle sensitize and important issues. In fact, it does not mean that those mentioned above deal with such issues all the time. It is, however, widely acknowledged even with Kenyan hip-hop followers that Tanzanian hip-hop music has more meaning than its Kenyan counterpart. There are a number of Tanzanian artists who sing about everyday issues like love, but even then their songs are appealing as they have good messages, are not vulgar, and can be enjoyed by those with conservative views on music and liberals alike. Examples of such songs include *Cinderella* by Ali Kiba, and what is currently referred to as the East African anthem, *Anita* by Matonya. In the former, the singer is telling his childhood friend who failed to accept and reciprocate his love for him then, but now wants to be with him, that it is late as he is already married to Cinderella who he loves very much. *Cinderella* was a very popular song in the region for quite some time. In *Anita*, Matonya talks about two lovers who stay apart, a fact that is having a toll on their love. The man is complaining because his lover (Anita) is away. She responds that he was traveling too much and leaving her lonely and that is why she took off. Behind, he goes through a lot of tribulations, including enduring ridicule from neighbors who want him to forger about her. However, since he loves Anita so much, he is willing to wait for her return. He is hopeful that she will come back and they will live together. The two songs are on love and have been well received because the message is well presented and the language is good. Furthermore, their video renditions can be watched by the whole family comfortably as they do not have obscene dancing styles, and the dressing is good too.

KENYAN HIP-HOP

There is less focus on important messages in hip-hop by Kenyan artists. Many talk about love, brag about their musical and sexual exploits (e.g. *swing* by Kleptomaniax), *and* other mundane issues and achievements. The few musicians who have come up with meaningful music include Eric Wainaina who, though in the group of young musicians, is not a hip-hop artist. His most popular track is *Nchi ya kitu kidogo* (a corrupt country) in which he castigates the country for turning out to be a country of corruption. According to him, by turning the vice into common practice, and accepting *kitu kidogo* (something small, a term used to refer to corruption) we are making our country a country of small people (watu wadogo). He then advises those who ask for bribes to take Limuru tea (chai ya Limuru) or a soda. *Chai* and soda are terms used in asking for bribes and Wainaina is advising the corrupt to take the real *chai* and soda instead of money.

In light of the prevalent corruption in Kenya, especially in the police force, the civil service and other government agencies, it was very gratifying to come a cross another track, titled "Sitoi kitu kidogo" (I am not paying any bribe) by jimw@t that dwells on the issue. Other than the popular Eric Wainaina's "Nchi ya kitu kidogo" (a corruption country), already mentioned above, this is the other piece that has tackled the vice directly. The singer has in fact, zeroed in on the police who have time and again been mentioned as the most corrupt in Kenya. In the video, the singer laments how he was accosted on his way from work and asked to part with *kitu kidogo* (something small), which is one of the many names used to refer to money given out in corruption environments. He adamantly refuses to pay claiming that he has committed no offense but the policemen tell him he is wrong and threaten to take him to jail. He sticks to his guns and is indeed taken to jail, where he however does not receive the dreaded treatment people talk about. The significance of the song in my view is the message that you can indeed stand up for your rights and refuse to co-operate with rogue officers. There is also a new release on corruption, titled *hatutaki kitu kidogo* (we don't want bribes) which is unique because it has been compose by a Kenyan police officer.

The Longombas have *vutapumz* (take a breath), a song that addresses HIV/AIDS. HIV is a major concern in East Africa, as is the case elsewhere. It should, therefore, have been a topic of concern to many artists in the region. While a number have taken it up, it is not in the expected vigor considering the enormity of the problem. In *vuta pumz,* the artists are warning their peers not to be too indulging. They are told not to turn every time someone with good (big) behinds passes by. They specifically warn that if they don't change their behaviors they will end up in Lang'ata Cemetery (a public cemetery in Nairobi). I believe that such a warning is necessary to the entire society and not the youth only. There is also a song by another musician which warns people against unprotected sex. The song, titled *Juala* aka *Manyake* (sheng for condom and a lot of meat) by Circute & Joel, equates a lady with meat and says that all sizes are available but proceeds to warn that you need a condom (referred to in the song as *juala*).The use of meat, and specifically in the plural sense, should be understood in the context of the traditional African concept of a woman's beauty. In the eyes of many people, the bigger in build the better.

As alluded to earlier, most Kenyan hip-hop artists do not play very educative and thought provoking songs. This section will give examples to illustrate that point. There is a song, *Kuhanya* (sheng for having multiple lovers) which seems to be praising infidelity and unfaithfulness by talking about how people in Nairobi like to engage in the practice. It would have been a different thing if the song then proceeded to indicate the dangers in the practice but that does not happen. Another song is *We Kamu* (you come) by Nonini. In the song, Nonini is basically asking a girl to come visit him in his room where he is waiting for her. That in itself is not a problem but it is what follows that is a little disturbing. To begin with, the singer advises the lady to lie to the mother that she is going to the library. She is urged not to come accompanied by any of her friends. He then does not hide his intentions at all as he informs her that he is so eager to meet her and immediately she gets to the door she should start undressing. Also indicated is the fact that he has no intention of using any protection as he states "nyama choma leo tutakula kavu" which translates to we shall "eat dry roast beef". This kind of song has a big potential to influence young girls negatively. Furthermore, in the current AIDS scenario in Kenya, 'eating roast beef dry' is not a wise thing to do at all.

In his other song, *Manzi wa Nairobi* (Nairobi ladies), he is basically talking about how good ladies of Nairobi are. He says they are the best in the world, they are very well dressed, and urges them to get up, dance and make noise. In the refrain of the song, there is a lady who says over and over again that "machali leo hamtalala" (guys you will not sleep today). Is it because they will be dancing with the ladies throughout the night or because of something else? Another song by Nonini titled *keroro* (beer) talks about excessive drinking but does not condemn it or even talk much about its effects, save for a mention of he fact that it makes some people abandon their family responsibilities. He says he is so intoxicated that all he sees in the club is beer and a girl seated at the counter. In the video he moves over to talk to her and among the things he does is ask her to stand up so that he can examine her. After doing so, he asks her the number of underwear she is wearing. The same partying message recurs in *Furahiday* (a corruption of Friday) used to mean a happy day. It talks about the party mood that engulfs Nairobi on Fridays, especially as the working day moves to a close. There are more vehicles on the roads, the dressing code changes as more ladies wear shorter skirts, and the phone networks get very congested. He even mentions that those who are married usually remove and hide their rings on Fridays. To his credit, in a song titled *Silazima* (it is not a must), Nonini talks about the need to be careful in relationships. He is telling a lady, who is otherwise very ready for sex, that they must not do it. He advises that they can go home and relax without doing anything. This is a good message since rushing into sex is one thing that puts many young men at risk. It is, therefore, gratifying to see an artist telling them that you can go out, have fun and go back home without necessarily engaging in careless sex.

There is a song by jimw@t called *under 18* which has a very good message that is dented by the use of inappropriate language at some point. It is a story of a young man who meets a girl in a club, who claims to be a third year student at the University of Nairobi. He buys her drinks after which they leave the club for a walk in the park. As they start getting intimate, policemen pounce and ask them for identification papers. The girl has none as she is actually under age. The young man is taken to jail and given 18 strokes by the officer. Upon his release he finds the girl waiting for him outside his house. Annoyed he tells her off for putting him in trouble but instead she suggests that they go home and have fun instead. He agrees - *"akasema twende nimpe manyundo, nikaweka tu, nikafyeka tu"* (she suggested I go give her some hammering, I did it, I just did it). The translation is being very soft because in the Swahili version, the words used are too direct and do not actually refer to love out of any respect and intimacy. It, instead, describes sex with very little concern or care. For his efforts, he is caught in the act by a police officer and ends up sentenced to serve some time. While it is a good warning to people who prey on young girls, it would have achieved much more if he had avoided sexually explicit lyrics.

LANGUAGE IN EAST AFRICAN HIP-HOP

East African hip-hop artists use different languages in their productions. In Tanzania, to a great extent they use Swahili. What's more, they often use standard Swahili. There are a few who mix Swahili and their first languages, and English but that is not a very common practice. Examples of such a mixture are found in Joni Walker and Ras Lion's song *Fani si Utani* (art is not a joke) who have mixed Swahili with Dholuo, as well as Mr. Ebbo who has some Maasai in *EE Emoti*. Kenyan hip-hop musicians use a mix of languages much more than their Tanzanian counterparts. There is more use of English, mixed with Swahili and *Sheng* (youth language used more in urban areas) as well. Several other Kenyan languages such as Dholuo, Kikuyu, Kamba, Giriama and others are often featured in such songs. Because of the dynamism of *Sheng*, it is difficult for many non *Sheng* speakers to follow some of what is said in a number of songs. In both countries, though more in Tanzania, there are many musicians who use standard Swahili which is more accessible to many people.

Many artists make an attempt to incorporate the Swahili poetry tradition of rhymes in their songs. An example is in the song *Kumekucha (it is day break)* by Professor Jay of Tanzania in collaboration with Nonini from Kenya. Here is a section of the song that does so.

Mapinduzi yamewadia, kwani huu ndio wakati
Nasimama kwenye ukweli, mpaka niione tamati
Tuko kimang'ati, siwezi kuvunja torati
Toka Bongo mpaka Kenya, nabeba bendera pati
(It is revolution time, for time is ripe
I stand on the side of truth until the very end
I carry the flag, from Tanzania to Kenya)
Shuka mpaka chini, hivi vita ndo madini
Usiseme umeona jini, ni profesa na Nonini
Niamini hakuna mapenzi tuko kazini
Sauti hizi zinatoka ghetto mpaka ubabini
(Go all the way down, this is an important fight
Do not mistake us for ghosts, it is Profesa and Nonini
Trust me as we are on duty and there are no jokes
These are voices from the ghetto on the way abroad)

In *Under 18*, jimw@t says:

weka mikono juu ya hewa
hata kama umelewa
ili wajue unaelewa
(Throw your hands in the air
Even if you are inebriated
How else will they know that you know?)

The same trend is evident in *Nyeti* (sensitive issues) and *Taarifa ya Habari* (news broadcast) by Wagosi wa Kaya and *Zali la Mentali* by Professor Jay and many others. Overall, the musicians have proved to be very creative in their use of language. There is a heavy presence of figures of speech like simile, metaphors, imagery and symbolism in their productions which serve to spice them up as works of art besides enriching their messages.

There in another aspect of language which, besides the themes tackled, is the reason for some of the criticism leveled against hip-hop artists. To start with, a number of them use vulgar language which makes it difficult for the music to be played in mixed audiences. A number of commentators have even complained to radio stations that play such music in their programs. When talking about the transformation of classical *taarab* music in Tanzania to the modern style which, according to some, has been polluted with obscenities, Lange (2002) attributed the change to popular taste. What is the situation with hip-hop? Is there any argument to support the artists' apparent ever present urge to be so explicit and somewhat obscene in their lyrics? There is no evidence of demand for such lyrics from the public. If anything members of the public have been complaining about the current situation as was the case in a discussion in the Tanzanian parliament in August 2008.

In Colombia, when record producers interested in the music of hip-hop artists approach them for a discussion of a possible deal; they often request them to "soften the denunciatory nature of their lyrics" (Dennis, 2008: 188) which are said to be offensive. Unfortunately, in East Africa, this is not the case, partly because of the mushrooming of individual producers who are eager to make a name and money (hence do not care much for the message of the songs), and probably because opposition from society is not strong enough. Do we need to come up with a policy with some sort of censorship in order to ensure that what is offered to us as music can be acceptable to a wider community?

Generally, hip-hop music is increasingly contributing to the development of Swahili language. As East African musicians of the 1960s and 1970s did, hip-hop artists are major players in the spread of the language. As a result, there are many people who have been drawn to the language through the music, which is a good result. Swahili speakers have also benefited from the music which deepens their understanding of the language due to the artists' creativity. Also, mainly because of its appeal to the youth, the music has now become a common feature in Swahili language classrooms in US and other western universities. The audio and video CDs have emerged as important parts of language teaching materials.

THE ROLE PLAYED BY HIP-HOP IN EAST AFRICA

With all its problems and shortcomings, hip-hop music has slowly started to play a role in sensitizing the youth on political participation and involvement in other issues that affect them in society. For instance, during the 2002 general elections in Kenya, the song *unbwogable* by Gidi Gidi Maji Maji was used by one of the political parties as a rallying call in their campaigns. During the 2007 electioneering period, many hip-hop musicians were involved in all major political meetings. They were utilized in attracting people to the meetings and entertaining them too. The most popular song in this regard was B*anjuka* by DNA. Musicians also talk to their fellow youths on various issues, like HIV/AIDS, crime, etc. A number of organizations have also been organizing concerts where the youth are not required to pay any entrance fee. These gatherings are then used as arenas for imparting various messages to the youth. For example, on December 6, 2008 a group of Kenyan and other top East African musicians converged in Nairobi for a music concert dubbed *Generation Jipange*. The show, also attended by Jamaican Dancehall king, Mr. Vegas, was organized to launch a movement by the same name, formed to address issues facing the youth in Kenya, including health, education, and economic issues.

Hip-hop has also contributed in tackling the high level of unemployment in the region. Many of the youth who would otherwise be hopeless, and end up in crime and other social ills, have a possible way out. A number have managed to make a decent living from music something that should be encouraged by all involved. Besides singing, music production has also created many employment opportunities although there is a big involvement by Asian businessmen in Tanzania. To some extent, the growth of hip-hop has greatly aided the rapid increase in the number of privately owned radio and television stations which thrive on entertainment programs that feature hip-hop artists. The music has also spiced up life in entertainment spots in the region. The amount of foreign music played in discos and pubs, as well as public transport vehicles is going down over time.

The value of music is Africa has been noted by none other than The World Bank. By noting the importance of music as an industry in Africa, the bank has launched a program in commercial music development. It states that, apart from minerals, music is the only area in which the continent has an opportunity to make money presently (Kirkegaard, 2002). The potential is clearly evident in East Africa. While local artists have not managed to break into the international arena to make big money, in a way hip-hop has become a source of employment to many young men who would otherwise be very miserable considering the state of our economies.

Despite all these, society seems to be divided in its position on hip-hop music in East Africa. While many, especially the youth and an increasing number of the adults, are embracing the music, there are some who are very critical of it. Indeed, the debate is ongoing. In the Tanzanian parliamentary debate aired on the Tanzania Broadcasting Corporation (TBC) TV on Tuesday the 26th August 26, 2008, a discussion ensured about the youth in Tanzania. The main issues being raised were their dress code, which was described as bizarre and largely unacceptable in the Tanzanian cultural context. The main culprits were *Bongo Flava* artists who were also accused of singing songs that have no moral teachings but are instead laden with sexually explicit lyrics. It was also stated that some of their videos are culturally inappropriate. However, Muhammed Seif Khatib, a minister in Tanzania, came to their defense. He said that their involvement in music is a source of employment that has helped many a youth leave their original lives in poverty. He also said that they are contributing in the development of Kiswahili language; and urged them on while advising adults who detest the music not to listen or watch it. The speaker of the national assembly, Mr. Sitta, caused laughter in the house by agreeing with the minister and enumerating the *Bongo Flava* musicians that he likes.

Those in opposition to the current crop of youthful musicians cite a number of reasons for their position. One major concern is the content of most of their songs. It is felt, especially, in the Kenyan situation that most of the music does not tackle themes that are valuable to society. Instead some of the songs are alleged to be encouraging bad practices.

Others insist that the music has lost direction. They call the hip-hop artists computer musicians who can't play a single musical instrument. This viewpoint is made worse by the realization that the musicians actually rely on play back in their live performances. What many people don't know is the fact that many of these musicians do not have any input in the beats of the songs they produce. Many buy the beats and then compose the songs while others actually buy both the beats and the lyrics. For people who were used to seeing a group of musicians playing different instruments while their colleagues belt out vocals, the current trend is not very appealing.

CONCLUSION

Popular music can have a very big impact in society. With regard to hip-hop in East Africa, it has a great potential to serve as a catalyst for social change, especially among the youth. It now has a following among many, including the elder members of society. If utilized to carry messaged that are thematically relevant to the problems and aspirations of society, it can contribute immensely to development. It is with that understanding that this article has discussed the state of hip-hop music in East Africa. Drawing from Kenya and Tanzanian hip-hop, it has attempted an analysis of the content tackled by several musicians. Besides the messages dealt with, a brief discussion of the language forms used by the musicians is carried out. While it is evident that hip-hop artists have managed to use language very creatively, there is a difference in terms of what Tanzanian and Kenyan artists predominantly talk about. Although this brief survey can't be said to present an exhaustive analysis of East African hip-hop, it clearly indicates that Tanzanian artists are more involved with relevant topics in comparison to their Kenyan counterparts. It is that kind of involvement that will make hip-hop play that significant role of serving as a catalyst for positive social change. Examples cited indicate that music can be very important in sensitizing the public for more involvement in politics and governance of public affairs. Unfortunately, a number of artists have not only failed to come up with music that provokes any serious thought, they have gone ahead to produce music that is offensive and largely meaningless to society.

However, a more detailed exploration of hip-hop, covering more artists, would be required before confidently drawing conclusions on the state of East African hip-hop with regard to the value of its messages. Language use is also a subject that calls for more research in order to bring out the rich features of language used by hip-hop artist in East Africa.

REFERENCES

Banda, T. K. (2008). *Youth converge on city for music show of the year. Saturday Nation* (p. 36). Nairobi, Kenya: Nation Media Group Ltd.

Big Youth. (2007). *Aluta Continua.* GMC Wasanii Promoters Ltd.

Bynoe, Y. (2004). *Stand and deliver: Political activism, leadership, and hip hop culture*. Brooklyn, NY: Soft Skull Press.

Circute and Joel. (2008). *Juala.* Retrieved on 30 September, 2008, from http://www.youtube.com/watch?v=ziF-HAATRGU

Cook, S. (2004). New technologies and language change: Toward an anthropology of linguistic frontiers. *Annual Review of Anthropology, 33*, 103–115. doi:10.1146/annurev.anthro.33.070203.143921

Corona, I., & Madrid, A. (2008). Introduction: The postnational turn in music scholarship and music marketing. In Corona, I., & Madrid, A. L. (Eds.), *Postnational musical identities: Cultural production, distribution, and consumption in a globalized scenario.* MD, USA: Lexington Books.

Corona, I., & Madrid, A. L. (2008). *Postnational musical identities: Cultural production, distribution, and consumption in a globalized scenario* (pp. 85–207). MD, USA: Lexington Books.

Dennis, C. (2007). *The Afro-Colombianization of hip-hop and discourses on authenticity. Banjuka.* Retrieved on 16 December, 2008, from http://www.youtube.com/watch?v=s1lBvP4BW0o

Firstenberg, S. (Dir.). (1984). *Breakin': Electric boogaloo.* Cannon Films Inc.

Gidi Gidi & Maji Maji. (2006). *Unbwogable.*

Jimw@t. (2006). Under 18. *Operation Fagia, 4.* Calif Records.

Jimw@t. (2008). *Sitoi Kitu Kidogo.* Retrieved on 20 September, 2008, from http://www.youtube.com/watch?v=e-L79xtsz40

Jones, S. (Ed.). (2002). *Pop music and the press.* Philadelphia, PA: Temple University Press.

Kiba, A. (2007). *Cinderella.* G-Records.

Kirkegaard, A. (2002). Introduction. In Palmberg, M., & Kirkegaard, A. (Eds.), *Playing with identities in contemporary music in Africa* (pp. 7–18). Uppsala, Sweden: Nordiska Afrikainstitutet.

Kitakufe, A. (2006). *The rise of East African hip-hop.* Retrieved on 20 September, 2008, from http://www.ugpulse.com/articles/daily/homepage.asp?ID=418#mat1

Kleptomaniax. (2005). *Swing.* Ogopa Deejays.

Kun, J. (2005). *Audiotopia: Music, race, and America.* Berkeley/ Los Angeles, CA & London, UK: University of California Press.

Lange, S. (in press). *Multipartysim, rivalry and Taarab in Dar es Salaam.* (pp. 165-180).

Lathan, S. (Dir.). (1984). *Beat street.* Orion Pictures.

Longombas. (2005). Vuta Pumz. *Chukua.* Ogopa Deejays.

Matonya and Lady Jaydee. (2007). Anita. *Siamini.* Mandugu Digital/41 Records.

Mr. Ebbo. (2006). *Kamongo.* Motika Records.

Mwangi, E. (2004). Masculinity and nationalism in East African Hip hop music. *Tydskrif Vir Letterkunde, 41*(2).

Nonini, J., & Qchilla. (2007). Kumekucha. *Aluta Continua.* GMC Wasanii PromotersLtd.

Ntarangwi, M. (2007). Hip-hop, westernization and gender in East Africa. In Njogu, K., & Maupeu, H. (Eds.), *Songs and politics in Eastern Africa.* East Lansing, MI: Michigan State University Press.

Professor Jay. (2003). *Mapinduzi Halisi.* GMC Wasanii Promoters Ltd.

Silberg, J. (Dir.). (1984). *Breakin'.* Cannon Films Inc.

Wachanga, D. N. (in press). Global media and information ethics: Challenges re-examined. In Yin, D. Y. (Ed.), *Global media convergence and cultural transformation: Emerging social patterns and characteristics*. Hershey, PA: IGI Global Publishing.

Wagosi wa Kaya. (2005). *Nyeti*. GMC Wasanii Promoters Ltd.

Wainaina, E. (2001). *Nchi ya Kitu Kidogo. Sawa sawa*. Christian Kaufmann.

Woka, J., & Lion, R. (2002). *Nani Alaumiwe?* MJ Records.

Youtube. (2007a). *Si Lazima.* Retrieved on 19 September, 2008, from http://www.youtube.com/watch?v=UiTR2Lt_fMw

Youtube. (2007b). *Manzi wa Nairobi*. Retrieved on 29 September, 2008, from http://www.youtube.com/watch?v=-AExNsJLKho

Youtube. (2007c). *Keroro*. Retrieved on 29 September, 2008, from http://www.youtube.com/watch?v=KIQcPm_btm0

Youtube. (2007d). *We Kamu*. Retrieved on 30 September, 2008, from http://www.youtube.com/watch?v=Ne2DXjIzOxk

Youtube. (2007e). *Furahiday*. Retrieved on 1 October, 2008, from http://www.youtube.com/watch?v=HlLOjXi9wDs

Section 3

Emerging Media, Global Politics, and Cultural Transformation

Chapter 14
Mediating Identity and Culture:
Nigerian Videos and African Immigrants in the U.S.

Adedayo Ladigbolu Abah
Washington and Lee University, USA

ABSTRACT

Using the media accessibility function from self-categorization theory, this study examines the role of the Nigerian video film in mediating the twin issues of culture and identity among African immigrants in the United States. Africans in diaspora constitute the majority of the transnational audience for Nigerian video films outside of Africa. Using a combined method of surveying and personal interviews, several African immigrants, their children, and friends living in the Dallas/Fort Worth area of Texas, USA were interviewed for their views on the role of the nascent Nigerian video industry in the way they sustain and straddle their multiple identities and culture in their society of settlement. Results indicate that most of the immigrants view the videos as affirmation of the values they grew up with and with which they still identify. This is in direct contradiction of professed cultural denigration they feel in their everyday professional lives in the United States. Most of the younger immigrants and first generation immigrants view these videos as a convenient way of accessing their Africanness as part of their multi-stranded identity and culture. Based on the expressed motivations for use and expressed outcome of use of the video-film, results indicate that the use of the video-film may have contributed to the accessibility of the African in diaspora label as a social category for this group of immigrants.

DOI: 10.4018/978-1-60960-591-9.ch014

MEDIATING IDENTITY AND CULTURE: NIGERIAN VIDEOS AND AFRICAN IMMIGRANTS IN THE U.S.

African Immigrants in the U.S.

Africa is a vastly diverse continent with several countries with approximately one billion people (http://www.prb.org/Publications/Datasheets/2009/2009wpds.aspx), and more than 2,000 languages spoken (www.nationsonline.org/oneworld/african_languages.htm). There are varying histories of colonization, civil strife, and political instability in each country. There are also differences in culture, ethnicity, traditions, religion, economic viability, and socio-cultural factors that have a significant impact on the needs of emigrants from each of these countries to the United States (Fanon, 1968). Despite the heterogeneity, sub-Saharan African immigrants often have similar needs based on a shared history and philosophy that includes "fundamental values and beliefs that permeate the continent of Africa, regardless of the diversity" (Young, 2003:166). Some of the shared history and philosophy emanated from the impact of colonialism on identity, including racism and discrimination that might be experienced in the new places of settlement. The experience of negotiating the conflicting messages between the African and Eurocentric worldviews cannot be overemphasized in the identity distortion of the African immigrant.

Of primary importance among the unique issues Africans face is that they have endured a long and complex history of colonialism and oppression, punctuated by the institution of enslavement by and a fight for freedom from Arab and European groups (Diop, 1974). The colonization period "led to Africa's loss of independence, loss of sovereignty, economic stagnation of the continent through colonialism, international humiliation, and the assumption of the inferiority of Africans and people of African descent" (Yansane, 1990:64). There are psychological and socio-psychological aspects of the historical legacy of colonialism and slavery relating to Africans that cannot be ignored (Appiah, 1992; Fanon, 1967).

However, African immigrants may be among the fastest growing immigrant populations in America. According to the 2000 US Census Bureau, there are over 1 million African foreign born in the United States in 2002-the 2010 census results for the population of foreign born Africans was not yet available at the time of this writing. According to the U.S. Census Bureau 2010 website (www.census.gov/prod/2010pubs/acs-11.pdf), a foreign-born is anyone who was not a U.S. citizen or U.S. national at birth. Non-U.S. citizens and those who have become citizens through naturalization are considered foreign born. More Africans have entered the United States voluntarily since 1990 than the total that disembarked as slaves prior to 1807, the year international slave trafficking was outlawed (Roberts, 2005). According to census data (as cited in Roberts, 2005, p. A1), the proportion of Black people living in the United States who described themselves as African-born more than doubled in the 1990s, and approximately 2 million U.S. residents identify their ancestry as sub-Saharan African.

African immigrants in the United States, much like other immigrants, often develop transnational ties to help nurture and ease the challenges of a dramatically different culture. These challenges may include negotiating ethnic and cultural identities in the confusing potpourri of race relations in the United States. They might also include economic marginalization as well as cultural denigration to some extent. Some of these challenges may be ameliorated by forming social networks to help tackle cultural denigration concerns having to do with socialization, gender roles and child rearing.

Many Africans are dynamic in their social construction of their identity in the United States while maintaining social and family ties in the home country. This study sought to assess the role of the nascent Nigerian video-film industry in the interplay of the multiple identities African immigrants create for themselves in the United States. The study also sought to examine the role of the videos in the coping mechanisms cultivated by this transnational audience to negotiate, maintain, straddle and balance the multiple identities created in the struggle to secure a sense of place and cultural identity in a 'foreign' land.

This balancing act often results in 'hybridity' (Bhabha 1994:4) of identities or an intersection of identities or multiples identities which could result in psychosocial stress (Mellow, van Slyck, & Eynon, 2003). Many Africans are border-crossers-building new identities in the United States while maintaining family ties in their home countries-and are often faced with uncertainty about the future (Rendon, 1996:20).

The social injustices of racism and discrimination that punctuate the life of minority populations in the United States as they negotiate their lives' objectives have the potential to influence self-worth and self-esteem. This feeling may be further complicated by the transition of coming from an African country where they may have been in the racial or ethnic majority. The professional lives may be cramped by additional obstacles faced by many U.S. racial and ethnic minorities, such as differential treatment for certain races in hiring practices (Bertrand & Mullainathan, 2004; Levitt & Dubner, 2005). From this perspective, it can be noted that many African immigrants will experience an intersection of multiple identities (Mellow, et al., 2003). Nwadiora (1996) wrote about the Nigerian immigrant experience. He contended that some immigrants experience adjustment issues because they assume a triple identity in the United States. First, African immigrants want to be identified as Africans. Due to the issues of racism and discrimination, American society does not clearly differentiate between the experiences of African immigrants and African Americans. Second, African immigrants must adjust to living and working in a predominately White culture as they simultaneously attempt to maintain their own cultural traditions and heritage. Last, African immigrants may struggle with the meaning of maintaining an African national identity versus adopting a Black American racial identity.

This hybrid perspective (Bhabha, 1994) is basically the combination of identities that immigrants need and use to successfully negotiate and adjust to their different socio-economic and political roles in their host countries. These roles range from parenting an America-born child, being a spouse (in America and sometimes to an American), being an employee, and being an American-born child of African immigrants, etc. In this process of hybridization, access to Nollywood videos is gained through globalization and transnationality. This study is therefore primarily concerned with the self-confessed role and impact, if any, of access to Nollywood video-films on the African immigrant experience in a particular location in the United States. Specifically, this study is concerned about how accessibility to the Nigerian video films as a subset of self-categorization theory has influenced group and individual identity among the African immigrant community in the Dallas/Ft. Worth area of Texas, USA.

Literature Review

The relative novelty of Nollywood as an industry and a subject of academic study has not affected the volume of available literature on the subject. Several scholars have addressed several aspects of the nascent industry. Early and ongoing scholarly works on the industry range from the history (Barrot, 2009; Dogbe, 2003; Haynes 1995, 2000, 2002, McCall 2004; Ukadike, 1994, 2000, 2003, forthcoming) social and religious issues of the videos (Adesokan 2004; 2006; 2007; Abah, 2008; 2009; Ogunleye, 2003a; 2003b; 2003c; Okome, 2000a; 2000b; 2004; 2005; 2007a; 2007b; Omoera, 2009; Ukah, 2003, 2004; Wendl, 2007) to the political and economic implications of the industry (Adesokan, 2009b; Ajibade, forthcoming; Ekwuazi, 2001; Haynes 2003, 2002, 2006, 2007; Marston, Woodward and Jones III 2007).

Other issues in Nollywood videos addressed by academic literature include the use of colonial language in the industry by Adejumobi (2002; 2003; 2004, 2007; 2008), where he investigated the use of English language by Nollywood and the types of audiences that the use of a 'foreign' language garners for the industry. Akpabio and Mustapha-Lambo (2008) used the viewpoint of the cultural imperialism hypothesis to interrogate the place of Nollywood videos vis-à-vis foreign films among Nigerian audiences and found that a large majority of the local audience have a favorable attitude towards the videos and a slight percentage actually prefer the videos over foreign films. Zajc (2009) argued that the culture of Nollywood not only challenges the global systems of video production and distribution, but 'also provides space for the articulation of public discourse by innovating the film language itself' (2009:65). Krings (2004; 2005; 2008; 2009), Brian Larkin (1997; 2000; 2002; 2003; 2004a; 2004b; 2008) and Abdalla (2004; 2010; forthcoming) are some of the scholars who have examined and commented extensively on the northern-Nigeria based and Muslim audience-oriented Kannywood branch of the home video phenomenon.

More proximate to the current study are several reception studies work on Nollywood. Notable among these is Adesokan (2009a) which addresses the reception of Nollywood videos among African immigrant communities and the global cultural markets. Adesokan (2009a) was primarily concerned with standardization and cultural translation of the videos outside the shores of Africa. Along the same lines, Ogundimu (2009) noted the popularity of the videos among African audiences both in Africa and globally. He makes the point that the cultural imperialism hypothesis no longer applies because audiences are active participants in the exportation of their culture globally rather than passive consumers of global, and sometimes, alien cultures. In one of his several works on the Nollywood industry, Okome 's (2007b) reception study on the sites of consumption of Nollywood is also relevant to the current study. Okome (2007b) critically examined the local sites of consumption of Nollywood videos and the relevance of these sites to the context of production and distribution of the videos. Becker (Forthcoming) argued that the consumption of these videos by university students and young professionals in post-apartheid southern Africa is part of the process of recreating identities of being consciously 'African' through the use of various forms of popular culture by these young adults. Becker (forthcoming) suggested that the videos serve as the link to global nodes of cosmopolitanism and cultural Pan-Africanism to which these young people yearn to belong. Esan (2008) conducted a similar study to the current one among UK-based audiences of Nollywood and found similar results-audiences watch the videos to learn a moral lesson, engage with the African society by proxy, and to instruct the younger generation about the 'home' culture. This current study fits snugly in the tradition of Nollywood works examining the places and spaces of reception of the video films.

Theoretical Assumptions

Nollywood films, as the Nigeria home videos are popularly called, is defined as ethnic media in this study as a mode of classification based solely on its utility among a group of immigrants. It is described as ethnic media solely in the sense in which Johnson (2010:108) defined ethnic media as "...communication channels that serve a particular cultural or racial group." In the words of Johnson (2010:109), ethnic media have sometimes contributed both to societal integration and reinforced cultural separatism that "resulted in cultural strife and political clashes." The pluralism function of the ethnic media in helping to preserve cultural heritage, language and identity, and providing a haven from unsavory stereotypes have been noted by several ethnic media researchers (Gutierrez, 1977; Riggins, 1992; Downing 1992; Subervi-Velez, 1994).

The recent attempt by Johnson (2010) to integrate the social identity perspective into the study of ethnic media has greatly influenced the theorization process of this study. According to Johnson, the two bodies of work that make up the social identity perspectives are the social identity theory and the self-categorization theory. While social identity theory "emphasizes that individuals describe self identity vis-à-vis other groups, rather than enact an innate, fixed identity, [s]elf-categorization theory describes how individuals assign themselves to particular groups and not to others" (2010:114). Fundamental to social-categorization theory is the socio-cognitive elaboration of the nature of social identity in the perception of self and others (Turner, Hogg, Oakes, Reicher & Wetherell, 1987). Social-categorization theory is therefore, "useful for the study of ethnic media because it is more focused on the social cognitive processes of how individuals are able to act as a group" (Johnson, 2010:114). The rationales for self-categorization theory include reduction of uncertainty and increasing social status through identification with a particular group (Turner et al., 1987). The role of the media, particularly the ethnic media, in generating and reinforcing the cultural social category cannot be overemphasized (Johnson, 2010). Johnson (2010) argued that four main self categorization concepts are especially relevant to ethnic media. One of these four is the accessibility of the social category to an individual to use in self-description. Media boost the accessibility of a social categorization due to their ubiquitous nature, but ethnic media in particular enhances the accessibility of a social category when it exists as a supplement or a replacement of general market media (Johnson, 2010). Other self-categorization concepts mentioned by Johnson include social norms, prototypes and fit. These are of minimal interest to this study. As noted above, this study is solely concerned about how accessibility to the Nigerian video films as a category of self-categorization theory has enhanced group and individual identity among the African immigrant community in the Dallas/Ft. Worth area of Texas, U.S.A.

Research Questions

In light of this background, this study seeks to answer the following research questions:

1. What is the significance of the Nigerian home videos in the diasporic experiences of African immigrants as transnational consumers of the products?
2. How is the Nigerian video-viewing experience connected or unconnected to the self-styled identity of African immigrants in the United States?
3. Is there a difference in the perspective of the African immigrant viewer of the videos and the domestic viewer in Africa?

Methodology

This researcher distributed 250 copies of a survey to African immigrants and people associated with African immigrants in the Dallas/Fort Worth area of Texas from July 22 –August 15, 2007. This location was chosen because of its moderate concentration of African immigrants and because this researcher has existing contacts within the community, a factor which eased the process of establishing trust. The surveys were distributed at places of worship, parochial organization meetings, and ethnic grocery stores. These were places often frequented by African immigrants. While copies of the survey were not distributed anonymously to an opportunity sample, each respondent completed the survey independently and was not required to indicate his or her identity on it. The respondents were given an envelope to enclose their responses and drop them off inside centrally located drop off boxes in the above named locations. Although respondent anonymity was strictly emphasized, respondents were given the option of identifying themselves by contact phone numbers if they would like to be contacted for face-to-face (FtF) interviews. One hundred twenty surveys were returned in which about 40 people indicated interest in a FtF interview. About 15 were randomly selected from the 40 while paying attention to gender and age distribution for balance. The FtF interviews were conducted over a period of 10 days with each lasting between 25 to 55 minutes.

From June 27 to July 14, 2008, this researcher visited Nigeria and interviewed an opportunity sample of civil servants, business men, market men and women, roadside entrepreneurs, and regular everyday Nigerians on the streets of Lagos, Oyo, Ibadan and Aba about their viewing habits regarding the Nigerian video-film. Forty-five interviews were conducted and recorded but due to a lot of background noise, about 20 of these were actually usable.

RESULTS

Immigrant Audiences

Based on those who answered the questions on gender, slightly more immigrant women (9% more) than men surveyed and interviewed watch the videos. Giving the missing value, this may or may not be an important gap statistically.

As seen in Table 1, about 88% of the respondents to the survey are from Africa. Out of this, about 76% are from Nigeria and the other 11% is distributed within other African countries. About 12.5% of the respondents who answered this question claim to be non-African. Among the respondents, 63.5% of respondents are married, 19% are single and the rest are widowed, divorced or separated. The group is fairly well –educated. About 47% have a degree from a 4-year college or some college education and 39% have some graduate work or a graduate degree. The age range for 63% of the respondents is 30-55 while 20% are between 13-29. Another 17% did not answer this question. Age is a private issue in some parts of Africa.

Table 1. Summary of respondents' demographical information

Xteristics	%	%	%	%	%	%	%	Mss	Total
Age	13-21 **8.7**	22-30 **11.5**	31-45 **24.0**	46-59 **36.5**	60-70 **1.0**	-	-	**18.3**	**100**
Gender	Male **39.4**	Female **48.1**						**12.5**	**100**
Education	some HS- **1.9**	Complet-ed HS **5.8**	some coll **19.2**	Grad-4yr coll **27.9**	some grad **8.7**	grad degree **28.8**	-	7.7	**100**
Employment	FTem-ployed **62.5**	PTem-ployed **11.5**	self empl out home **5.8**	notem-ployed by choice **2.9**	Not employed looking **1.0**	NA **6.7**	-	**9.6**	**100**
Income	30,000 **12.5**	30-49,999 **12.5**	50-65,000 **13.5**	65,001-75,000 **11.5**	75,001-99,999 **12.5**	100,000-150,000 **11.6**	150,000+ **3.8**	**22.1**	**100**
Nationality	Nigeria **76.9**	Ghana **1.9**	Cameroun **1.9**	Liberia **1**	Kenya **1.9**	Other **3.8**	-	**12.5**	**100**
Marital	Single **19.2**	Married **63.5**	Divorced **1.9**	Separated **1**	Widowed **1**	–		**13.5**	**100**

As seen in Table 2, most of the videos (64.4%) are either rented from ethnic grocery stores, borrowed from friends, brought or mailed from Nigeria. A significant 29% of the people watch by swapping with friends. Only one person rented online. There is limited use of online resources to access the videos. Online pay per view (PPV) channels such as www.izogn.com are not very popular as sources for the movies from this particular immigrant group. But this group is also not using the internet to buy other movies either. Nearly 86% get their American movies through pay-per-View (PPV), movie channels, renting and buying from store. They basically use the same mechanisms they use to get their American movies as sources for the Nigerian movies as well. When asked about the lack of use of online services during the FtF interview, one of the subjects stated that she didn't trust online sources with her credit cards.

About 60% have been watching Nigerian videos for several years, 19% in the last couple of years, 10% don't recall how long they've been watching, and the rest have watched anywhere from the last few months to the last one year.

As seen in Table 3, in terms of frequency of watching the videos, 26% watch as infrequently as about 4 times a year, 9% watch daily, 16% watch 2-3 times a week, 18% watch weekly and another 18% watch anywhere from monthly to bi-monthly as self-reported. For these respondents, the videos do not seem to interfere with their consumption of American movies. While 35.5% said they watch American movies more than once a week in the theater and at home, only 25% said they watch the Nigerian videos more than once a week. However, about 5% have never watched an American movie and another 26% have never been to a movie theater to watch an American movie, only 2% said they rarely watch the Nigerian videos. It would seem that the Nigerian videos are not the main source of filmic entertainment for this group, but one in a variety of choices available to them.

Table 2.

Variable	%	%	%	%	%	%Missing	Total
Source (Nollywood)	Rent from store **37.5**	Swap with friends **28.8**	Buy from store **17.3**	Rent online **1.0**	Buy from Nigeria **9.6**	**5.8**	**100**
Source (Hollywood)	Rent from store **28.8**	Swap with friends **4.8**	Buy from store **20.2**	Rent/Buy online **1.9**	PPV/Movie channels **36.5**	**7.7**	**100**
Genre Nollywood	Drama **35.6**	Action **10.6**	Tradoepics **11.5**	Religion **13.5**	No preference **25.0**	**3.8**	**100**
Language Preferred	English **66.3**	Yoruba **18.3**	Hausa **1.9**	Ibo **0**	No Preference **10.6**	**3**	**100**
History of watch	Several yrs **59.6**	Last 2yrs **19.2**	Last 1 yr **2.9**	few months **6.7**	Don't **recall** 9.6	**1.9**	**100**
(N)Mode: Alone	Always **4.8**	Almost always **4.8**	Sometimes **56.7**	Rarely alone **16.3**	Never alone **12.5**	Mss= **2.9** NA= **1.9**	**100**
(N)Mode: w/ spouse	Always **11.5**	Almost always **15.4**	Sometimes **36.5**	Rarely **12.5**	Never **2.9**	Mss=**2.9** NA=**18.3**	**100**
(N) Mode: w/ children	Always **6.7**	Almost always **7.7**	Sometimes **28.8**	Rarely **21.2**	Never **12.5**	Mss=**1.9** NA=**21.2**	**100**

Table 3.

Variable	%	%	%	%	%	%	%	%	%	%
Freq. of watching (Nollywood)	Rarely **1.9**	3-4 x/yr **26.0**	Bi-monthly **6.7**	Monthly **11.5**	Weekly **18.3**	2-3x/wk **16.3**	Daily **8.7**	Other **8.7**	Missing **1.9**	Total **100**
Freq. of watching American (Theater)	Never **26**	Less than 4x/ yr **44.2**	2ce/ month **13.5**	Monthly **4.8**	Weekly **3.8**	More than 1/ wk **3.8**	-	-	**3.8**	**100**
Freq. of watching American (Home)	Never **4.8**	Less than 4x/ yr **25**	2ce/ month **14.4**	Monthly **11.5**	Weekly **9.6**	More than 1/ wk **31.7**	-	-	**2.9**	**100**
Length stay in US	Less 1 yr. **1.0**	1-2 yrs **2.9**	2-5 yrs **6.7**	+5-10 yrs **15.4**	+10-15 years **17.3**	+15-20 yrs **7.7**	20+ yrs **26.0**		**23.0**	**100**
Last visit to Africa	Less 1 yr.ago **2.9**	More than 1 yr **1.9**	2-4 yrs **34.6**	+5-10 yrs **15.4**	+10-15 years **3.8**	+15 **3.8**	Never-been **8.7**		**28.8**	100

Ten percent of the respondents have lived in the United States for 5 years or less, 40% have lived in the U.S. for more than 5 years but less than 21 years, and 26% have lived in the U.S. for over 21 years (this also includes non-Africans who have lived in the U.S. all their lives and immigrants' adult children who were born in the U.S.). Another 23% chose not to answer this question. Forty one percent have visited Africa in the last 4 years or less, another 20% in the last 5 to 15 years, 8% have never been to Africa, and 29% would not answer this question. The length of time an immigrant has lived in the U.S. did not connect with whether or not they watch the videos.

The most preferred genre is drama at nearly 36%, followed by religion-based videos at 13.5%. A quarter of the people surveyed did not have any preference when it comes to the genre of the videos. The most favored language for the videos is English as these are more accessible to the first generation immigrants whose understanding of the local languages may be limited as well as the non Africans or Africans from other parts of Africa or of a different ethnicity. The videos are consumed sometimes alone, with spouses and with children (see Table 2).

In order of ranking, (there were 12 choices) the top reasons why the African immigrant respondents watch the movies are:

1. To feel connected to home culturally (85% ranked this in their top 5 reasons)
2. See the new developments in the social scene in Africa (77% ranked this in their top 5)
3. The plot represents my own worldviews (77% ranked this in their top 5 reasons)
4. Remind me of the values I grew up with (70 percent ranked this in top 5 reasons)
5. Teach my children what Africa is all about (66% ranked this in their top 5)
6. Other: Fun, entertainment, comparison with American movies (56% rank this in top 5 reasons)

Other options on the list were:

- See how other Africans abroad live (59% checked this as one of their top 10 reasons)
- See changes in expectations for Africans abroad (50% checked this in top 10)
- See how other Africans live in other parts of Africa (40% in top 10)

- See how Nigerians live in Nigeria (55% check in top 10)
- See new economic developments (43% check in their top 10)

During the face-to-face (FtF) interviews, however, most of the respondents said their top reasons for watching the videos are:

Desire to learn a moral lesson: This was a common reason tendered by most adult male and female interviewees. This was somehow reflected in the survey in the amalgamation of feelings of being connected to home culturally and the movies representing the values the respondents felt they grew up with. Some of them noted that story-telling in the villages was about teaching morals and when they started watching television, most of the cultural performances, even the stage performances were about teaching morals. Therefore, they expect to learn when they watch the movies. When asked during a FtF interview why this is so important, a woman in Arlington, Texas stated that oftentimes when she watches an American movie, she cannot tell what the moral of the story was and she liked watching Nigerian movies because they always have a moral lesson to teach through the story.

The desire to see trends in clothing style, architecture, modern infrastructures and social norms is another reason for watching the movies: Mrs Yetunde* of Ft. Worth, Texas mentioned that she liked watching the movies because she can look at some of the developments that have taken place all over the country. When asked to name some of the developments, she answers:

Mrs. Yetunde: You know, the new styles of clothing, new house designs and so on. It makes you feel that Nigeria is developing in some ways and that people are evolving in their fashion sense. The houses are so beautiful, more beautiful than the houses we have here. It makes me think of what kind of house I would have if I lived in Nigeria and the kind I would like to build there.

Nostalgia for Nigerian culture: Most of the interviewees seem to live vicariously through the videos. The videos remind them of their imagination of 'home' compared to 'here' and their desire for return at some point in the future when things get better back home. It is this use of the videos that largely fulfils part of what Shobat and Stam (1996) noted as the empowering nature of media as a means for immigrant communities dealing with feelings of cultural alienation and geographical displacement. During the interview, this particular answer resurfaces for many of the interviewees partly as an indication of the doubts and cultural complexity of the immigrant or transnational experience. The desire for cultural authentication is often sated by watching these videos. Mrs. F. in the Garland area of Texas who wishes to remain anonymous noted that, the videos make her feel that she is not crazy for having cultural expectations of her husband and her children. She loves watching the videos so much that her children bought a TV/DVD player combo for her on her birthday so she could watch the videos in her room lying on her bed. She noted that she uses the videos to teach her children the cultural expectations for children in Africa. For example, the reason why every older African they know is an aunt or an uncle is that it is culturally rude in most of West Africa to call older people by their first names, and it is too distant to call them Mr. this or Mrs. that. She, like most African immigrant women, tells her children to call all their family friends uncles and aunties.

Mrs F: When I can get my daughter or son to watch with me, I tell them, see Africa, see, it is not ugly, it is not naked people running around, children have respect for parents, and you are never too old to be caned by your parents. This is the Africa where I was raised. This is my culture.*

She noted how gratified she was when she overheard her daughter talking to an American friend who had come home with her about Africa.

Her daughter told the friend of the Africa where people have Mercedes Benz cars and big houses and lots of food. She recollects how proud she was that her daughter who at the time had not visited Africa was able to "remove the shame" about the perception of Africa by Americans.

The feeling that Africa is culturally denigrated in America through the media and in common perception by Americans combined with issues of racial inequality in America is a pervasive one for some of the interviewees. It is a source of considerable pride to some of them that they are not "African-American" and to make sure that their children understand the difference. Another lady in her fifties from the same Church as Mrs. F noted this fact. Mrs. T is a vocational nurse at a nursing home. Prior to coming to America, she had obtained certification in teaching from a College of Education in Nigeria. She went to a community college in the Dallas/ Fort Worth area to obtain certification in vocational nursing after doing several odd jobs she considered beneath her qualifications.

Mrs. T: I like watching the movies because they make me feel good. You know you are from somewhere when you watch the movies. All the bad things of the day will be forgotten. The people who treat you badly outside (the home?) are all forgotten. They think you are black like the blacks here, they don't know that you know who you are. I tell my children, watch the videos with me, learn about your culture, so people outside (the home) would know you are not just another black person. You have a culture; you are not like them, these Americans with no culture.*

Mrs. T and several interviewees like her, underscores the role of the videos to transnational and immigrant communities in constructing and contextualizing their identities vis-à-vis their country of residence. This is a theme related to construction of identity in a highly racialized American society that makes few distinctions between ethnicity and race (Msia Clark, 2009). Despite the fact that Blacks in America are made up of different ethnic groups within a black racial identity, the politics of race in America often force Black people constituting different ethnic groups to embrace an African-American identity that is based only on the values and experiences of one ethnic group (Clark, 2009).

This theme is related to the notion of a desire to maintain a particularized identity in the midst of trans-nationality and a globalized world, and is emphasized by Gupta and Ferguson (1992):

The irony of these times, however, is that as actual localities and places become ever more blurred and indeterminate, the ideas of cultural and ethnically distinct places become perhaps even more salient. It is here that it becomes more visible how imagined communities (Anderson, 1983) come to be attached to imagined places, as displaced people cluster around remembered or imagined homelands, places, or communities in a world that seems increasingly to deny such firm territorialized anchors in their actuality... Territoriality is thus re-inscribed at just the point when it threatens to be erased (pp.10-11).

Other themes that emerged from the interviews include uses such as:

1. Movies as education tools to teach first generation immigrants about their ethnic history
2. Movies teach non-Nigerian Africans about some of the similarities and differences in African beliefs and culture.
3. More men who watch the videos also watch American movies while more women who watch the videos do not watch American movies.
4. Length of immigration is a not a significant or indicator of interest in the movies.
5. Fosters belief in the reality or close-to-reality nature of the videos.

6. Viewers expressed concern about immodesty in dressing by young actresses.

The FtF interview with 1st-generation immigrants and immigrants who came here as children produced different reasons for their interests in the movies. The themes that emerged from the interviews with some of the first-generation immigrants and immigrants who came here as minors are that the videos:

- Provide cultural affirmation of their ethnic identity that is significantly different from the Africa of American mainstream media (plagued by wars or famine) and the Africa of National Geographic (exotic people and animals). An Africa they can associate with.
- Provide a better understanding of cultural values that parents try to impart to them.
- Provide opportunities to learn more about African culture.
- Provide pride in ethnic background.

This group is also:

- More likely to complain about quality of the video production
- More concerned about treatment of women
- More concerned about polygamous unions

Here's a couple of excerpts from two young women who are first generation African immigrants. Neither of them has been to Africa and they are both college students.

Seun*(23): I like watching the videos. Uhm, I watch with my friends some time and I also watch with my mom.*

Interviewer: *Are your friends Africans or children of Africans too?*

Seun: *Most of them that I watch the movies with are children of Africans just like my parents. At first, we used to laugh at some of the things they say and do because we didn't understand. But when I watch with my Mom, she talks about those things, like genuflecting and deference to parents, you know, stuff like that. She said they are cultural. Me and my friends though, we like the houses and some of the clothes. It's like a different Africa. Where people are not starving like in the news and they are actually very rich people and nice cars and nice houses. We like that.*

Interviewwer: *So the nice houses and the nice cars are the reason you and your friends watch the movies?*

Seun: *Not just that. I think they also help us understand some of the weird things our parents do. Like my mum always yelling that we have no respect and asking me to kneel down for people. You know, cultural stuff like that. When you watch the videos, you kind of try to understand that it is the culture that made them like that.*

Tatiana* is 21 and still lives at home with her mother and grandmother while she attends a 2-year college. Her parents are divorced. Her father is originally from Nigeria and her mother is from Cameroun. She says she rarely watches the video except when she's visiting other African families and joins them if she finds them watching. She says her grandmother would sometimes borrow some and watch with her mom, and if she's home, she might watch with them. Also in the home are four other children who are all younger than Tatiana and most of them do not watch the movies. One of the younger children is Tisha*, 19, who is an African-American young woman being fostered by Rebecca, Tatiana's mom.

Tatiana: *I don't watch a lot of them except if I just kind of run into them at friends houses or mom and grandma. I have watched more since*

Tisha came here. She likes them. She thinks she can learn about Africa from the videos.

Interviewer: *How do you feel about the videos? Do you like them the few times you've watched them?*

Tatiana: *Well, I really don't remember any of the story lines and I don't even understand the plots half the time. But I like the houses and the clothes and the way all the children are very deferent to the older folks.*

Interviewer: *What about you Tisha? What do you like about the videos?*

Tisha: *I really like the clothes. They are even prettier than the ones you see at parties (celebrations) here. They are so colorful and loud and I wish I could get clothes like that.*

Interviewer: *Apart from the clothes, is there anything about the videos and the stories that you like or don't like?*

Tisha: *Well, me and Tatiana sometimes watch with mom and grandma and they talk too loud that I don't understand what they are saying. But I still like them because I didn't know Africa had such nice houses and modern roads, like big cars, kind of like America. You know, uhm, they never show us that kind of Africa in school. Just sick people and hungry people. The videos show a lot of beautiful people in Africa and nice, modern Africa.*

Nigerian Audiences

Among the audiences in Nigeria interviewed,

- Generally more women than men admit watching the videos.
- The highly-educated /elite class interviewed admit that they do not watch the videos as often as their spouses, children and other significant others; they admit that they do not watch the movies as often because they are preoccupied with work and it takes too long to watch the movies. They also have more criticisms about the quality of the videos.
- The class of people mentioned above admits to buying a lot of the videos for their significant others who watch and they occasionally would watch with the family members.
- The Nigerian audiences express vehement cross-ethnic, cross-gender disapproval of sexual content in the videos because most parents watch with their children regardless of the ratings on the videos.
- Cross-ethnic and cross-gender expression of serious concern about immodesty in young actresses' mode of dressing in the movies.
- Belief in the reality or close-to-reality nature of the videos more prevalent especially among the less educated.
- Strong preference for the videos over foreign movies.
- Moral teachings and a strong expectation to learn a strong reason for watching the videos.

DISCUSSION

The general themes that emerged from the research are: among the African immigrants in the Dallas/ Ft. Worth area who watch Nollywood, their reasons for watching are complex and not often clear. Entertainment with a purpose is one big reason and the purpose range from identity affirmation to cultural affirmations. This group of immigrants define themselves vis-à-vis the larger American mainstream society by their values. They watch these movies because the movies affirm those values and these values are the representations of their culture and sense of self. This confirms

Johnson's (2010) declaration that 'a cultural group without its own ethnic media lacks one possible powerful ethnic symbol..leading to potentially diminished group identity..[and] less potential for behavioral empowerment' (pp.116). Based on their own proclamations, the accessibility to, and use of the Nigerian video-films by this particular immigrant group can therefore be said to contribute to an increased group identity and their self-categorization as a distinct group of Africans living in America.

This is a rather interesting notion among a group of people who mostly voluntarily left their original cultures to seek a different way of life somewhere else. A group, who despite being inundated with all forms of global media content and images in the general market media, still yearns for the 'local.' Most of the interviewees are not exiled and are therefore, able to return if they choose to. Most of them harbor the hope of returning 'one day' when things get better in their different countries.

One of the surprising elements of the research is that both the immigrant group and the domestic group in Nigeria, across different ethnic groups, express the desire to learn a moral lesson as a major reason for watching the videos. This is a stated reason for watching the video by most of the respondents regardless of their location and transnationality. Another surprising element is that the individuals interviewed in Nigeria, across cultures, decry the globalized mode of dressing employed by young women in the movies as indecent. This sentiment was echoed by both men and women in Nigeria but more vehemently by men. Within the immigrant group, however, the men interviewed never mentioned this as a factor but the same sentiment was echoed by the women. One explanation might be that the immigrant men have been desensitized through exposure to global media and is therefore unaffected by this.

In addressing the research questions, the first one asks:

What is the significance of the Nigerian home videos in the diasporic experiences of African immigrants as transnational consumers of the products? Scholars in migration studies have emphasized the multiple linkages that immigrants use to "forge and sustain simultaneous multi-stranded social relations that link together their societies of origin and settlement" (Schiller, Basch, & Blanc, 1995:48). Relevant to this multi-stranded social relation between societies of origin and settlement is the role of electronic media and new technologies of communication. The mobility and flow of media products contest the notion of a geographic boundary of cultures. The use of media products to reconstruct the transnational identity of migrants has been documented by scholars (Appadurai, 1996; Chambers, 1994; & Naficy, 1993, 1999).

The role of the Nigerian videos in the diasporic experience of the African immigrants fits snugly in the attempt to forge and maintain simultaneously, the multi-stranded social relations of their host society and their societies of origin. In the self-professed words of some of these immigrants, the top five uses of these videos is to feel connected to 'home' culturally; to see new developments in the social scene in Africa, to watch drama plots that represent their worldviews; to be reminded of the values they grew up with, to teach their children about Africa, and for fun and entertainment. The concepts of here and home contextualized this study. Almost all the immigrants spoken to, regardless of which part of Africa they originally came from, referred to Africa as 'home' and America as 'here.' The only group of people who called Africa by name was the first generation immigrants and those who were married to Africans or friends to Africans.

It is also important to note that these immigrants did not seem to be living for the day they can return 'home.' Actually, most of them seemed to be living a somewhat contented lifestyle in the United States, but there is a sense of deprivation and longing to remain connected to Africa, and this is where the videos seemed to come in. It is also important to note that these people are not exiles, refugees or in any way, prevented from or isolated from visiting home. They mostly seem to take a good advantage of this opportunity as the average year of last visit of 39% of them is three years. They also consume American movies with the same-sometimes more-ferocious appetite with which they consume the African videos. It would seem then, that the consumption of these videos is just one more way in which African immigrants and trans-nationals mediate their identities and experiences as immigrants and trans-nationals. Overall, while most of the immigrants surveyed use the videos as part of their identity reconstruction, it is by no means, the only way they define themselves as Africans in America given their use of other American filmic offerings.

The second research question asks how the Nigerian video-viewing experience is connected or unconnected to the reconstructed identity of African immigrants in the United States? Based on the responses from the survey and the interviews, it would seem that viewing these videos is connected to the self-styled identity of African immigrants in the Dallas./ Ft. Worth area of the United States. In an attempt to make a distinction between themselves and African-Americans as a cultural group, viewing these videos seemed to act as evidence of their distinctiveness from other minority groups.

Appiayong, a 52 year old Ghanaian man with a graduate degree who was interviewed puts it best when he said:

Appiayong: *It is good to see Africans trying to tell stories about Africa, showing Africans in their locality. They are speaking in English but it is the English that Americans don't understand because they have no context for it. It is like speaking a language that only you understand. It is seeing certain acts and knowing what they mean without being told. You know what I'm saying? It is the feeling that only an African can fully understand what is going on sometimes. It is not a black thing or a white thing; it is simply an African thing. You know sometimes, I don't even see those things until some of my friends asks a question about what they mean or why someone is doing it, and this is a Black friend, it is cultural I guess.*

Interviewer: *But how is that experience different from someone from another country watching an American movie for the first time and not knowing all the cultural markers?*

Appiayong: *Well, I guess it is almost the same. But American movies, you know, you know Schawznegger is going to kill everyone and be victorious, and yes, there are cultural markers there too but I just think that for me anyway, watching the Nollywood movies is a different experience. It is a confirmation of another worldview, another world, where certain values are more important than the individual, something like that.*

The consumption of these videos is not necessarily a rejection of American values by these immigrants, but a confirmation and valuation of other worldviews and belief systems that may not be mainstream, that may even be imperfect but for which they feel ownership and a sense of belonging. These videos, it would seem, help these immigrants affirm these worldviews and values without denying their co-existing enthusiastic embrace of American values as well.

Regarding the question of whether there is a difference in the perspective of the African immigrant viewer of the videos and the domestic viewer in Africa? The audiences in Nigeria, in different ethnic settings, all expressed pride in the videos. Some of the people interviewed are barely literate and they express the strongest beliefs in the ability of these videos to express not only their worldviews, but recreate their lived experiences. The civil servants and business men and women among the local audience in Nigeria are not so enthusiastic about the ability of the videos to tell their own lived experiences but they gave positive ratings to the videos. Except for the women, the more educated audiences in Nigeria are also more likely to say they do not watch the videos regularly but they buy them regularly for their wives, girlfriends and children.

This is a distinct difference in the experience of the immigrant audience and the local audience. The immigrant audiences talk generally about the worldviews, social scenes and teaching tools as the value of the videos. The audiences in Nigeria are more vociferous about the reflection of their lived everyday experiences in these videos.

Another significant difference is in the language barrier and cultural cues. Most of the people interviewed in Lagos, Oyo and Ibadan grew up on the Yoruba travelling theater and stage plays, and the ones in the East grew up on 30 minute to 1 hour TV dramas. These videos, according to most of them, give them a little bit of control over their own entertainment. Due to infrequent power supply, and the synchronous nature of television programming, they have limited control over their enjoyment and fulfillment from those dramas. The videos, however, give them control and power over their enjoyment. Some of them have power generating plant they use to watch the videos. Some can choose when they want to watch the videos, and when power supply returns, they can continue watching. This is unlike the experience of watching a television programming that is permanently interrupted by a power failure. Some of them claim to buy and watch American movies as well as the Nigerian videos, more of the women buy and watch Nollywood movies than buy or watch American movies.

More of the men in the Nigerian audience interviewed were also vehemently opposed to the mode of dressing of young women in the Nigerian movies. This is often described as aping the white cultures in their mode of dressing. This is considered un-African by the majority of men interviewed in Nigeria without regard to their ethnic affinities.

CONCLUSION

Identity is defined by Castells as "people's source of meaning and experience" (Castells, 1997:6). The type of identity recreation or hybrid identity that the consumption of these videos help create for African immigrants is similar to the " less formally organized expressions of identity as in those involved in the local consumption preferences" (Tomlinson, 2003:270). This is the kind of identity formation that is fostered and anchored by globalization of communication technologies rather than destroyed by globalization. In this sense, this type of identity formation is a complex, multifaceted, "particular, modern way of socially organizing and regulating cultural experiences" (Tomlinson, 2003:273). These groups of immigrants, who are in a sense, deterritorialized, have been enabled through the globalization of media products to diminish the importance of geographical location to the flow of their cultural experience in a sense (Tomlinson 2003; 1999; Garcia-Canclini, 1995) through the use of this particular ethnic media product. In conformity to the accessibility concept of the self-categorization theory proposed by Johnson (2010), these video-films made the social category of Africans living in America more salient to this group of immigrants and their offsprings whose physical ties to Africa may be more tenuous.

The end result for the African immigrants studied is, in some sense, "the emergence of a 'hybrid' cultural identity as a consequence, both of the multicultural constitution of modern nation-states, and of the emergence of transnational forms of popular culture" (Tomlinson 2003:276; Pieterse 1995; Werbner and Modood 1997). This seems to be the function of the consumption of these videos, at least for the African immigrants interviewed. In the end, these videos seem to have provided for the African immigrants studied, the ability to "experience and express, without contradiction, both attachments to the nation, multiethnic allegiances and cosmopolitan sensibilities" (Tomlinson 2003:276).

This multiple and complex sense of identity was also similar to the experience of the domestic audiences interviewed in Africa. They express pride and a sense of ownership in the stories told by the Nollywood videos while acknowledging their avid consumption of other global media products simultaneously.

While there are certain similarities in the experiences derived by different audiences of these videos, these particular immigrant audiences are more likely to distinctively use the videos to negotiate and manage their multiple identities in a foreign land. The transnational audiences of these videos have multiple levels of use and derive different gratifications from the movies distinct from domestic audiences. The videos are more than mere entertainments for African audiences both domestic and transnational. They serve interesting roles in helping African immigrants straddle their multiple identities and in mediating their numerous and hybrid cultural experiences.

It is important to point out that this study is limited in its use for generalized statements about all immigrant and domestic audiences. While the findings are similar to the findings from similar studies (e.g. Esan, 2008). The sample size is rather small and that is why the study was augmented with face-to-face interviews. The respondents are also quite different. Due to immigration and need to find good-paying jobs, most of the respondents in the Dallas/Ft. Worth area of Texas who participated in this study are fairly well-educated and therefore self-select (they have to be able to read, write and comprehend the questions in the survey) for the study. The respondents in Nigeria are a mix of highly-educated (bachelor's degree and above), fairly educated (high school diploma and some college) and barely educated (less than high school diploma). The samples are therefore not comparable. It is also worthy of note that the highly educated participants in Nigeria are more likely to spend less time watching the video, even though they do a lot of purchasing of the videos for their significant others and family members. This class of people are also more likely to complain about the quality of the videos. Participants in the DFW area of Texas were also eliminated from participating in the survey if they claim to not watch the videos.

Despite these limitations however, it is important to note how different immigrant groups are using these videos and what gratifications are being experienced from the use. The results of this study compare favorably with other results obtain from other immigrant groups in other parts of the world who use the videos. The accumulation of these results may advance the process of making generalizations about African immigrant groups and their consumption pattern regarding the videos.

*Names have been changed or modified to protect the identity of the interviewees.

REFERENCES

Abah, A. L. (2008). One step forward, two steps backward: African women in Nigerian video-film. *Communication, Culture & Critique, 1*(4), 335–357. doi:10.1111/j.1753-9137.2008.00027.x

Abah, A. L. (2009). Popular culture and social change in Africa: The case of the Nigerian video-film industry. *Media Culture & Society, 31*, 731–748. doi:10.1177/0163443709339456

Adamu, A. (2010). The muse's journey: Transcultural translators and the domestication of Hindi music in Hausa popular culture. *Journal of African Cultural Studies, 22*(1), 41–56. doi:10.1080/13696810903488579

Adamu, A. (forthcoming). Private sphere, public Wahala: Gender and delineation of intimisphare in Muslim Hausa video films. In Forster, T., & Nyamnjoh, F. (Eds.), *Negotiations of culture: Perspective from Africa*. Basel, Switzerland: Centre for African Studies.

Adamu, A. U., Yusuf, M. A., & Umar, F. J. (Eds.). (2004). *Hausa home videos: Technology, economy and society*. Kano, Nigeria: Centre for Hausa Cultural Studies/Adamu Joji Publishers.

Adejunmobi, M. (2002). English and the audience of an African popular culture. *Cultural Critique*, *50*, 74–103. doi:10.1353/cul.2002.0001

Adejunmobi, M. (2003). Video film technology and serial narratives in West Africa. In Ogunlcye, F. (Ed.), *African video film today* (pp. 51–68). Manzini, Swaziland: Academic Publishers.

Adejunmobi, M. (2004). Foreign languages, local audiences: The case of Nigerian video film in English. In Palaver, V. (Ed.), *Imaginations of the local and non-native languages in West Africa* (pp. 101–130). Clevedon, UK & Buffalo, NY: Multilingual Matters.

Adejunmobi, M. (2007). Nigerian video film as minor transnational practice. *Postcolonial Text, 3*(2), 1-16. Retrieved from http://Jounials.sfu.ca/pocol/ index php/ pcl/ anicle/ vicw/548/405

Adejunmobi, M. (2008). Technorality, literature, and vernacular literacy in twenty-first-century Africa. *Comparative Literature*, *60*(2), 164–185. doi:10.1215/-60-2-164

Adesokan, A. (2004). Loud in Lagos: Nollywood videos. *Wasaftri, 19*(43), 45–49.

Adesokan, A. (2006). Issues in the new Nigerian cinema. *Black Camera, 21*(1), 6-8, 11.

Adesokan, A. (2007). A revolution from below: The aesthetics of West African videofilms. *Uka, 21*, 60–67.

Adesokan, A. (2009a). Excess luggage: Nigerian films and the world of immigrants. In Okpewho, I., & Nzegwu, N. (Eds.), *The new African diaspora* (pp. 401–422). Bloomington, IN: Indiana University Press.

Adesokan, A. (2009b). Practising 'democracy' in Nigerian films. *African Affairs*.

Ajibade, B. (forthcoming). Nollywod videos: Is a western audience possible? In Krings, M., & Okome, O. (Eds.), *Nollywood and beyond: Transnational dimensions of an African video film industry*.

Akpabio, E., & Mustapha-Lambe, K. (2008). Nollywood films and the cultural imperialism hypothesis. *Perspectives on Global Development and Technology*, *7*, 257–270. doi:10.1163/156914908X370683

Appadurai, A. (1996). *Modernity at large: Cultural dimensions of globalization*. Minneapolis, MN: University of Minnesota Press.

Appiah, K. A. (1992). *In my father's house: Africa in the philosophy of culture*. New York, NY: Oxford University Press.

Barrot, P. (Ed.). (2009). *Nollywood: The video phenomenon in Nigeria*Taylor, L., Trans.). Bloomingtonm, IN: Indiana University Press.

Becker, H. (Forthcoming). Nollywood in urban southern Africa: Nigerian video films and their audiences in Cape Town and Windhoek. In Krings, M., & Okome, O. (Eds.), *Nollywood and beyond: Transnational dimensions of an African video film industry*.

Bertrand, M., & Mullainathan, S. (2004). Are Emily and Brandon more employable than Latonya and Tyrone? Evidence on racial discrimination in the labor market from a large randomized experiment. *The American Economic Review*, *94*, 991–1013. doi:10.1257/0002828042002561

Bhabha, H. (1994). Introduction: Location of cultures. In *The Location of culture* (pp. 1–18). New York, NY & London, UK: Routledge.

Castells, M. (1997). *The power of identity, vol. II of The information age: Economy, society and culture*. Oxford, UK: Blackwell.

Chambers, J. (1994). *Migrancy, culture, identity*. London, UK: Routledge.

Clark, M. K. (2009). Questions of identity among African immigrants in America. In Okpewho, I., & Nzegwu, N. (Eds.), *The new African Diaspora* (pp. 255–269). Bloomington, IN: Indiana University Press.

Diop, C. A. (1974). *The African origin of civilization*. Westport, CT: Lawrence Hill Books.

Dogbe, E. (2003). Warped identities: Dress in popular West African video films. *Africa Identities*, *1*(1), 95–117. doi:10.1080/1472584032000127905

Downing, J. D. H. (1992). Spanish-language media in the greater New York region during the 1980s. In Riggins, S. H. (Ed.), *Ethnic minority media: An international perspective* (pp. 256–275). Newbury Park, CA: Sage.

Ekwuazi, H. (2001). Political values and the Nigerian film. *Matatu*, *23-24*, 275–282.

Esan, O. (2008). Appreciating Nollywood: Audeinces and Nigerian films. *Participations: Journal of Audience and Reception Studies*, *5*(1).

Fanon, F. (1967). *Toward the African revolution*. New York, NY: Grove Press.

Garcia-Canclini, N. (1995). *Hybrid cultures: Strategies for entering and leaving modernity*. Minneapolis, MN: University of Minneapolis Press.

Glick Schiller, N., Basch, L., & Blanc, C. S. (1995). From immigrant to transmigrant: Theorizing transnational migration. *Anthropological Quarterly*, *68*(1), 48–63. doi:10.2307/3317464

Gupta, A., & Ferguso, J. (1992). Beyond culture: Space identity, and the politics of difference. *Cultural Anthropology*, *7*(1), 6–23. doi:10.1525/can.1992.7.1.02a00020

Gutierrez, F. (1977). Spanish-language media in America: Background, resources, history. *Journalism History*, *4*(2), 34–68.

Haynes, J. (1995). Nigerian cinema: Structural adjustments. *Research in African Literatures*, *26*(3), 97–119.

Haynes, J. (1997). In Harrow, K. (Ed.), *African cinema: Postcolonial and feminist readings* (pp. 155–190). Lawrenceville, NJ: African World Press.

Haynes, J. (2000). Introduction. In Haynes, J. (Ed.), *Nigerian video films* (pp. 1–36). Athens, OH: Ohio University Press.

Haynes, J. (2002). Devaluation and the video boom: Economics and thematics. In Guyer, J. I., Denier, L., & Agbaje, A. (Eds.), *Money struggles and city life: Devaluation in Ibadan and other urban centers in southern Nigeria. 1986-1996* (pp. 207–217). Portsmouth, NH: Heinemann.

Haynes, J. (2003). Mobilizing Yoruba popular culture: Babangida must go. *Africa*, *73*(1), 122–138. doi:10.3366/afr.2003.73.1.77

Haynes, J. (2006). Political critique in Nigerian video films. *African Affairs*, *105*(421), 511–533. doi:10.1093/afraf/adi125

Haynes. J. (2007). Nnebue: The anatomy of power. *Film International*, *5.4*(28), 30-40.

Johnson, M. A. (2010). Incorporating self-categorization concepts into ethnic media research. *Communication Theory*, *20*(1), 106–125. doi:10.1111/j.1468-2885.2009.01356.x

Krings, M. (2004). From possession rituals to video dramas: Some observations of dramatis personae in Hausa performing arts. In Adamu, Y., Uba Adarou, A., & Faruk Jibrin, U. (Eds.), *Hausa home videos: Technology, economy, and society* (pp. 162–170). Kano, Nigeria: Center for Hausa Cultural Studies.

Krings, M. (2005). Muslim martyrs and pagan vampires: Popular video films and the propagation of religion in northern Nigeria. *Postscripts: The Journal of Sacred Texts and Contemporary Worlds*, *1*(2-3), 185–205.

Krings, M. (2008). Conversion on screen: A glimpse at popular Islamic imaginations in northern Nigeria. *Africa Today*, *54*(4), 45–68. doi:10.2979/AFT.2008.54.4.44

Krings, M. (2009). *Glocalization from below:Intercultural transcriptions and Muslim hegemony in Nigerian video culture*. Retrieved on September 22, 2009, from http://www.gps.uni-annover.de/ ts/ pdfs/ Krings.pdf

Larkin, B. (1997). Indian films and Nigerian lovers: Media and the creation of parallel modernities. *Africa*, *67*(3), 406–440. doi:10.2307/1161182

Larkin, B. (2000). Hausa dramas and the rise of video culture in Nigeria. In Haynes, J. (Ed.), *Nigerian video films* (pp. 209–241). Athens, OH: Ohio University Press.

Larkin, B. (2002). The materiality of cinema theaters in northern Nigeria. In Ginsburg, F. D., Abu-Lughod, L., & Larkin, B. (Eds.), *Media worlds: Anthropology on new terrain* (pp. 319–336). Berkeley, CA: University of California Press.

Larkin, B. (2003). Itineraries of Indian cinema: African videos. Hollywood, and global media. In Shohat, E., & Sham, R. (Eds.), *Multiculturalism, postcolonialism, and transnational media* (pp. 170–192). New Brunswick, NJ: Rutgers University Press.

Larkin, B. (2004a). Degraded images, distorted sounds: Nigerian video and the infrastructure of piracy. *Public Culture*, *16*(20), 289–314. doi:10.1215/08992363-16-2-289

Larkin, B. (2004b). From Majigi to Hausa video films: Cinema and society in northern Nigeria. In Adamu, Y., Uba Adarou, A., & Faruk Jibrin, U. (Eds.), *Hausa home videos: Technology, economy, and society* (pp. 46–53). Kano, Nigeria: Centre for Hausa Cultural Studies & Adamu Joji Publishers.

Larkin, B. (2008). *Signal and noise: Media, infrastructure and urban culture in northern Nigeria*. Chapel Hill, NC: Duke University Press.

Levitt, S. D., & Dubner, S. J. (2005). *Freakonomics: A rogue economist explores the hidden side of everything*. New York, NY: Morrow.

Marston, S. A., Woodward, K., & Jones, J. P. III. (2007). Flattening ontologies of globalization: The Nollywood case. *Globalizations*, *4*(1), 45–63. doi:10.1080/14747730701245608

McCall, J. C. (2004). Nollywood confidential: The unlikely rise of Nigerian video film. *Transition*, *95*, 98–109.

Mellow, G. O., van Slyck, P., & Eynon, B. (2003). The face of the future: Engaging in diversity at LaGuardia Community College. *Change*, *35*(2), 10–17. doi:10.1080/00091380309604088

Naficy, H. (1993). *The making of exile cultures: Iranian television in Los Angeles*. Minneapolis, MN: University of Minneapolis Press.

Naficy, H. (Ed.). (1999). *Home, exile, homeland: Film media and the politics of place*. New York, NY: Routledge.

Nederveen Pieterse, J. (1995). Globalization as hybridization. In Featherstone, M. (Eds.), *Global modernities* (pp. 45–68). London, UK: Sage.

Nwadiora, E. (1996). Therapy with African families. *The Western Journal of Black Studies*, *20*, 117–124.

Ogundimu, F. (2009). African video, film cinema, and cultural repackaging in the diaspora. In Okpewho, I., & Nzegwu, N. (Eds.), *The new African Diaspora* (pp. 387–399). Bloomington, IN: Indiana University Press.

Ogunleye, F. (2003a). Christian video film in Nigeria: Dramatic sermons through the silver screen. In Ogunleye, F. (Ed.), *African video film today* (pp. 105–128). Manzini, Swaziland: Academic Publishers.

Ogunleye, F. (2003b). Female stereotypes in the Nigerian video film: A case for re-socialization. *Humanities Review Journal*, *3*(2), 1–14.

Ogunleye, F. (2003c). That we may serve him without fear: Nigerian Christian video film and battle against cultism. *International Journal of Humanistic Studies*, *2*, 16–27.

Okome, O. (2000a). Naming suffering and women in Nigerian video films: Notes on the interview with Emem Isong. *Hdunode: Calabar Journal of Humanities*, *3*(1), 43–56.

Okome, O. (2000b). Onome: Ethnicity, class, gender. In Haynes, J. (Ed.), *Nigerian video films* (pp. 148–164). Athens, OH: Ohio University Press.

Okome, O. (2004). Women, religion and the video film in Nigeria. *Film International*, *7*(1), 4–13. doi:10.1386/fiin.2.1.4

Okome, O. (2005). *From the periphery of the visual space: Women and the video film in Nigeria. Film in Africa: A reader*. NEH Africa Cinema Institute Project.

Okome, O. (2007a). The message is reaching a lot of people: Proselytizing and video films of Helen Ukpabio. *Postcolonial Text, 3*(2), 1-20. Retrieved on March 18, 2009, from http://journals,sfu.ca/ pocol/ indcx.php/ pct/ article/ view/ 750/419

Okome, O. (2007b). Nollywood: Spectatorship, audience and the sites of consumption. *Postcolonial Text, 3*(2), 1-21. Retrieved on March 18, 2009, from http://Joumals.sfu.ca/ pocol/ index. php/ pct/ article/ view/ 763/425

Omoera, O. S. (2009). Video film and African social reality: A consideration of Nigeria Ghana block of West Africa. *Journal of Human Ecology (Delhi, India)*, *25*(3), 193–199.

Rendon, L. (1996). Life on the border. *About Campus*, *1*, 14–20.

Riggins, S. H. (1992). The promise and limits of ethnic minority media. In Riggins, S. H. (Ed.), *Ethnic minority media: An international perspective* (pp. 276–787). Newbury Park, CA: Sage.

Roberts, S. (2005, February 21). More Africans enter U.S. than in days of slavery. *The New York Times*, p. A1.

Shohat, E., & Stam, R. (1996). From the imperial family to the transnational imaginary: Media spectatorship in the age of globalization. In Wilson, R., & Dissanayake, W. (Eds.), *Global, local: Cultural production and the transnational imaginary* (pp. 145–171). Durham, NC: Duke University Press.

Subervi-Velez, F. A. (1994). Mass communication and Hispanics. In Padilla, F. (Ed.), *Handbook of Hispanic cultures in the United States: Sociology* (pp. 304–357). Houston, TX: Arte Publico.

Tomlinson, J. (1999). *Globalization and culture*. Cambridge, UK: Polity Press.

Tomlinson, J. (2003). *Globalization and cultural identity*. Retrieved on October 15, 2008, from http://www.polity.co.uk/ global/ pdf/ GTReader2eTomlinson.pdf

Turner, J. C., Hogg, M. A., Oakes, P. J., Reicher, S. D., & Wetherell, M. S. (1987). *Rediscovering the social group: A self-categorization theory*. Oxford, UK: Basil Blackwell.

Ukadike, F. (1994). *Black African cinema*. Berkeley, CA: University of California Press.

Ukadike, F. (2000). Images of the reel thing: African video-films and the emergence of a new cultural art. *Social Identities*, *6*(3), 237–240. doi:10.1080/13504630050137615

Ukadike, F. (2003). Video booms and the manifestations of first cinema in Anglophone Africa. In Guneratnc, A. R., & Dissanayakc, W. (Eds.), *Rethinking third cinema* (pp. 127–143). New York, NY: Routledge.

Ukadike, F. (forthcoming). Nollywood history criticism: Rethinking African film discourse. In Krings, M., & Okome, O. (Eds.), *Hollywood and beyond: Transnational dimensions of an African video film industry*.

Ukah, A. (2003). Advertising god: Nigerian Christian video-films and the power of consumer culture. *Journal of Religion in Africa. Religion en Afrique*, *33*(2), 203–231. doi:10.1163/15700660360703141

Ukah, A. (2004). Pentecostalism, religious expansion, and the city: Lessons from the Nigerian bible belt. In Probst, P., & Spittler, G. (Eds.), *Between resistance and expansion* (pp. 415–442). Berlin, Germany: LIT Verlag.

Wendl, T. (2007). Wicked villagers and the mysteries of reproduction: An exploration of horror movies from Ghana and Nigeria. *Postcotonial Text, 3*(2), 1-21. Retrieved from http://Journals.sfu.ca/ pocol/ index php/ pct/ anicle/ vicw/529/420

Werbner, P., & Modood, T. (Eds.). (1997). *Debating cultural hybridity*. London, UK: Zed Books.

Yansane, A. Y. (1990). Cultural, political, and economic universals in West Africa. In Asante, M. K., & Asante, K. W. (Eds.), *African culture: The rhythms of unity* (pp. 39–68). Trenton, NJ: Africa World Press.

Young, C. (2003). Assimilation and social change dynamics in African and African American communities. *The Western Journal of Black Studies*, *27*, 164–175.

Zajc, M. (2009). Nigerian video film cultures. *Anthropological Notebooks*, *1*(15), 65–85.

Chapter 15
The Role of Mass Mediated Messages and Cultural Identity with Cross-Cultural Communication Failures Resulting from Flawed U.S. Military Policy in Iraq

Jim Schnell
Ohio Dominican University, USA

ABSTRACT

This chapter addresses mass media and cultural identity dynamics that impact U.S. military policy in the execution of the Iraq war and related cross-cultural communication issues that have evolved during the course of the war. I completed primary levels of Professional Military Education (PME), and this context has provided me with ample opportunity to become familiar with military doctrine, and, after the start of the Iraq war in 2003, observe how primary tenets of established military theory have been consistently ignored. At times it has almost been as if the U.S. invasion/liberation of Iraq has been executed as a case study for future students to review and learn how not to address such a challenge.

The inability to consider, let alone plan for, cross-cultural ramifications has been a central communication failure that has proven tremendously problematic. The mass media, via traditional channels and new communication technologies, have been keen to report on these matters which, in turn, have modified public understanding of the matters being reported. Mass media messages have been vital in creating, and impacting, this phenomenon. This phenomenon has been equally true among traditional media channels and the new communication technologies.

DOI: 10.4018/978-1-60960-591-9.ch015

INTRODUCTION

As the U.S. continues with military operations in Iraq the implications of shallow doctrine and subsequently vague mission objectives has been given thorough coverage in mass media. This reporting has, in turn, impacted U.S. execution of the war. Thus, mass media coverage conveys the story and has become part of the story it is reporting. The speed of reporting associated with the new communication technologies has made the new communication technologies especially relevant in this regard. Thus, there is a communicative nexus between mass media and war. This chapter will describe variables that compose and perpetuate that nexus.

BACKGROUND

On June 1, 2007 I retired from the U.S. Air Force at the rank of Colonel after 30 years service in the reserves and on active duty. My entire career was in the intelligence community, with the final 14 years spent as an Air Force attache in China. Prior to that I was assigned to Special Operations Command/Pacific. Given my civilian background in higher education I was involved with military educational opportunities. I graduated from Squadron Officers School, Air Command & Staff College, Air War College (these courses are commonly referred to as Professional Military Education) and completed many other specialized courses. In the closing years of my military career, after the start of the Iraq war in 2003 in particular, I had the opportunity to speak at Army War College, serve as an adjunct faculty member at Joint Special Operations University and assist with war games exercises at the Pentagon (as recently as March, 2007) and Air War College (as recently as May, 2007).

Since retiring from the military I have continued to serve via various civilian opportunities that have been made available to me. Such service has included duty as: 1) Senior Research Fellow at Air University, 2) Senior Research Analyst at the Urban Warfare Analysis Center, 3) Political/Social Science Advisor at HQ SOCOM (Headquarters, Special Operations Command) and 4) Lecturer to MARSOAG (Marine Special Operations Advisor Group). At present I am the Lead Social Scientist at the Army Culture and Foreign Language Management Office.

This provided me with ample opportunity to become familiar with military doctrine and, after the start of the Iraq war in 2003, observe how primary tenets of established military theory were consistently ignored. At times it has almost been as if the U.S. invasion/liberation of Iraq was executed as a case study for future students to review and learn how not to address such a challenge. The inability to consider, let alone plan for, cross-cultural ramifications has been a central communication failure that has proven tremendously problematic. The mass media have been keen to report on these matters which, in turn, has modified public understanding of the matters being reported on.

I completed Air Command & Staff College in 1998 and Air War College in 2000. Air Command & Staff College (intermediate service school) completion is required for promotion from major to lieutenant colonel and Air War College (senior service school) is required for promotion from lieutenant colonel to colonel. Each branch of the military has these intermediate and senior service schools. Communication processes are consistently recognized as being primary concerns in these courses. My completion of both courses, being so close to September 11, 2001 and the March, 2003 start of the Iraq war, has allowed me to observe the U.S. military actions in relation to various areas of U.S. military theory. Some of this military theory conveyed inaccurate speculations but some of the speculations and assumptions have been incredibly accurate.

All works quoted in this report are from sources used in Professional Military Education course readings. The only exception is Fiasco: The American Military Adventure in Iraq (Ricks, 2006) and a fall, 2007 CNN.com article quoting retired Army Lt. General Ricardo Sanchez. The Ricks book is used to compare and contrast U.S. military doctrine and theory against actual practices evidenced in Iraq as reported by Ricks.

In 1997 Russell Travers predicted "that the United States is on the threshold of a protracted military breathing space. During this period, U.S. leaders will be confronted with less risk of large-scale conflict than has been the case since the end of World War I" (Travers, 1997, p. 98). This speculation was obviously proven inaccurate as a result of the events of September 11, 2001. Hindsight provides an enhanced position for interpreting what has transpired since Travers wrote those words. However, for the most part, the Professional Military Education readings tend to be accurate and relevant in the assessments offered.

Charles Dunlap (1998) posited "Future adversaries may wage asymmetrical warfare by combining available low-tech equipment with a culturally oriented strategy.... an enemy could use a civilian airliner loaded with explosives to launch an attack on a high value target" (p. 4). He goes on to say "the kind of asymmetrical warfare adversaries may wage is not that which seeks to actually defeat U.S. or Western military forces, but rather that which assaults the psyche and will of the populations whose political support is required by Western democracies to sustain military operations" (p. 5).

Most of the Professional Military Education readings lay a foundation for understanding, among other things, under what conditions the United States should commit military troops and, when those troops are committed, considerations for guiding the deployment of troops. What follows is a general overview of some of the more common tenets. That is followed by representative excerpts from Fiasco (Ricks, 2006) that, in addition to the facts of the war thus far, illustrate how many of these tenets have not only been disregarded but, in many cases, have been directly contradicted.

Department of Defense (DoD) Joint Publication 3-0, Doctrine for Joint Operations (1993, v. 4) states there are three criteria for establishing a 'good' political objective for war. "The first criterion is that the objective must be easily understood.... The second is that the objective can be turned into a crusade that appeals to our moral sense....The third is that the objective must represent a self-interest perceived to be worth the cost of the war....Unless the objective is considered important and vital, Americans are hesitant to commit lives and resources on a large scale or for a long term to attain it" (page 4).

Drew and Snow (1988) stress the importance of national security objectives in Making Strategy: An Introduction to National Security Processes and Problems. They state the initial task of the strategist is to define the national security objectives that support an intended operation. "If the objectives are ill defined, inconsistent, or unsupported by some degree of national consensus, the strategist's function becomes exceedingly difficult" (p. 4). They go on to stress that, without attention to this detail, the result will most likely be a decline in American national will and military morale (p. 6).

Former Secretary of Defense William Perry (2002) clarifies three elements to be considered regarding the use of force. The political element: "a judgment as to the nature of the interests at stake and whether the use or the threat of use of military force is the most appropriate way to protect those interests." The military element: "a judgment as to the capability of the U.S. military forces to achieve our goal and the probable losses entailed." The ethical element: "a judgment as to whether achieving our goals by military force is in keeping with America's fundamental respect for human life" (pp. 1-2). He states there are three different cases in which we may employ military power: "The first category is when our vital national interests are threatened. Our second category is when important, but not vital, national interests are threatened. The third category is when a situation causes us deep humanitarian concern" (p. 4). Perry goes on to say, regarding national political objectives, that "accurate identification and evaluation of national interests and the political objectives that support those interests are vital if war is to serve a rational purpose" (p. 20).

A caveat to this position by Michael Howard is acknowledged. Howard (1979) stresses "The military instrument is not equally applicable to the pursuit of all national political objectives. Use of military power, even in war, may be inappropriate for some objectives and may even be counterproductive for others" (p. 15).

Perry cautions against fighting wars with limited political objectives as we are doing in Iraq. "The differences between limited and unlimited objectives can be crucial in determining popular levels of support for a war, especially in a democracy. For example, Americans generally have supported unlimited political objectives in war. However, limited objectives, such as those in Korea and Vietnam have been far less popular" (Perry, 2002, p. 20). Similarly, Ambassador David Passage (1998) asserts "A truly democratic government has no business undertaking schemes that arouse genuine hostility among a significant percentage of its citizens; such schemes should be left for private institutions or others" (p. 30).

ISSUES, CONTROVERSIES, PROBLEMS

An overall theme that is consistently reinforced in professional military education centers on clarity. Before deploying U.S. troops the administration should have clearly stated goals that are well defined and can be understood, achievable, measurable and assessable. An exit strategy should be established and there should be public support. William Snyder (1995) stresses this line of thinking when he explains the importance of military strategy. He indicates "military strategy is a broad concept which includes a military objective and a plan for achieving that objective by means of military resources" (p. 3).

At this point insights about the "friction" of war, offered by Carl von Clausewitz, have considerable relevance in that they highlight the crucial consequence of unplanned variables. "So in war, through the influence of an infinity of petty circumstances....Friction is the only conception which in a general way corresponds to that which distinguishes real war from war on paper.... This enormous friction, which is not concentrated, as in mechanics, at a few points, is therefore everywhere brought into contact with chance, and thus incidents take place upon which it was impossible to calculate, their chief origin being chance" (von Clausewitz, 1962, pp. 46-47). In the contemporary vernacular of today's contemporary soldier---"shit happens."

As a result of the aforementioned friction, and related variables, the original mission planning can spin toward unintended directions and the consequences can be daunting. These unintended issues often lead to "mission creep" whereby the original mission that was planned in detail grows into a plethora of unexpected sub-missions that need to be addressed. A sub-mission can become larger than the original mission. Such has been the situation in Iraq. The original primary mission was to oust Saddam Hussein. This was accomplished in short order. However the aftermath, which was barely planned for (i.e. there was no exit strategy), has dragged on since March, 2003.

As a result we find the U.S. Army, which had planned for a "shock and awe" invasion of Iraq to oust Saddam Hussein (which they successfully carried out), mired down with the overarching goal of trying to simmer a civil war among warring factions. Donald Snow, former professor at the Army War College, warns of this very phenomenon.

Do not think of peacekeeping and peace-enforcement either as an extension of what the Army does or as parallel and compatible missions. The

Army has limited experience in peacekeeping.... Significant involvement in peace-enforcement and peacekeeping operations will require the Army to modify the way it does business.... Peace-enforcement will require some fairly basic changes in the way the Army prepares for war. (Snow, 1993, pp. 5-6).

It is within this context that the findings from Fiasco: The American Military Adventure in Iraq (Ricks, 2006) can be best illuminated. The book by itself is provocative and exceedingly relevant given the loss of life, on all sides of the issue, that occurs on a daily basis. That the reality it presents, regarding how the war is being executed, contradicts how our professional military education system prepares our military leaders to fight wars underscores the degree of confusion that exists within the U.S. government on this matter. The U.S. military, particularly the Army, is being given orders to fight a war that it is not prepared to fight. Significant in this scenario, according to Ricks, is a lack of awareness regarding cross-cultural communication dynamics (Ricks, 2006, p. 238). Add to that the continual mass media reporting of the war, gains and losses, and the resulting public mind that evolves.

His highlighting of cross-cultural concerns is couched within other themes in the book. Such topics include focus on the role of military doctrine, the Bush rationale for military intervention in Iraq, problems associated with strategy, confusion regarding the notion of counterinsurgency, misuse of the U.S. Army, cross-cultural issues and parallels with U.S. involvement in the Vietnam War. The relevance of mass media reporting of these topics is inherent in our understanding of them.

Ricks (2006) opens his study with an overarching interpretive foundation that he builds from.

President George W. Bush's decision to invade Iraq in 2003 ultimately may come to be seen as one of the most profligate actions in the history of American foreign policy. The consequences of his choice won't be clear for decades, but it already is abundantly apparent in mid-2006 that the U.S. government went to war in Iraq with scant solid international support and on the basis of incorrect information—about weapons of mass destruction and a supposed nexus between Saddam Hussain and al Qaeda's terrorism—and then occupied the country negligently. (p.1) He builds from this broad assertion throughout the book.

A central premise of the book is that Bush's venture into Iraq has veered off course regarding longstanding military doctrine. In June 2002 Bush gave a speech at West Point where "he made preemption the national strategy—an astonishing departure from decades of practice and two centuries of tradition. Henceforth, the United States was prepared to attack before threats became full-fledged.... Bush stated 'If we wait for threats to fully materialize, we will have waited too long'" (Ricks, 2006, p. 38). Earlier that year, during his State of the Union address in January 2002 that was widely disseminated via mass media venues, Bush had acknowledged Iraq, Iran and North Korea as rogue states. "States like these and their terrorist allies constitute an axis of evil, arming to threaten the peace of the world" (Ricks, 2006, p. 35). As Ricks notes, "The first speech had done the targeting—that is, stated the goal. Then the West Point speech provided the doctrinal, or intellectual, rationale for doing it" (Ricks, 2006, p. 39).

Ricks (2006) proceeds to interpret the rationale used by Bush to defend his use of the military in this manner.

The Bush administration offered three basic rationales for U.S. intervention in Iraq: the threat it believed was posed by Saddam's WMD; the supposed nexus it saw between Saddam Hussein's government and transnational terrorism; and the need to liberate an oppressed people. In the spring of 2004, the first two arguments were undercut by the official findings by the same government that had invaded Iraq, and the third was tarred by the revelation of the Abu Ghraib scandal. (p. 375)

Thus, according to Ricks, Bush is defined as being doctrinally bankrupt and operating from dubious premises. The implication being that such a scenario is a recipe for tragic results.

Much of the Ricks book lays out, in vivid detail, the absence of meaningful strategy that is an outgrowth of faulty doctrine and misguided rationale for military intervention in Iraq.

Strategy was seen (by the Bush administration's national security team) as something vague and intellectual, at best a secondary issue, when in fact it was the core of the task they faced.... An overly simplistic conception of the war led to a cascading undercutting of the war effort: too few troops, too little coordination with civilian and governmental/non-governmental agencies and too little allotted time to achieve success. A confused strategy can be every bit as lethal as a bullet. (Ricks, 2006, p. 185)

Representative Ike Skelton sent a letter to Bush in which he invoked the advice of the highly regarded Prussian military theorist Carl von Clausewitz "to remind the White House of the requirement in war 'not to take the first step without considering the last step.'" That is, don't go into war without an exit strategy. Similarly he quoted Chinese military philosopher Sun Tzu "To win victory is easy; to preserve its fruits, difficult" (Ricks, 2006, p. 59).

This description of vacant strategy implies that the U.S. Army is one of the victims of U.S. military intervention in Iraq. What much of the Army leadership tried to caution against, and that we can see clearly in hindsight, is that there was not a clear overarching strategy in place prior to the invasion. "Speed didn't kill the enemy—it bypassed him. It won the campaign, but it didn't win the war, because the war plan was built on the mistaken strategic goal of capturing Baghdad, and it confused removing Iraq's regime with the far more difficult task of changing the entire country" (Ricks, 2006, pp. 127-128). Ricks goes on to convey example after example where this strategic void has led to tragic results.

"The effect was that the U.S. occupation in its very nature violated the fundamental military principle of unity of command—that is, having one person in charge of the effort, so that all hands have a common goal and work together toward it....Chain of command—of all the problems in Iraq, this is the biggest problem" (Ricks, 2006, pp. 179-180). Ricks explains that a casualty of the lacking chain of command is the absence of an understandable commander's intent. "The U.S Army divisions operated like fingers without an operational hand or strategic arm to guide them.... it works only if (they're) guided by a larger strategy that coordinates each unit's actions. In military shorthand, that direction is called the commander's intent. (General) Sanchez didn't provide it" (Ricks, 2006, p. 225).

The Abu Ghraib prison situation reflects one of the more pronounced examples of how the U.S. Army subsequently has become a train wreck waiting to happen. The tragedy of the Abu Ghraib abuses can be attributed to a large degree to the lack of strategy. "As the need for actionable intelligence arose, the realization dawned (among U.S. commanders) that pre-war planning had not included planning for detainee operations." It is bizarre to think this inevitability was not considered (Ricks, 2006, p. 291). The mass mediated visual images of Abu Ghraib prison abuses will haunt U.S. legitimacy in the region for years to come in relation to cultural identity violations. Internet based technologies and related new communication technologies had significant impact on the distribution of such messages.

Thus, U.S. military intervention in Iraq is basically summarized as being thoroughly misguided because, among other things, without sound strategy it is left vacillating in prevailing political winds of the day. "Unless we ensure we have coherency in our policy, we will lose strategically....I do not believe we had a clearly defined war strategy, end state and exit strategy before we commenced our invasion" (Ricks, 2006, p. 362). Multiple misunderstandings have flowed from this situation. A key misunderstanding deals with a missing awareness of what a counterinsurgency is and how to combat it.

The Bush administration has been very slow to grasp the U.S. military is in the throes of a counterinsurgency. "The Iraqi people were the prize in this fight, not the playing field....(the) classic counterinsurgency strategy holds that the objective is first gain control of the population and then win their support" (Ricks, 2006, p. 250). Many problems can flow from a lack of awareness in this area. "The key to counterinsurgency is focusing on the people, not on the enemy.... Every time you treat an Iraqi disrespectfully, you are working for the enemy....One of the keys to winning a counterinsurgency is to treat prisoners well, because today's captive, if persuaded to enter politics, may become tomorrow's mayor or city council member" (Ricks, 2006, pp. 420-21).

There is no doubt that the U.S. holds the upper hand regarding weaponry however correct use of that weaponry is key. "Firepower must be viewed very differently (in a counterinsurgency) than in regular war. A soldier fired upon in conventional war who does not fire back with every available weapon would be guilty of dereliction of his duty; the reverse would be the case in counterinsurgency warfare, where the rule is to apply the minimum of fire" (Ricks, 2006, p. 266). This confusion about what kind of war we are fighting obviously places the U.S. soldier in a very difficult situation regarding how one is to proceed and react.

Kalev Sepp is a retired Special Forces officer who earned his Ph.D. in history at Harvard and became a professor at the Naval Postgraduate School. Sepp drafted a short paper that distilled the lessons of fifty-three counterinsurgency campaigns in the twentieth century, with an eye to identifying the characteristics of those that had won and those that hadn't....Sepp's chart of the nine unsuccessful characteristics reads like a summary of the U.S. occupation in 2003-04. These were his hallmarks of failure: 1) primacy of military direction of counter-insurgency; 2) priority to kill-capture enemy, not on engaging population; 3) battalion-size operations as the norm; 4) military units concentrated on large bases for protection; 5) Special Forces focused on raiding; 6) adviser effort a low priority in personnel assignment; 7) building, training indigenous army in image of U.S. Army; 8) peacetime government processes; 9) open borders, airspace, coastlines. (Ricks, 2006, pp. 393-394)

Ricks portrays resulting frustrations among U.S. military personnel and implies such frustrations are understandable. One of the ways for soldiers to function in such an environment is depicted via them being able to remove themselves from that environment, even if it is counter to best practices for fighting a counterinsurgency.

Since the 2003 Iraq invasion quality of life issues for U.S. soldiers have been stressed more and more. In order to keep a volunteer force relatively happy and willing to come back for third and perhaps fourth tours, the Pentagon had to provide a high quality of life for its people. But counterinsurgency doctrine says that the only way to win such a campaign is to live among the people. One of the nine hallmarks of failure identified by Kalev Sepp was 'military units concentrated on large bases'—and that was precisely the new force posture of the U.S. military....He knew that classic counterinsurgency theory held that troops must live among the people as much as possible, developing a sixth sense of how the society works. (Ricks, 2006, pp. 417-418)

U.S. military involvement in Iraq is painted as classic misuse of the U.S. Army. During my March, 2007 military tour at the Pentagon I noticed many copies of the Ricks book being read. The Pentagon library had many copies prominently displayed. It will no doubt be useful reading in future case study analysis for how U.S. leaders should not proceed. "What you see here is an unconventional war fought conventionally....In other words, U.S. forces were fighting hard, and might even be able eventually to claw their way to victory, but they were working far harder and less productively than necessary....They were pounding the square peg of the U.S. Army into the round hole of Iraq" (Ricks, 2006, p. 214).

As such the U.S. Army is praised and the Army leadership is not blamed for the prevailing circumstances. "The cause of the turmoil among the troops wasn't the quality of the commanders, but rather the disconnect between what the Army was designed to do and what it actually found itself doing....We've got a military designed to fight big wars, and it's constantly fighting small wars" (Ricks, 2006, p. 310). The reader can easily grasp the predicament this presents for the soldier trying to carry out his/her duty.

Army Chief of Staff Eric Shinseki appeared on Capitol Hill February 25, 2003 and presented his pessimistic view of the proposed invasion of Iraq. Among his key points was that the postwar force should be considerably larger than the wartime force. This put him in opposition to the plan being put forth by Secretary of Defense Donald Rumsfeld (Ricks, 2006, p. 96) and, as a result, Shinseki resigned. He is one of the few senior leaders who can be admired for transparent integrity.

There are numerous pockets of resulting problems that can be categorized compartmentally. Communication scholars can easily point to issues associated with cross-cultural communication misunderstandings. Colin Powell warned against military intervention into Iraq. Powell asserted to Bush "You are going to be the proud owner of twenty-five million people.... You will own all their hopes, aspirations and problems." (Ricks, 2006, p. 48). Powell also later resigned in frustration.

Important cultural norms were trampled as the Army sought to achieve their objectives. "Entering the private space of the house where the women and children were, then tying up and interrogating (i.e, humiliating) the man in the house in front of his family, the premier cultural value of family honor was violated....You've created a blood debt when you do that" (Ricks, 2006, p. 238). Such scenarios have played out over and over again during the U.S. military occupation of Iraq.

Making enemies when we should be making friends has created a disaster for the U.S. "Each Iraqi owed it to himself and his family to decide whether it made more sense to cooperate with us or to cooperate with somebody else, the insurgents. Unfortunately, because of our incompetence, more and more Iraqis have made the decision that their interests don't lie with us" (Ricks, 2006, pp. 325-326). This is depicted as one of the cascading failures. The mass mediated visual images of the Abu Ghraib abuses hurt our cause immensely in this regard.

Marine four-star General Anthony Zinni, who led the 1998 Desert Fox raids on Iraq, has been vocal in his opposition to the 2003 U.S. invasion of Iraq. "The more I saw, the more I thought that this was the product of the neocons who didn't understand the region and were going to create havoc there....U.S. soldiers would wind up paying for the mistakes of Washington policy makers" (Ricks, 2006, pp. 87-88). This lack of cross-cultural understanding among senior U.S. leaders is recognized as being an acute failure that will haunt us for years to come. The parallels with U.S. involvement in Vietnam during the 1960's-1970's are many.

In fall, 2003 General Zinni addressed the U.S. Naval Institute and the Marine Corps Association in Crystal City, Virginia (near the Pentagon). "It kills me when I hear of the casualties and the sacrifice that's being made, especially because the casualties are being suffered because some policy wonk back here had a brain fart of an idea of a strategy" (Ricks, 2006, p. 242). Zinni closed by cautioning that Iraq was beginning to feel like the Vietnam War. He received a standing ovation.

In May 2004 a Washington Post/ABC poll indicated a majority of Americans felt the war in Iraq was not worth the losses incurred. "General Zinni came to a similar conclusion. 'I have seen this movie,' he said in April 2004. 'It was called Vietnam'" (Ricks, 2006, p. 362). Zinni has periodically presented such views on the Sunday morning political talk shows. When a prominent retired Army four-star general conveys fervent opposition to U.S. involvement in Iraq on national television this makes a strong impression with the viewing public.

What Ricks presents would stand alone as interesting and engaging reading even if it were fiction. That it is dealing with a factual situation, and so many lives are being destroyed in the wake of this scenario, makes it an imperative topic for discussion. The implications for the future are stunning.

In January 2005, the CIA's internal think tank, the National Intelligence Council, concluded that Iraq had replaced Afghanistan as the training ground for a new generation of jihadist terrorists... .Iraq is likely to dominate American foreign policy for years....What happens in Iraq will influence the fate of the Middle East For generations to come, with a profound impact on our own national security. (Ricks, 2006, pp. 430-431)

SOLUTIONS AND RECOMMENDATIONS

One is left wondering what, amid the staggering losses in Iraq, has been gained? Ricks searches for an answer to this question and concludes "Yet inside all these problems there lay a major victory for President Bush and his plan to transform the Middle East. Like it or not, the U.S. government through his actions has been tethered to Iraq and to the region around it as never before. Under him, the U.S. military has carried out its first ever occupation of an Arab nation.... The stakes are simply too high to let Iraq become a sanctuary for anti-U.S. terrorists" (Ricks, 2006, p. 433).

We in the U.S., individually and collectively, are left to draw our own conclusions of what this represents for the nation. It is a sensitive topic and one that might easily be avoided given the discomfort associated with it. Given what I've observed I am not opposed to U.S. involvement in Iraq, and the region overall if we have national security interests at risk, but I do oppose military intervention if there is no coherent plan in place that stresses goals that are defined, understandable, achievable, measurable, and assessable and if there is no exit strategy end game and public support. Otherwise, it seems to be a recipe for disaster for all concerned.

FUTURE RESEARCH DIRECTIONS

On March 17, 2007 the second largest anti-war protest against the Pentagon occurred as a means to register dissent against U.S. military involvement in Iraq (the largest had occurred 40 years earlier as a statement against U.S. involvement in Vietnam). The protesters gathered on the mall at the Vietnam Veterans Memorial and proceeded across the Memorial Bridge and onto the Pentagon. I was working a war game exercise that day at the Pentagon but arranged my duty day so I could go to the mall and then observe the march as it proceeded to the Pentagon. I was surprised at the number of counter-protesters who were at the mall to oppose the anti-war protesters. I was also surprised at the degree of feeling that both constituencies expressed. There were some fairly ugly confrontations.

As we enter the coming years, and U.S. involvement in Iraq continues, it is my guess we will see more expressions of dissent and support regarding U.S. military involvement in the region. The role of communication processes in such expressions, and the understanding of such expressions, will continue to be pivotal. Similarly the role of communication processes will be pivotal regarding the U.S. military functioning in Iraq. There will continue to be effective practices and ineffective practices. Regarding the latter (effective practices), the Human Terrain System is worth noting. This innovative approach teams civilian anthropologists with Army brigades to advise military leaders regarding cross-cultural dynamics that can enhance mission effectiveness. It has been widely embraced.

CONCLUSION

All the while, however, the impact of an absent overall strategy continues to be felt and the loss of life (Iraqi and U.S.) continues to mount. In October, 2007 Lt. General Ricardo Sanchez, coalition commander in 2003-04, labeled the Iraq war as "a nightmare with no end in sight....(and) the Iraq war plan from the start was catastrophically flawed, unrealistically optimistic.... (involving) partisan politics that have prevented us from devising effective, executable and supportable strategies" (Sanchez, 2007). Hopefully lessons from the present will be heeded in the future.

REFERENCES

Department of Defense. (1993). *Joint publication 3-0, doctrine for joint operations*. Washington, D.C.: Department of Defense.

Drew, D. M., & Snow, D. M. (1988). *Making strategy: An introduction to national security processes and problems. Maxwell AFB*. AL: Air University Press.

Dunlap, C. (1998). *Preliminary observations: Asymmetrical warfare and the Western mindset, challenging the United States symmetrically and asymmetrically: Can America be defeated? Carlisle Barracks*. PA: U.S. Army War College.

Howard, M. E. (1979). *Restraints on war: Studies in the limitations of armed conflict*. New York, NY: Oxford University Press.

Passage, D. (1998). *Africa: New realities and U.S. policy. Special Warfare, 11(3)*. Fort Bragg, NC: JFK Special Warfare Center.

Perry, W. (2002). *The ethical use of military force*. International Studies: National and International Security. Maxwell Air Force Base, AL: Squadron Officer College (Air University).

Ricks, T. E. (2006). *Fiasco: The American military adventure in Iraq*. New York, NY: Penguin Books.

Sanchez, R. (2007, October 17). Iraq War a nightmare with no end in sight. *CNN*. Retrieved from www.cnn.com

Snow, D. M. (1993). *Peacekeeping, peacemaking and peace-enforcement: The U.S. role in the new international irder*. Carlisle Barracks, PA: Strategic Studies Institute (U.S. Army War College).

Snyder, W. P. (1995). Strategy: Defining it, understanding it and making it. Maxwell Air Force Base, AL: Air War College (Air University).

Travers, R. (1997). A new millennium and a strategic breathing space. *The Washington Quarterly*, *20*(2), 97–114. doi:10.1080/01636609709550243

von Clausewitz, C. (1962). *On war*. Chicago, IL: Regnery Press.

Chapter 16
New Media and Hegemonic Discourse in Pakistan

Saman Talib
Beaconhouse National University, Pakistan

Saadia Gardezi
Lahore School of Economics, Pakistan

ABSTRACT

Cell phones in the Sahara, Internet in Burma, cable television in Pakistan: modern media developments are finally touching the far flung reaches of Earth. With the arrival of these innovations come myriad uses, effects, and consequences. Once again the debate on the emancipatory utility of these media is renewed as these media reach deeper into global society. This chapter states that the effects of the 'new media' may not be what we expect with respect to the democratization of society. The analysis shows that ideological hegemony is not only alive and well, but might even thrive in these new media. To scrutinize these effects, the chapter focuses on the impact of Pakistani cable television on identity and culture discourse through the lens of three topics: identity and the new communication technologies; ideology, power and the new media; and new media and new popular cultures.

INTRODUCTION

Cell phones in the Sahara, Internet in Burma, cable television in Pakistan, modern media developments are finally touching the far flung reaches of Earth. However, it needs to be accepted that the rate of diffusion of different media varies across the globe. While many developed nations have had television since the 1940s and cable television since the 1960s, other nations still do not even have cable television. This chapter starts with a consideration of what qualifies as 'new media' for a large portion of the world. It is argued that in a country like Pakistan, cable television qualifies as a new medium just as the Internet and its many innovations are in Western nations. The discussion then moves on to the effects of the popularization of cable television in Pakistan. The chapter will look at the impact of cable television on identity and culture discourse through the lens of three topics: Identity and the new communication tech-

DOI: 10.4018/978-1-60960-591-9.ch016

nologies; Ideology, power and the new media; and New media and new popular cultures.

Haythornthwaite defines new media as "the implementation of any medium that is new to the adopting unit" (2002, p.386). Keeping this definition in mind, cable television can be seen as a new medium in Pakistan. Cable television has been a late arrival to Pakistan primarily due to the lack of availability of infrastructure investment. Prior to the arrival of cable television, Pakistanis had two options; a handful of state-sponsored television channels or multiple 'foreign' channels accessed through satellite dishes. The latter option, due to high personal cost, was limited to only a very few households in Pakistan. This all changed with the arrival of cable television five years ago. Similar to the explosive growth of the Internet in Western countries, cable television channels multiplied rapidly in Pakistan. In the last 3 years alone, there has been a 132% growth in the number of channels. The result of this rapid expansion has been that the government has been able to exert little influence on the regular broadcast of all except the news channels. The audiences, long used to a controlled diet of state sponsored fare, are now beset by a multiplicity of viewing choices. However, the question to ask is that does this ever increasing menu of cable channels enhance the marketplace of ideas in Pakistani society? Or are the channels merely pouring old wine into new bottles when it comes to identity, power and cultural discourse?

Looking at cable television from the perspective of identity, over the last few years, cable television has become the battleground where an identity for Pakistan is being forged. Citizens and audiences are surrounded on all sides by different identity pulls through the multiplicity of unmonitored and uncensored programming available on cable. The traditional 'elite discourses' have again gained the loudest voice in this new medium – this includes religious elites, social elites, and political elites. Of course, this tension is further exacerbated by indiscriminate and massive exposure to foreign broadcast channels. This struggle for precedence in dominant discourse is made more complex by the post-colonial and partition trauma of the country. Sixty years after gaining independence from the British Raj and separating from India, there is still a raging debate as to what constitutes the 'identity' of Pakistan or a Pakistani. In this vacuum of a historic identity, many 'groups' have stepped into the vacuum to birth an identity for Pakistanis. With the arrival of cable television, the public discourse is inevitably co-opted by the privileged few because they control both the means of production and the access to media. The effect is that of magnifying the voice of elites and magnifying the voicelessness of the 'other'. Of course, there are some channels attempting to challenge this trend but the result is pillarization rather than dialogue.

This then leads us to the consideration of how propagation of ideology is subservient to who already holds the power to define discourse as cable television emerges as the predominant medium in Pakistan. One way of reinforcing elite hegemonic discourse is through the use of local 24-hour news channels. These are some of the most prolific types of channels, numbering 15 at last count. A few of these 24-hour news channels are also among the most watched channels on cable. The predominance of certain stories and perspectives on these cable news channels leads to the monopolization of public discourse. This is affected through agenda setting, in that some issues are given prominence while others are ignored. Though, superficially, there are programs and channels that espouse the 'common man', but again these are drowned out by higher funded, partisan channels. Also, the method of coverage favours the elite, since topics are treated in a manner that is episodic and reductionist. Furthermore, since the 24-hour news channel format is so new and requires a large influx of both human resources and information, the producers in the channel have to depend on the most readily available resource. Therefore, the politicians, experts, and even the talk show hosts are from the same social strata. Finally, through years of state-sponsored news on the broadcast media, the public has become habituated to talking about certain issues. These then are again the most popular topics and perspectives on the all-news cable channels. There is little attempt to broach new or less popular topics.

With the arrival of new media comes the inevitable emergence of new forms of popular culture. The creation and propagation of popular culture is another example of how elite discourse maintains its position through new media. Depiction of culture in Pakistan cable television comes though music channels, drama and entertainment channels, as well as religious channels. Music channels frequently use English language, which is the elite mode of communication. The exclusiveness of the use of English language in regular communication in privileged classes is exacerbated due to low literacy rate as well as low international mobility for most of the population. The music videos also often depict scenarios that would appeal to a certain strata of society. The clothing, the language, and the cultural depictions are mostly from a limited population. Before the arrival of cable television, the government channel PTV tried to maintain equal artistic representation of different demographic components of society. On the commercial channels, the majority of music is of the western inspired pop and rock variety.

In entertainment channels, dramas and variety shows are the most common types of programming. Within these types of programs there is a tendency to minimize coverage or depiction of tough social issues. Focus is mostly placed on luxurious lifestyles, foreign locations, and over-the-top outfits. The result is that the image of the ideal life is again seen as materialistic and power-driven. The reason given for the abundance of such shows is that this is what people expect to and want to watch. However, traditionally, it is a very different kind of television drama that was popular in Pakistan. Also, the features and images shown most frequently on television is not that of the majority of the population, but of a chosen few. Beyond this, there is an almost absolute absence of programming showing ethnic or religious minorities and their issues or celebrations.

The discussion in this chapter will further illustrate the view that "the use of centralized media systems, particularly broadcasting, for hegemonic control and as an ideological tool by national elites to propagate a sense of national and regional identity" (McMillin, 2007, p.77). Media of course in Pakistan is not a single focused, ideology controlling mechanism. Its control over ideology exists in subtle ways, and its impact on national identity is fragmented and reflected in the identity crisis of the citizen consumers. National media systems serve as vessels to disseminate ideologies of power to establish and maintain the relations of domination (McMillin, 2007). Radio and television act as conspirators to establish and maintain the dominant codes of nation (Hall, 1980).

NEW MEDIA, IDENTITY AND STATUS QUO ELABORATION

Many of the new media studies can be seen from the perspective of the competing theories of technical determinism and social shaping. Technical determinism posits that technology shapes social habits, whereas social shaping contests that it is society that dictates the development of technology. Speaking of the Internet, Nicholas Garnham in Emancipation, the Media, and Modernity suggests that the current optimism about the information revolution draws from the tradition of technological determinism. He suggests that these theories draw on traditions emerging from McLuchan and Walter Benjamin, and through him Marx. Garnham says that a "productivist romanticism now surrounds hypertext and the world-wide web and underpins the constantly forecast but constantly postponed death of the book" (Garnham, 2000). In the social shaping tradition "science, and thus the technologies built upon it, is like religion or ideology, simply a set of beliefs as to what counts as truth, that is developed, negotiated and sustained as part of the cultural repertoire of a given social group" (Garnham, 2000). Garnham's claims about a new medium and its uses in any society stand true of cable television in Pakistan. There is a contest on what determines the uses of the new medium, the penetration and use of this new medium is contingent both upon the perceived social utility of the technology as well as the ability of the hardware delivery of cable television to increasingly diverse communities.

Both schools of thought have their limitations. Social determinism is limited by the fact that much of the trajectory of the development of most technologies is determined before the technology becomes widely available. So, "The key technological decisions in designing the system will have necessarily been taken before there is a single consumer. Thus, how those decision are taken and within what interplay of power and interest becomes a crucial question" (Garnham, 2000). Thus before the social influence begins, certain features have already been locked into the capability of the technology. This can be seen by the distribution and programming content of media in Pakistan. Diffusion of Cable television requires an active participation and impetus from both the federal government of Pakistan as well as from the government of each of the provinces. Until recently cable television grew unchecked and organically, with a few providers wiring buildings in densely populated cities and routing satellite television programming through these cable lines. The national regulatory body PEMRA has thus far limited itself to issuing licenses and regulating the content for the many 24 hour news channels available over cable. However, each province has control over the introduction and viewing of cable in its region, and in 2000, on of the provinces, NWFP, banned cable television on the insistence of religious proponents (BBC, 2000).

One of the theories that highlight the limitations of technology determinism is the techno-economic paradigm. There are three salient aspects of this paradigm. First, according to this paradigm, technological innovation itself is driven by economic concerns. Second, it takes a while for the new technology to penetrate into the social system as it has to first overcome the inertia created by previous infrastructure investments. Finally, different societies may have varying suitability for a certain technology. Perhaps, the best approach to studying new media use is to "focus on the social – political, economic, cultural, and legal – factors that shape technology, and reflect the social and institutional context in which they emerge" (Docter & Dutton, 1999). These political, economic, cultural and legal issues will be considered in detail in the following sections. Suffice it say that these multiple issues are creating a picture of cable television in Pakistan that is unique to its own history, political milieu, current search for identity and emerging personality. For example, even in the case of governmental restrictions and regulations to the contrary, cable operators continue to provide channels and shows that are ostensibly banned from viewing.

One of the theories utilized in the communication field that supports this attitude towards technology is the Diffusion of Innovations theory. The earliest example of the study of diffusion in political communication can be found in "The People's Choice" by Lazarsfeld, Berelson, and Gaudet (1968). In this study, the authors refuted the hypodermic needle theory of media by offering the two step approach. This methodology emphasized the significance of opinion leaders in the spread of information from the mass media to the public. Recent studies have applied the diffusion of innovation theory to such aspects of media and political communication as elections, political campaigns, and general media effects (Swanson & Mancini, 1996; Valente & Davis, 1999; Bryant & Thompson, 2002).

The theory was formalized by Everett Rogers in 1962 and has found widespread application and acceptance across several fields of research. What

Rogers states in this theory is that "Diffusion is the process in which an innovation is communicated through certain channels over time among the members of a social system" (2003, p.5). He emphasizes the importance of considering the four elements of the definition: innovation, communication through channels, time, and the social system.

Rogers (2003) points out that another factor of particular theoretical and methodological significance is the concept of technology clusters. That is, often in innovations, it is difficult to clearly delineate the boundaries of a specific innovation. In some cases, innovation can lead to the occurrence of a package of closely related technologies. In such cases, it is key to study the diffusion of these interrelated technologies simultaneously, which lends support to Bennett's (2004) suggestion that scholars should study new media technology rather than merely the Internet. This is particularly true because in developing countries, like Pakistan, the introduction of what is generally defined as new media, The Internet and its many applications, has arrived before cable television. However, the introduction of the Internet did not have as measureable an effect as cable television due to the high initial cost of access as well as the limited literacy levels of the population. As opposed to the Internet, cable television has had a significant impact in Pakistan because a large percent of the households already own television sets. In addition, the audio visual nature of cable television removes the literacy barrier that the Internet presents.

New media, such as cable television in Pakistan is also shaped by the focus of society and history as it spreads through any community. The ideological and cultural hegemonic elements are thus passed through the medium intact. McMillin (2007) proposes in her book is that current institutions that exist in the superstructure of post-colonial societies are a continuation of colonial institutions. She looks at a variety of studies done in third world 'tri-continent' countries that lend support to the 'change in guard' thesis, that in post-colonial states, the mechanism for subjugation is still the same; only the controlling party has changed. A cultural continuity exists in the mechanism of hegemonic control. A change in draconian rule and emancipation of the subaltern rarely occurs with a change in guard (McMillin, 2007, p. 134). Media is thus an institution that maintains the status quo of the elements of the superstructure.

Since Pakistan is a post-colonial nation, it inherits media use traditions transmitted from its colonial historic roots. During colonialism the focus of mass media systems was to increase the administrative efficiency of the government (McMillin, 2007, p. 71). Today this can still be seen to hold, following the thesis that there has been a change in ruler but not a change in rules. Colonial rulers wanted to create and maintain the basic conditions of their rule (McMillin, 2007, p.71). They did this by reorganizing existing hierarchical structures. Once the power structure is set up, you can put anything into it, like the variable constituent elements of nationalism for example (Young, 2001, p.173). This is to say that values and institutions had to be constructed and then enforced onto traditional society, and this strategy has been adopted by postcolonial governments.

Shohat and Stam (1994) describe ways in which colonial racism played out and was portrayed in the media. First by projecting a lack of modesty and material resources in the colonized population; second by a hierarchy, in placing for example Europeans over non-Europeans or modernism over traditionalism; third by placing the protagonist at a distant from colonial struggles and thus removing empathy between the races; and lastly by a systematic devalorization of the ways of the colonized. Today in the context of ruling class hegemony though media, these messages still exist. Portrayed by the focus on top level political drama by 24 hours news channels and not on the minorities or grass root issues. The producers of the news and their subject of focus, the politicians, bureaucrat and individual actors, all belong to the same social strata of the countries elites, or to put it crudely, to the same income bracket. This becomes a way new media distances itself from the masses, yet has a historical and ideological hold over them due to a long standing medium of communication between national events and Pakistan's population. 'Devalorization' in the post-colonial context happens not in active opposition to dissenting views and values of society but by the cooption of the masses into believing that the content and views of the media reflect the views of society, and the silence of new or resisting voice due to this intellectual cooption. Centralized control of media, even accompanied by private investment means a centralized control over content. For instance, in post colonial contexts television relies on existing gender, class and caste hierarchies to produce discourses of progress, technology and innovation. These hierarchies place gender or minorities as objectives of rescue or ridicule and identify upper-class males as progenitors of a new nation (McMillin, 2007, p.89).

While on the surface there is an accepted discourse propagated for decades through official media, PTV and Pakistan Radio, that Pakistan is a religiously defined state. However, there always has and does exist draws towards competing identities. Traditionally there has always existed a contest between Islamic-secular identity and national-regional-ethnic identities. However, with the arrival of new media this quest for identity has expanded to include western-Arab-south Asian identity as well as linguistic identity.

There long been a dispute as to the very reasons for the creation of the country itself. The history books teach that Pakistan was created based on a 'two nation theory' with the objective of the creation of a homeland for the Muslims of the Indian subcontinent. The explanation generally offered is that the Muslims of pre-partition India faced constant persecution and discrimination and Pakistan was created as a means of offering them a homeland where they would be safe and prosperous. This overarching view of Pakistan as an Islamic state has traditionally led to laws and government actions that have institutionalized discrimination again religious minorities as well as women. Such laws have reinforced the culture of Patriarchy that already had existed in Pakistan. In addition, there has also been a pull for citizens of the country towards identifying with the nation or retaining their regional, ethnic identity (Rais, 2007).

As to the religious identity of Pakistan, "the role of religion is not a settled issue, which greatly impacts the statecraft, the status and rights of minorities, and the larger question of internal peace and security" (Rais, 2007, p.111). This is evidenced by the Blasphemy laws as well as Hadood Ordinance laws which have both been widely criticized as being discriminatory against religious minorities and women, respectively. This tension then seeps through the society as citizens try to negotiate the terrain between religiosity and secularism. Pakistan is home to people who run the range from fundamentalists to moderates within just the Muslim community. This contest is also reflected in the content offered up on cable television.

Underlying the overarching discourse of national religious identity, is the fact that Pakistan was created as a Federal state. The federation consists of four provinces – Sindh, Punjab, Balochistan, and North West Frontier Province – and two territories – Islamabad and Federally Administered Tribal Areas. Each constituent is home to distinct populations with separate languages and differing cultures. Historically, "the question of cultural identity and representation has been an issue of contention in Pakistani society since the country's creation in 1947" (Syed, 2008, p. 318).

Recently, this identity crisis in Pakistan has been further complicated by the introduction of the Internet and cable television. As these influences have made inroads into society, a new identity of 'modernity' and 'traditionalism' has seeped into the society. Here modernity is seen to mean adherence to western values, such as speaking in

English, consuming English language programming, and more liberal stances on dating and marriage. An additional influence, made more obvious through the media is the confounding of the Islam with Arabic culture. What this means is that people who want to portray themselves as devoutly religious and/or traditional begin wearing Arabized clothing like Abayas and Jilbabs.

Identity and Hegemonic Discourse

Looking the history of dictatorships in Pakistan, Pakistan Television Corporation (PTV) was established in 1964 by first Martial Law Administrator, General Ayub Khan as a state controlled broadcast television channel. PTV served, and it can be argued that it still serves, the purpose of state propaganda. Khan (2009) argues that when in the 1980's western countries were gradually developing in to 'Media-Centered Democracies', Pakistan was under military rule with General Zia-Ul-Haq. Though this to an extent inhibited an earlier emergence of a free media it set the tone for a future media that would be more vocal and would oppose non-democratic politics. Are the increasingly right-wing, pro-democracy cable channels born out of a resistance to authoritarian regimes?

The case of Pakistan's media certainly implies so and the idea of hegemony can explain this. Military rule in Pakistan was coercive. Media controlled though force could bend public opinion to its advantage but it could never influence people on an intellectual level. Which is why the consensual, co-optive power of the free and seemingly 'neutral' media, that is not a puppet like PTV, comes in and produces a hegemonic discourse, rather than just a dominant discourse for Pakistan's masses. The only lasting hegemonic effect of direct control by the government or military has been the fostering of a Muslim nationalist streak that is detached from ethnic or regional identity. Thus we can talk of the current situation of identity in Pakistan where the media in Pakistan since the 1960's has adequately promoted a general ideology that is Muslim and nationalistic. Today with the increase in information and images with over 25 cable news channels there is a change in trend from the production of a postcolonial identity of being Pakistani, to an identity that is fragmented. The identity projected by the media is still Pakistani, but what that this identity means is open to political debate, ethnic conflict, international relations with USA, regional migration, that is, the sort of discourse that is abundant cable news channels. To confound further the issue of identity formation, audiences in Pakistan are not homogenous. Pakistan as a country has strong regional and ethnic divides. Yet the news watching population is subjected to a somewhat homogenous format of information projection. Instead of discussion on ethnic issues of migration from Afghanistan, displaced tribes in northern Pakistan for example, media is seen to focus on political conflict at the top level and a criticism of US policy on Pakistan.

Recent trends in news channels that guide identity include the discussion on the Kerry Luger bill which has raised much anti-American and anti-West sentiment in the Pakistani people. The Pakistani government has always been pro-America, yet the people are not. Television discourse makes its way to public discourse and there is an irrational angst present against the US. Much of it is a reflection of the popular talk show hosts. As a journalist writes, 'We would rather have the US conduct all its actions in accordance with the wishes of our mullahs, media magnates and anchors' (Habib, 2009). Another example is of the drone attacks by the US in Pakistan to get rid of Taliban militancy. The issue is not that the drone attacks are bad, but of sovereignty, without talk of what sovereignty means, or to what extent US military intervention an advantage, or about military resources and goals. The same attacks would be useful if conducted by Pakistan and not the US, that is, a problem of possession and ownership. As mentioned earlier, content is limited to a one sided argument where there is no room for counter arguments. For example, pro-American journalists are absent from the discourse. We can't even know if they exist.

The uniformity of content should bring about a uniformity of opinion and thus a uniformity of identity. Yet class cleavages and ethnic divides are so strong, and the monopolized subject discussed on TV is so contentious that consolidation of identity has been impossible. The media may have hegemony over public discourse, but this discourse is unable to cause change in its audiences and reinforces the problems that it discusses. Content produced is subject to the hegemonic discourse itself. The perceived resistance of these channels to corruption in the political system, or their pro-democracy stance is not resistance in the emancipatory sense. This kind of resistance only reinforces the hegemonic discourse onto the society.

IDEOLOGY, POWER AND THE NEW MEDIA

This sort of an analysis of mass media borrows heavily form Marxist cultural theory that takes Gramsci's theory of hegemony and Althusser's (1971) concept of media being an ideological state apparatus that reproduces dominant ideologies. According to Stuart Hall (1996), discourses are a group of statements that provide a language that can present a particular type of knowledge about a subject. For the purpose of this study, this description of discourse is an apt way to look at the content produced by news TV channels. Discourses are produced through language and practices. A discourse includes ways of talking and on cable news channels in Pakistan we can even pinpoint the kinds of subjects and setting that produce and constrain dominant discourse. These include US policy for aid and development, US policy for Afghanistan and Pakistan, India and security, democracy and national politics.

McMillin (2007) is of the view that the symbiotic relationship between government and private owned media has led to a shift in goals from less developed countries from social development to capitalist expansion altering the identity of the audiences form citizens to consumer. Though this is valid to some extent, the citizenry of Pakistan depends on private news channels for all dissemination of information. It is true that cable television channels in Pakistan are part of a capitalist expansion and are profit driven corporations, yet the effect of the information produced influences the identity of the audience and the ideology of the nation, regardless of what economic or political motive is behind the production of content on television. Pakistan has hybrid citizen-consumers, as identity and national ideology has much to do with the viewing of these new cable channels. With this shift come more viewing choices and thus can make the audience feel more empowered and freer due to the presence of alternatives to state run broadcasting (McMillin, 2007). Though the quantity of television channels expands, content may or may not. Freedom of choice narrowly means just more access to information. As in the case of cable television in Pakistan, the expansion has led to more channels but a homogenizing of content. All of which is increasingly right wing and other than hourly news, the most popular form of programming is the talk shows with one anchor who is not usually neutral arguing with three guests from the political arena.

For Stuart Hall et al. (1978) the mass media reproduces interpretations which serve the interests of the ruling class yet the media is also 'a field of ideological struggle'. This allows for the media to be relatively autonomous (Woollacott, 1982). The media claims to be the voice of public opinion to reinforce the dominant viewpoint and garner consent, or at least silence from its audiences. Control of the media of course does not mean direct government control of the media. This is clarified by our structuralist analytical framework that explains how media has some autonomy from direct class influence. Looking to neo-Marxist cultural theory, we can speak of the state or government as part of the societal superstructure that is created by classes and social groups. The case of Pakistan illustrates this, for example media in time of the dictatorship of General Musharraf after 1999 becoming increasingly free and remaining pro-democracy.

Employing a structuralist framework of analysis, it can be argued that the hegemonic discourse being produced by in the superstructure, of which national media is a part, is not in the direct control of the ruling classes, but historically present out of its construction by motives of efficient colonial rule, and motives of a development of a modern nation state. These issues of control are not obvious of course, but provide an important picture of the way television in Pakistan has hegemony over ideology and ideas, and secondly how the identity of the media and its messages is linked to the messages and conflicts between the ruling elites. Thus it is useless to talk of specific interest groups and political actors and direct control of one political party over one TV channel. This is not how it is in Pakistan, the superstructure in which media exists in Pakistan is to a degree autonomous of direct instrumental control by specific actors. But like any media, various TV channels, various talk shows, various anchors have specific ideological dispositions that impact the identity of the masses. Since the coming of cable television in Pakistan, the number of news channels has grown from one in 2000 to 22 in 2009. These are all domestically produced channels, most of which report in the Urdu language. Two of these are in English, and there four channels who broadcast in regional languages.

The hegemony of the media on discourse in Pakistan comes from the structure of society which includes post-colonial goals for modernization. For example, cable news channels during the time of resident Musharraf branded him a dictatorship and were pro-democracy. Today, when we have a democratically elected President, Asif Zardari, his actions are interpreted as being un-democratic in talk show discussion (an observed trend on GEO News channel). This points to firstly the fact that the media is not under direct control by the ruling politicians. Secondly there is an ideological bias in the media to non-democratic polices, to corrupt politicians, to a criticism of some political parties more than others, to a nit picking of government policies. Thirdly that this illustrates a hegemony of discourse because it is issues like "what the president did yesterday" and "who the chief minister of Punjab sacked today", that news anchors repeat a dozen times a day on 24/7 format news channels. All other issues, all other forms of discourse, dissent, or any voice other than that of a handful of politicians that revolve around the channels and a couple of human rights activists (who mostly talk only of the politicized judiciary issue) is missing. Information to be made public is set by the TV channel producers and with this hegemony over content and a single minded focus on the politics of the ruling elite's news channels in Pakistan makes other content silent.

One of the biggest trends on cable news channels is that of the talks show program that have a deep structural bias towards anti-Americanism and right wing politics. There is usually some debate that is part of the content of the talk show. The anchor mediates between the public and the politicians, or between the guests that are invited. Due to the rising popularity and long running of some of these shows, these anchors are now self styled experts and usually give their own opinions on policy issues and political debates thus moving from the role of a facilitator of discussion to active participants. This observation is generally accepted by critics of the media in Pakistan in academic studies (Khan, 2009), and news paper journalism[1]. Khan (2009) writes that even in generally criticized programs, like *Jawabdeh* (meaning 'The Giving of Answer') on Geo TV anchors tend to act as judges in their tone and scripts, while the political leadership is presented there as a guilty party. Such examples are abundant.

On 8th August, 2009, Waheed Hussain an anchor on the channel *Waqt News* interrupted a guest discussing the $5.3 billion in aid that Friends of Democratic Pakistan[2] and had promise and said "You mean the foes of Pakistan". The same host went on to call the Kerry Luger bill[3] the "Kerry Looter" bill. Yet with their flaws, these programs are the most popular programs as well as being deviants from ideas of impartiality and reality. The sparks that fly on television between the invited politician and hosts, are usually what make the show more popular. There is a lack of access to books and magazines that provide impartial and well researched journalism of important events to local audiences thus the television is the cheapest alternative. Added to that the talks shows and style of journalism has great entertainment value for audiences. Not only are the discussions on talks show entertaining but there is a dearth to good quality alternative programming in the Urdu language on television solely for entertainment purposes

The purpose of new channels is information and entertainment, giving the discourse produced more room for influence over audiences and ideology. Impartiality of the content obviously has an impact on the ideology and worldviews of the masses that are also anti-American and suspicious of any US policy about this region of South Asia. Even with such clear biases these news channels never declare publicly their political affiliations. Unlike western media CNN or FOX who are open with their affiliation with particular viewpoints. The acts of being neutral and reporting the truth is firmly adhered to.

Added to partiality, the other issue is that reporting practices of news channels where facts are often made up, are wrong, or are reported without verification. Most of this is attributed to the competition between the dozens of news channels to be the first to report an issue of an event. There are many examples that illustrate the contradictions between actual facts and reported facts. Talat Hussain, one of the most influential and popular talk show hosts in Pakistan during an interview with Senator Hillary Clinton in 2009 accused the US of giving too little aid to Pakistan. To present a comparison he cited a figure of $700 million rent of a military base the US has in Kyrgyzstan. Senator Clinton replied that the figure was near $50 million (which is the actual per year figure), to which Hussain replied that $50 million must be the monthly rent. For the news media in Pakistan there seems to be minimum accountability. With its hegemony over information, there is no way for an audience to know to verify it. Thus with the control of information in the hand of private players with no checks on them, the line between reality and television content becomes blurred. Anchors and show hosts have the power to construct discourse in whatever way is best suited to personal biases i.e. a means of hegemonic control.

NEW MEDIA AND NEW POPULAR CULTURES

With the new media come new forms of popular culture. As a new medium spreads through society it offers new opportunities to share popular pre-existing cultural texts. However, the as these traditional texts are shared they are often also repurposed and transformed by the new media. In addition, the added possibilities of a new medium give rise to new forms of popular culture and opens up new avenues for these popular cultural artefacts. However, in most developing cultures the 'keepers' or propagators of popular culture are most often the state and elite. While popular culture is created by and consumed by the common man, the means of sharing and financial support have been traditionally provided by the Government or wealthy patrons. So, traditionally, while elements of popular culture could arise from the populace it required the interest of Government and patrons for its perpetuation and distribution, which subjugates popular culture to the tastes of the wealthy.

Under Gramsci's ideas of hegemony, we can see subjugation as cooperation. Cooption works "primarily by inserting the subordinate class into the key institutions and structures which support the power and social authority of the dominate order. It is, above all, in these structures and relations that a subordinate class lives its subordination" (Hall, 1976, p. 39). Another way to look at this is through the concept of 'cultural patrimony' or the measures of the state to sustain power relations. Cultural patrimony maintains structures of social difference between different social groups who have differentiated access to production and distribution of products. The "symbolic capital of subaltern groups has a place, but one that is subordinate, secondary, or on the margins of the hegemonic institutions and apparatuses" (Cancilini, 1995, p. 136). Ideas and their production by subalterns have no value unless it is co-opted into the patrimonial private structure that has a degree or expertise in production and distribution. This allows us to think of some audience agency, where the messages of the subalterns get absorbed by the system, are rewritten and become part of the dominant discourse.

Cable television in Pakistan has grown exponentially over the last few years, and much of this growth has taken on the nature of the Wild West. There has been little in the way of oversight or regulation by the government in the development and broadcast channels. PEMRA, the regulatory body responsible for oversight of broadcasting in Pakistan, was set up in 2007. Even now, the legislation is limited to granting licenses and not much else. The one restriction PEMRA did impose was on the broadcast of cable news channels during the Emergency declared by Pervaiz Musharraf. Otherwise, the birth of channels is determined largely by the ability of the licensee to afford a license and turn a profit from their venture. This has meant that there are no real standards as to what should be broadcast, what constitutes entertainment, or even what obscenity is. This has meant that audiences, especially those financially and socially disadvantaged, are often exposed indiscriminately to programming that is not audience driven or even audience sensitive.

This is further complicated by the factor that the production and broadcast facilities of Pakistani produced channels are located exclusively in the urban centers of Karachi and Lahore. A few channels are broadcast directly from Dubai. As television production and broadcast is a relatively new field in Pakistan and there are only three university level institutions that offer training in this field, most of the production is done by amateurs. These amateur producers, actors, and other creatives have no grounding in media theory or audience studies. Little is offered in the way of professional or on-the-job training. What this results in is entertainment that is produced solely based on the producer's perceptions of the marketability and appeal of a concept. This combination of urban concentration, lack of professional training, and ignorance of audience preferences has led to programming that is subjugated to the ideals and self-image of the social, political, and cultural elite.

Entertainment channels in Pakistan can be broadly categorized as: Local Urdu channels, Local regional language channels, English Language channels, and Indian channels. There are more than 40 Local Urdu channels broadcasting various kinds of entertainment genre programming in the national language, Urdu. Local regional language channels consist of channels offering a mix of entertainment programming in Punjabi, Sindhi, Balochi, and Pushto. English language channels offer a variety of popular western branded channels such as Discovery, Nickelodeon, and MTV. Indian channels consist of soap opera and entertainment news offerings. Both the English and Indian channels are not localized for the Pakistani audience but are in fact satellite feeds downloaded from neighbouring regions.

All four of these regional and language types of channels can further be divided into News, Islamic, Music, Movies, Cooking, and Variety channels. News channels, the Urdu versions of which were discussed earlier, include CNN, Fox News, Sky News, and Aljazeera in its English language offerings. The Islamic channels provide 24 hour programming focused on Islam and its teachings. Music channels are made up of local and international music channels which broadcast in Urdu and English languages. Local movie channels basically provide a fare of bootlegged versions of Bollywood and Hollywood movies. Movie channels sourced from abroad include such international brands as HBO and Filmax. There are two cooking channels offering a variety of cooking shows in the Urdu language. Variety channels include channels from the Urdu, Regional Language, English Language, and Indian offerings.

This globalization of content is having an interesting effect on the rise and metamorphosis of popular culture in Pakistan. Structural imperialism theory suggests references that the world can be divided into core and peripheral nations. Core nations hold a position of power and influence in economic and cultural aspects. Peripheral nations, on the other hand, are of the nature of the satellites or moons revolving around a core nation. Moreover, within each country, there are core and peripheral populations. Within a certain country the core population wields ideological, financial and political power over the peripheral population. In Pakistan the religious, political and social elites can be seen as the core population, whereas the remaining population holds the place of peripherals. The theory also states that the core populations in each country are more closely aligned and connected with the core nations. For core Pakistanis, three countries can be seen as holding core status: United States, India, and Saudi Arab. These countries have come to define not only the divergent identities of parts of the core populations, but also their divergent productions and texts of popular culture.

If this theory of structural imperialism is applied to the emergence of cable television and popular culture in Pakistan, we can see certain trends emerging. Traditional homespun popular culture is being quickly replaced by culture that is heavily influenced by these three core countries. For example, while the predominant form of music available ten years ago was based on traditional forms of Ghazal and folk songs, the majority of music available today is a westernized version of these very forms of music. Influences from the core nations are carried over in two major ways: Localization and Hybridization.

Localization is the process by which the concept of a program or a format that has proven successful in one country is transported into another by making changes to the original to better fit the indigenous culture, language and tastes, American Idol being adapted into localized versions, the most popular of which have been the Indian Idol and Pakistan Sangeet Icon. Perhaps the most prominent example of this phenomenon is MTV and its numerous localized versions such as MTV Middle East, MTV Pakistan, and MTV India. Hybridization happens when a program or television format is adopted almost completely into a new country with changes that create a new form consisting of elements from the original and others from the local culture. Soap Operas and Telenovelas are by far the most popular forms of hybridized programming. Each country has its own version, but each version is a mix of the conventions of these programs merged with the traditional storytelling mores of the local culture.

The effect of this kind of localization and hybridization of foreign programs can be seen as creating new vehicles for the dominant discourse to be disseminated in the form of entertainment programs to the masses. The novelty of the cable television combined with the explosive growth has given rise to an insatiable rise in appetite for a variety of television programs. Since there is a dearth of creative personnel trained in the field of entertainment, often young professionals are hired in urban areas to generate ideas for television shows. These young people mostly belong to the core elite and often turn to foreign programming as inspiration for new programming. While there is nothing wrong in itself to create localized or hybridized programming, but the way the foreign influences are combined with the producers` sensibilities and interpretation of local culture results in products that support the status quo. Three specific types of programming can be used as examples: Soap operas channels, Music channels, and Islamic channels. Television Drama driven channels, heavily influenced by the soap operas produced in India, have come to represent the faces and ideals of the core elite, while diminishing the peripheral population to the status of the other. Music channels have become the home of the core youth, depicting their aspiration to a United States inspired lifestyle. Islamic Channels tend to align themselves with Saudi Arabian influences and propagate the image and discourse emerging from the religious core elite.

Television drama in Pakistan has a long and illustrious history and has been one of the mainstays of entertainment in the country. Dramas created by the state television PTV were rivals to the most popular movies of their times and were much more critically acclaimed. However, the great appeal of these dramas also lay in their resonance with the broader audience and their lives. Two examples of these kinds of dramas which are still being talked about and watched on DVD today are Waris and Dhoop Kinarey. Waris was a show that focused on the atrocities done by the feudal lords in society and Dhoop Kinarey focused on the daily struggles of the urban middle class. Such dramas were produced under cultural and social policy targets established by the government in power at that time. Thus, the dramas focused on such political debates as the disadvantages of feudalism and the constraints in lifestyle of the middle class.

The television dramas produced today depend heavily on hybridization of conventions established in soap operas, which include hyperbole, politics of family, and wealth. Two of the most popular current dramas can serve to illustrate this trend. The first is Vasal and the second, is Noorbano. Vasal is a drama based on the life of a young woman who is married to a Pakistani man settled in the US. The story follows the struggles this young woman faces in adjusting to the new culture and being separated from her family. Noor bano depicts the life of a woman of the same name and how she negotiates life as an orphan living with a wealthy relative.

The storylines as well as the visual cues in these dramas are geared towards the economic and social elites. This leads to the repetition of ideas and themes in these dramas that would have little resonance with the peripheral populations. The dramas often focus on the ridiculously dramatic lives of beautiful and privileged urban populace. The peripheral population, if it ever does make an appearance in these shows, is always depicted as the bereft, needy and other. Thus the dramas continue to privilege the ideals, lifestyles and concerns of the core. The peripheral population is incidental to the lives of the core and, ostensibly exist only to support and underscore the dramas of the privileged.

A similar progression can be seen on Music channels broadcast through cable television in Pakistan. These channels have become the domain of the young and attractive people belonging to the social elite. Pakistani popular music so far could have been divided into two main genres: ghazal and folk. Ghazals were songs with a focus on high poetry and refined singing. Folk music reflected a variety of bard styles of singing with themes drawn from religion and traditional tales. Each style of singing depended most heavily on live performances and the highest values was placed on refinement of singing skills. These skills take years to develop and perfect, therefore, most of the famous singers shown on the government television were of middle age and not much attention was placed on their looks or fashion style. However, this popular culture was controlled by the Government and wealthy patrons, because they were the ones who could fund public performances.

With the arrival of cable television, there has been a proliferation of music channels which have grown to nine 24-hour channels. The visual nature of the medium and the localization of the MTV format have led to the predominance of western-style music videos and VJ (Video Jockey) shows. These channels are predominantly showcasing music that is a fusion of ghazal and folk traditions, pop, rock and jazz. Culturally, it can be seen as aspiration media, that is, it displays the aspirations of the younger core class. What is seen most commonly is westernized clothing, scenes of love relationships at various stages, dance sequences, and opulent settings. However, these cannot be said to represent the aspirations of the vast majority of the population, even the youth that belong to the periphery. The lyrics of these songs are also primarily about love, acquisition, rebellion and teen angst. Little is offered in the way of consideration of subaltern culture, aspirations, or concerns.

The Islamic channels are a brand new phenomenon born with the popularization of cable television. They were initially regarded as providing a counterbalance to all the westernized and secular influences emerging out of the other entertainment channels. Here in the Islamic Channels we have television channels dedicated to 24 hour programming based on religious teachings and traditions. Each channel portrays particular perspectives on Islam, perspectives that are already form the dominant discourse in terms of religion. So the personalities and their associated discourses not only persist with the introduction of the new media, they are now enhanced by video and ubiquity. So, where as an 'Islamic Scholar' would have had only the opportunity to address a congregation of only a few thousand for a few hours and only the most well know were allowed a few hours occasionally on PTV, now they have regular broadcasts that address millions.

The additional complicating factor is that these channels tend to confound the religion with Arab visual cultural ques. QTV, a channel from ARY, has many 'scholars' and programs with people in Arabized costumes. The only women's program, Bazm e Khwateen, actually has women wearing Middle Eastern outfits. Another factor that is noticeable is that on Peace TV, which is actually a Satellite channel, broadcast from India, most of the programs are pre-recorded and no interactive. This is also true of QTV. So the interpretation and transmission of religious views is uni-directional and not subject to public questioning. A glaring example of discrimination occurred when Dr. Amir Liaquat Hussain (Asian Human Rights Commission, 2008) was accused of inciting violence when he declared that Ahmadis, a religious minority, were Wajib ul Qatal (valid for murder). This declaration was quickly followed by murders of members of this religious minority. Noticeably missing are also any channels representing minority religions such as Hindus and Christians. So these religious channels not only offer a megaphone effect to traditional religious discourse, they further silence any opposing views by the very absence of the voices of the already disadvantaged.

CONCLUSION

New communication technologies have proven to be a boon for many developing countries. They have been substantiated to be quite an ally for development purposes as well as for information gathering. Still, it cannot be stated without reservation that these new technologies have reached their emancipatory or even democratizing potential. The reason for this is that the primary socio-technical decisions rest with the traditional elites in a society. The media dynamics created by these initial decisions are then reinforced by the reality that production and programming influence is held by elites also.

The chapter uses three perspectives to explore these dynamics as regards to cable television in Pakistan. First, the interaction between new media diffusion dynamics and identity politics in Pakistan was discussed to illustrate how the newness of a medium does not isolate it from the effects of ideological hegemony that already exists in society. Second, 24-hour news channels are used to consider how a seemingly free press can still become a vehicle for hegemonic discourse. Finally, the discussion turns to the fact that the predominance of the elite viewpoint is not limited to prior forms of content, but that it permeates the popular culture formats to come.

REFERENCES

Akhmad, A. (2010, February 21). Dumb questions invite dumb answers. *The News International.*

Althusser, L. (1971). *Lenin and philosophy, and other essays.*

Asian Human Rights Commission. (2008, September 10). *Two persons murdered after an anchor person proposed the widespread lynching of Ahmadi sect followers.* Retrieved March 21, 2010, from http://www.ahrchk.net/ ua/ mainfile.php/ 2008/2999/

BBC. (2000). *Pakistan province bans cable television.* Retrieved March 21, 2010, from http://news.bbc.co.uk/ 2/hi/ south_asia/ 800515.stm

Cancilini, N. G. (1995). *Hybrid cultures: Strategies for entering and leaving modernity.* Minneapolis, MN: University of Minnesota Press.

Garnham, N. (2001). *Emancipation, the media, and modernity.* Oxford, UK: Oxford University Press.

Habib, E. (2009, November 12). Tragedy of the talk shows. *The Daily Times.*

Hall, S. (1980). Cultural studies and the centre: Some problematics and problems. In Hall, S. (Ed.), *Culture, media, language: Working papers in cultural studies.* London, UK: Hutchison.

Hall, S. (1996). The West and the rest: Discourse and power. In Hall, D. H. S. (Ed.), *Modernity: An introduction to modern societies.* Malden, MA: Blackwell Publishers.

Hall, S. A. (1976). *Resistance through rituals: Youth subcultures in post-war Britain.* London, UK: Hutchison.

Hall, S. C. (1978). *Policing the crisis.* London, UK: Macmillan.

Haythornthwaite, C. (2002). Strong, weak, and latent ties and the impact of new media. *The Information Society*, 385–401. doi:10.1080/01972240290108195

Khan, M. A. (2009). The mediatization of politics in Pakistan: A structural analysis. *Pakistaniaat: A Journal of Pakistan Studies, 1*(1).

McMillin, D. C. (2007). *International media studies.* Oxford, UK: Blackwell Publishers.

Paracha, F. N. (2009, April 19). Smokers'corner: The creeping malaise. *Dawn News.*

Rais, R. (2007). Identity politics and minorities in Pakistan. *South Asia: Journal of South Asian Studies*, 111-125.

Shohat, E. A. (1994). *Unthinking Eurocentrism: Muticulturalism and the media.* New York, NY: Routledge.

Syed, J. (2008). The representation of cultural diversity in Urdu-language newspapers in. *South Asia: Journal of South Asian Studies*, 317-347.

Woollacott, J. (1982). Messages and meanings. In Gurevitch, M. T. (Ed.), *Culture, society and the media.* London, UK: Methuen.

Young, R. J. (2001). *Postcolonialism: An historical introduction*. Oxford, UK: Blackwell Publishers.

ENDNOTES

1 Paracha (2009) at *DAWN News*, and Akhmad (2010) at *The News International*.

2 Friends of Democratic Pakistan is a US led group of countries that aims to support democratic governments in Pakistan to consolidate democracy and to helping economic development. Some members of the group are United Kingdon, China, Italy, Turkey, Canada and France.

3 The Kerry Luger Bill is a non military aid package offered to Pakistan for a period of 5 years

Chapter 17
The World Narrow Web: Internet Content Regulation in South Korea

Siho Nam
University of North Florida, USA

ABSTRACT

The inauguration of the Lee Myung-Bak administration in 2008 signaled a crisis for Internet-driven participatory, democratic public culture in South Korea. One of the most visible effects was immediately found in the administration's repressive media policy. A series of anti-democratic regulations, grounded in both conservative and neo-liberal philosophies, was implemented to control and tame civic participation, public deliberation, and identity politics on the Internet. Firstly, in light of this, this chapter summarizes certain main debates regarding the role of the Internet in promoting or hindering democracy. Next, it takes up the recent spate of regulations regarding Internet content in Korea to shed critical light on how the Internet is reconfigured as a new site of cultural politics. Finally, it advocates anonymity as a constitutional free speech right and ascertains that anonymity in cyberspace contributes to, rather than impairs, the quality of public culture and democracy.

The cure for the ailments of democracy is more democracy (John Dewey, The Public and Its Problems, 1927, p. 240).

INTRODUCTION: SETTING THE CONTEXT

By many accounts, the Republic of Korea or South Korea (hereinafter, Korea) is considered to be one of the most connected countries in the world, with over 90 percent of households served by broadband Internet connection. According to the International Telecommunication Union's Information and Communication Technology (ICT) Development Index (IDI),[1] Korea was ranked third in the world as of 2008. Further, it was the only non-European country listed among the top ten countries (ITU, 2010). Korea has witnessed the emergence of new forms of civic, alternative media and online democracy over the past two decades

DOI: 10.4018/978-1-60960-591-9.ch017

as a result of its highly advanced Internet and telecommunication infrastructure. This has indeed helped the relatively young Korean democracy to further progress toward a more substantive democracy.[2] The efflorescence of participatory, democratic Internet culture has thus taken various manifestations such as free speech portals, public knowledge sites, and online citizen journalism (e.g., *OhmyNews.com,* see Cheon, 2004).

However, the arrival of the new conservative Lee Myung-Bak government in 2008 has posed a serious challenge to the continual experimentation of online deliberative democracy in Korea. The new government has been trying to suppress the radical potentials of the Internet by implementing a series of anti-democratic Internet policies, including Internet Real Name Policy (IRNP). This is probably the most resisted kind of policy by netizens in Korea and abroad. This much debated and dubious policy mandates that users of major websites – those that have more than 100,000 visits per day – should register with their legal names and national resident registration numbers[3] (Fitzpatrick, 2008). Not surprisingly, Reporters Without Borders (RWB) in its 2009 publication entitled *Internet Enemies*, listed Korea as a democracy under surveillance because of its "some disproportionate measures to regulate the Net" (p. 2). The same organization's international rankings of press freedom for Korea have dropped sharply from the 39th position in 2007 to the 47th and 69th positions in 2008 and 2009, respectively (Press Freedom Index, 2007, 2008, and 2009). Notwithstanding its world leader status in Internet penetration, the OpenNet Initiative (2007) observes, "[Korean] citizens do not have access to a free and unfiltered Internet. The state imposes a substantial level of filtering for a free and democratic society. It requires ISPs to block sites on government lists and fosters a culture of self-censorship…." via ambiguous and questionable laws (p. 4).

Under the conservative political regime, Korean Internet is now reduced to a narrowly defined and politically sanitized space of daily entertainment media only for celebrity/political gossip, gaming, dating, social networking, gambling, etc., while it is regulated exclusively from an economic point of view as a new engine for elevating Korea as a global information technology powerhouse in the post-Fordist global economy. In fact, this alleged "compressed economic development model" had been used as an ideological and moral justification for the suppression of citizen resistance and democratic values during Korea's rapid, yet questionable economic development after the Korean War (Nam, 2008). Consequently, Internet regulation and policy are now being reconfigured into two noticeable directions. On the one hand, legal and economic restrictions are being loosened for the sake of telecommunication market deregulation. Alternatively, on the other hand Internet content regulation is being tightened to control democratic participation and civic resistance. Currently, the regulatory framework and language overtly attempt to separate between the political, the cultural, and the economic parameters. Thus, the Korean citizens' identities are increasingly confined to consuming subjects, while not being allowed to use the Internet as a political space to act against the present government's anti-democratic impulses.

Appalled by this unjustifiable retreat from democracy, citizens and media activist groups have expressed strong opposition to the current administration's Internet policy. Undeniably, the Internet has become a new site of cultural politics and hegemonic struggle among the various governmental, regulatory agencies, civic organizations, netizens, and even domestic and global Internet powerhouses such as Google and YouTube Korea. It is within this context that this chapter aims to present a critique of the Lee government's anti-democratic and shortsighted Internet regulation and policy. In so doing, it seeks to critically examine how the Internet is regulated as a politically sanitized public space in Korea, and to uncover the actual anti-democratic consequences of repressive Internet content regulation. The chapter starts out by discussing certain main theoretical debates over the ways by which new media, like the Internet, help or impede democracy. The chapter weaves together a host of illustrative cases that have caused a great deal of public debates in recent times. Thus, it seeks to elucidate how free speech restrictions placed on the Internet help forge a politically sanitized cyberspace and undermine the democratic foundations of virtual public spheres. In addition, the article sheds light on the notion of anonymity, defending it not only as a constitutional free speech right, but also as a discursive democratic practice. The main argument favoring anonymity on the Internet is that it essentially helps to enhance, rather than degrade, the quality of public culture and democracy.

Therefore, certain main issues and questions for the chapter include: What roles does the Internet play in promoting or hindering civic participation, public culture, and democracy? What are the political and ideological implications of the Internet for relatively new democracies like Korea, where traditional media institutions are largely conservative and anti-democratic? Why and how are the current administration's new Internet policy and regulations anti-democratic? What forms and directions do hegemonic struggles take among the Korean government, domestic, global Internet powerhouses, Internet users, and civic organizations?

THEORETICAL BACKGROUND: INTERNET AND DEMOCRACY

Certainly, there is a lack of consensus on the nature of the relationship between new media like the Internet and democracy. Thus, it is often presented in the form of a theoretical fissure between a rosy depiction of Internet-driven participatory democracy and a rather pessimistic view of declining communitarian democracy. The main debate essentially pins down to the question of "whether the new media are a force for fragmentation or cohesion in society" (McQuail, 2005, p. 140). The use of computer networking and online civic participation for overcoming obstacles in the way of democracy has been discussed widely. Newly coined terms such as electronic democracy (or e-democracy), online democracy, and teledemocracy point to the theoretical postulation that relates the democratic potential of the Internet and other electronic media to the actual practice of democracy. Alternatively, some social critics focus on the proliferation of new media, including the Internet, which ironically erodes the democratic foundation of the society. According to them (e.g., Putnam, 2000; Sunstein, 2001), an advancement in the communication technologies does not necessarily correlate to an advancement in democracy. Instead, it could pose a grave threat to the very notion of democracy which requires an exchange of different ideas and voluntary participation in various communities.

For people who emphasize the democratic potential of the Internet, the Internet serves as a corrective force to the problems of an elite-controlled and sluggish democracy where citizen participation is limited and the means of public communication are controlled by a few heavyweight media oligopolists. One of the most influential early proponents of virtual community and its radically different ways of organizing human interaction and community formation would be Howard Rheingold. Throughout his work (1993, 2002), he noted that the Internet and the virtual community defy the status quo because of the decentralized structure of networking and that they have a real potential to augment collective intelligence and ultimately realize a new form of social life and democracy. McQuail (2005) also presents a similar observation that online users are collectively building their own norms and behavioral rules (i.e., netiquette). Further, it was also observed that "online communities have the added advantage of being in principle open and accessible." (p. 149).

The Internet's contribution to democracy is often theoretically validated in light of Harbermas's conception of public sphere. Habermas (1991) developed the normative notion of the public sphere where citizens could exchange views on matters of importance to the common good, so that public opinion could be formed. Central to his understanding of the significance of the public sphere lies in the process of public opinion formation, which must take the form of rational critical debate.[4] For instance, Poster (2001) attempts to find a possibility of reshuffling the conventional political order in the wake of expanding the cyberspace. In an attempt to avoid the baseless utopian hype that has characterized much of the early public debate concerning the future of cyberdemocracy, Poster delineates what justly distinguishes the Internet from other preceding forms of media and then updates Habermas's public sphere. He notes that, whereas Habermasian public sphere is "a homogeneous space of embodied subjects in symmetrical relations, pursuing consensus through the critique of arguments and the presentation of validity claims" (pp. 181-182), cyberspace is essentially marked by its heterogeneity. Even though Poster (2001) admits that both bureaucratic and commercial forces may threaten the Internet's transformative capacity, it is still capable of realizing the postmodern democratic politics. Thus, the Internet can bring about the diminution of hierarchy in authority, gender, race, and class. Accordingly, cyberspace has been conceived of as a space in which all individuals, including traditionally marginalized groups strive to expand the public sphere by multiplying, opening, and extending political discussion. Similarly, Saco (2002) expands an insight for asserting that cyberspace must be viewed as a "produced social space," that successfully confounds the ordering conventions of our physical spaces. Within this particular framework, she stresses upon the significance of the Internet as a social space in developing civic-mindedness and the politics of collective action. From a journalistic perspective, Sifry (2009) analyzes how the Internet opens up a possibility of new participatory democracy, observing that "[t]he low-cost, high-speed, always-on Internet is changing the ecology of how people consume and create political information" (p. 43).

Unquestionably, there is considerable empirical evidence for championing the Internet as both a democratic and a democratizing force. For example, Roh Moo-Hyun's election as the president of Korea in 2002 was principally credited to the power of the Internet in mobilizing net generation's participation in politics, thereby increasing the actual voting turnouts. Accordingly, he was dubbed "the world's first Internet president" (Watts, 2003). It is also true that the Internet has helped serve as an alternative source of information, and has aided independent and often radical journalism in authoritarian states such as Singapore and Malaysia (George, 2003). Further, the Internet's role in helping organize new forms of social movements on a global scale has also been observed (Bennett, 2003; Gill, 2000). Citizen journalism and social networking sites like Twitter and Facebook have recently vividly exemplified how the democratic potential of new media can be materialized by effectively disseminating real-time news, stories, pictures, and even videos of the evolving revolt against the Iranian government over the election fraud. In reality, these empirical cases attest to the fact that the Internet can challenge the sycophantic roles of traditional media in emerging democracies.

However, despite the democratic potential in realizing a two-way, deliberative, and participatory democracy, the Internet's actual contribution may be easily refuted by certain stark realities such as fragmented public opinion, prevailing cyber crimes, and ruthless mercantilists bombarding consumers with unsolicited promotional messages. Regarding these anti-democratic propensities and effects of the Internet, at least three different criticisms can be made. Firstly, Sunstein (2001) argues that the Internet and other types of hyper-personal media tend to retribalize the society, rather than revitalize democracy. According to him, the Internet has a polarizing effect on the communal life and democracy at large. The traditional forms of mass media help generate a certain level of shared culture and consensus on social issues. However, the Internet tends to create a socio-political condition in which a highly personalized use of the medium leads to sharp divisions of individuals and social groups on the basis of race, gender, class, religion, and most importantly, political orientation. These divisions themselves pose a serious threat to democracy, when viewed from his communitarian perspective. Sunstein (2001) conceives democracy as "[requiring] both a range of common experiences and unanticipated, unchosen exposures to diverse topics and ideas" (p. 205). Thus, democracy could be in serious danger by an egocentric medium, the Internet, in which each individual possesses the power to filter what he or she wants to use. In this scenario, social fragmentation and mutual misunderstanding are likely to prevail on the Internet.

Secondly, from a critical political economic perspective, the Internet largely helps to create and maintain a digital marketplace, connecting systems of capitalist economic production and distribution without undermining the unequal economic and political power relations (Schiller, 2000). McChesney (1999) and Chester (2007) also observe that the new media have become a significant anti-democratic force in the United States, instead of providing a foundation for freedom and democracy. According to them, the ongoing media mergers and anti-regulatory environment in the media and the telecommunications industries prove that the main beneficiaries of the information revolution are a handful of global media giants. Furthermore, contradictory to the theorization of the Internet as a qualitatively different participatory medium, the actual Internet usage may easily negate such hypothesis. For example, most users tend to remain only spectators and do not actively participate in content creation. Only a small fraction of Internet users contribute content and approximately 1% of users account for all Internet action (Nielsen, 2006).

Thirdly, many critics note that the Internet has become more like a panopticon. The online activities of individuals are increasingly carefully monitored via various advanced, but often unapproved surveillance technologies by both the government and the corporation. Earlier, Gandy (1993) called this the transformation of personal information and privacy into a commodity. His acute criticism might be no exaggeration, when considering the fact that the majority of websites today collect personal data without the informed consent of the users. In addition, even routine search results can be stored and retrieved by the myriad of data mining companies. Further, in the aftermath of the 9/11 attacks, the government has been given a series of anti-democratic, unconstitutional rights to monitor the individuals' Internet activities (Kellner, 2004). Panopticon refers to an architectural design devised by Jeremy Bentham aimed at creating a sense of being watched all the time. This new social control technique becomes most effective when the watched internalizes the sense of constant monitoring by the watcher. Foucault's acute insight is quite revealing: "Hence the major effect of the Panopticon: to induce in the inmate a state of conscious and permanent visibility that assures the automatic functioning of power" (Foucault, 1995, p 201).

Having reviewed certain main debates over the Internet's contribution to democracy, it should be noted that a communication technology's role ought to be understood in a context that is conditioned, if not determined, by a mixture of socio-political constraints of the given society. Thus, the degree to which the Internet realizes its democratic potential will greatly depend upon the institutional or regulatory arrangements that fashion both the structure and content of the Internet. Looking back at the history of media and communication technologies, no new medium of that time proved to be democratic in itself. Nor did a utopian society envisioned at the dawn of the new medium take place. Therefore, the question whether the Internet in itself is good or bad for democracy is neither productive nor intuitive. The rightful questions to ask are how, to what extent, and under what conditions the Internet can contribute to democracy. Hence, there arises the need for a non-deterministic, context-specific, and cultural politics-centered approach in exploring the intersections and tensions between the Korean Internet and the online democracy.

In this context, Raymond Williams's (1990) cultural materialist view of technology-society nexus and Brian Winston's (1998) intuitive notions of "supervening social necessity" and "suppression of radical potentials of a new technology" could prove to be useful. Williams (1990) asserts that technology is best understood as an intervention into the relationship between the human agency and a fabric of social, cultural, and political systems. Similarly, Hill (1989) notes that "technological change... is not, by itself, productive of social change. Instead, the direction of change is a product of the *particular alignment* between the technological possibilities and the society and culture that exists" (p. 33, emphases added). According to Winston (1998), any new invention of the communication technology arises out of specific social needs (e.g., cable for better reception of broadcast signals and a prototype of the Internet for military readiness) aimed at solving existing problems. Thus, the adoption of an invention depends largely on the degree to which the technology meets the social need, whether it is technological, economic, political, or cultural. On the other hand, the "law of suppression" indicates that whenever a new technology poses threats to the existing social order, technological standard, or cultural value, it tends to be suppressed by various legal, technological, discursive, and economic means. Consequently, an extent to which a certain new communication technology is adopted by society is determined as a result of an interaction between the "historical brakes" (law of suppression) and the "accelerators" (social necessities) (pp. 5-9).

Seen from this theoretical perspective, it is evident that as new forms of civic participation and radical journalism on the Internet pose threats to the status quo, various legal, economic, and moral measures are introduced by authoritarian states to suppress and tame the radical potential of the Internet. For emerging democracies, the Internet is now reconfigured within the logic of economic globalization and telecommunication liberalization. However, it is prevented from fully developing its radical democratic potentials. For example, the Chinese government overtly justifies various measures of Internet content regulation or even censorship. While the Chinese economy enjoys its continual, phenomenal growth, the Internet as a means of free speech and democratic participation is facing a troublesome future in the face of tight government controls. Specifically, the Chinese propaganda department has mandated that "no website has the right to gather news on its own account; websites can only publish news collected by the tightly controlled traditional media. Even reports from some traditional media outlets are not allowed to be reproduced online" (Li, 2008). Furthermore, advanced economies like Canada also regulate the Internet content, albeit in a more subtle way. Fraser (2007) examines how the Canadian Internet policy functions as "civilizing discourse," "creating model citizens for the information age" (p. 201). These examples clearly illustrate what Winston (1998) theorized with his own original concept of the suppression of the radical potential of a new communication technology.

INTERNET CONTENT REGULATION IN KOREA: CASES AND MEASURES

As discussed in the previous section, neither a utopian nor a dystopian view is adequate in examining the Internet's volatile contribution to the actual democracy at work. A context-specific, cultural politics-based approach is required to examine the various intersections between new media and social change from a non-deterministic perspective. Within this theoretical framework, it may be useful to present a brief history of alternative, citizen-driven media in the context of Korea's democratic transition that started in the late 1980s.

The turbulent history of democracy in Korea was never easy, and it was not until 1987 that Korea began a democratic transition. During the several decades of military rule, Korean media were locked in the state-led economic development model. Thus, their main goal was the advancement of the national economic interests and propagating the government's developmental economic doctrines, while effectively preventing dissident voices from being developed. "In return for their loyalty, the military government provided preferential treatment to the oligopoly of established media companies, which played a critical role in supporting the authoritarian rule" (Shim, 2002, p. 338). Regarding this treacherous state-media alliance, the late 1980s witnessed the emergence of a media reform movement in the form of a nation-wide campaign to boycott the television reception fee. Angered by the obedient and sycophantic content of the government-controlled Korea Broadcasting System (KBS), many concerned citizens, civic activist groups, and religious organizations organized a series of public campaigns and large-scale street demonstrations (Kim, 2001).

After the victory of the June Protest in1987, which forced the then president Chun Doo-Whan to accept the direct election of a president, the media reform movement gained a real momentum, leading to some significant fruition. It would be almost impossible to talk about Korean democracy without highlighting the birth of a citizen-funded national daily newspaper, *Hankyoreh*, which roughly translates into English as "one nation." Many journalists were previously imprisoned and expelled from their jobs for their dissident reporting and commentary. These people united to experiment with an alternative news media, which was free from both the state and the market in the newly created and expanded democratic space. More than 3,000 people including public intellectuals, democratic movement leaders, and progressive religious leaders joined hands and devised a blueprint for a commercial influence-free, truly independent national newspaper for expediting democratic transition and deepening Korean democracy. Citizens across the country were equally enthusiastic in supporting this ambitious vision and in helping to get the daily started (Cheon, 2004). The idea of a citizen-funded, independent daily came to fruition in May 1988. Its distribution covered half-million copies of the first publication nationwide. Through the 1990s, the paper grew as one of the most influential national newspapers. However, after it decided to print advertisements, its left-leaning, progressive editorial tone has become muted, in part. Nevertheless, this unique historical experimentation clearly shows how civil society is capable of not only undermining the media status quo, but also more importantly helping to realize an alternative news media.

This kind of democratic experiment never halted in Korea. Consequently, it resulted in the much talked about citizen-driven online journalism, *OhmyNews.com* in 2000 (Chang, 2005). Having the motto of "the citizen is the journalist," its news stories are mostly supplied by citizen journalists who come from various backgrounds and do not have much formal journalism education, or any previous actual journalistic experience. This was a first of its kind in the world in the history of journalism. After the phenomenal success of this alternative journalism model, several similar news websites, albeit with different political orientations, mushroomed in Korea. The explicit political nature of the Korean Internet distinguishes it from that of other countries. However, it is also true that the Internet, largely owing to the enormous popularity of online gaming across generations, has probably become the most influential and lucrative entertainment industry in Korea. Not only do politicians, government officials, educators, and public figures maintain their web presence, but they also interact with citizens on their shared interests and matters of importance. In fact, organizing and mobilizing the civil movement by virtue of the Internet has become a natural part of Korean democracy in the 21st century.[5]

The arrival of the new right-wing, conservative Lee Myung-Bak administration in 2008 signaled a considerable retreat from the Internet-driven participatory democracy in many regards. Unlike his predecessor, Roh Moo-Hyun, the current president Lee Myung-Bak was predominantly supported by the older generations, and he often demonstrates his overt disdain for dissident public opinion on the Internet. As a matter of fact, he promptly implemented strong and often anti-democratic media policy and regulations to help fashion his own vision of politically clean and economically competent media environment, including the creation of the consolidated regulatory body (Korea Communications Commission, KCC) to oversee all media and telecommunications sectors, and the appointment of his long-time political advisor, Choi See-Jung to the chair of the commission. As mentioned above, the Internet Real Name Policy (IRNP) would probably be one of the most blatant attempts to control and tame the Internet users' free expression and exchange of ideas, especially when they aim to criticize the current government's dubious economic and cultural policies. It must be noted that the verification of one's identity online can be traced back as early as 2001 during the Kim Dae-Jung administration. However, a full-scale, strict legal verification of one's identity was never implemented until recently. The new IRNP that was passed in April 2009 mandates that every popular Internet portal (e.g., Yahoo.co.kr) and government site (e.g., president.go.kr) having more than 100,000 daily visitors should verify the user's identity by his/her legal name and resident registration number (South Korean YouTube users, 2009). Of course, this caused an elevated level of discontent and resentment among the netizens, civil liberty groups, and web companies including global IT powerhouses like Google. Initially, YouTube Korea showed explicit opposition to the new regulation. However, they later decided to comply with the IRNP in the fear of losing one of the most lucrative Internet markets in the world.

Under the Lee government, though the Internet content is increasingly under heavier regulation, it remains as the main economic engine for Korea as a global IT powerhouse. The three recent cases may suffice in unearthing the grounds and the hidden political purposes for which the Internet content is increasingly exposed to anti-democratic surveillance. Firstly, during the course of media reform movements, the citizens targeted a right-wing, ultra-conservative daily, *Chosun Ilbo* (or *Chosun* Daily) that previously served as the military dictators' propaganda machine and continued to help manufacture conservative public opinion today. The citizens organized both online and offline anti-*Chosun* movements, using their freedom of speech, assembly, and association guaranteed by the Korean Constitution. As part of their creative tactics, they demanded that major advertisers like Samsung and Hyundai stop advertising on *Chosun Ilbo* and spreaded their campaign messages widely on the Internet. Shockingly, a considerable number of people got arrested for their so-called illegal, disruptive behavior that was thought to damage the financial foundation of the daily. Knowing that the daily played a large role in helping Lee to get elected as the president, his government did not take the Lawyers for a Democratic Society's warning seriously: "Boycotting should be allowed as it is based on consumer rights and freedom of speech, which are guaranteed by the Constitution of Korea" (Kim, 2008).

Secondly, the Minerva case was probably one of the most vivid examples of anti-democratic Internet content regulation in Korea. Minerva was a pseudonym for a self-educated economic analyst who periodically published his unconventional, witty writings on major Korean portals and bulletin boards from 2008. Though his economic forecasts were rather gloomy, they still accurately predicted the impending meltdown of the global financial markets and the housing crisis. For example, his writing predicted the collapse of Lehman Brothers five days before it was reported by the mainstream U.S. media, recording more than 40 million hits in a week (Kirk, 2009). When Korea faced a currency crisis in 2009 reminding its people of the 1997 International Monetary Fund (IMF) crisis, President Lee and the government officials publicly criticized Minerva for spreading false information about the health of the Korean economy and the level of foreign debts. Subsequently, he was arrested for harming national interests. Later he was proven innocent and released on the ground that his writings did not "intentionally harm the public interest" (Ramstad, p. 3). In a sense, Minerva's main crime was "that of embarrassing the government" by disseminating unofficial knowledge (Kirk, 2009). He attended a community college and was mostly self-educated in economics. President Lee's contempt for unofficial, online information is clear: "[W]e have to guard against 'infodemics,' a phenomenon in which inaccurate, false information is disseminated; prompting social unrest that spreads like an epidemic" (cited in Fitzpatrick, 2008).

Thirdly, another unfortunate case also warrants a critical evaluation. A nationally loved actress, Cho Jin-Sil (the Korean equivalent of Julia Roberts or Meg Ryan) committed suicide within a week after her fellow actor, Ahn Jae-Hwan killed himself because of his financial meltdown. After her suicide, the mainstream media, politicians, and many people almost immediately blamed the Internet for spreading rumors about her, especially of her being a loan shark and pressing Ahn to pay back a $2 million debt (Choe, 2008). In the aftermath of this accident, the Korean Congress was very quick in enacting a new policy called the Cyber Insult or the Cyber Defamation Law. A congressman, Hong Joon-Pyo said, "Internet space in our country has become the wall of a public toilet" (cited in Choe, 2008) and asked for stricter penalties against those who post malicious or defamatory comments on the Internet. This is not only problematic, but also unconstitutional as it can be used to prosecute a citizen for posting legitimate, investigative writings about public officials, corrupt politicians, and businesses. According to the law, even without the consent of the plaintiff, the law enforcement agencies can go ahead to delete the posted message and prosecute the person who posted the malicious message. Together, the above mentioned three cases illustrate clearly how the Internet content regulation, under the current conservative government is reconfigured as a way to suppress the dissident voices on the Internet.

What these three cases illuminate and what they imply for Korean democracy is no small matter. The current administration is making attempts to transform the Internet into a political panopticon, wherein dissident voices and anti-government citizen mobilizations are under constant surveillance. The Internet is progressively becoming a politically sanitized space in which the citizens are increasingly internalizing the fear of monitoring. Thus, they are voluntarily softening or forgoing their politically dissident voices. In this context, the notion of deliberative democracy is in danger. Although there is a wide array of theoretical approaches to deliberative democracy, one commonality is that democracy needs citizens who are well informed and willing to participate in the formal political process, as well as the daily public culture and democratic practices. As Mouffe (2000) argues, democracy develops through the contestation of different ideas, which requires free access to and the use of public communication networks like the Internet. If Korea is to intensify its democracy not only in terms of formal, institutional arrangements, but also in terms of public culture, identity politics, and citizen participation, the current administration's approach to Internet content regulation is by no means justifiable. Despite its possible negative effects, citizen participation via the Internet is increasingly important to the articulation of political beliefs, the enhancement of public deliberation, and ultimately, the practice of democratic citizenship. The Internet must remain free of government content regulation, not because free Internet automatically guarantees substantial democracy, but rather political content regulation could pose serious, empirical threats to the development of participatory, plural democracy.

IMPLICATIONS FOR DEMOCRACY: ANONYMITY AS A FREE SPEECH RIGHT

Having examined some of the most representative cases of Internet content regulation in Korea, this section proceeds to discuss the implications that anonymity has for democracy. However, it does not aim to delve into this question from a formal legal analysis or a techno-practical approach. Instead, the main focus is whether anonymity is a friend or foe of democracy at the normative or theoretical level. Nevertheless, it may still be useful to determine whether a mandate to reveal the user's identity for online communication by legal force is constitutional or not. Articles 17, 18, and 21 of the Korean Constitution[6] clearly guarantee the Korean citizens' freedom of speech including the freedom to remain anonymous. Korean Constitution Article 18 states, "The privacy of correspondence of no citizen shall be infringed." Article 17 also advocates anonymity in light of privacy, affirming that "the privacy of no citizen shall be infringed." Further, Article 21 states that "all citizens shall enjoy freedom of speech and the press, and freedom of assembly and association. Licensing or censorship of speech and the press, and licensing of assembly and association shall not be recognized."

Several important court cases in the United States have well documented anonymity as a free speech right or a basic human right. Although the right to speak and receive information anonymously can be breached by a justifiable legal procedure, anonymity is considered an essential free speech right protected by the First Amendment. The basis for this reasoning was: "[I]dentification and fear of reprisal might deter perfectly peaceful discussions of public matters of importance" (*Talley v. California,* 1960, p. 65) and "the decision in favor of anonymity may be motivated by fear of economic or official retaliation, by concern about social ostracism or merely by a desire to preserve as much of one's privacy as possible" (*McIntyre v. Ohio Elections Commission,* 1995, pp. 341-342). Regardless of the actual motivation, the U.S. Supreme Court holds that "anonymity is a shield from the tyranny of the majority ... to protect unpopular individuals from retaliation – and their ideas from suppression – at the hand of an intolerant society" (*McIntyre v. Ohio Elections Commission,* 1995, p. 357). This line of reasoning is well in fact in line with Mouffe's (2000) conception of radical, pluralist democracy where democracy is not defined in terms of the ruling by the majority, but by the contestation of differing ideas, opinions, and ideologies.

However, Sunstein (2001) rightfully noted, that the Internet,, can promote free speech by hate groups in isolated enclaves. Accordingly, he proposed a policy reform to correct the misuse of the Internet as a ruthless free speech outlet. Besides Sunstein's well-reasoned warning, negative externalities of free, anonymous exchange of ideas undoubtedly exist on the Internet. Such phenomena are named as trolls, tipsters, sock puppets, cyber stalking, cyber bullying, and cyber defamation. Consequently, the array of such cases served as the Lee government's raison d'être for implementing stricter and more enforceable legal measures including the IRNP. However, the question to ask is: Would restricting anonymous comments reduce malicious cyber behaviors? Although this may look like a legitimate question at first, it tends to divert attention from the essential point and is a mere knee-jerk response to the bigger social problems. Even from a comparative perspective, it can be simply argued that the level of vitriolic cyber behavior is not particularly devastating in Korea.

Subsequently, a more important and fundamental question to ask is: What are the effects that restrictions on anonymity have on free speech and democracy at large? Akdeniz (2002) presents an excellent overview of why and how anonymity serves larger goals of democracy in the society. Essentially it helps to create a public sphere where ideas are evaluated on merit, and not on social castes (Weicher, 2007). This is in line with Habermas's envisioning of an ideal public sphere where the process and outcomes of democratic deliberation are not influenced by hierarchy along the lines of class, gender, race, and age, but instead by merit of the argument. The Internet can help to realize a new form of public sphere that can help deepen "democracy through open, reciprocal communication within liberatory praxis" (Green, 1999, p. xi). From this viewpoint, one can then expand the Habermasian public sphere "as a range of institutionalized processes of public communication" (Barnett, 2003, p. 81) rather than an actual physical site of social relations. The notion of a virtual public sphere should be actively sought after recognizing the legitimacies of the multiple modes of communication and identities. Anonymity is also theoretically justified from a deliberative or discursive democracy perspective. As Dryzek (2002) analyzed empirically, deliberation across the social class, race, and gender is possible via the means of electronic networks and can produce positive effects on democracy as a whole. Balkin (2004) extends the democratic potential of free speech on the Internet by elaborating on the Internet's contribution to the development of democratic culture. According to him, political expression and dissent via new communication technologies, such as social networking sites, blogs, and forums have become essential in realizing both political and cultural democracy: "Freedom is participation. Freedom is distribution. Freedom is interaction.... Dissent is central to this conception of free speech, for dissent is cultural as well as political" (p. 44). According to him, dissent should involve all the features of liberty including anonymity.

The notion of anonymity has also been studied in the organizational context (Kahai, 2009). According to Kahai's empirical study, anonymity does not necessarily promote anti-social behavior. Overall, it can be rightfully argued that the IRNP is a panoptic political scheme aimed at policing the questioning mind. The notion of deliberative democracy would be most severely affected by the authoritarian state's tighter Internet content regulation (Dahlberg, 2001). The mentality "If you are innocent, why would you try to conceal your identity?" in essence considers netizens to be potential criminals. The same logic must be applied to the government: If they do not plan to suppress the democratic formation of public opinion and new forms of dissent, why do they specifically target some of the most visited portals and electronic forums?

The democratic benefits of free speech including the freedom to remain anonymous far outweigh its negative externalities. The problems of online democracy should not be fixed by reversing the wheel of history, to a time when there was no Internet. Even on a practical level, the IRNP is neither effective nor efficient. Even if users do something online without revealing their legal names, they can still be traced because the law enforcement authorities can easily follow their "digital bread crumbs" (Anderson, 2009) In fact, today it is almost impossible to remain completely anonymous because of the advancements (or retreats) made in the online surveillance technologies. It should also be noted that a legitimate place for government intervention and regulation, such as fostering an environment for fair market competition, extending universal service, and closing the digital divide certainly exist. Even freedom of speech requires progressive government efforts to extend the technological access and democratize the means of public communication. As there are no visible signs of the Lee government's interests and concerted efforts in these important matters, its strong Internet content regulation drive remains politically suspicious and unwarranted.

CONCLUSION

Today Korea faces a retreat from democracy. If the current spate of anti-democratic Internet content regulation continues, online citizen journalism and alternative media would be the greatest victims. In addition, the new forms of civic participation would in most likelihood lose their vitality. As discussed, the Lee government's new legal and policy initiatives designed for a tighter Internet content regulation are not only unconstitutional, but also against human rights. Accordingly this article questioned and problematized the recent spate of Internet content regulation in Korea as an insidious political operation, and not a natural moral response to the vitriolic cyber crimes. More specifically, the IRNP creates an invisible system of surveillance and social control to tame resistive political debates and discourses on the Internet and ultimately aims at creating docile, obedient netizens. It has been demonstrated that the de facto Internet censorship is simply not compatible with democracy.

Consequently, it was argued that freedom of speech must be protected on the Internet, including the freedom not to reveal one's identity by force. Regardless of certain negative remnants of anonymous online behaviors, anonymity as a free speech right needs to be valued and protected because of its contribution in fostering democratic public spheres. Unquestionably, anonymity in itself is not the goal of democracy. Nevertheless, it is important as it can contribute significantly to creating a space where diverse, and at times, confronting ideas are exchanged without the fear of reprisal and threats to enhance the quality of democratic deliberation and its outcomes. Thus, anonymity should not be confined to a narrowly defined privacy issue. Instead it must be seen as a necessary condition for free expression and democratic public culture. As the Lee government continues to control the flow of ideas on the Internet, the netizens' voluntary, collective resistance against it to realize the real World "Wide" Web of free speech and democratic culture in Korea and beyond also continues.

REFERENCES

Akdeniz, Y. (2002). Anonymity, democracy, and cyberspace. *Social Research*, *69*, 223–237.

Anderson, N. (2009, March 31). Pulling back the curtain on anonymous Twitters. *Law & Disorder: Tech law and policy in the digital age.* Retrieved February 19, 2010 from http://arstechnica.com/ tech-policy/ news/ 2009/ 03/ pulling-back-the-curtain-on-anonymous-twitterers.ars

Balkin, J. M. (2004). Digital speech and democratic culture: A theory of freedom of expression for the information society. *New York University Law Review*, *79*(1), 1–58.

Barnett, C. (2003). *Culture and democracy: Media, space, and representation*. Tuscaloosa, AL: University of Alabama Press.

Bennett, W. L. (2003). New media power: The Internet and global activism. In Couldry, N., & Curran, J. (Eds.), *Contesting media power: Alternative media in a networked world* (pp. 17–37). Lanham, MD: Rowman & Littlefield.

Chang, W.-Y. (2005). Online civic participation, and political empowerment: Online media and public opinion formation in Korea. *Media Culture & Society*, *27*(6), 925–935.

Cheon, Y.-C. (2004). Internet newspapers as alternative media: OhmyNews in South Korea. *Media Development*, *51*(1), 28–33.

Chester, J. (2007). *Digital destiny: New media and the future of democracy*. New York, NY: The New Press.

Choe, S.-H. (2008, October 13). Korean star's suicide reignites debate on Web regulation. *The New York Times*.

Dahlberg, L. (2001). Democracy via cyberspace. *New Media & Society*, *3*, 157–177.

Dryzek, J. S. (2002). *Deliberative democracy and beyond: Liberals, critics, contestations*. New York, NY: Oxford University Press.

Fitzpatrick, M. (2008, October 9). South Korea wants to gag the noisy Internet rabble: The leaders of the most wired country on Earth are seeking to curb online anonymity and debate, with laws that many say will fail. *The Guardian*, p. 6.

Foucault, M. (1995). *Discipline & punish: The birth of the prison* (Sheridan, A., Trans.). New York, NY: Vintage Books.

Fraser, N. (2007). Creating model citizens for the information age: Canadian Internet policy as civilizing discourse. *Canadian Journal of Communication*, *32*, 201–218.

Gandy, O. H. Jr. (1993). *The panoptic sort: A political economy of personal information*. Boulder, CO: Westview.

George, C. (2003). The Internet and the narrow tailoring dilemma for Asian democracies. *Communication Review*, *6*, 247–268.

Gill, S. (2000). Toward a postmodern prince? The battle in Seattle as a moment in the new politics of globalization. *Millennium: Journal of International Studies*, *29*(1), 131–140.

Green, J. (1999). *Deep democracy: Community, diversity, and transformation*. New York, NY: Rowman & Littlefield.

Habermas, J. (1991). *The structural transformation of the public sphere: An inquiry into a category of bourgeois society* (Burger, T., Trans.). Cambridge, MA: MIT Press.

Hill, S. (1988). *The tragedy of technology: Human liberation versus domination in the late twentieth century*. London, UK: Pluto Press.

International Telecommunication Union. (2010). *Measuring the information society 2010*. Retrieved January 25, 2010, from http://www.itu.int/ ITU-D/ ict/ publications/ idi/ 2010/ Material/ MIS_2010_without %20annex %204-e.pdf

Internet Enemies. (2009). *Reporters without borders*. Retrieved February 27, 2010, from http://www.rsf.org/ IMG/ pdf/ Internet_enemies_2009_2_.pdf

Kahai, S. S. (2009). Anonymity and counternormative arguments in computer-mediated discussions. *Group & Organization Management*, *34*, 449–478.

Kellner, D. (2004). The media and the crisis of democracy in the age of Bush-2. *Communication and Critical. Cultural Studies*, *1*(1), 29–58.

Kim, J.-E. (2001). *Journalists in peril: The Korean case*. People's Coalition for Media Reform. Retrieved February 27, 2010, from http://www.pcmr.or.kr/ english/9.html

Kim, S. (2000). *The politics of democratization in Korea: The role of civil society*. Pittsburgh, PA: University of Pittsburgh Press.

Kim, S.-H. (2008, June 24). Hundreds admit to pushing boycott. *Korea Herald*.

Kirk, D. (2009, January 27). Financial blogger's arrest tests Korea's progress on human rights. *The Christian Science Monitor*. Retrieved February 23, 2010, from www.csmonitor.com/ World/ Asia-South.../ p04s02-wosc.html

Lee, Y. (2008, October 26). S. Korean actress' suicide fuels cyber-insult debate. *Los Angeles Times.* Retrieved February 21, 2010, from http://articles.latimes.com/ 2008/ oct/26/ world/ fg-net26

Li, D. (2008, July 8). China's leaders, the media and the Internet. *Open Democracy: Free Thinking for the World.* Retrieved February 28, 2010, from http://www.opendemocracy.net/ article/ china-s-leaders-and-the-internet

McChesney, R. (1999). *Rich media, poor democracy: Communication politics in dubious times.* Urbana, IL: University of Illinois Press.

McIntyre v. Ohio Elections Commission, 514 U.S. 334 (1995).

McQuail, D. (2005). *Mass communication theory* (5th ed.). Thousand Oaks, CA: Sage.

Mouffe, C. (2000). *The democratic paradox.* New York, NY: Verso.

Nam, S. (2008). The politics of compressed development in new media: A history of Korean cable television, 1992-2005. *Media Culture & Society, 30*(5), 641–661.

Nielsen, J. (2006). *Participation inequality: Encouraging more users to contribute.* Retrieved February 21, 2010, from http://www.useit.com/ alertbox/ participation_inequality.html

OpenNet Initiative. (2007). *South Korea.* Retrieved from February 27, 2010, from http://opennet.net/ sites/ opennet.net/ files/ south_korea.pdf

Poster, M. (2001). *What's the matter with the Internet?*Minneapolis, MN: University of Minnesota Press.

Press Freedom Index. (n. d.). *Reporters without borders for press freedom.* Retrieved February 23, 2010, from http://www.rsf.org/ en-classement1003-2009.html

Putnam, R. (2000). *Bowling alone: The collapse and revival of American community.* New York, NY: Simon & Schuster.

Ramstad, E. (2009, April 21). Leading the news: Korean judge finds blogger not guilty. *The Wall Street Journal Asia,* p. 3.

Rheingold, H. (1993). *The virtual community: Homesteading on the electronic frontier.* Reading, MA: Addison-Wesley.

Rheingold, H. (2002). *Smart mobs: The next social revolution.* New York, NY: Perseus.

Saco, D. (2002). *Cybering democracy: Public space and the Internet.* Minneapolis, MN: University of Minnesota Press.

Schiller, D. (2000). *Digital capitalism: Networking the global market system.* Cambridge, MA: MIT Press.

Shim, D. (2002). South Korean media industry in the 1990s and the economic crisis. *Prometheus, 20,* 337–350.

Sifry, M. L. (2009). A see-through society: How the Web is opening up our democracy. *Columbia Journalism Review, 47*(5), 43–47.

Stein, L. (2008). Speech without rights: The status of public space on the Internet. *Communication Review, 11,* 1–23.

Sunstein, C. (2001). *Republic.com.* Princeton, NJ: Princeton University Press.

Talley v. California, 362 U.S. 60 (1960).

Watts, J. (2003, February 24). World's first Internet president logs on: Web already shaping policy of new South Korean leader. *Guardian.* Retrieved February 27, 2010, from http://www.guardian.co.uk/ technology/ 2003/ feb/ 24/ newmedia.koreanews

Weicher, M. (2007). [Name withheld]: Anonymity and its implications. *Proceedings of the American Society for Information Science and Technology*, *43*, 1–11.

Williams, R. (1990). *Television: Technology and cultural form*. New York, NY: Routledge.

Winston, B. (1998). *Media technology and society: A history: From the telegraph to the Internet*. New York, NY: Routledge.

ADDITIONAL READING

Anderson, D. M., & Cornfield, M. (Eds.). (2003). *The civic web: Online politics and democratic values*. Lanham, MD: Rowman & Littlefield.

Balkin, J. M., Grimmelmann, J., Katz, E., Kozlovski, N., Wagman, S., & Zarsky, T. (Eds.). (2007). *Cybercrime: Digital cops in a networked environment*. New York: New York University Press.

Bennett, W. L., & Entman, R. (Eds.). (2001). *Mediated politics in the future of democracy*. New York: Cambridge University Press.

Calhoun, C. (Ed.). (1993). *Habermas and the public sphere*. Boston: MIT Press.

Castells, M. (1996). *The rise of the network society*. Malden, MA: Blackwell.

Colby, D. (2005). Toward a new media autonomy. *Communication Law and Policy*, *10*(4), 433–476.

Dahlberg, L. (2007). The Internet, deliberative democracy, and power: Radicalizing the public sphere. *International Journal of Media and Cultural Politics*, *3*(1), 47–64.

Dahlgren, P. (2005). The Internet, public spheres, and political communication: Dispersion and deliberation. *Political Communication*, *22*, 147–162.

DeNardis, L. (2009). *Protocol politics: The globalization of Internet governance*. Boston: MIT Press.

Downing, J. (2001). *Radical media: Rebellious communication and social movements*. Thousand Oaks, CA: Sage.

Eko, L. (2001). Many spiders, one Worldwide Web: Towards a typology of Internet regulation. *Communication Law and Policy*, *6*(3), 445–484.

Fish, E. (2009). Is Internet censorship compatible with democracy? Legal restrictions of online speech in South Korea. *Asia Pacific Journal on Human Rights and the Law*, *10*, 43–96.

George, C. (2003). The Internet and the narrow tailoring dilemma for Asian democracies. *Communication Review*, *6*, 247–268.

Gomez, J. (2004). Dumbing down democracy: Trends in Internet regulation, surveillance and control in Asia. *Pacific Journalism Review*, *10*(2), 130–150.

Hacker, K. L., & van Dijk, J. (Eds.). (2000). *Digital democracy: Issues of theory and practice*. Thousand Oaks, CA: Sage.

Kalathil, S., & Boas, T. C. (2003). *Open networks, closed regimes: The impact of the Internet on authoritarian rule*. New York: Carnegie Endowment for International Peace.

Kluver, R., & Banerjee, I. (2005). Political culture, regulation, and democratization: The Internet in nine Asian nations. *Information Communication and Society*, *8*(1), 30–46.

Kobayashi, T., Kenichi, I., & Miyata, K. (2006). Social capital online: Collective use of the Internet and reciprocity as lubricants of democracy. *Information Communication and Society*, *9*, 582–611.

Lessig, L. (1999). *Code and other laws of cyberspace*. New York: Basic Books. *Code version 2.0* is available for download at http://pdf.codev2.cc/Lessig-Codev2.pdf.

Lessig, L. (2002). *The future of ideas: The fate of the commons in a connected world.* New York: Vintage Books. Available for download at http:// thefutureofideas.s3.amazonaws.com/ lessig_FOI. pdf.

Lessig, L. (2004). *Free culture: The nature and future of creativity.* New York: Penguin Books. Available for download at http://www.free-culture. cc/ freecontent/.

Norris, P. (2001). *Digital divide: Civic engagement, information poverty, and the Internet worldwide*. New York: Cambridge University Press.

Poster, M. E. (n. d.). Cyberdemocracy: Internet and the public sphere. Retrieved February 23, 2010 from http://www.hnet.uci.edu/ mposter/ writings/ democ.html.

Stein, L. (2006). *Speech rights in America: The First Amendment, democracy, and the media.* Chicago: University of Illinois Press.

Taubman, G. (1998). A not-so World Wide Web: The Internet, China, and the challenges to non-democratic rule. *Political Communication, 15*, 255–272.

KEY TERMS AND DEFINITIONS

Alternative Media: Media organizations that aim to promote independent, critical voices, while being controlled by neither the government nor the media conglomerate.

Anonymity: The freedom not to reveal one's identity in cyberspace.

Citizen Journalism: The production and distribution of news via various channels including the Internet by citizen journalists who are not members of established, mainstream news media outlets.

Free Speech: Freedom to express one's opinion without fear of reprisal.

Internet Content Regulation: Legal and regulatory restrictions placed on the Internet for the purpose of controlling what, where, and how something can be communicated.

Internet Democracy: Advancement of democracy via civic participation, public deliberation, and debates on the Internet.

Internet Real Name Policy: A form of Internet regulation in which users are required to resister for websites with their legal names and other identifying information to be able to fully participate in online communication including posting/replying messages to forums.

Pseudonymity: The use of a fictitious name in cyberspace.

Virtual Public Sphere: An imaginary space/ community on the Internet where people convene to express their opinions and to exchange ideas on matters of social importance.

ENDNOTES

1. The index measures the level of information and communication technology advancement in 159 countries by utilizing various socio-economic data sets in three subcomponents (i.e., access, use, and skills) of ICT.
2. For a detailed historical account of Korean democracy, see Kim (2000). His book presents an excellent history of Korean democracy in the 20th century, focusing on the rise of civil society, and roles played by democratic movements in the late 1980s.
3. This is a Korean equivalent of the U.S. Social Security Number (SSN), but it is more strictly enforced legally, and is used widely for various legal, educational, and economic purposes and activities.
4. For an excellent anthology on Jurgen Habermas's notion of public sphere, see Calhoun, C. (Ed.). (1993). *Habermas and the public sphere.* Boston: MIT Press.

5 For a nice overview of citizen journalism and online democracy in Korea, refer to Ronda Hauben's *The rise of netizen democracy,* which is available at http://www.columbia.edu/~rh120/other/misc/korean-democracy.txt.

6 The English version of Korean Constitution is available on the Constitutional Court of Korea website via http://english.ccourt.go.kr/home/att_file/download/Constitution_of_the_Republic_of_Korea.pdf.

Chapter 18
An Agent for Change:
The Internet is Setting New Agendas in China

Jiafei Yin
Central Michigan University, USA

ABSTRACT

China became the largest Internet user in the world with 420 million of its citizens connected to the new media by June 2010. This chapter investigates the social conditions and ways in which new communication technologies are transforming the politics, culture, and the society in China through analyses of uses of the Internet, differing roles played by the traditional and the new media, Internet regulations in the country, and cases catapulted to the national media spotlight by the online community, and through contrasts with the roles new communication technologies play in Western and African societies. The chapter also attempts to explore the implications of these transformations.

INTRODUCTION

There is no doubt that the Internet, as a new and instantaneous mass communication as well as interpersonal communication tool, is transforming the whole world, east or west, north or south, regardless of differences in political, economic or media systems. In well-developed democratic societies, it has threatened the survival of established mainstream media outlets by providing the public with a channel for accessing free media content, siphoning away streams of advertising revenue, and giving rise to "citizen journalism". In the Middle East, it has helped bring about democratic movements across the Arab world and beyond, toppling governments in Tunisia and Egypt, bringing on a raging civil war in Libya and mass demonstrations in Algeria, Yemen, Bahrain, and even Saudi Arabia and Iran. Elsewhere in Africa, new communication technologies have become new tools of domination in reinforcing prevailing ideologies (Wachanga, 2011), as they tend to be in the hands of the ruling class or the privileged. In China, the adoption of the Internet, an open medium that defies total control, repre-

DOI: 10.4018/978-1-60960-591-9.ch018

sents a unique and fascinating experiment with a communist media system.

The Internet entered China in 1994, sixteen years after China opened up to the rest of the world following almost three decades of isolation from the west. It is the right medium at the right time for China. No other mass medium caught up in China as fast as the Internet has. Radio and television entered average Chinese homes decades after they became household items in the West. Despite breathtaking economic, social and cultural developments in China since the start of the reforms, China's political system – the communist system – has remained largely intact. China has, thus, become this odd combination of mostly laisser-faire private and capitalist economic system with one-party communist political system, which, according to predictions by Western scholars, will never work because of the incompatibilities of the two systems. Capitalism, at least in its early stages, is supposed to thrive on free competition predicated on the free flow of labor, capital, products, and information while communist governments closely monitor and heavily regulate such flows. In China after years of economic reforms, the opening up of the media sector seems to be the last intransigent frontier. Then came the Internet, the one unstoppable communication channel that has the capability of reaching millions instantaneously anywhere in the world. It may help negotiate the incompatibilities of the political and economic systems in China.

This is a fascinating experiment because it is unprecedented and because the outcome of such an experiment cannot be predicted. Already the impact of the Internet has caught many Party and government officials in China by surprise and prompted top state and Party leaders to log on to the virtual world to chat with ordinary citizens in order to maintain a government presence on the net. The Internet is bringing about revolutionary and irreversible changes to the news media in China and to the way the country is governed.

This chapter tries to answer research questions, including what kind of political, social and cultural impact the Internet is having in China, how the Internet brings about those changes, and what the implications of those changes in the Chinese society are in comparison to the impact of the Internet in Western societies. It will also provide a context for examining the controversies surrounding the regulations and control of the Internet in China.

The research in this chapter focuses on the analysis of nationally-known media cases, uses of the Internet, role of the new and old media, the laws and regulations concerning the Internet, and the factors that made the acceptance of the Internet in China possible. Statistical data for this study was drawn from the Statistical Reports on Internet Development in China, which are summaries of biannual national surveys of Internet use in the country by the China Internet Network Information Center (CNNIC), annual reports by the Chinese Academy of Social Sciences (CASS) on the Analysis of Internet Public Opinion in China, and 2009 China Journalism Yearbook.

BACKGROUND

Pre-Internet Days: Official Media Domination

Before the opening up of China in late 1970s, the media system in China was rather rigid, consisting only of official newspapers, television and radio stations. Crime was not news, and police was not a beat. There was very little variety in news coverage. Listening to the Voice of America, the U.S. government-funded radio, was a crime during the "cultural revolution."

The opening up of China changed much of that once the control of the media was relaxed. Since then media outlets in China have proliferated and the media market has substantially expanded. Before the reforms, most of the cities in the country had only one Party paper, one official television station and one official radio station. In contrast, by the end of 2008, there were 1,943 newspapers and 9,549 periodicals published in China ("China," 2009). There were no television cable services before the reforms. By the end of 2008 cable subscribers reached 160 million households. News coverage has become much more lively and diverse. However, before the adoption of the Internet, the official media still controlled the coverage of important news events, which made suppression of sensitive news events relatively easy. There were few alternative news sources.

There were also few outlets for public opinion. In fact, public opinion did not seem to exist because there was no gathering place for it and because there seemed to be no need for it either. The opinion and editorial pages of official newspapers were filled with the voices of the Party and government. There was little discussion or debate on important Party or government policies, only interpretations and justifications of those policies. There were no controversial issues in China because nothing was controversial. The Party already determined what was right and what was wrong. Public opinion was limited to "letters to the editor" on the opinion pages. Communication model for the Chinese society was mostly top-down – information was flowing one way from the Party and the government to the people while the people had little avenue to having their voice heard. This communication model confirms Altschull's theory, "In all press systems, the news media are agents of the people who exercise political and economic power." (Altschull, 1995, p. 440) With little political clout, the public had no voice.

There were only a few brief exceptions when public opinion briefly entered the mainstream media or was tolerated by the government. Before the crackdown on the "rightists" in 1957, intellectuals in China were encouraged to criticize Party and government policies and to engage in free debate of ideas, hence the "let hundred flowers bloom campaign." In 1978 at the start of the reforms, the Xidan Democracy Wall, known as the "Beijing Spring," symbolized new freedom of speech before it was shut down by the government. The last time the news media was allowed to reflect the sentiment of public opinion was in 1989 when college students in Beijing, who was later joined by Beijing residents and even government employees, staged massive demonstrations in the Tiananmen Square, demanding end to corruption.

Chinese journalism textbooks define the role of the news media as a bridge between the Party and the people, meaning the media should publicize the Party's policies and plans while relaying people's concerns to the Party. In practice, all too often the news media were preoccupied with promoting Party's policies. Thus, they became one-way bridge. Just as Altschull (1995) theorized, "Press practices always differ from theory (p. 441)." Even in the People's Daily it was hard to find people's voice. Public opinion polls and surveys are only a recent practice in China.

The Right Medium at the Right Time: Internet's Fast Growth and Uses

The arrival of the Internet in China in 1994 changed the communication model in China. In the national drive to modernize China, the Chinese government decided to adopt the new communication technology and allow China to become a part of the wired international community. Without opening up China, the introduction of the Internet in the country could have been delayed for decades. The reforms speeded up its introduction, and now information and public opinion on the Internet are pushing the reforms to the next level.

Efforts at bringing the Internet to China started in 1987 when a group of Chinese scientists sent an e-mail to their German counterpart with the title, "Across the Great Wall we can reach every corner in the world." ("The Internet Timeline," 2003, 1). Two years later the Chinese Research Network was connected to the German Research Network. Ironically, the initial request by the Chinese Academy of Sciences (CAS) to be connected to the

Internet was denied in 1992 because of political concerns that many U.S. government departments are connected to the Internet. Two years later CAS officials repeated the request, which finally won approval by the U.S. National Science Foundation. Once embraced by the government, Internet has experienced phenomenal growth in China. With a monthly fee, millions of Chinese have logged on. And China is never the same again. By June 2010, there were a total of 420 million Chinese who had access to the Internet, ranking China the largest Internet user in the world. The potential for further growth is considerable as Chinese Internet users accounted for only 32 percent of the total population compared with the United States where about 77 percent of the population has access to the Internet and Japan at 78 percent ("Top 20," 2010). About 46 percent of the Internet users in China were mobile Internet users ("Statistical Report," 2009). There were also 180 million bloggers in China (Qin, 2010). By June 2009, the total number of Chinese domain names had reached 16 million and the number of Chinese websites 3 million ("Statistical Report"). The rapid growth rate was attributed to heavy government investment, continued lowering of the prices for Internet access, improvements in the quality of technology, and expansion of Internet content and applications, such as mobile blog, mobile video and mobile television. The Internet is just one new communication technology embraced by the Chinese government. In 2007 the People's Daily, the official Party paper, started its mobile phone distribution all over the country.

In China the majority of Internet users are young people with the age group 10-19 accounting for 33 percent, and as high as 83.5 percent of the Internet users in China are under the age of 39 ("Statistical Report"). The largest group of net users are students at 32 percent. Contrary to popular belief, neither the income level nor the educational level of net users is high with 67.3 percent of them having only a junior high or senior high school education, and the largest group 23.5 percent earning less than 500 yuan (73 US dollars) a month. Despite Western media's spotlight on government crackdowns on Internet cafés in China, 80 percent of the users access Internet from home.

The Internet has become a very important source for news and public opinions in China with 78.7 percent of the users checking out news online, second only to those who tune in to online music at 85.5 percent ("Statistical Report"). As high as 54 percent of the users host their own blogs, and about 30 percent of the users participate in discussions at online forums. As social networking tools, the Internet connects high proportions of users through instant messaging (72 percent) and e-mails (55 percent). The Internet also serves as a research tool with 69 percent of the users taking advantage of the search engines.

A Review of Theories and Research

Media's Agenda-Setting Role

One of the most salient effects of the news media is its agenda-setting role. According to Severin and Tankard (2001), "The agenda-setting function of the media refers to the media's capability, through repeated news coverage, of raising the importance of an issue in the public's mind (p. 219)."

Norton Long (1958) wrote in an article in the *American Journal of Sociology*, "In a sense, the newspaper is the prime mover in setting the territorial agenda. It has a great part in determining what most people will be talking about, what most people will think the facts are, and what most people will regard as the way problems are to be dealt with (p. 260)."

When research findings fluctuate between limited and powerful effects of the media, media scholars seem to have agreed on one conclusion drawn by Bernard Cohen (1963) – the media may not be successful in telling people what to think, but it is stunningly effective in telling its readers what to think about (p.13).

On the Chinese media scene, who is setting the news agenda is changing because of the adoption of the Internet, which is faster and wider in its reach in disseminating news and often publishes news stories first shunned by the traditional news media. No traditional media outlets can compete with the Internet in their power to reach millions of readers at the same time. The interactive nature of the Internet allows the readers to post their views on what is published, turning one story into so many more stories.

The Spiral of Silence

Another theory that is relevant to explaining the changes the Internet is bringing to China is the spiral of silence advocated by Elizabeth Noelle-Neumann (1993). The Internet is bringing the Chinese people out of that spiral by providing them with access to public opinion forums, which were non-existent in pre-Internet days (Yin, 2005).

Noelle-Neumann contended that the mass media do have powerful effects on public opinion and that those effects have been underestimated. According to Noelle-Neumann, three characteristics of mass communication – its accumulation, ubiquity and consonance – combine to produce powerful effects on public opinion, which is formed through a process called the spiral of silence. People form impressions on a controversial issue based on the distribution of public opinion. They try to discover whether they are in the majority and whether public opinion is swinging to their side. If they sense they are in the minority or if they think public opinion is not on their side, they tend to remain silent on the issue. The more they remain silent, the less other people are aware or supportive of their particular point of view, and the more they remain silent.

According to Noelle-Neumann, willingness to speak out on issues is affected mostly by perception of the climate of opinion – if the climate of opinion is unfavorable to a person, that person will remain silent for fear of isolation. The media also plays an important role in the spiral of silence because they are the source to which people look to find the spectrum of public opinion.

Contending Perspectives on the New Media in China

Guobin Yang (2009) approached the topic from a sociological perspective and focused on the "multidimensional interactions" that both enabled and constrained online activism in China. The interactions were between online activism and state power, culture, the market, civil society, and transnationalism. Yang's conclusion was that Chinese online activism was a generalized response to the consequences of Chinese modernity, and a countermovement rooted in material grievances and the identity crisis as a result of dramatic social change. He believed online activism gave rise to the civil society in China and was renegotiating political power. He also found that the radicalization of online activism was in direct proportion to the degree of outside influences.

Yongnian Zheng (2008) tried to answer the question of whether progress in science and technology would lead to political liberalization and even democracy in China. He made the distinction that democracy required changes in the political structure and, thus, argued that the new media brought about liberalization in China, but not democracy. He predicted that a more realistic scenario of regime change in China would be a change from gradual political liberalization, in the process of which the state and social forces would transform each other through their increasingly intensive interactions over Internet-mediated platforms.

A more pessimistic evaluation on the impact of the new media in China was provided by Yongming Zhou (2005) in his comparison of the arrival of telegraphy in the late Qing Dynasty with the arrival of the Internet in China in 1990s. Zhou argued that even though attitudes toward the adoption of the new technologies were very different between the Qing officials who strongly resisted telegraphy and Chinese leaders today who embraced the Internet, their controls of the technologies were very similar. Because of the authoritarian nature of the governments then and today, Zhou contended that "public sphere" as defined by Habermas never existed in the Chinese society because there were no free expressions. Zhou saw the view that technologies were either inherently oppressive or automatically emancipating as a "fallacy of technological determinism". He contended that many of the early enthusiasts of the Internet had abandoned their rosy expectations for China.

THE INTERNET IS TRANSFORMING THE LARGEST COMMUNIST MEDIA SYSTEM

The arrival of the Internet in China has ended traditional media's monopoly in news coverage and for the first time in decades provided a public forum where people are connected with each other and with the rest of the world and where people are able to comment on government policies and public affairs. The Internet has opened up information flow within the Chinese society and changed the way the government runs its business by making the top-down communication model obsolete. The power of online public opinion has attracted government leaders to the Internet, where everyone has a mike.

The Internet Becomes the Silence Breaker

Since 1949 when the Chinese Communist Party came to power, China has mostly banned unofficial or underground news publications. Coverage of important or sensitive news events often comes from a couple of official news sources such as the Xinhua News Agency and the People's Daily, which set the tone on those news events. In the days pre-dating the Internet in China, if official media outlets did not cover a news event, there were few alternative news sources for the public. Therefore, it was relatively easy for the government to manage news flow or block the publication of a news event. The Internet, given its open nature, rendered news blackout a thing of the past and has posed strong challenges to the official media and local governments. As unofficial media outlets, commercial websites in China are not allowed to engage in newsgathering; they can only reprint news stories from sanctioned news sites. But their public forums and blogs are open communication channels, which have published many breaking news events turned down by the conservative traditional media and pushed the traditional media to follow online leads and do follow-up reporting when an online story caught wide public attention and became a big story. In this ironic twist, the unofficial medium, the Internet, has become a news leader and news agenda setter instead of playing second fiddle to mainstream media outlets.

According to the 2010 Bluebook on China's Social Development and Forecast published by the Chinese Academy of Social Sciences (CASS), of the 77 hot news topics covered in the Chinese media in 2009, 23 appeared on the Internet first, which accounted for about 30 percent of the total (Zhu et al, 2009). Most of the news events that broke on the Internet were exposés, such as official corruption and abuse of power, mass protests and fraud. These cases were not limited to 2009 either.

One such online story, which attracted intense publicity in 2009 concerned attempted police cover-up. One night in May in Hangzhou in East China, twenty-year-old driver Hu Bin was racing down the street with two of his friends in their own cars. Hu hit and killed twenty-five year old Tan Zhuo, who was crossing the street. Some eyewitnesses claimed that Tan was sent five meters into the air and twenty meters away ("70 km," 2009). Most of the mainstream media outlets remained silent on the news the next day. In contrast, public forums on the Internet exploded with information and response. Within one hour of the tragedy, the news, photos and Hu's vehicle license plate were posted online. One eyewitness photo showed Hu's friends gathered at the scene of the accident, smiling and smoking in total disregard of a human life just lost. One post expressed the concern that money might prevent justice being served to these rich kids. Within hours, the online post received more than 2,000 hits ("70 km," 2009) and responses reached 140,000 (Zhu et al, 2009).

Under pressure, Hangzhou police held a press conference claiming that the speed of the racing car was 70 km (43.75 miles) per hour, which would allow Hu to keep his driver's license. The police also insisted that there was no evidence that Hu's car was refitted for racing as suspected at online forums. The public was convinced that the police was covering up the truth, as it was impossible to send someone that high at that speed. Public outrage online was instantaneous and intense. Overnight "70 km" became the euphemism for police cover-ups nationwide. Students at the Zhejiang University, where the victim graduated, held a candlelight vigil along the street where the tragedy took place and wrote a letter to the mayor of the city, demanding the truth. The letter soon found its way to the Internet, catapulting it to a major national news event despite initial silence from the traditional media.

The pressure of national outrage prompted the mayor to respond. He issued a statement, emphasizing the need to punish the perpetrators according to the law so as to end such tragedies and illegal racing in the streets. A more thorough police investigation followed, which revealed that the car was refitted for racing and that the speed of the car was between 84 km (52.5 miles) and 101 km (63 miles). The police had to issue apologies to the public. The case ended in compensation for the victim's family and a three-year sentence for Hu Bin, the driver. Without the Internet and its popular forums breaking the silence, the truth could have been suppressed, and the privileged could have escaped justice, as the racing car problem had been plaguing communities for quite some time.

Another widely publicized case of police abuse of power exposed on the Internet in 2009 occurred in the metropolis of Shanghai, where traffic police launched a crackdown on unlicensed taxis in the city. Because of difficulties in obtaining evidence, police hired undercover agents and had these agents impose on innocent drivers, some of whom were kind enough to oblige pleas for help to give a "helpless" stranger a ride. Once these agents arrived at their destinations, they would plant cash in a car, and a group of police officers would suddenly jump onto the scene and arrest the driver for driving an "illegal taxi." The crackdown lasted several years in Minhang District, Shanghai, and netted more than 50 million yuan (7.35 million U.S. dollars) in fines and confiscated cash (Zhou, 2009). The online community called such law enforcement as "fishing expeditions."

One victim, Zhang Jun, was ambushed in one such "fishing expedition" when he thought he was giving a man with "stomach ache" a ride. Frustrated with such injustice, Zhang, who was fined 10,000 yuan (1,470 US dollars), posted his story online, appealing for help (Zhu et al, 2009). The post immediately attracted nationwide outrage as such "fishing" tactics were being used elsewhere in the country. Condemnations and advice poured in. The most empowering part of the online post was that it connected Zhang with a dozen other victims of such cases. The Shanghai judiciary department learned of the post and got involved. Eventually Zhang was cleared, and the fines were returned to him. What was more important was that the Internet not only helped Zhang but other victims of similar cases. One month after Zhang was charged, eighteen-year-old Sun Zhongjie was framed the same way and attempted suicide, protesting the charges against him and trying to prove his innocence. Without the Internet, Zhang and Sun would have been helpless victims, fighting hopeless battles on their own against the powerful local police.

Other exposés concerned professional negligence of duty, malpractices and cover-ups in the workplace. One of the top ten hot topics on the Internet in China in 2009, according to the China Daily, was that a five-month-old baby died in a hospital in Nanjing also in East China while the doctor on duty was playing an online game and missed the opportunity to save the baby. Initial hospital investigation concluded that the doctor was writing his thesis that night (Hu, 2009). But one post online claimed that the doctor was playing online games. The article attracted the attention of the provincial health department, which launched an independent investigation. The department had its computer experts check the doctor's computer for evidence and confirmed that the doctor was playing online games at the time. Eventually the doctor admitted to game playing and had his practice license revoked.

The Internet Becomes an Opinion Leader

The Internet in China today not only breaks the silence of the traditional media in covering touchy news topics and events; it also provides open forums for public opinion, unavailable before the days of the Internet. Some forums even debated such sensitive topics as the need for the reform of the political system in the country (Zhu at al, 2008). Others have become very popular and command national attention. In the Chinese society today, hot media topics often come from popular websites and online forums instead of from traditional media outlets, which are tasked with "guiding public opinion." In contrast, information online is often user-generated. Without the Internet, traditional media had a dominant voice in the Chinese society. The arrival of the Internet has added voices to the Chinese society. Online public opinion, when focused and intense, can be a very powerful tool for social change by influencing the outcome of controversial news events as it often generates much-needed publicity (Yin & Gao, 2004).

One of the most prominent cases that showcased the power of the Internet occurred in 2003 when a college student Sun Zhigang was beaten to death by local police in Guangzhou in South China for failing to show his identification. At the time China had a law stipulating that if anyone failed to produce his identification upon request by law enforcement, he would be treated as homeless and could be detained. One month after the tragedy was published in a local paper, nothing happened. But once it was posted on the Internet, the news spread like wildfire. Public outrage online eventually helped bring justice to the case and ended the discriminatory law on the homeless.

In the 2009 CASS study, the authors examined top twenty hot news topics from five online forums, including Tianya.cn/bbs, club.kdnet.net, bbs1.people.com.cn, bbs.sina.com.cn, and club.china.com. Sixteen of the twenty topics recorded more than five thousand original posts, not including follow-up posts, while five of the sixteen topics registered more than ten thousand posts. Most of the topics were concerning conflicts between the powerful and the powerless, and most of the posts were calling for the protection of citizens' rights, supervision of official power, and the promotion of moral behaviors. The topic with the highest number of posts, 25,133 in total, was about the case of a young female hotel waitress, Deng Yujiao. According to Deng, two male guests, both local officials in their forties, allegedly forced her to provide "special services" and beat her when she refused to oblige. After repeated futile protests and attempts to resist them, Deng stabbed one to death and injured another with a pocketknife. The police detained Deng and sent her to a local hospital for mental examination by medical experts. Online forums exploded with postings. The majority supported Deng's self defense and demanded that she be released from the hospital. The whole proceedings of the case were under very close public scrutiny online, bringing transparency in the justice system.

In today's China, it is becoming increasingly difficult for the police to hide its misconduct from online forums and blogs. For example, when one young man died after eleven days in jail, the police report concluded that he died after crashing into a wall while playing a game of hide and seek with fellow inmates. The public was not convinced of the police report, and the term "hide and seek" became a hot search term online. Public pressure forced the police to reveal the truth that the young man was beaten to death by his bullying fellow inmates.

For government officials who still had not realized the power of the Internet, they were surprised to find themselves at the center of unwanted online attention. In Central China's Henan Province, the planning bureau of the city of Zhengzhou set aside a large piece of land for building low-priced housing for low-income families (Ren & He, 2009). But local villagers found out later that the land was used for the construction of high-priced villas by land developers. One villager filed complaints but to no avail. When cornered by persistent reporters, the deputy director of the bureau, Lu Jun, angrily demanded: "Are you going to speak for the Party or for the common men?" Overnight that rhetorical question became the most quotable quote on the Internet. The story received 5,385 original posts online (Zhu et al, 2009) while the quote was queried 15.8 million times on Google.com and 280,000 times on Baidu.com, the most popular search engine in China.

One of the five popular online forums studied is the forum of the official media—the People's Daily. Even there unorthodox views were expressed. One post by He Dechuan (2010) critically reviewed export subsidies provided by the Chinese government. The post sharply criticized the irrationality of government's policies by comparing the high prices of products in China, a developing country, with lower prices of the same products in developed countries, such as the United States. It cited Western media in praising China for its contribution to the world economy and argued that such policies ended in low income and high prices in China and high income and low prices in rich countries such as the United States. It ridiculed such policies as "robbing the poor to aid the rich." Such critical posts were not likely to be published in the pages of the People's Daily, which still mostly reflects the Party's voice. But this was one example that even the official media had to adapt to a new media environment in the age of the Internet.

The biggest difference the online forums made in 2009 was perhaps concerning the regulation of the Internet itself when massive online objections pressured the government to reverse its directive on requiring the installation of the screening software Green Dam in all computers sold in China. In May the Chinese Ministry of Industry and Information Technology issued a directive, requiring all PC manufacturers install the filtering software, Green Dam, in all PCs sold in China by July 1 (Cui, 2009). Officials said the software was intended to block out pornographic and violent materials for the protection of minors. Some American researchers said the software would also block politically sensitive content (Wolchok et al, 2009). Online opposition to the directive was massive with 9,016 original posts recorded at the five online forums (Zhu et al, 2009), protesting the government's move. The majority of the posts came from PC users, and some from PC makers both domestic and foreign. A University of Michigan study found "serious security vulnerabilities" in the software and concluded that any web site a Green Dam user visited can take control of the PC (Wolchok et al, 2009).

One month after the directive was issued, a postponement of the launch of the software was announced, and by August Minister of Industry and Information Technology Li Yizhong had to admit to the public that the implementation of the Green Dam was "not well thought out" (Cui, 2009). He explained that it was never the ministry's intention to demand that the software be pre-installed. "PC users were to have the final say about whether to use the software or not," Li said. "We will listen to the public's views before issuing a new directive on Green Dam." And to the delight of PC makers and users, Li did not set a new deadline for the implementation of the software except that the software would be installed on computers in schools and Internet cafés. Without the Internet, public objections to the directive would be rather weak as there would be no gathering place for them.

Based on the three annual studies conducted for the Chinese Academy of Social Sciences (Zhu et al, 2007, 2008, 2009), it is apparent that public opinion not only starts to play an important role in the public life in China, but also is growing louder every year. In 2007, the most popular topic at online forums, a fake news photo of extinct South China Tigers, generated 3,076 original posts at three top online forums (Zhu et al, 2007). In 2008, the most debated topic online was the real estate bubble. Original posts at five popular online forums reached 10,176 (Zhu et al, 2008). And in 2009 the case of Deng Yujiao discussed earlier in the section attracted 25,133 original online posts at five popular forums (Zhu et al, 2009). With the increasing number of Internet users in China, the impact of online forums is snowballing.

Online forums also provide an opportunity for the general public to participate in public discourse on sensitive topics traditionally reserved for the élites. In 2007 and 2008 discussions among the intelligentsia and Party officials on the need of reforming the political system to match the reform of the economic system were spread to the Internet. Articles such as "Democracy is a good thing," and concepts such as "government for the people and by the people" and "universal values" appeared on the Internet and attracted wide attention and discussions (Zhu et al, 2007 & 2008). It is revolutionary to have debates about the reform of the political system in the Chinese society even though such debates are only limited to online forums.

The Internet Has Become the Social Bellwether

Given its unprecedented power in drawing public attention to news events and issues, the Internet has almost single-handedly re-directed public attention from government policies to concerns of the people. Chinese President Hu Jintao fully realized the impact of the Internet. He reminded government officials that whether they could cope with the Internet was "a matter that affects the development of socialist culture, the security of information, and the stability of the state." ("President," 2007) He urged government officials to become more knowledgeable about the Internet and to improve their ability to administer the Internet in order to "better guide public opinions" on the net. The official Xinhua News Agency described the Internet in China in 2009 as "rocked by public opinion and people's political views." ("2009 flashback," 2010) Xinhua acknowledged that the Internet had become an important tool for all levels of government officials in China to assess public opinions and that the top leadership in China had started to pay unprecedented attention to online public opinion ("China's," 2009).

Widely publicized in the international media, the live online conversations between Chinese President Hu Jintao and net users in 2008 highlighted the impact of the Internet in China. When asked what he read on the web, Hu responded that he read news on the web and checked out net users' thoughts, opinions, and suggestions for the government ("Hu," 2008). When asked specifically whether he reviewed suggestions and proposals from net users on the web, He said, "We pay great attention to suggestions and advice from our netizens. ... The web is an important channel for us to understand the concerns of the public and assemble the wisdom of the public." ("Hu," 2008, ¶ 11) More than three hundred questions were asked, including those concerning the fall of the stock market, inflation, and corruption (Spencer, 2008). Other questions were about democracy, Taiwan, Hu's salary, and when he was going to root out corruption (Page, 2008). The online chats lasted about twenty minutes. In a traditionally hierarchical society such as China, public conversations between top leaders of the country and ordinary citizens rarely occur. The fact that it happened online indicated the level of importance the top government leaders attached to public opinions online and the government's determination to have a presence on the net.

Chinese Premier Wen Jiabao followed the lead of Hu in carrying on live online conversations with net users. More than 150,000 questions were posted online for Wen ("China's," 2009), most concerning issues such as unemployment, social security, health care, and corruption. Wen focused on combating official corruption by emphasizing the importance of transparency in government affairs and the need to make officials' assets public ("Premier," 2009). He said he always believed that the public had a right to know what the government was doing and a right to criticize government policies ("China's," 2009). Wen attributed the cause of government corruption to its unchecked and excessively concentrated power. He acknowledged that many government officials had too much power. To fight corruption he believed that the most important thing to do was to establish a good system with less centralized and unchecked power ("Premier," 2009).

The one thing that Chinese leaders and the people can agree on is the importance of online public opinion. An online survey conducted in 2009 by the People's Daily, the National School of Administration and the Renmin University found that 69 percent of the net users in China believed that the Internet provided an effective means for public officials to gauge public opinion and that net users had high hopes of promoting democracy in China through the Internet ("New highlights," 2009). Within the central government, fourteen ministries have set up message boards, e-mails and blogs on the Internet to gather public opinion, and they have also replied to online messages and appointed network spokesmen. According to the People's Daily Online, its message board for local leaders gathered about 420,000 messages in 2009. Altogether thirty-eight provincial Party secretaries and governors, and more than ninety municipal officials publicly replied to the messages. As a result, more than seven thousand problems were resolved.

The new wisdom, which goes against the age-old traditional wisdom, is that silence is not gold and that prompt response from the government may help prevent a crisis ("Evaluation," 2009). To improve the efficiency of local governments and prevent the outburst of protests, the People's Daily graded local governments' responses to online grievances concerning ten incidents occurred in 2009. The highest score was given to the city of Chengdu in Southwest China, where a bus blaze killed twenty-eight passengers and injured dozens of others ("Evaluation," 2009). Public outrage online was immediate and strong with speculations about the possible cause of the fire spreading by the minute. Within three days of the incident, the municipal government held five news conferences, answering questions in detail. The grading used timeliness and transparency as the two important criteria in evaluating government responses.

Officials in China who fail to realize the power of the Internet do so at their own peril. Local officials in the city of Shishou in Central China was given a warning for their failure to respond quickly when online rumors was spreading fast about the death of a cook ("Evaluation," 2009). The incident generated more than four thousand original posts at five popular online forums and was listed among the twenty most debated topics online in 2009 (Zhu et al, 2009). Detached from the virtual world, local officials remained silent for eighty hours after the incident, fueling rampant rumors about the cause of death, which eventually led to riots in the streets. An autopsy conducted by experts from the Ministry of Public Security ruled the death a suicide.

In China the Internet is often the first and final outlet for public grievances against the government. Slow government response to pent-up public frustrations resulted in riots not only in Shishou but also elsewhere in the country, such as Southwest China's Weng'an County, Guizhou Province, where the death of a female middle school student triggered a violent riot against the local government. The incident was listed among the 20 hot topics online in 2008 (Zhu et al, 2008). The local government first gave very little information and later claimed that the student committed suicide by jumping from a bridge while one of her friends was doing push-ups nearby. The official version of the incident was widely ridiculed by the online community. Rumors that the girl was raped and murdered led to a riot of 30,000 local residents ("Guizhou," 2008). An army of 1500 police officers was sent in, and about two hundred rioters were arrested ("China ordered," 2008). Local officials labeled the rioters as "evil forces" challenging the government and launched media campaigns against them. The Guizhou provincial government intervened. It fired the police chief ("Police," 2008) and warned that the term "evil forces" could not be used at will to describe an upset public (Zhu, 2008). Guizhou officials pointed to public grievances against rampant crimes in the county as the real cause of the riot. About six hundred to eight hundred cases of crime occurred in the county every year, and half of them were never solved ("Guizhou," 2008). Theft, robbery and fighting were reported to be frequent occurrences. Following the incident, the central government instituted a pubic survey, assessing the performance of government officials and giving the people a voice in deciding the promotions of government officials (Wu, 2008), which was unprecedented in China and a landmark in reforming the government by bringing in public input.

The bite of the Internet was so real that one local official in northwestern Shaanxi Province was nostalgic about the days without the Internet when the government could have the final say on anything (Zhu et al, 2009). Another local official, one magistrate in Lingfen city, Shanxi Province, lamented, "Being a government official in Shanxi has become a high-risk occupation. If you're not careful, you can land in jail. Now our job is like dancing on an egg. The days when our positions were secure no matter what are gone forever." (Zhu et al, 2008, ¶ 35)

DISCUSSIONS

The Internet is transforming the Chinese society—its culture, its traditional media, and its government— by creating in China a better informed, more vocal and more engaged public, a more scrutinized and responsive government, and a more open society. This transformation is made possible with the help of the constant supply of news and views on the Internet, which provides user-generated information in real time and to a much wider audience than the traditional media ("Statistical report," 2009). The Internet could not have arrived in China at a better time than 1994 when the debate over the nature and speed of the economic reforms within the top leadership of the Party was mostly over, and the government was eager to join the international community in the development of new communication technologies. The government sought the arrival of the Internet, and the Internet has changed the face of the government, which now has a populist premier and a leadership that is much more responsive to the people today than anytime in the history of the Party.

The economic reforms in the country have also produced a business environment conducive to the introduction and development of the Internet. With the dismantling of government policies favoring the official media such as government subsidies and compulsory subscriptions, commercial media such as the Internet have had a chance to expand and thrive. The more cautious traditional media have left a huge information gap between supply and demand, which has created a perfect opportunity for the growth of the Internet.

The opening-up of China and its economic reforms have exposed the Chinese society to new ideas and values from the outside world. The younger generation is particularly receptive to new ideas and trends. In China the Internet is a medium of the young people with 83.5 percent of the net users under the age of thirty-nine. Only with a more tolerant, open, and less judging cultural environment can public forums and personal blogs in China overcome a reserved traditional culture and flourish.

New Media, New Agenda

In China the Internet is competing with the traditional media in setting the agenda for news and government policies. While the traditional media's primary source of information is still the government, the Internet carries a substantial amount of user-generated content, thus re-directing public attention from the cares of the government to the concerns of the people. Repeated exposures of high-profile cases of official corruption and police abuse of power, as the annual lists of online hot topics have shown, have highlighted the extent of official corruption and the need for the reform of the political system. Official corruption was not new in China. The student demonstrations at the Tian'anmen Square in 1989 were triggered by official corruption and lack of democracy. Comparatively speaking, official corruption was on a much smaller scale back then than it is today. In 1999 the number of leading officials at various levels of the government who were disciplined by the Party increased 45.5 percent over 1993 while those sentenced by the law increased 471 percent (Zheng, 2008). With the pressure of public opinion on the Internet, the government has to be getting more serious about fighting official corruption. In 2006 the government started to require leading officials to declare their family assets. In 2010 the government expanded those rules to include the declaration of the employment conditions of their family members and pledged to strengthen the implementation of those rules ("CPC pushes," 2010). Public frustration on the Internet over official corruptions prompted the Chinese premier to address the need for the reform of the political system and unchecked concentrated government power. And online forums have enabled the participation of the public in such debate and added urgency to the resolution of the problem.

Other online topics directed the public attention to something positive in the Chinese society. Following the massive earthquake in Southwest China's Sichuan Province in 2008, online public opinion testified to the quick response of the government in trying to rescue quake victims and contrasted the quick government response with the one more than thirty years ago when a devastating earthquake hit Northern China's city of Tangshan in 1976. Back then there was very little information available about the deadly earthquake. Some posts also compared government response with U.S. government's response to hurricane Katrina (Zhu et al, 2008). Such discussions highlighted the issues of government responsibility, accountability and efficiency during major crisis. Online forums also paid tribute to the high professional standards and love of the teachers, some of whose bodies, when dug out, were shown bending forward with arms around their students. At a time when many Chinese lamented the pursuit of money and self-interest following the economic reforms, the focus on the heroic deeds of the teachers gave new impetus to the discussions of moral issues, such as the value of life.

Whether positive or negative, the new agendas were less created at government meetings than at the online forums, which have a unique capability in attracting immediate and intense public reaction to a news event or an issue. Given the fact that 78.7 percent of the net users tune in to online news, it is little wonder that the Internet has become a strong contender in setting the agendas in the country.

New Media, More Voices and More Public Participation

New communication technologies have ended the monopoly of the traditional media in releasing information. They have enabled the vast Chinese population to spiral out of its silence and provided a channel for the helpless to have their voice heard. On the Internet the Chinese have discovered that their voice can be powerful and can make a difference. Now they can express their doubts about the official version of an event and even overturn whitewashed official versions of events and get to the truth. Encouraged by the differences it has made, the public has become more engaged in social affairs and public life, forcing government officials to be more responsive to the concerns of the people. When it comes to government policies, it is no longer government business, it becomes the people's business, and the people demand a say in it. It is refreshing that the Chinese society can now hear diverse voices and has public discourse on government performance. Now government websites also solicit public suggestions in solving social problems. These are the unmistakable budding signs toward a democratic government—government of the people, by the people and for the people—even though the path to that goal is still long.

Online forums have brought about real actions, too. Following the earthquake in Sichuan, the websites of four traditional media outlets including the People's Daily, Xinhua News Agency, China News Agency, and China Central Television released about 123,000 pieces of news about the quake, while four top commercial websites—sina.com, sohu.com, NetEase and Tencent—together published 133,000 stories about the quake ("The Internet," 2009). The eight websites received 11.6 billion hits on the news with 10.6 million responses to the news. They played an important role in raising awareness and donations, delivering aid and helping quake victims find their missing family members. Disaster relief has become people's concern, not just the responsibilities of the government. The Internet has not only given the people a voice, it also motivates people to get involved.

New Media, New Pressures

With citizen journalism, every individual citizen has the potential of becoming a watchdog of the government. After the earthquake in Sichuan, rumors were circulating on the Internet that tents donated by the China Red Cross were sold at exorbitant prices to quake victims (Zhu et al, 2008). Teams were then dispatched to audit donations and supervise the work of the China Red Cross, which published on its website a detailed table listing all the donations and how they were used. The Internet has helped increase the transparency in disaster relief.

As the Chinese economy keeps expanding and the national wealth accumulating, there are more opportunities for government officials to seek and take bribes. Power corrupts. Absolute power corrupts absolutely. In China with concentrated government power, opportunities abound for government officials to become corrupt. Many of them are often oblivious to the concerns of the people. There are laws and regulations binding them, but none of them seem to be effective enough to stem the tide of widespread corruption.

Under such circumstances, the Internet has become a very effective tool in exposing official corruption and abuse of power. Without newsroom gatekeepers, the Internet has provided a wide-open field for the people to voice their grievances and expose the wrongdoings of government officials. The public through the Internet has become the ubiquitous watchdog, leading many investigations of controversial cases. Online forums may not always be able to get to the truth, but it has started a welcome trend of asking tough questions and demanding answers. When government officials start to see the vulnerability of their positions and power, change becomes possible.

Online public scrutiny can also be very effective in exposing professional fraud. One photo about the Qinghai-Tibet railway, which won the 2006 top ten CCTV news photo award, was exposed to be a fake after it was posted online and examined by the online community ("CCTV," 2008). Another fake photo of South China Tigers was exposed by a net user, who found it to be very similar to an old painting of the tigers (Bristow, 2007). Both BBC and U.S. magazine Science reported the hoax. The online watchdogs never stop watching.

New Media, New Culture

The Internet itself has become a cultural phenomenon. Getting on the Internet has become a popular pastime, as what becomes a hot topic online will soon be popular offline. The China Daily compiles its annual list of "top ten hot topics on Internet" ("Top ten," 2009). The CASS annual reports also had a list of catchphrases coined and made popular online. Most of these phrases were mocking corrupt officials while the rest put a humorous spin on daily events in people's lives. The Internet has become the go-to place for young people, the trendsetter.

The open nature of the Internet is liberalizing the conservative Chinese culture by promoting creativity and individuality among the younger generation. One five-minute music video attracted more than 300,000 hits and thousands of requests to become a friend of the singer (Yi & Chen, 2009). In the online world, "I" is replacing "we" when people have to identify themselves. Han Han, a twenty-seven-year-old racecar driver, novelist and a blogger, has become the outspoken voice of China's younger generation by being different (Wu, 2009). Personal blogs on the Internet have added diverse voices to the Chinese society, challenging traditional Chinese cultural values. Sex has been a taboo topic in the Chinese culture. However, one magazine columnist kept an online diary about her sex life and developed a following of 110,000 readers ("Sex writer," 2003). Her weblog caused a huge stir among the public and cost the columnist her job. As the majority of the net users in China are young, the Internet seems to be widening the generation gap.

New Media, New Challenges

Given the unique open nature of the Internet and its power, there is no doubt that new communication technologies pose unique challenges to the Chinese government in trying to regulate it. The government's often stated rationale for regulation is to protect the young from pornography and violence, as 33 percent of the net users in China are between the ages of ten and nineteen ("Statistical report," 2009). Periodically the government launches campaigns to crack down on online porn or violence in video games. For example, in 2009 the government arrested 5400 people accused of involvement in online porn and deleted 9000 pornographic websites ("China nabs," 2010). Such campaigns are mostly supported by parents and teachers, who are also concerned about school-age children's addiction to the Internet. In China there are rehabilitation camps where parents send their children for the treatment of the addiction even though scientists worldwide are still debating about the scientific definition of Internet addiction.

Other areas of regulations go beyond pornographic and violent material. The media in China are instructed to promote national security and unity. Along those lines eleven categories of materials are banned online according to the 2005 administrative rules, including information concerning national security and unity, national honor and interest, social order and stability, and information concerning superstition, terrorism, pornography and violence ("Administrative rules," 2005). Any calls for illegal rallies and activities are banned also. Those are very broad categories, which have attracted many critics. To implement these rules the government holds service providers responsible for removing banned material and demands these companies to practice "self discipline" (Dombey & Hille, 2010). As a result, many Internet service providers have to install filtering software to keep out banned materials, some of which, critics argue, are politically sensitive information. Violators are fined according to the rules. But China's firewall is not airtight. There are plenty of ways to get around the firewall (Stone & Barboza, 2010). Besides, Western news websites are readily accessible inside China, such as American broadcasters CNN, which was boycotted by the Chinese online community over its perceived biased coverage of the riot in Xinjiang in 2009, the Voice of America (VOA), U.S. government-funded radio, Fox News, ABC, NBC, CBS, and American newspapers such as the New York Times, the Washington Post, the Christian Science Monitor, and European wire services including Reuters and AFP. Inside China, headlines such as "China using detention to silence dissent" from VOA (Ide, 2010) and "China can't both pull and push the yuan" from Reuters ("China can't," 2010) can be easily found online.

One challenge for Chinese Internet service providers is to catch up with new government regulations. Without a press law or much experience in a more open media environment, the government's regulations of the media are constantly changing. During the 2008 Beijing Olympics, the government's experiment with relaxing rules concerning the operations of foreign media companies caught worldwide media attention. Under the new rules, foreign correspondents no longer had to obtain approval before they visited places outside major metropolises or interviewed ordinary Chinese. That experiment was an attempt to further open up China.

The riots in Tibet and Xinjiang presented a sharp contrast in the way the government dealt with the media. During the riot in Tibet in 2008, the government banned Western media from the area, resulting in much criticism from the international community and perceived biased reporting in the Western media. With new and mobile communication technologies news blackout was no longer possible. So when riot broke out in Xinjiang one year later, the government allowed western journalists to stay in the area to continue their reporting. However, the government cut off the region's access to the Internet during the riot when it discovered that rioters planned their actions through public forums on the Internet. By early 2010 the region's access to the Internet was gradually restored.

The Internet is a new medium, and the government's experiment with it has just started. Even though the path of the future development of the Internet in China is unpredictable, the direction of that development can only lead to more openness on the Internet and in the Chinese society.

New Media, Old Media in the East and West

In the most developed Western and Asian societies such as Britain, Germany, the United States, Japan and South Korea, where the percentage of Internet users hovers around 80 percent, the new media have seriously eroded the market of the traditional media. In 2009 alone, quite a few newspapers in the United States had to stop print editions entirely and move online such as the Christian Science Monitor, or combine copy desks such as the Booth Newspapers, or downsize the editorial staff such as the New York Times. Even in Japan, which claims the top five spots of the largest newspapers in the world, new communication technologies are drawing readers away from the traditional media; the same is happening in South Korea. Readers, especially younger ones, are moving online, so are the advertisers. The U.S. newspaper industry was hit particularly hard as the economic recession also results in reduced advertising spending. Even some print journalists are saying that the print is dying. In the West, the press is already free. What the Internet provides is convenience of delivery and more individualism in the form of citizen journalism and personal blogs.

In Africa, new communication technologies are reinforcing ideological hegemony (Wachanga, 2011).

In contrast, new communication technologies are rejuvenating the traditional media in China by creating a new and more competitive media environment. The new media raised the bar, and the old media had to catch up. In many cases the new media and the old media follow up on each other's reporting to lend more media power to a controversial news event (Zhu et al, 2009). In China, traditional media outlets are still very powerful as only 32 percent of the 1.3 billion Chinese has access to the Internet. With pressures from the new media, old media have to adapt from their traditional role of speaking mostly for the government to a more populist role of speaking for the people as well. That is a revolutionary shift in the Chinese communication model.

CONCLUSION

Tom Friedman of the New York Times said, "China's going to have a free press. Globalization will drive it." (Nordlinger, 2007) The Internet is the one truly global media whereas media giants, including BBC and CNN, are only national media with a global reach. Whether Google.cn is based in Beijing or Hong Kong, it will not change the fact that the Internet is bringing about an irreversible revolution in the Chinese society, where the balance of information power between the government and the people is shifting. The Internet has provided a platform for publishing hidden information and people's voice.

The Internet has singlehandedly redirected the flow of information in China from a top-down vertical flow to a ubiquitous horizontal flow, which spreads public-generated information and ideas. Like everything else in China, the Internet is still expanding, and regulations of it are still changing. But no matter how they change, there is no turning back once people's mind has been opened. A wired China is a new China in the making.

REFERENCES

Altschull, J. H. (1995). *Agents of power: The media and public policy* (2nd ed.). New York, NY: Longman Publishers.

Associated Press. (January 1, 2010). China nabs 5,400 people for online porn in 2009. Retrieved March 12, 2010, from http://abcnews.go.com/ Technology/ wireStory? id=9459979

Bristow, M. (December 5, 2007). Row over China tiger photo fakes. *BBC*. Retrieved February 2, 2010, from http://news.bbc.co.uk/ 2/ hi/ asia-pacific /7128383.stm

Chengdu Evening News. (February 16, 2008). CCTV-awarded photo was determined a fake: The photographer admitted that the photo was doctored. Retrieved February 2, 2010, from http://news.sina.com.cn/ c/ p/ 2008-02-16/ 034914948587.shtml

China Daily. (November 24, 2003). Sex writer creates a stir, gains a following. Retrieved March 27, 2010, from http://www.chinadaily.com.cn/ en/ doc/ 2003-11/24/ content_284039.htm

China Daily. (February 28, 2009). Premier Wen talks online with public. Retrieved January 31, 2010, from http://www.chinadaily.com.cn/ china/ 2009-02/28/ content_752 2765.htm

China Daily. (Dec. 24, 2009). Top 10 hot topics on Internet 2009. Retrieved February 21, 2010 from http://www.chinadaily.com.cn /china/ 09ten-topnews/ 2009-12/24/ content_919 6462.htm

China Digital Times. (February 28, 2009). China's Wen makes Internet debut. Retrieved March 9, 2010, from http://chinadigitaltimes.net/ 2009/ 02/ chinas-wen-makes-internet-debut/

China Internet Network Information Center. (2003). *The Internet Timeline of China* 1987-1996. Retrieved January 21, 2010, from http://www.cnnic.net.cn

China Internet Network Information Center. (2009). *Statistical report on Internet development in China.* Retrieved January 21, 2010, from http://www.cnnic.net.cn

China Internet Network Information Center. (May 18, 2009). *The Internet Timeline of China* 2008. Retrieved January 21, 2010, from http://www.cnnic.net.cn

China Journalism Yearbook. (2009). Institute of Journalism and Communication under the Chinese Academy of Social Sciences. Beijing: China Journalism Yearbook Publisher.

Chinese, B. B. C. (June 30, 2008). China ordered Weng'an rioters to turn themselves in. Retrieved from http://news.bbc.co.uk/ chinese/ simp/ hi/ newsid_7480000/ newsid_7480700/ 7480702.stm

Cohen, B. C. (1963). *The press and foreign policy*. Princeton, NJ: Princeton University Press.

Cui, J. (Aug. 14, 2009). Green Dam launch not handled well. *China Daily.* Retrieved March 7, 2010, from http://www.chinadaily.com.cn/ cndy/ 2009-08/ 14/ content_8568194.htm

He, D. C. (Feb. 19, 2010). *CCTV New Year skit revealed the truth about China's exports.* Retrieved February 22, 2010, from http://bookmark.people.com.cn/ toViewBookmark.do? id=126265

Hille, K., & Dombey, D. (January 15, 2010). Beijing demands companies show 'self-discipline.' *Financial Times,* London. Retrieved January 26, 2010, from http://www.ft.com/ cms/ s /0/ 26b045c2-0176-11d f-8c54-00144 feabdc0.html

Hu, Y. Q. (Nov. 13, 2009). *Baby dies while doctor plays game online.* Retrieved February 21, 2010, from http://www.chinadaily.com.cn/ cndy/ 2009-11/ 13/ content_8962291.htm

Ide, W. (2010). China using detention to silence dissent. *Voice of America.* Retrieved August 12, 2010, from http://www1.voanews.com/ english/ news/ asia/ China-Increasingly-Using-Prison-Sentences-and-Detention-to-Silence-Dissent-10044 2599.html

Internet World Stats. (2009). *Top 20 countries with the highest number of Internet users*. Retrieved February 12, 2010 from http://www.internetworldstats.com/ top20.htm

70km (欺实马). (2009). Retrieved January 31, 2010 from http://www.hudong.com/ wiki/ %E6% AC% BA%E5 %AE%9E %E9%A9 %AC

Lang, G. E., & Lang, K. (1983). *The battle for public opinion: The President, the press, and the polls during Watergate*. New York, NY: Columbia University Press.

Long, N. E. (1958). The local community as an ecology of games. *American Journal of Sociology, 64*, 251–261. doi:10.1086/222468

Noelle-Neumann, E. (1993). *The spiral of silence: Public opinion – our social skin* (2nd ed.). Chicago, IL: University of Chicago Press.

Nordlinger, J. (April 2, 2007). China dreamin'. *National Review Online*. Retrieved March 14, 2010, from http://nrd.nationalreview.com/ article/ ?q= NjdmMmI zOWZhM 2NkMjA5 Y2EyZDJm NDY 3MD AzN2 VkM2Y=

Page, J. (June 24, 2008). Elder brother Hu Jintao makes Internet debut. *Times Online.* Retrieved January 31, 2010, from http://www.timesonline.co.uk /tol/ news/ world/ asia/ article41983 10.ece? print=yes & randnum= 1151003209000

People's Daily. (December 24, 2009). New highlights in China's democratic politics 2009. Retrieved January 31, 2010, from http://english.peopledaily.com.cn/ 90001/ 90780/ 91342/ 6851678.html

People's Daily Online. (June 20, 2008). Hu Jintao talks to netizens via People's Daily Online. Retrieved January 31, 2010, from http://english.peopledaily.com.cn/ 90001/ 90776/ 90785/ 6433952.html

Qin, J., Liu, Q., & Ding, Q. (2010). Ministry refutes US claims China restricts Internet. *China Daily.* Retrieved January 26, 2010 from http://www.chinadaily.com.cn/ china/ 2010-01/23/ content_9365824.htm

Ren, L. P., & He, Y. (2009). *Land for low-priced housing turned into land for villas while a reporter was questioned by an official, "Who are you speaking for?"* Retrieved January 26, 2010, from http://news.sina.com.cn/ c/ 2009-06-17/ 0751158 03313s.shtml

Reuters. (Aug. 11, 2010). China can't both pull and push the yuan. Retrieved August 12, 2010, from http://blogs.reuters.com/ columns/ 2010/ 08/11/ china-cant-both-pull-and-push-the-yuan/

Severin, W. J., & Tankard, J. W. Jr. (2001). *Communication theories: Origins, methods, and uses in the mass media.* New York, NY: Longman.

Spencer, R. (June 20, 2008). China: Internet debut for leader Hu Jintao. *Telegraph.* Retrieved January 31, 2010, from http://www.telegraph.co.uk/ news/ worldnews/ asia/ china/ 2164637/ China-Internet-debut-for-leader-Hu-Jintao.html

Stone, B., & Barboza, D. (January 15, 2010). Scaling the digital wall in China. *The New York Times.* Retrieved January 19, 2010, from http://www.nytimes.com/ 2010/01/16/ technology/ internet/ 16evade.html

Top 20 countries with the highest number of Internet users. (June 2010). *Miniwatts Marketing Group.* Retrieved March 12, 2011 from http://www.internetworldstats.com/top20.htm

Wolchok, S., Yao, R., & Halderman, J. A. (2009). *Analysis of the Green Dam censorware system.* Computer Science and Engineering Division, the University of Michigan. Retrieved March 8, 2010, from http://www.cse.umich.edu/~jhalderm/ pub/gd/

Wu, C. (September 30, 2009). Han Han: finding happiness by being different. *The Daily Telegraph.* Retrieved January 31, 2010, from http://www.telegraph.co.uk/ news/worldnews/ asia/ china/6244112/ Han-Han-finding-happiness-by-being-different.html

Wu, J. (July 15, 2008). Public poll to help judge work of officials. *China Daily.* Retrieved January 28, 2010, from http://www.chinadaily.com.cn/ cndy/ 2008-07/15/ content_6846121.htm

Xinhua News Agency. (January 24, 2007). President Hu Jintao asks Chinese officials to better cope with Internet. Retrieved January 31, 2010, from http://news.xinhuanet.com/ english/ 2007-01/24/ content_5648674.htm

Xinhua News Agency. (January 13, 2010). CPC pushes Party leaders to report family information to stem corruption. Retrieved January 31, 2010, from http://news.xinhuanet.com/ english/ 2010-01/13/ content_12804650.htm

Xinhua News Agency. (January 22, 2010). 2009 Flashback: Public opinion and political views rocked the Internet in China. Retrieved January 28, 2010, from http://news.xinhuanet.com/ comments/ 2010-01/21/ content_12848338.htm

Xinhua News Agency. (July 26, 2009). Evaluation tells Chinese officials to listen to online opinions and act fast. Retrieved February 11, 2010 from http://english.people.com.cn/ 90001/90776/ 90785/ 6709971.html

Xinhua News Agency. (July 4, 2008). Guizhou officials admit residents harbored grievances. Retrieved January 31, 2010, from http://www.chinadaily.com.cn/ china/ 2008-07/04/ content_6817910.htm

Xinhua News Agency. (July 4, 2008). Police chief sacked after violent protest in Guizhou. Retrieved January 31, 2010, from http://www.chinadaily.com.cn/ china/2008-07/04/ content_6820613.htm

Xinhua News Agency. (September 25, 2005). Administrative rules for providing Internet information services. Retrieved March 14, 2010, from http://news.xinhuanet.com/ politics/ 2005-09/25/ content_3538899.htm

Yang, G. (2009). *The power of the Internet in China: Citizen activism online*. New York, NY: Columbia University Press.

Yi, L., & Chen, C. (September 30, 2009). How the Internet is shaping China's future. *The Daily Telegraph*. Retrieved January 31, 2010, from http:// www.telegraph.co.uk/ news/worldnews/ asia/ china/ 6244019/ How-the-internet-is-shaping-Chinas-future.html

Yin, J. (2005). *Spiraling out of silence: The Internet is influencing news agenda in China*. Presentation at the 14th AMIC Annual Conference: Media and Society in Asia: Transformations and Transition, Beijing, China, July 18-21, 2005.

Yin, J., & Gao, J. (2004). *The one truly global media in China: The Internet is transforming the largest Communist country*. Presentation at the Conference on Communication & Globalization, Center for Global Media Studies, Washington State University, Seattle, Washington, July 16-17, 2004.

Zheng, Y. (2008). *Technological empowerment: The Internet, stat, and society in China*. Stanford, CA: Stanford University Press.

Zhou, K. L. (Oct. 16, 2009). *Investigation into fishing law enforcement: Fines and confiscations totaled more than 50 million yuan*. Retrieved February 21, 2010 from http://china.huanqiu.com/ roll/ 2009-10/ 604442.html

Zhou, Y. (2006). *Historicizing online politics: Telegraphy, the Internet, and political participation in China*. Stanford, CA: Stanford University Press.

Zhu, H. X., Hu, J. C., & Sun, W. T. (2007). *2007 report on the analysis of Internet public opinion in China*. Published by the Chinese Academy of Social Sciences. Retrieved February 18, 2010, from http://media.people.com.cn/ GB/ 22114/ 52789/ 115416/6852601.html

Zhu, H. X., Shan, X. G., & Hu, J. C. (2008). *2008 report on the analysis of Internet public opinion in China*. Published by the Chinese Academy of Social Sciences. Retrieved February 18, 2010 from http://www.china.com.cn/ aboutchina/ zhuanti/ 09zgshxs/ content_17100922.htm

Zhu, H. X., Shan, X. G., & Hu, J. C. (Dec. 22, 2009). 2009 report on the analysis of Internet public opinion in China. *2010 Bluebook on China's Social Development and Forecast*. Published by the Chinese Academy of Social Sciences. Retrieved January 31, 2010, from http://yq.people.com.cn/ htmlArt/ Art392.htm

ADDITIONAL READING

Brady, A. (2009). *Marketing dictatorship: Propaganda and thought work in contemporary China*. Lanham, MD: Rowman & Littlefield.

Chase, M. (2002). *You've got dissent! Chinese dissident use of the Internet and Beijing's counter strategies*. Santa Monica, CA: RAND Corporation.

Cheung, A. S. Y. (2009, April). A study of cyber-violence and Internet service providers' liability: Lessons from China. *Pacific Rim Law & Policy Journal, 18*(2), 323–346.

Fallows, J. (2010, March). Cyber Warriors. *Atlantic Monthly, 305*(2), 58–60, 62–63.

Foroohar, R., et al. (2010). It's China's world: We're just living in it. *Newsweek*. v. 155 no. 12 (March 22 2010) 36-9.

Gaudin, S. (2010, April 5). Google, China play game of cat and mouse. *Computerworld, 44*(7), 8.

Google. (2010, March/April). China in Internet scuffle. *Information Management Journal, 44*(2), 6.

Gross, D. (2010). Google vs. China: China's silicon ceiling. *Newsweek* v. 155 no. 4 (January 25 2010) 38-9.

Han, G. K. (2009, November). Starbucks is forbidden in the Forbidden City: Blog, circuit of culture and informal public relations campaign in China. *Public Relations Review, 35*(4), 395–401. doi:10.1016/j.pubrev.2009.07.004

Hvistendahl, M. (2009, May). The China Syndrome. *Polar Science, 274*(5), 60–65.

Li, X. (2010). Google and the cyber infiltration. *Philosophy and public policy quarterly.* v. 30, no. ½ winter/spring 2010, 13-17.

Lyons, D. (2010). Google vs. China: You can't fight the future. *Newsweek* v. 155 no. 4 (January 25 2010) 40.

Madden, N. (2009, November 2). 20 blogs marketers to China should be reading. *Advertising Age, 80*(37), 10.

Madden, N. (2010, January 18). Whether it stays or goes, Google wasn't winning in China anyway. *Advertising Age, 81*(3), 2.

Mengin, F. (2004). *Cyber China: Reshaping national identities in the age of information*. Hants, U.K.: Palgrave Macmillan.

Nathan, A. J. (2009, November/December). The Power of the Internet in China. *Foreign Affairs (Council on Foreign Relations), 88*(6), 172.

Pan, P. L. (2009, September). Online strategic communication: A cross-cultural analysis of U.S. and Chinese corporate websites. *Public Relations Review, 35*(3), 251–253. doi:10.1016/j.pubrev.2009.04.002

Polumbaum, J. (2008). *China ink: The changing face of Chinese journalism*. Lanham, MD: Rowman & Littlefield Publishers.

(2010, April). Present and future of Internet banking in China. *Journal of Internet Banking and Commerce., 15*(1), 1A–10A.

Scotton, J. F. (2010). *New media for a new China.* Hoboken, NJ: Wiley-Blackwell.

So, S., & Westland, J. C. (2010). *Red wired, China's Internet revolution*. Singapore: Marshall Cavendish Business.

Tai, Z. (2006). *The Internet in China: Cyberspace and civil society.* New York, NY: Routledge.

Talbot, D. (2010, May/June). China's Internet Paradox. *Technology Review, 113*(3), 62–67.

(2010, February). Taunting the dragon. (2010). [International edition]. *The Institutional Investor, 35*(1), 7–8.

(2009, Summer). Vicarious and contributory liability for Internet host providers: Combating copyright infringement in the United States, Russia, and China. *Wisconsin International Law Journal., 27*(2), 363–408.

Wachanga, D. N. (2011). Ed. *Cultural Identity and New Communication Technologies: Political, Ethnic and Ideological Implications.* Hershey, PA: IGI Global.

Wacker, G. (2007). *China and the Internet*. Oxford, U.K.: Taylor & Francis.

Xiong, J. A., et. al., (2009) Chinese e-government information resources and services: a survey report. *Chinese Librarianship* no. 28 (December 1 2009).

Zakaria, F. (2010). Google vs. China: Clash of the titans. *Newsweek*. v. 155 no. 4 (January 25 2010) 34-6.

Zhao, Y. (1998). *Media, market, and democracy in China: Between the party line and the bottom line*. Champaign, IL: University of Illinois Press.

Zhao, Y. (2008). *Communication in China: Political economy, power, and conflict*. Lanham, MD: Rowman & Littlefield Publishers.

KEY TERMS AND DEFINITIONS

CNNIC: China Internet Network Information Center

CASS: Chinese Academy of Social Sciences

CCTV: Chinese Central Television

Xinhua: China's official news agency

CPC: Communist Party of China

Firewall: website blocking software system

Green Dam: porn screening software

Netizens: members of the online community

Compilation of References

Abah, A. L. (2008). One step forward, two steps backward: African women in Nigerian video-film. *Communication, Culture & Critique*, *1*(4), 335–357. doi:10.1111/j.1753-9137.2008.00027.x

Abah, A. L. (2009). Popular culture and social change in Africa: The case of the Nigerian video-film industry. *Media Culture & Society*, *31*, 731–748. doi:10.1177/0163443709339456

Abdi, J., & Deane, J. (2008, April). *The Kenyan 2007 elections and their aftermath: The role of media and communication*. BBC World Service Trust. Retrieved February 17, 2010 from http://downloads.bbc.co.uk/worldservice/trust/pdf/kenya_policy_briefing_08.pdf

Adam, L., & Wood, F. (1999). An investigation of the impact of Information and Communication Technologies in sub-Saharan Africa. *Journal of Information Science*, *25*(4), 307–318. doi:10.1177/016555159902500407

Adamolekun, W. (1996). The information superhighway and traditional communication: Where we stand. *Africa Media Review*, *10*(2), 22–36.

Adamu, A. (2010). The muse's journey: Transcultural translators and the domestication of Hindi music in Hausa popular culture. *Journal of African Cultural Studies*, *22*(1), 41–56. doi:10.1080/13696810903488579

Adamu, A. U., Yusuf, M. A., & Umar, F. J. (Eds.). (2004). *Hausa home videos: Technology, economy and society*. Kano, Nigeria: Centre for Hausa Cultural Studies/Adamu Joji Publishers.

Adamu, A. (forthcoming). Private sphere, public Wahala: Gender and delineation of intimisphare in Muslim Hausa video films. In Forster, T., & Nyamnjoh, F. (Eds.), *Negotiations of culture: Perspective from Africa*. Basel, Switzerland: Centre for African Studies.

Adejunmobi, M. (2002). English and the audience of an African popular culture. *Cultural Critique*, *50*, 74–103. doi:10.1353/cul.2002.0001

Adejunmobi, M. (2008). Technorality, literature, and vernacular literacy in twenty-first-century Africa. *Comparative Literature*, *60*(2), 164–185. doi:10.1215/-60-2-164

Adejunmobi, M. (2003). Video film technology and serial narratives in West Africa. In Ogunlcye, F. (Ed.), *African video film today* (pp. 51–68). Manzini, Swaziland: Academic Publishers.

Adejunmobi, M. (2004). Foreign languages, local audiences: The case of Nigerian video film in English. In Palaver, V. (Ed.), *Imaginations of the local and non-native languages in West Africa* (pp. 101–130). Clevedon, UK & Buffalo, NY: Multilingual Matters.

Adejunmobi, M. (2007). Nigerian video film as minor transnational practice. *Postcolonial Text, 3*(2), 1-16. Retrieved from http://Jounials.sfu.ca/ pocol/ index php/ pcl/ anicle/ vicw/548/405

Adesida, O. (2001). *Governance in Africa: The role for information and communication Technologies*. Retrieved June 18, 2008, from http://www.comminit.com/en/node/221626/307

Adesokan, A. (2004). Loud in Lagos: Nollywood videos. *Wasaftri*, *19*(43), 45–49.

Adesokan, A. (2007). A revolution from below: The aesthetics of West African videofilms. *Uka*, *21*, 60–67.

Adesokan, A. (2009a). Excess luggage: Nigerian films and the world of immigrants. In Okpewho, I., & Nzegwu, N. (Eds.), *The new African diaspora* (pp. 401–422). Bloomington, IN: Indiana University Press.

Adesokan, A. (2006). Issues in the new Nigerian cinema. *Black Camera, 21*(1), 6-8, 11.

Adesokan, A. (2009b). Practising 'democracy' in Nigerian films. *African Affairs*.

African Review of Business and Technology. (2005). Genset sales continue to soar: The world wants more diesel generator sets, and manufacturers are hard pressed to supply. *Entrepreneur*. Retrieved on May 13, 2010, from http://www.entrepreneur.com/tradejournals/pub/gen sale.html

Ajibade, B. (forthcoming). Nollywod videos: Is a western audience possible? In Krings, M., & Okome, O. (Eds.), *Nollywood and beyond: Transnational dimensions of an African video film industry*.

Akaki, S. (January 20, 2008). Will 'Balaalo' and land define the 2011 elections? *The Sunday Monitor Online*. Retrieved January 20, 2008 from http://monitor.co.ug

Akdeniz, Y. (2002). Anonymity, democracy, and cyberspace. *Social Research, 69*, 223–237.

Akhmad, A. (2010, February 21). Dumb questions invite dumb answers. *The News International*.

Akiwumi, A. (1998). *Land clashes commission report*. Nairobi.

Akpabio, E., & Mustapha-Lambe, K. (2008). Nollywood films and the cultural imperialism hypothesis. *Perspectives on Global Development and Technology, 7*, 257–270. doi:10.1163/156914908X370683

Alemneh, D., & Hastings, S. (2006). Developing the ICT infrastructure for Africa: Overview of barriers to harnessing the full power of the Internet. *Journal of Education for Library and Information Science, 47*(1), 26–38. doi:10.2307/40324333

Alemneh, D. G., & Hastings, S. K. (2006). Developing the ICT infrastructure for Africa: Overview of barriers to harnessing the full power of the internet. *Journal of Education for Library and Information Science, 47*(1), 4–16. doi:10.2307/40324333

Alexander, G. (2008). Channeling better images. *Black Enterprise, 39*(5), 70.

Althusser, L. (1971). Ideology and ideological state apparatuses. In *Lenin and Philosophy and other essays*. London, UK: New Left Books.

Althusser, L. (1971). *Lenin and philosophy, and other essays*.

Altschull, J. H. (1995). *Agents of power: The media and public policy* (2nd ed.). New York, NY: Longman Publishers.

Among, B., & Tebajjukira, M. (March 3, 2008). Museveni says 9,000 sq miles claim false. Retrieved March 3, 2008, from http://www.newvision.co.ug/D/8/13/614763

Andanda, P., & Cook, L. J. (2007). *Majengo HIV/AIDS research case: A report for GenBenefit*. Retrieved on 19th June, 2007, from www.uclan.ac.uk/genbenefit

Anderson, P. (1964). Origins of the present crisis. *New Left Review, 23*, 24–53.

Anderson, N. (2009, March 31). Pulling back the curtain on anonymous Twitters. *Law & Disorder: Tech law and policy in the digital age*. Retrieved February 19, 2010 from http://arstechnica.com/ tech-policy/ news/ 2009/ 03/ pulling-back-the-curtain-on-anonymous-twitterers.ars

Anduzia, A. (2009). The Internet, election campaigns and citizens: State of affairs. *Quaderns del CAC, 33*, 5–12.

Appadurai, A. (1996). *Modernity at large: Cultural dimensions of globalization*. Minneapolis, MN: University of Minnesota Press.

Appadurai, A. (1993). Disjuncture and difference. In During, S. (Ed.), *The cultural studies reader*. London, UK & New York, NY: Routledge.

Appel, L. (2003). White supremacy in the movement against the prison-industrial complex. *Social Justice (San Francisco, Calif.), 30*(2), 81–88.

Appiah, K. A. (1992). *In my father's house: Africa in the philosophy of culture*. New York, NY: Oxford University Press.

Apter, D. (1967). *The political kingdom in Uganda* (2nd ed.). Princeton, NJ: Princeton University Press.

Aseka, E. (2009). Media consumerism and cultural transformation. In Njogu, K., & Middleton, J. (Eds.), *Media and identity in Africa* (pp. 19–35). Edinburg University Press for the International African Institute.

Aseka, E. M. (1989). *A political economy of Baluyia, 1900-1964*. Unpublished Ph.D Thesis, Kenyatta University, Kenya.

Asian Human Rights Commission. (2008, September 10). *Two persons murdered after an anchor person proposed the widespread lynching of Ahmadi sect followers*. Retrieved March 21, 2010, from http://www.ahrchk.net/ ua/ mainfile.php/ 2008/2999/

Associated Press. (January 1, 2010). China nabs 5,400 people for online porn in 2009. Retrieved March 12, 2010, from http://abcnews.go.com/ Technology/ wireStory? id=9459979

Bako, P. (2009). Walking the talk: Are land evictions in Uganda in line with human rights standards? Masters thesis. Retrieved March 13, 2011 from http://upetd.up.ac.za/thesis/available/ etd-08032010-145221/unrestricted/ dissertation.pdf

Balkin, J. M. (2004). Digital speech and democratic culture: A theory of freedom of expression for the information society. *New York University Law Review*, *79*(1), 1–58.

Ball-Rokeach, S. J. (1985). The origins of individual media-system dependency: A sociological framework. *Communication Research*, *12*(4), 485–510. doi:10.1177/009365085012004003

Ball-Rokeach, S. J. (1998). A theory of media power and a theory of media use: Different stories, questions, and ways of thinking. *Mass Communication & Society*, *1*(1/2), 5–40. doi:10.1207/s15327825mcs0101&2_2

Ball-Rokeach, S. J., & Defleur, M. L. (1976). A dependency model of mass-media effects. *Communication Research*, *3*, 3–21. doi:10.1177/009365027600300101

Ball-Rokeach, S. J., Power, G. J., Guthrie, K. K., & Waring, H. R. (1990). Value-framing abortion in the United States: An application of media system dependency theory. *International Journal of Public Opinion Research*, *2*(3), 249–273. doi:10.1093/ijpor/2.3.249

Banda, T. K. (2008). *Youth converge on city for music show of the year. Saturday Nation* (p. 36). Nairobi, Kenya: Nation Media Group Ltd.

Banda, F., Mudhai, O. F., & Tettey, W. J. (2009). Introduction: New media and democracy in Africa-a critical interjection. In Mudhai, O. F., Tettey, W. J., & Banda, F. (Eds.), *African media and the digital public sphere* (pp. 1–20). New York, NY: Palgrave Macmillan.

Banks, A. J. (2006). *Race, rhetoric and technology: Searching for higher ground*. New Jersey: NCTE Press.

Baran, S. J. (2009). *Introduction to mass communication: Media literacy and culture*. New York, NY: McGraw-Hill.

Barnett, C. (2003). *Culture and democracy: Media, space, and representation*. Tuscaloosa, AL: University of Alabama Press.

Barrot, P. (Ed.). (2009). *Nollywood: The video phenomenon in Nigeria*Taylor, L., Trans.). Bloomingtonm, IN: Indiana University Press.

Batibo, H. M. (1992). The fate of ethnic languages in Tanzania. In Brenzinger, M. (Ed.), *Language death: Factual and theoretical explorations, with special reference to East Africa* (pp. 85–98). Berlin, Germany: Mouton de Gruyter.

Baylies, C., & Bujra, J. (2000). *AIDS, sexuality and gender in Africa: The struggle continue*. London, UK: Routledge.

Bayne, S. (2008). *Post-election violence in Kenya: An assessment for the UK government*. London, UK: DFID Kenya-UK Government.

BBC. (2000). *Pakistan province bans cable television*. Retrieved March 21, 2010, from http://news.bbc.co.uk/ 2/hi/ south_asia/ 800515.stm

Becher, A. & Basedau. M. (2008). *Promoting peace and democracy through party regulation? Ethnic party bans in Africa*. German Institute of Global and Area Studies (GIGA) Working Papers No. 66.

Beck, R. M. (2006). Popular media for HIV/AIDs prevention? Comparing two comics; *Kingo* and the *Sara* communication initiative. *The Journal of Modern African Studies*, *44*(4), 513–541.

Becker, L. B., & Whitney, D. C. (1980). Effects of media dependencies: Audience assessment of government. *Communication Research*, *7*(1), 95–120. doi:10.1177/009365028000700105

Becker, H. (Forthcoming). Nollywood in urban southern Africa: Nigerian video films and their audiences in Cape Town and Windhoek. In Krings, M., & Okome, O. (Eds.), *Nollywood and beyond: Transnational dimensions of an African video film industry*.

Bennett, W. L. (2003). New media power: The Internet and global activism. In Couldry, N., & Curran, J. (Eds.), *Contesting media power: Alternative media in a networked world* (pp. 17–37). Lanham, MD: Rowman & Littlefield.

Berger, P., & Luckman, T. (1966). *The social construction of reality: A treatise in the sociology of knowledge*. New York, NY: Doubleday.

Berger, G. (2005). Powering African newsrooms: Theorising how Southern African journalists make use of ICTs for newsgathering. In Berger, G. (Ed.), *Doing digital journalism: How Southern African newsgatherers are using ICTs* (pp. 1–14). Grahamstown, South Africa: High Way Africa.

Bertrand, M., & Mullainathan, S. (2004). Are Emily and Brandon more employable than Latonya and Tyrone? Evidence on racial discrimination in the labor market from a large randomized experiment. *The American Economic Review*, *94*, 991–1013. doi:10.1257/0002828042002561

Bhabha, H. (1994). Introduction: Location of cultures. In *The Location of culture* (pp. 1–18). New York, NY & London, UK: Routledge.

Bhuiyan, S. I. (2006, September 5). Impact of new media technology on society. *Daily Star*.

Big Youth. (2007). *Aluta Continua*. GMC Wasanii Promoters Ltd.

Bijker, W. E. (1995). *Of bicycles, Bakelite and bulbs: Towards a theory of socio-technical change*. Cambridge, MA: MIT Press.

Bitugi, E. M. (2007). Taaluma ya tafsiri katika vyombo vya habari nchini Kenya; matatizo, changamoto na mustakabali wake. In K. Njogu (Ed.), *Kiswahili na Elimu Nchini Kenya.* (pp. 158-170). Nairobi, Kenya: Twaweza Communications.

Black, J. (2001). Semantic and ethics of propaganda. *Journal of Mass Media Ethics*, *16*(2-3), 127–137.

Blackledge, A., & Pavlenko, A. (2001). Negotiating identities in multilingual contexts. *The International Journal of Bilingualism*, *5*(3), 243–257. doi:10.1177/13670069010050030101

Boal, A. (1979). *Theater of the oppressed*. New York, NY: Theater Communication Group.

Boal, A. (1992). *Games for actors and non-actors*. New York, NY: Routledge.

Boal, A. (1998). *Legislative theater: Using performance to make politics*. London, UK: Routledge.

Bomer, B. (2007). How information shapes political institutions. In D. Graber (Ed.), *Media power in politics*. (pp. 8- 18). Washington, DC: CQPress.

Bordens, K. S., & Abbot, B. B. (2008). *Research design and methods: A process approach* (p. 165). Boston, MA: McGraw-Hill.

Bosire, E. K. (2010). *Access and use of Information Communication Technologies (ICTs) for rural poverty reduction in Uasin Gishu, Kenya.* (Unpublished PhD thesis). Edoret, Kenya: Moi University.

Bourdieu, P. (1991). *Language and symbolic power*. Cambridge, MA: Harvard University Press.

Bourgault, L. (1995). *The mass media in sub-Saharan Africa*. Bloomington/ Indianapolis, IN: Indiana University Press.

Bourgault, L. M. (1995). *Mass media in sub-Saharan Africa*. Bloomington, IN: Indiana University Press.

Bratton, M., Chikawana, A., & Sithole, T. (2005). Propaganda and public opinion in Zimbabwe. *Journal of Contemporary African Studies*, *23*(1). doi:10.1080/0258900042000329466

Brint, S. (2001). Gemeinschaft revisited: A critique and reconstruction of the community concept. *Sociological Theory*, *19*(1), 1–23.

Bristow, M. (December 5, 2007). Row over China tiger photo fakes. *BBC*. Retrieved February 2, 2010, from http://news.bbc.co.uk/ 2/ hi/ asia-pacific /7128383.stm

Bucholtz, M. (1999). Why be normal? Language and identity practices in a community of nerd girls. *Language in Society, 28*, 203–223. doi:10.1017/S0047404599002043

Buganda Kingdom. (2004). *Buganda culture*. Retrieved June 12, 2008, from http://www.buganda.com/culture.htm

Bunyoro-Kitara Kingdom. (2007). *The Bunyoro-Kitara kingdom*. Retrieved December 14, 2009 from http://www.bunyoro-kitara.com/history.html

Busoga Kingdom. (2004). *Obwa kyabazinga bwa Busoga*. Retrieved June 10, 2010, from http://www.busoga.com/facts.php

Busoga Kingdom. (2007). *Evolution of obwa kyabazinga bwa Busoga*. Retrieved February 12, 2009 from http://www.busoga.com/history.php

Butagira, T. (May 25, 2008). Democracy, media under threat. Retrieved May 25, 2008 from http://allafrica.com/stories/200805270901.html

Byam, L. D. (1999). *Community in motion: Theatre for development in Africa*. Westport, CT: Bergin & Gorvey.

Bynoe, Y. (2004). *Stand and deliver: Political activism, leadership, and hip hop culture*. Brooklyn, NY: Soft Skull Press.

Byrnes, R. M. (1992). The society and its environment. In R. M. Brynes (Ed.), *Uganda: A country study (*chapter 2). Retrieved from http://lcweb2.loc.gov/cgi-bin/query2/r?frd/cstdy:@field(DOCID+ug0025)

Campbell, C. (2003). *Letting them die: Why HIV/AIDS prevention programs fail*. Bloomington, IN: Indiana University Press.

Campbell, C., Nair, Y., Maimane, S., & Nicholson, J. (2007). Dying twice: A multi-level model of the roots of AIDS stigma in two South African communities. *Journal of Health Psychology, 12*, 403. doi:10.1177/1359105307076229

Cancilini, N. G. (1995). *Hybrid cultures: Strategies for entering and leaving modernity*. Minneapolis, MN: University of Minnesota Press.

Carmody, P. (2009). *A new socio-economy in Africa: Thintegration and the mobile phone revolution.* Retrieved February 4, 2010 from https://www.tcd.ie/iiis/documents/discussion/pdfs/iiisdp279.pdf

Carroll, W., & Hackett, R. (2006). Democratic media activism through the lens of social movement theory. *Media Culture & Society, 28*(1), 83–104. doi:10.1177/0163443706059289

Castells, M. (1997). *The power of identity, vol. II of The information age: Economy, society and culture*. Oxford, UK: Blackwell.

Central Broadcasting Services (CBS). (2008). *Ettaka Lyaffe Ligenda*. Retrieved June 12, 2008 from http://www.radiocbsfm.com

Chaffee, S. H. (Ed.). (1975). *Political communication*. Beverly Hills, CA: Sage Publications.

Chambers, J. (1994). *Migrancy, culture, identity*. London, UK: Routledge.

Chang, W.-Y. (2005). Online civic participation, and political empowerment: Online media and public opinion formation in Korea. *Media Culture & Society, 27*(6), 925–935.

Chari, T. (2009). Ethical challenges facing Zimbabwean media in the context of the Internet. *Global Media Journal: African Edition, 3*(1), 1-34.

Cheeseman, N. (2008). *The Kenyan elections of 2007: An introduction. Journal of Eastern African Studies, 2(2)*. London, UK: Routledge.

Chengdu Evening News. (February 16, 2008). CCTV-awarded photo was determined a fake: The photographer admitted that the photo was doctored. Retrieved February 2, 2010, from http://news.sina.com.cn/ c/ p/ 2008-02-16/034914948587.shtml

Cheon, Y.-C. (2004). Internet newspapers as alternative media: OhmyNews in South Korea. *Media Development, 51*(1), 28–33.

Chepkonga, S. (2002). *Helping Kenya extend access to ICTs to rural areas (Acacia II). Communications Commission of Kenya (CCK)*. Nairobi, Kenya: Kenya Bureau of Statistics.

Chester, J. (2007). *Digital destiny: New media and the future of democracy*. New York, NY: The New Press.

China Daily. (Dec. 24, 2009). Top 10 hot topics on Internet 2009. Retrieved February 21, 2010 from http://www.chinadaily.com.cn /china/ 09tentopnews/ 2009-12/24/ content_919 6462.htm

China Daily. (February 28, 2009). Premier Wen talks online with public. Retrieved January 31, 2010, from http://www.chinadaily.com.cn/ china/ 2009-02/28/ content_752 2765.htm

China Daily. (November 24, 2003). Sex writer creates a stir, gains a following. Retrieved March 27, 2010, from http://www.chinadaily.com.cn/ en/ doc/ 2003-11/24/ content_284039.htm

China Digital Times. (February 28, 2009). China's Wen makes Internet debut. Retrieved March 9, 2010, from http://chinadigitaltimes.net/ 2009/ 02/ chinas-wen-makes-internet-debut/

China Internet Network Information Center. (2003). *The Internet Timeline of China* 1987-1996. Retrieved January 21, 2010, from http://www.cnnic.net.cn

China Internet Network Information Center. (2009). *Statistical report on Internet development in China.* Retrieved January 21, 2010, from http://www.cnnic.net.cn

China Internet Network Information Center. (May 18, 2009). *The Internet Timeline of China* 2008. Retrieved January 21, 2010, from http://www.cnnic.net.cn

China Journalism Yearbook. (2009). Institute of Journalism and Communication under the Chinese Academy of Social Sciences. Beijing: China Journalism Yearbook Publisher.

Chinese, B. B. C. (June 30, 2008). China ordered Weng'an rioters to turn themselves in. Retrieved from http://news.bbc.co.uk/ chinese/ simp/ hi/ newsid_7480000/ newsid_7480700/ 7480702.stm

Choe, S.-H. (2008, October 13). Korean star's suicide reignites debate on Web regulation. *The New York Times.*

Circute and Joel. (2008). *Juala*. Retrieved on 30 September, 2008, from http://www.youtube.com/watch?v=ziFHAATRGU

Clark, J. P. (1999). Ibadan. In Gerald, M., & Beier, U. (Eds.), *Penguin book of modern African poetry* (p. 256).

Clark, M. K. (2009). Questions of identity among African immigrants in America. In Okpewho, I., & Nzegwu, N. (Eds.), *The new African Diaspora* (pp. 255–269). Bloomington, IN: Indiana University Press.

Clifford, J. (1992). Travelling cultures. In Grossberg, L., Nelson, C., & Treichler, P. (Eds.), *Cultural studies*. London, UK & New York, NY: Routledge.

Cohen, R. (1978). Ethnicity; problem and focus in anthropology. *Annual Review of Anthropology*, *7*, 379–403. doi:10.1146/annurev.an.07.100178.002115

Cohen, D. W. (1970). A survey of interlacustrine chronology. *Journal of African History*, *11*(2), 177–202. doi:10.1017/S0021853700009932

Cohen, B. C. (1963). *The press and foreign policy*. Princeton, NJ: Princeton University Press.

Collins, P. H. (1991). Learning from the outsider within: The sociological significance of Black feminist thought. In *Beyond methodology: Feminist scholarship as lived research* (pp. 35–59). Indiana Press.

Constitution of the Republic of Uganda. (1995). Retrieved November 31, 2009, from http://www.ugandaonlinelawlibrary.com/files/ constitution/constitution_1995.pdf

Cook, S. (2004). New technologies and language change: Toward an anthropology of linguistic frontiers. *Annual Review of Anthropology*, *33*, 103–115. doi:10.1146/annurev.anthro.33.070203.143921

Cook, S. (2004). New technologies and language change: Toward an anthropology of linguistic frontiers. *Annual Review of Anthropology*, *33*, 103–115. doi:10.1146/annurev.anthro.33.070203.143921

Corona, I., & Madrid, A. L. (2008). *Postnational musical identities: Cultural production, distribution, and consumption in a globalized scenario* (pp. 85–207). MD, USA: Lexington Books.

Corona, I., & Madrid, A. (2008). Introduction: The postnational turn in music scholarship and music marketing. In Corona, I., & Madrid, A. L. (Eds.), *Postnational musical identities: Cultural production, distribution, and consumption in a globalized scenario*. MD, USA: Lexington Books.

Creese, G., & Kambere, E. N. (2004). What colour is your English? *The Canadian Review of Sociology and Anthropology. La Revue Canadienne de Sociologie et d'Anthropologie*, *40*(5), 565–573.

Creswell, J. W. (1998). *Qualitative inquiry and research design: Choosing among five traditions*. Thousand Oaks, CA: Sage Publications.

Crystal, D. (2001). *Language and the Internet*. Cambridge, UK: Cambridge University Press.

Cui, J. (Aug. 14, 2009). Green Dam launch not handled well. *China Daily.* Retrieved March 7, 2010, from http://www.chinadaily.com.cn/ cndy/ 2009-08/ 14/ content_8568194.htm

Dahlberg, L. (2001). Democracy via cyberspace. *New Media & Society*, *3*, 157–177.

Danmole, H. O. (2010). From the global to the local: The media and Islam in Nigeria, 1979-2007. In Adebayo, A. G., Adesina, O. C., & Olaniyi, R. O. (Eds.), *Marginality and crisis: Globalization and identity in Contemporary Africa* (pp. 89–196). New York, NY: Lexington Books.

Davis, T. C., & Postlewait, T. (2003). *Theatricality: Theater and performance theory*. Cambridge, UK: Cambridge University Press.

De Jager, C. (2007). SABC gives Africa a voice in the news. *Variety*, *407*(10), 18.

DeFleur, M. L., & Ball-Rokeach, S. (1989). *Theories of mass communication* (5th ed.). White Plains, NY: Longman.

Dennis, C. (2007). *The Afro-Colombianization of hip-hop and discourses on authenticity. Banjuka*. Retrieved on 16 December, 2008, from http://www.youtube.com/watch?v=s1lBvP4BW0o

Denton, R. E., & Woodward, G. C. (1990). *Political communication in America*. New York, NY: Praeger.

Denzin, N. K., & Lincoln, Y. S. (Eds.). (1994). *Handbook of qualitative research*. Thousand Oaks, CA: Sage.

Department for International Development. (2010). *Debt relief in Uganda*. Retrieved February 26, 2010, from http://www.dfid.gov.uk/Media-Room/News-Stories/ 2010/Debt-relief-in-Uganda/

Department of Defense. (1993). *Joint publication 3-0, doctrine for joint operations*. Washington, D.C.: Department of Defense.

Deuze, M. (2003). The Web and its journalisms: Considering the consequences of different types of news media online. *New Media & Society*, *5*(2), 203–230. doi:10.1177/1461444803005002004

Dialogue Kenya. (2007). *The Kenya national human rights commission report on post election violence*. Retrieved December 2, 2009 from http://www.dialoguekenya.org/docs/Agreement-commission-on-post-Election-violence.pdf

Diamandaki, K. (2003). Virtual ethnicity and digital diaporas: Identity construction in cyberspace. *Global Media Journal, 2*(2).

Diane, C., & Meier, P. (2009). *New technologies in emergencies and conflicts: The role of information and social networks*. Washington, DC & London, UK: UN Foundation-Vodafone Foundation Partnership.

Diop, C. A. (1974). *The African origin of civilization*. Westport, CT: Lawrence Hill Books.

Dogbe, E. (2003). Warped identities: Dress in popular West African video films. *Africa Identities*, *1*(1), 95–117. doi:10.1080/1472584032000127905

Dopke, S. (2000). The interplay between language-specific developments and cross linguistic influence. In Dopke, S. (Ed.), *Cross-linguistic structures in simultaneous bilingualism* (pp. 79–104). Amsterdam, The Netherlands & Philadelphia, PA: John Benjamins.

Downing, J. D. H. (1992). Spanish-language media in the greater New York region during the 1980s. In Riggins, S. H. (Ed.), *Ethnic minority media: An international perspective* (pp. 256–275). Newbury Park, CA: Sage.

Drew, D. M., & Snow, D. M. (1988). *Making strategy: An introduction to national security processes and problems. Maxwell AFB*. AL: Air University Press.

Driscoll, W. (2002). Fire for our children. CARE International/Kenya special report. Retrieved on January 1, 2006, from http://www.careusa.org/newsroom/special reports/aids/fire.asp

Drolet, K. (2009). *Cell phone culture*. Retrieved October 7, 2009, from http://media.www.thecurrentonline.com/home

Dryzek, J. S. (2002). *Deliberative democracy and beyond: Liberals, critics, contestations*. New York, NY: Oxford University Press.

Dube, M. W., & Kanyoro, M. (Eds.). (2004). *Grant me justice: HIV and gender readings of the Bible*. Maryknoll, NY: Orbis Books.

Dubois, N. (2002). *A sociocognitive approach to social norms*. New York, NY: Routledge.

Dunlap, C. (1998). *Preliminary observations: Asymmetrical warfare and the Western mindset, challenging the United States symmetrically and asymmetrically: Can America be defeated? Carlisle Barracks*. PA: U.S. Army War College.

Dunn, A. (2005). Participatory communication in malaria control: Why does it matter? *Findings, 4*, October.

Dyson, M. E. (2006). *Is Bill Cosby right? Or has the black middle class lost its mind?*Perseus Books Group.

Ebo, B. M. P. (2002). *Old wine in new wine bottle: The Internet and the techolonization of Africa*. Retrieved February 20, 2008, from http://www.arts.uwa.edu.au/MotsPluriels/MP2002be.html

Economist. (2010, January 30). *A world of connections: A special report on social networking*. (pp. 3-20).

Edugie, A., Mosanya M. E., Amajoh C., Otsemobor O., Ezedinachi E. N. U., Afolabi B. M.,... Aaron, K. O. (2004). *Nigeria roll back malaria consultative mission: Essential actions to support the attainment of the Abuja targets*. Nigeria RBM Country Consultative Mission, Final Report, 07 March.

Edwards, J. (1985). *Language, society, and identity*. Oxford, UK: Basil Blackwell.

Egbunike, N. A. (2009a). *Peering into the future: The fate of newspapers in a digital world*. Unpublished Seminar, Department of Communication and Language Arts, University of Ibadan.

Egbunike, N. A. (2009b). Power play. *Interpress Newsletter, 34*(3-4).

Eko, L. (2001). Steps toward Pan-African exchange: Translation and distribution of television programs across Africa's linguistic regions. *Journal of Black Studies, 31*, 365–379. doi:10.1177/002193470103100308

Ekwuazi, H. (2001). Political values and the Nigerian film. *Matatu, 23-24*, 275–282.

Elischer, S. (2008). *Ethnic coalitions of convenience and commitment: Political parties and party systems in Kenya*. German Institute of Global and Area Studies (GIGA) Working Papers No. 68.

Emenari, O. G. (2003). The Internet as a challenge to broadcasting and allied media. In Onyeka, N. (Ed.), *Communication and national development*. Onitsha, Nigeria: African Links Books.

Esan, O. (2008). Appreciating Nollywood: Audeinces and Nigerian films. *Participations: Journal of Audience and Reception Studies, 5*(1).

European Union, Election Observation Mission. (2008). *Kenya: Final report general elections*. 27 December 2007. European Union.

Fanon, F. (1967). *Toward the African revolution*. New York, NY: Grove Press.

Fantini, A. (1995). Introduction-language, culture and worldview: Exploring the nexus. *International Journal of Intercultural Religion, 19*(2), 143–153. doi:10.1016/0147-1767(95)00025-7

Farr, A. (2004). Whiteness visible: Enlightenment racism and the structure of racialized consciousness. In G. Yancy (Ed.), What White looks like: African-American philosophers on the whiteness question (pp. 143-157). New York, Y: Routledge.

Ferguson, R. (2004). *The media in question*. New York, NY: Arnold Publishers.

Firstenberg, S. (Dir.). (1984). *Breakin': Electric boogaloo*. Cannon Films Inc.

Fishman, M. (1980). *Manufacturing the news*. Austin, TX: University of Texas Press.

Fishman, J. (1991). *Reversing language shift: Theoretical and empirical foundations of assistance to threatened languages*. Clevedon, UK: Multilingual Matters.

Fitzpatrick, M. (2008, October 9). South Korea wants to gag the noisy Internet rabble: The leaders of the most wired country on Earth are seeking to curb online anonymity and debate, with laws that many say will fail. *The Guardian*, p. 6.

Fleur, M. L. (1973). *Theories of mass communication* (2nd ed.). Seattle, WA: Washington State University.

Flis, H., Miller, N., Senker, P., & Wyatt, S. (2000). *Technology and in/equality: Questioning the information society*. New York, NY: Routledge.

Folarin, B. (1998). *Theories of mass communication: An introductory text*. Ibadan, Nigeria: Stirling-Horden Publishers Ltd.

Fontein, J. (2009). Anticipating the tsunami: Rumours, planning and the arbitrary state in Zimbabwe. *Africa: The Journal of the International African Institute*, *79*(3), 1–21. doi:10.3366/E0001972009000862

Foucault, M. (1995). *Discipline & punish: The birth of the prison* (Sheridan, A., Trans.). New York, NY: Vintage Books.

Fourchard, L. (2002). *The case of Ibadan, Nigeria*. Ibadan, Nigeria: Institut Francais de Recherche en Afrique (IFRA), University of Ibadan. Retrieved May 13, 2010 from http://www.ucl.ac.uk/dpuprojects/Global_Report/pdfs/Ibadan.pdf

Frank, M. (1995). *AIDS-education through theatre*. Bayreuth African Studies, Bayreuth.

Fraser, N. (2007). Creating model citizens for the information age: Canadian Internet policy as civilizing discourse. *Canadian Journal of Communication*, *32*, 201–218.

Friedman, T. L. (2006). *The world is flat: A brief history of the twenty-first century* (3rd, ed.). New York: Farrar, Straus and Giroux.

Gandy, O. H. Jr. (1993). *The panoptic sort: A political economy of personal information*. Boulder, CO: Westview.

Garcia, O. (2009). *Bilingual education in the 21st century: A global perspective*. Wiley-Blackwell.

Garcia-Canclini, N. (1995). *Hybrid cultures: Strategies for entering and leaving modernity*. Minneapolis, MN: University of Minneapolis Press.

Garnham, N. (2001). *Emancipation, the media, and modernity*. Oxford, UK: Oxford University Press.

Gastil, J. (2008). *Political communication and deliberation*. Los Angeles, CA: Sage Publications.

Gathanju, D. (2010). Kenya builds digital villages. *EContent*, *33*(4), 18–22.

Gathanju, D. (2007). *Safaricom introduces mobile money transfer*.

George, C. (2003). The Internet and the narrow tailoring dilemma for Asian democracies. *Communication Review*, *6*, 247–268.

Giddens, A. (1990). *The Consequences of Modernity*. Cambridge, UK: Polity.

Gidi Gidi & Maji Maji. (2006). *Unbwogable*.

Giles, H., & Powersland, P. (1975). *Speech style and social evaluation*. London, UK: Academic Press.

Giles, H., Taylor, D., & Bourhis, R. (1973). Towards a theory of interpersonal accommodation through speech: Some Canadian data. *Language in Society*, *2*, 177–192. doi:10.1017/S0047404500000701

Giles, H., Copeland, N., & Copeland, J. (1991). Accommodation theory: Communication, context, and consequence. In Giles, H., Coupland, J., & Coupland, N. (Eds.), *Developments in applied sociolinguistics*. Cambridge, UK: Cambridge University Press. doi:10.1017/CBO9780511663673.001

Gill, S. (2000). Toward a postmodern prince? The battle in Seattle as a moment in the new politics of globalization. *Millennium: Journal of International Studies*, *29*(1), 131–140.

Githinji, P. (2006). Bazes and their shibboleths: Lexical variation and Sheng speakers' identity in Nairobi. *Nordic Journal of African Studies*, *15*(4), 443–472.

Githiora, C. (2002). Sheng: Peer language, Swahili dialect or emerging Creole? *Journal of African Cultural Studies*, *15*(2), 159–181. doi:10.1080/1369681022000042637

Glasgow University Media Group. (1980). *More bad news*. London, UK: Routledge.

Glick Schiller, N., Basch, L., & Blanc, C. S. (1995). From immigrant to transmigrant: Theorizing transnational migration. *Anthropological Quarterly*, *68*(1), 48–63. doi:10.2307/3317464

Goffman, E. (1967). On face-work: An analysis of ritual elements in social interaction. In Jaworski, A., & Coupland, N. (Eds.), *The discourse reader*. London, UK: Routledge.

Goldstein, J., & Rotich, J. (2008). *Digitally networked technology in Kenya's 2007 post election crisis*. Harvard, MA: Berkman Publishers.

Graber, D. A. (1981). Political language. In Nimmo, D., & Sanders, K. R. (Eds.), *Handbook of political communication* (pp. 195–223). Beverly Hills, CA: Sage.

Gramsci, A. (1971). *Selections from prison notebooks*. London, UK: Lawrence & Wishart.

Gramsci, A. (1971). *Selections from the prison notebooks* (Hoare, Q., & Smith, G. N. (Trans. Eds.)). New York, NY: International Publishers.

Grant, A. E., Guthrie, K. K., & Ball-Rokeach, S. (1991). Television shopping: A media system dependency perspective. *Communication Research*, *18*(6), 773–798. doi:10.1177/009365091018006004

Green, J. (1999). *Deep democracy: Community, diversity, and transformation*. New York, NY: Rowman & Littlefield.

Grice, H. (1975). Logic and conversation. In Jaworski, A., & Coupland, N. (Eds.), *The discourse reader*. London, UK: Routledge.

Grossberg, L. (2006). *Media making: Mass media in a popular culture*. Thousand Oaks, CA: Sage.

Gumperz, J. J. (1982). *Discourse strategies*. London, UK: Cambridge University Press. doi:10.1017/CBO9780511611834

Gumucio, D., & Tufte, T. (Eds.). (2006). *Communication for social change anthology*. New York, NY: Communication for Social Change Consortium.

Gupta, A., & Ferguso, J. (1992). Beyond culture: Space identity, and the politics of difference. *Cultural Anthropology*, *7*(1), 6–23. doi:10.1525/can.1992.7.1.02a00020

Gustainis, J. J. (1990). Demagoguery and political rhetoric: A review of the literature. *Rhetoric Quarterly*, *20*(2), 155–161.

Gutierrez, F. (1977). Spanish-language media in America: Background, resources, history. *Journalism History*, *4*(2), 34–68.

Gwyn, R. (2002). *Communicating health and illness*. Thousand Oaks, CA: Sage Publications.

Habermas, J. (1991). *The structural transformation of the public sphere: An inquiry into a category of bourgeois society* (Burger, T., Trans.). Cambridge, MA: MIT Press.

Habib, E. (2009, November 12). Tragedy of the talk shows. *The Daily Times*.

Hall, S., Critcher, C., Jefferson, T., Clarke, J., & Roberts, B. (1978). *Policing the crisis: Mugging, the state, and law and order*. London, UK: Routledge & Kegan Paul.

Hall, S. A. (1976). *Resistance through rituals: Youth subcultures in post-war Britain*. London, UK: Hutchison.

Hall, S. C. (1978). *Policing the crisis*. London, UK: Macmillan.

Hall, S. (1982). The re-discovery of ideology. In Bennett, J. C., & Woollcott, J. (Eds.), *M. Gurevitch, T* (pp. 56–90). London, New York: Methuen.

Hall, S. (1980). Cultural studies and the centre: Some problematics and problems. In Hall, S. (Ed.), *Culture, media, language: Working papers in cultural studies*. London, UK: Hutchison.

Hall, S. (1996). The West and the rest: Discourse and power. In Hall, D. H. S. (Ed.), *Modernity: An introduction to modern societies*. Malden, MA: Blackwell Publishers.

Hamilton, K. (2002). UVA lateest in string of black face incidents. *Black Issues in Higher Education*, *19*(22), 10.

Harding, F. (2003). Africa and the moving image: Television, film and video. *Journal of African Cultural Studies*, *16*(1), 69–84. doi:10.1080/1369681032000169276

Hatar, A. (2002). Theatising AIDS in rural communities: Lessons learnt. *Utafiti*, 4.

Hatar, A. (2005). Are we answering the right questions? Challenges in participatory development communication among the Barbaig in Northern Tanzania. *Tanzania Journal of Population Studies and Development, 1*.

Haynes, J. (1995). Nigerian cinema: Structural adjustments. *Research in African Literatures*, *26*(3), 97–119.

Haynes, J. (2003). Mobilizing Yoruba popular culture: Babangida must go. *Africa*, *73*(1), 122–138. doi:10.3366/afr.2003.73.1.77

Haynes, J. (2006). Political critique in Nigerian video films. *African Affairs, 105*(421), 511–533. doi:10.1093/afraf/adi125

Haynes, J. (2002). Devaluation and the video boom: Economics and thematics. In Guycr, J. I., Denier, L., & Agbaje, A. (Eds.), *Money struggles and city life: Devaluation in Ibadan and other urban centers in southern Nigeria. 1986-1996* (pp. 207–217). Portsmouth, NH: Heinemann.

Haynes, J. (1997). In Harrow, K. (Ed.), *African cinema: Postcolonial and feminist readings* (pp. 155–190). Lawrenceville, NJ: African World Press.

Haynes, J. (2000). Introduction. In Haynes, J. (Ed.), *Nigerian video films* (pp. 1–36). Athens, OH: Ohio University Press.

Haynes. J. (2007). Nnebue: The anatomy of power. *Film International, 5.4*(28), 30-40.

Haythornthwaite, C. (2002). Strong, weak, and latent ties and the impact of new media. *The Information Society*, 385–401. doi:10.1080/01972240290108195

Hazen, K. (2002). Identity and language variation in a rural community. *Language, 78*(2), 240–257. doi:10.1353/lan.2002.0089

He, D. C. (Feb. 19, 2010). *CCTV New Year skit revealed the truth about China's exports.* Retrieved February 22, 2010, from http://bookmark.people.com.cn/toViewBookmark.do? id=126265

Hermida, A., & Thurman, N. (2007). *Comments please: How the British news media is struggling with user-generated content.* Paper presented at the 8th International Symposium on Online Journalism, University of Texas, Austin, USA.

Higginbotham, E. B. (1992). African-American women's history and the metalanguage of race. *Signs, 17*(2), 25–274.

Hill, S. (1988). *The tragedy of technology: Human liberation versus domination in the late twentieth century.* London, UK: Pluto Press.

Hill Collins, P. (2000). Black feminist thought knowledge, consciousness, and the politics of empowerment (rev. 10th anniversary ed.). New York, NY: Routledge.

Hille, K., & Dombey, D. (January 15, 2010). Beijing demands companies show 'self-discipline.' *Financial Times,* London. Retrieved January 26, 2010, from http://www.ft.com/cms/s/0/26b045c2-0176-11df-8c54-00144feabdc0.html

Hintereder, P. (2009). Media moments. In *Magazin-deutschland; The New World of Media, 3.*

Hirsch, F. S. (2008). *Putting hate speech in context: Observations on speech, power, and violence in Kenya.* Paper prepared by George Mason University.

Hooks, b. (1989). Talking back: Thinking feminist, thinking black. Boston, MA: South End Press.

Howard, M. E. (1979). *Restraints on war: Studies in the limitations of armed conflict.* New York, NY: Oxford University Press.

Hu, Y. Q. (Nov. 13, 2009). *Baby dies while doctor plays game online.* Retrieved February 21, 2010, from http://www.chinadaily.com.cn/ cndy/ 2009-11/ 13/ content_8962291.htm

Human Rights Watch. (2008). *Ballots to bullets: Organized political violence and Kenya's crisis of government.* Retrieved from http://hrw.org/reports/2008/Kenya0308/kenya0308web.pdf 37

Hyden, G., Leslie, M., & Ogundimu, F. (2002). *Media and democracy in Africa.* New Brunswick, NJ: Transaction Publisher.

Hyden, G. (Ed.). (2003). Media and democracy in Africa, Uppsalla: NordiskaAfrikainstitutet

Ibelema, M., & Bosch, T. (2009). Sub-saharan Africa. In De Beer, A., & Merrill, J. (Eds.), *Global journalism: Topical issues and media systems* (pp. 293–336).

Ide, W. (2010). China using detention to silence dissent. *Voice of America.* Retrieved August 12, 2010, from http://www1.voanews.com/ english/ news/ asia/ China-Increasingly-Using-Prison-Sentences-and-Detention-to-Silence-Dissent-10044 2599.html

International Monetary Fund (IMF). (March, 2008). *Statement by IMF Staff Mission to Uganda.* Press Release No. 08/58, March 21. Retrieved June 18, 2008, from http://www.imf.org/external/np/ sec/pr/2008/pr0858.htm

International Parliamentary Union. (2009). *Women in national parliaments*. Retrieved October 25, 2009, from http://www.ipu.org/wmn-e/classif.htm

International Telecommunication Union. (2010). *Measuring the information society 2010*. Retrieved January 25, 2010, from http://www.itu.int/ ITU-D/ ict/ publications/ idi/ 2010/ Material/ MIS_2010_without %20annex %20 4-e.pdf

Internet Enemies. (2009). *Reporters without borders*. Retrieved February 27, 2010, from http://www.rsf.org/ IMG/ pdf/ Internet_enemies_2009_2_.pdf

Internet World Stats. (2009). *Top 20 countries with the highest number of Internet users*. Retrieved February 12, 2010 from http://www.internetworldstats.com/ top20.htm

ITU. (2009). *Manual for measuring ICT access and use by household and individuals*. Geneva, Switzerland: ITU Telecom World.

Jackson, N. A., & Lilleker, D. G. (2009). Building an architecture of participation? Political parties and Web. 2.0 in Britain. *Journal of Information Technology & Politics*, *6*, 232–250. doi:10.1080/19331680903028438

Jensen, M. (2001). Afribox, telecenters, cybercafés: ICT in Africa. *ICT in Africa*, *1*, 96–109.

Jimw@t. (2006). Under 18. *Operation Fagia, 4*. Calif Records.

Jimw@t. (2008). *Sitoi Kitu Kidogo*. Retrieved on 20 September, 2008, from http://www.youtube.com/ watch?v=e-L79xtsz40

Johnson, M. A. (2010). Incorporating self-categorization concepts into ethnic media research. *Communication Theory*, *20*(1), 106–125. doi:10.1111/j.1468-2885.2009.01356.x

Jones, S. (Ed.). (2002). *Pop music and the press*. Philadelphia, PA: Temple University Press.

Jones, J. (2004). The impairment of empathy in goodwill Whites for African Americans. In Yancy, G. (Ed.), *What White looks like: African-American philosophers on the Whiteness question* (pp. 65–86). New York, NY: Routledge.

Kahai, S. S. (2009). Anonymity and counter-normative arguments in computer-mediated discussions. *Group & Organization Management*, *34*, 449–478.

Kaid, L. L. (2004). Introduction and overview of the field. In Kaid, L. L. (Ed.), *Handbook of political communication research* (pp. xiii–xviii). Mahwah, NJ: Lawrence Erlbaum Associates, Publishers.

Kalichman, S. E., Simbayi, L. C., Jooste, S., Toefy, Y., Cain, D., Cherry, C., & Kagee, A. (2005). Development of a brief scale to measure AIDS-related stigma in South Africa. *AIDS and Behavior*, *9*(2). doi:10.1007/s10461-005-3895-x

Kaloki, T. (2010). *Kenya accounts for 7% of Africa mobile industry*. Retrieved March 17, 2010, from http://bizcommunity.co.ke/Article/111/78/43445.html

Kalyango, Y. Jr. (2009). Political news use and democratic support: A study of Uganda's radio impact. *Journal of Radio and Audio Media*, *16*(2), 200–215. doi:10.1080/19376520903279399

Kalyango, Y. Jr, & Eckler, P. (2010). Media performance, agenda building, and democratization in East Africa. *Communication Yearbook*, *34*, 355–389.

Kalyango, Y., Jr., & Winfield, B. (2009). Rhetorical media framing of two First Lady candidates across cultures. *Global Media Journal, 8*(15), 1-16. American Edition.

Kamwangamalu, N. (1998). 'We-code', 'they-codes' and 'codes-in-between': Identities of English and codeswitching in post-apartheid South Africa. *Multilingua*, *17*(2/3), 277–296. doi:10.1515/mult.1998.17.2-3.277

Kang, N., & Kwak, N. (2003). A multilevel approach to civic participation. *Communication Research*, *30*(1), 80–106. doi:10.1177/0093650202239028

Kang, J. (2000). Cyber-race. *Harvard Law Review*, *113*(5), 1131–1209.

Karan, K., Gimeno, J. D. M., & Tandoc, E. (2009). The Internet and mobile technologies in election campaigns: The Gabriella Women's Party during the 2007 Philippine elections. *Journal of Information Technology & Politics*, *6*, 326–339. doi:10.1080/19331680903047420

Karan, K., Tandoc, E., & Gimeno, J. D. M. (2009). The case of party-list winners and new media technologies in the 2007 Philippines elections. *Quaderns del CAC, 33*, 87–94.

Karatzogianni, A. (2006). *The politics of cyberconflict*. Routledge research in information technology and society.

Karunakaran, C. (2006). *Personal email describing systemic Whiteness* (Harper, B., Ed.). New York.

Kasoma, F. P. (1996). The foundations of African ethics (Afriethics) and the professional practice of journalism: The case of society-centred media morality. *Africa Media Review, 10*(3), 93–116.

Kawoya, V., & Makokha, J. S. (2009). The case for Kiswahili as a regional broadcasting language in East Africa. *The Journal of Pan African Studies, 2*(8), 11–35.

Kellner, D. (2004). The media and the crisis of democracy in the age of Bush-2. *Communication and Critical. Cultural Studies, 1*(1), 29–58.

Kembo-Sure, & Ogechi, N. O. (2006). Language planning and language reform in Kenya. In Kembo-Sure, S. Waitiki & N. O. Ogechi (Eds.) *Language planning for development in Africa.* (pp. 37 – 54). Eldoret, Kenya: Moi University Press.

Kenya Information Guide. (2009). *Kenya population*. Retrieved October 25, 2009, from http://www.kenya-information-guide.com/kenya-population.html

Kenya National Commission on Human Rights. (2007). *Still behaving badly: Second periodic report of the election-monitoring project*. Nairobi, Kenya: Kenya National Commission on Human Rights.

Kenya Population Census. (1999). Retrieved January 16, 2010 from http://www.minority-Rights.org/download.php?id=147/html

Kershaw, B. (1992). *The politics of performance: Radical theater as cultural intervention*. New York, NY: Routledge. doi:10.4324/9780203412282

Khan, M. A. (2009). The mediatization of politics in Pakistan: A structural analysis. *Pakistaniaat: A Journal of Pakistan Studies, 1*(1).

Kiai, M. (2008). Speech, power, and violence: Hate speech and the political crisis in Kenya. In *Proceedings of the International Professional Communication Conference*. Nairobi, Kenya: Kenya National Commission on Human Rights.

Kiba, A. (2007). *Cinderella*. G-Records.

Kim, S. (2000). *The politics of democratization in Korea: The role of civil society*. Pittsburgh, PA: University of Pittsburgh Press.

Kim, J.-E. (2001). *Journalists in peril: The Korean case*. People's Coalition for Media Reform. Retrieved February 27, 2010, from http://www.pcmr.or.kr/ english/9.html

Kim, S.-H. (2008, June 24). Hundreds admit to pushing boycott. *Korea Herald*.

King, P. (1997). The press, candidate images, and voter perceptions. In McCombs, M., Shaw, D. L., & Weaver, D. (Eds.), *Communication and democracy: Exploring the intellectual frontiers in agenda-setting theory* (pp. 29–40). Mahwah, NJ: Lawrence Erlbaum Associates.

Kirk, D. (2009, January 27). Financial blogger's arrest tests Korea's progress on human rights. *The Christian Science Monitor*. Retrieved February 23, 2010, from www.csmonitor.com/ World/ Asia-South.../ p04s02-wosc.html

Kirkegaard, A. (2002). Introduction. In Palmberg, M., & Kirkegaard, A. (Eds.), *Playing with identities in contemporary music in Africa* (pp. 7–18). Uppsala, Sweden: Nordiska Afrikainstitutet.

Kitakufe, A. (2006). *The rise of East African hip-hop*. Retrieved on 20 September, 2008, from http://www.ugpulse.com/articles/daily/homepage.asp?ID=418#mat1

Kiwanuka, M. S. M. (1968). Bunyoro and the British: A reappraisal of the causes for the decline and fall of an African kingdom. *Journal of African History, 9*, 603–619. doi:10.1017/S0021853700009063

Kiwanuka, M. S. M. (1971). *The kings of Buganda*. Nairobi, Kenya: East African Press.

Kiwere, N. (January 10, 2008). Land saga brewing widespread agitation in Uganda. Retrieved on March 13, 2011 from http://www.ugandanlink.com/politics/land-saga-brewing- widespread-agitation-in-uganda.html

Kleptomaniax. (2005). *Swing*. Ogopa Deejays.

km (□□□). (2009). Retrieved January 31, 2010 from http://www.hudong.com/ wiki/ %E6% AC% BA%E5 %AE%9E %E9%A9 %AC

KNCHR & KHRC. (2006). *Behaving badly: Deception, chauvinism and waste during the referendum campaigns*. Nairobi, Kenya: KNCHR & KHRC.

Kolko, B. E., Nakamura, L., & Rodman, G. B. (2000). *Race in cyberspace*. New York, NY: Routledge.

Koloko, B., Nakamura, L., & Rodman, G. (2000). *Race in cyberspace*. New York, NY: Routledge.

Kramsch, C. (2001). *Language and culture*. Oxford, UK: Oxford University Press.

Krauss, R., & Chiu, C. (1998). Language and social behavior. In Gilbert, D., Fiske, S., & Lindzey, G. (Eds.), *Handbook of social psychology* (4th ed., *Vol. 2*, pp. 41–88). New York, NY: McGraw-Hill.

Kreiss, D. (2009). Developing the good citizen: Digital artifacts, peer networks, and formal organizations during the 2003-2004 Howard Dean campaign. *Journal of Information Technology & Politics*, *6*, 281–297. doi:10.1080/19331680903035441

Krings, M. (2005). Muslim martyrs and pagan vampires: Popular video films and the propagation of religion in northern Nigeria. *Postscripts: The Journal of Sacred Texts and Contemporary Worlds*, *1*(2-3), 185–205.

Krings, M. (2008). Conversion on screen: A glimpse at popular Islamic imaginations in northern Nigeria. *Africa Today*, *54*(4), 45–68. doi:10.2979/AFT.2008.54.4.44

Krings, M. (2004). From possession rituals to video dramas: Some observations of dramatis personae in Hausa performing arts. In Adamu, Y., Uba Adarou, A., & Faruk Jibrin, U. (Eds.), *Hausa home videos: Technology, economy, and society* (pp. 162–170). Kano, Nigeria: Center for Hausa Cultural Studies.

Krings, M. (2009). *Glocalization from below: Intercultural transcriptions and Muslim hegemony in Nigerian video culture*. Retrieved on September 22, 2009, from http:// www.gps.uni-annover.de/ ts/ pdfs/ Krings.pdf

Kukubor, K. B. (2006). *Ethnicity, discrimination and national integration*. Accra, Ghana: Daily Graphic.

Kun, J. (2005). *Audiotopia: Music, race, and America*. Berkeley/ Los Angeles, CA & London, UK: University of California Press.

Kuoega, J. (2003). Camfranglais: A novel slang in Cameroon schools. An account of the intricate hybrid used by young people in West Africa. *English Today*, *19*(2), 23–29.

Kupe, T. (2004). An agenda for researching African media and communication contexts. *Ecquid Novi: South African Journal for Journalism Research*, *25*(2), 353–356. doi:10.3368/ajs.25.2.353

Labelle, R. (2005). *ICT policy formulation and e-strategy development: A comprehensive guidebook*. Retrieved June 8, 2010 from http://www.apdip.net/publications/ ict4d/ict4dlabelle.pdf

Labov, W. (1972). *Sociolinguistic patterns*. Philadelphia, PA: University of Pennsylvania Press.

Labov, W. (1972). *Language in the inner city: Studies in Black English vernacular*. Philadelphia, PA: University of Pennsylvania Press.

Lang, G. E., & Lang, K. (1983). *The battle for public opinion: The President, the press, and the polls during Watergate*. New York, NY: Columbia University Press.

Lange, S. (in press). *Multipartysim, rivalry and Taarab in Dar es Salaam*. (pp. 165-180).

Larkin, B. (1997). Indian films and Nigerian lovers: Media and the creation of parallel modernities. *Africa*, *67*(3), 406–440. doi:10.2307/1161182

Larkin, B. (2004a). Degraded images, distorted sounds: Nigerian video and the infrastructure of piracy. *Public Culture*, *16*(20), 289–314. doi:10.1215/08992363-16-2-289

Larkin, B. (2008). *Signal and noise: Media, infrastructure and urban culture in northern Nigeria*. Chapel Hill, NC: Duke University Press.

Larkin, B. (2004b). From Majigi to Hausa video films: Cinema and society in northern Nigeria. In Adamu, Y., Uba Adarou, A., & Faruk Jibrin, U. (Eds.), *Hausa home videos: Technology, economy, and society* (pp. 46–53). Kano, Nigeria: Centre for Hausa Cultural Studies & Adamu Joji Publishers.

Larkin, B. (2002). The materiality of cinema theaters in northern Nigeria. In Ginsburg, F. D., Abu-Lughod, L., & Larkin, B. (Eds.), *Media worlds: Anthropology on new terrain* (pp. 319–336). Berkeley, CA: University of California Press.

Larkin, B. (2000). Hausa dramas and the rise of video culture in Nigeria. In Haynes, J. (Ed.), *Nigerian video films* (pp. 209–241). Athens, OH: Ohio University Press.

Larkin, B. (2003). Itineraries of Indian cinema: African videos. Hollywood, and global media. In Shohat, E., & Sham, R. (Eds.), *Multiculturalism, postcolonialism, and transnational media* (pp. 170–192). New Brunswick, NJ: Rutgers University Press.

Latham, K. (2007). SMS, communication, and citizenship in China's information society. *Critical Asian Studies, 39*(2), 295–314. doi:10.1080/14672710701339493

Lathan, S. (Dir.). (1984). *Beat street*. Orion Pictures.

Le Page, R., & Tabouret-Keller, A. (1985). *Acts of identity: Creole-based approaches to language and ethnicity*. Cambridge, UK: Cambridge University Press.

Lee, Y. (2009). Internet election 2.0? Culture, institutions, and technology in the Korean presidential elections of 2002 and 2007. *Journal of Information Technology & Politics, 6*, 312–325. doi:10.1080/19331680903050085

Lee, Y. (2008, October 26). S. Korean actress' suicide fuels cyber-insult debate. *Los Angeles Times*. Retrieved February 21, 2010, from http://articles.latimes.com/ 2008/ oct/26/ world/ fg-net26

Levert, L., & Mumma, O. J. (Eds.). (1995). *Drama and theatre: Communication in development*. Nairobi, Kenya: KDEA.

Levine-Rasky, C. (2002). Critical/relational/contextual: Toward a model for studying Whitness. In Levine-Rasky, C. (Ed.), *Working through Whiteness: International perspectives* (pp. 319–352). New York, NY: SUNY.

Levitt, S. D., & Dubner, S. J. (2005). *Freakonomics: A rogue economist explores the hidden side of everything*. New York, NY: Morrow.

Li, D. (2008, July 8). China's leaders, the media and the Internet. *Open Democracy: Free Thinking for the World*. Retrieved February 28, 2010, from http://www.opendemocracy.net/ article/ china-s-leaders-and-the-internet

Loges, W. E., & Ball-Rokeach, S. J. (1993). Dependency relations and newspaper readership. *The Journalism Quarterly, 70*(3), 602–614.

Long, N. E. (1958). The local community as an ecology of games. *American Journal of Sociology, 64*, 251–261. doi:10.1086/222468

Longombas. (2005). Vuta Pumz. *Chukua*. Ogopa Deejays.

Lords, E. (2001). Free speech becomes primary issues in Auburn blackface case. *Black Issues in Higher Education, 18*(22), 11.

Louw, E. (2001). *The media and cultural production*. London, UK: Sage.

Low, D. A., & Pratt, R. C. (1960). *Buganda and British overrule, 1900-1955: Two studies*. Nairobi, Kenya: Oxford University Press.

Luganda, M. M. (2001). Magazeti ya Kiswahili na maendeleo nchini Kenya. In Shitemi, N. L. (Eds.), *Kiswahili: A tool for development- a multidisciplinary approach* (pp. 197–205). Eldoret, Kenya: Moi University Press.

Lull, J. (2002). *Media, communication, culture: A global approach*. New York, NY: Columbia University Press.

Lynch, G. (2006). Negotiating ethnicity: Identity politics in contemporary Kenya. *Review of African Political Economy, 33*(107), 49–65. doi:10.1080/03056240600671282

MacGregor, K. (2003, August 18). Africa's satellite effect: They are creating the first Pan-African networks—and hits. *Newsweek (International Edition), 39*.

Machill, M., & Beiler, M. (2009). The importance of the Internet for journalistic research. *Journalism Studies, 10*(2), 178–203. doi:10.1080/14616700802337768

Maddox, A. H. (2003). Blackface, friend chicken, watermelon and a lynching - it's OK. *New York Amsterdam News, 10*, 13.

Maganda, F. (2006). A hybrid identity among the Sukuma in the Tanzanian Africa Inland Mission. *Southern African Review of Education (SARE) with Education with Production (EWP), 12*(1), 31-43.

Makoloo, M., & Ghai, Y. (2008). *Kenya: Minorities, indigenous peoples and ethnic diversity*. Nairobi, Kenya: Minority Rights Group International.

Makunja, G. S. (1995). Sarufi na msamiati wa Kiswahili katika vyombo vya habari. In Mlacha, S. A. K. (Ed.), *Kiswahili na Vyombo vya Habari* (pp. 14–24). Dar es salaam, Tanzania: TUKI.

March 07, 2008, from http://www.ugandaobserver.com/new/index.php

Marcus, E. (2005). *Meat market: Animals, ethics, and money*. New York, NY: Brio Press.

Marczyk, G., DeMatteo, D., & Festinger, D. (2005). *Essentials of research design and methodology*. Hoboken, NJ: John Wiley & Sons, Inc.

Marston, S. A., Woodward, K., & Jones, J. P. III. (2007). Flattening ontologies of globalization: The Nollywood case. *Globalizations*, *4*(1), 45–63. doi:10.1080/14747730701245608

Maseruka, J., & Musoke, C. (September 30, 2009). *Kabaka Mutebi, Museveni meet*. Retrieved October 5, 2009, from http://www.newvision.co.ug/D/8/12/696342

Mason, J. (1996). *Qualitative researching*. London, UK: Sage.

Mason, J. (2006). Mixing methods in a qualitatively driven way. *Qualitative Research*, *6*(1), 9–25. doi:10.1177/1468794106058866

Masuku, J. (2009, October 20). Letter from Zimbabwe: Digital tv taking over. *Radio Netherlands Worldwide*. Retrieved from http://www.rnw.nl/english

Matonya and Lady Jaydee. (2007). Anita. *Siamini*. Mandugu Digital/41 Records.

Matsiko, G. (May 2, 2008) Press freedom remains elusive in Uganda. Retrieved May 2, 2008, from http://allafrica.com/stories/ 200805010675.html

Matu, P. M. (2009). Newspaper reporting of the run-up to the 1997 Kenyan general election. *Journal of Intra-African Studies*, *1*, 37–53.

Mbiriyamveka, J. (2009, September 19). Studio 263 Back! *The Herald*. Retreived from http://www.allAfrica.com/stories

Mbogo, M. (2010). The impact of mobile payments on the success and growth of micro-business: The case of M-Pesa in Kenya. In Iraki, F. K. (Ed.), *Journal of Language, Technology & Entrepreneurship in Africa (JOLTE), 2(1)*.

Mbugua wa, M. (2003). *Identity politics in Nairobi Matatu folklore*. Unpublished doctoral dissertation. Jerusalem, Israel: Hebrew University.

McCabe, J. (2009). Resisting alienation: The social construction of Internet communities supporting eating disorders. *Communication Studies*, *60*(1), 1–16.

McCall, J. (2007). *Viewer discretion advised: Taking control of mass media influences*. Lanham, MD: Rowman and Littlefield.

McCall, J. C. (2004). Nollywood confidential: The unlikely rise of Nigerian video film. *Transition*, *95*, 98–109.

McChesney, R. (1999). *Rich media, poor democracy: Communication politics in dubious times*. Urbana, IL: University of Illinois Press.

McClendon, J. H. (2004). On the nature of Whiteness and the ontology of race. In Yancy, G. (Ed.), *What White looks like: African-American philosophers on the Whiteness question* (pp. 211–225). New York, NY: Routledge.

McIntyre v. Ohio Elections Commission, 514 U.S. 334 (1995).

McMillan, S. J. (2006). Coming of age with the internet: A qualitative exploration of how the Internet has become an integral part of young people's lives. *New Media & Society*, *8*(1), 73–95. doi:10.1177/1461444806059871

McMillin, D. C. (2007). *International media studies*. Oxford, UK: Blackwell Publishers.

McNair, B. (2007). *An introduction to political communication* (4th ed.). New York, NY: Routledge.

McQuail, D. (2007). *McQuail's mass communication theory*. London, UK: Sage Publications Ltd.

McQuail, D. (2005). *Mass communication theory* (5th ed.). Thousand Oaks, CA: Sage.

Mda, Z. (1993). *When people play people: Development communication through theatre*. London, UK: Zed Books.

Media Monitoring Project Zimbabwe. (2007). *Free the airwaves-your right to the freedom of choice*. Retrieved from http://www.mmpz.org.zw

Mellow, G. O., van Slyck, P., & Eynon, B. (2003). The face of the future: Engaging in diversity at LaGuardia Community College. *Change*, *35*(2), 10–17. doi:10.1080/00091380309604088

Merrill, J. C., & Lowenstein, R. L. (1971). *Media, messages, and men: New perspectives in communication*. New York, NY: McKay.

Messer, D. E. (2002). *Breaking the conspiracy of silence: Christian churches and the global AIDS crisis*. Minneapolis, MN: Fortress Press.

Meyerhoff, M. (1998). Accommodating your data: the use and misuse of accommodation theory in sociolinguistics. *Language & Communication*, *18*, 205–225. doi:10.1016/S0271-5309(98)00006-8

Meyer-Weitz, A. (2005). Understanding fatalism in HIV/AIDS protection: The individual in dialogue with contextual factors. *African Journal of AIDS Research*, *4*(2), 75–82. doi:10.2989/16085900509490345

Miller, D., & Slater, D. (2000). *The Internet: An ethnographic approach*. Oxford, UK: Berg.

Mkilifi, M. (1972). Triglossia and Swahili-English bilingualism in Tanzania. *Language in Society*, *1*(2), 197–213. doi:10.1017/S004740450000049X

Mlacha, S. A. K. (Ed.). (1995). *Kiswahili na Vyombo vya Habari*. Dar es salaam, Tanzania: TUKI.

Mlama, P. M. (1991). *Culture and development: The popular theatre approach in Africa*. Uppsala, Sweden: Nordiska Afrikainstitute.

Mohochi, S. (2000). Kiswahili na vyombo vya habari nchini: Matatizo katika matumizi ya lugha. In Njogu, K. (Eds.), *Kiswahili katika Karne ya Ishirini na Moja (Kiswahili in the 21st Century)* (pp. 147–157). Cape Town, South Africa: Centre for Advanced Study of African Society.

Mokwetsi, J. (2005, July 24). Satellite tv proves a hit. *The Standard*. Retrieved from http://thestandard.co.zw/entertainment/14296.html

Montero, M. D. (2009). Political e-mobilisation and participation in the election campaigns of Ségolène Royal (2007) and Barack Obama (2008). *Quaderns del CAC*, *33*, 27–34.

Mouffe, C. (2000). *The democratic paradox*. New York, NY: Verso.

Moyo, D. (2007). Alternative media, diasporas and the mediation of the Zimbabwe crisis. *Ecquid Novi: African Journalism Studies*, *28*(1&2), 81–105. doi:10.3368/ajs.28.1-2.81

Moyo, D. (2009). Citizen journalism and the parallel market of information in Zimbabwe's 2008 elections. *Journalism Studies*, *10*(4), 551–567. doi:10.1080/14616700902797291

Moyo, L. (2009). Repression, propaganda, and digital resistance: New media and democracy in Zimbabwe. In Mudhai, O. F., Tettey, W. J., & Banda, F. (Eds.), *African media and the digital public sphere* (pp. 57–71). New York, NY: Palgrave Macmillan.

Moyo, J. (2004, November). *The role of the media in politics*. Paper presented at the Department of Journalism and Media Studies, National University of Science and Technology, Bulawayo, Zimbabwe.

Mpiranga, F. (1995). Kiswahili katika vyombo vya habari Rwanda. In Mlacha, S. A. K. (Ed.), *Kiswahili na Vyombo vya Habari* (pp. 93–123). Dar es salaam, Tanzania: TUKI.

Mr. Ebbo. (2006). *Kamongo*. Motika Records.

Mudhai, F. (2004). Researching the impact of ICTs as change catalysts in Africa. *Ecquid Novi: South African Journal for Journalism Research*, *25*(2), 313–335. doi:10.3368/ajs.25.2.313

Mudhai, F. O., & Nyabuga, G. (2001, September). *The Internet: Triumphs and trials for journalism in Kenya.* Paper presented at the Annual Highway Africa Conference, Rhodes University. Retrieved June 24, 2004 from http://www.highwayafrica.org.za/presentations/55.doc

Muhumuza, R. (May 28, 2008). Traditional institutions have no place – Museveni. Retrieved May 28, 2008, from http://allafrica.com/stories/ 200805280331.html

Mukhongo, L. L. (2009). The Kenyan media: Confronting state terror and atrocities. *Journal of Communication and Language Arts. Special Issues on Media and Democracy*, *3*, 26–43.

Mulama, J. (2008, December 24). *Kenya press freedom: Going, going, gone*. Retrieved February 14, 2010, from http://ipsnews.net/news.asp?idnews=45214

Mulira, P. (November 6, 2006). Demarcation of Uganda led to present problems. *The New Vision Online*. Retrieved November 6, 2006, from http://www.newvision.co.ug/D/8/20/530625

Mulira, P. (October 7, 2009). Buganda must outgrow the culture of nostalgia. Retrieved December 14, 2009 from http://www.newvision.co.ug/ D/8/20/697029

Muloki dies at 87. Retrieved December 12, 2008, from http://majimbokenya.com/home/2008/09/01/ rest-in-peace-kyabazinga-wako-muloki-dies-at-87/

Mumma, O. J., Mwangi, E., & Odhiambo, C. (Eds.). (1998). *Orientation of drama, theatre, and culture*. Nairobi, Kenya: KDEA.

Muriuki, G. (1977). *A history of the Agikuyu*. Nairobi, Kenya: Oxford University Press.

Musau, P. M. (2001). Freeing the airwaves or cultural enslavement: The case of media in Kenya. In Shitemi, N. L. (Eds.), *Kiswahili: A tool for development- a multi-disciplinary approach* (pp. 36–42). Eldoret, Kenya: Moi University Press.

Musiyiwa, M. (2008). The mobilization of popular music in the promotion of national unity in Zimbabwe. *Journal of Music Research in Africa*, *5*(1), 11–29.

Musoke, C. (May 21, 2008). Cabinet discusses state of media. Retrieved May 21, 2008, from http://www.newvision.co.ug/D/8/12/629236

Musoke, C., & Bita, G. (August 25, 2008). Museveni visits Kyabazinga in hospital. Retrieved March 13, 2011 from http://www.newvision.co.ug/ D/8/12/646353

Mute, L. (2008). Kenya: Legislation, hate speech, and freedom of expression. Retrieved from http://allafrica.com/stories/200810240331.html

Mutula, S. M. (2004). Making Botswana an information society: Current developments. *The Electronic Library*, *22*(2), 144–153. doi:10.1108/02640470410533407

Muwanga, D. (June 18, 2008). Uganda gets tractor assembly plant. Retrieved June 18, 2008, from http://www.newvision.co.ug/D/ 8/13/634339

Muyita, S. & Bareebe, G. (September 1, 2008). Rest in peace: Kyabazinga Wako

Mwalongo, R. (2008, January 26). Spreading the word of hate in Kenya. *Guardian*.

Mwangi, E. (2004). Masculinity and nationalism in East African Hip hop music. *Tydskrif Vir Letterkunde*, *41*(2).

Mwanje, R. (2008, June 15). Buganda refuses to be smeared. Retrieved June 15, 2008 from http://allafrica.com/stories/ 200806160517.html

Mwanje, R., & Senkabirwa, A. M. (January 5, 2008). Police grills Buganda officials over land debate. *The Monitor Online*. Retrieved January 5, 2008 from http://monitor.co.ug/

Mwansoko, H. J. M. (1995). Dhima ya vyombo vya habari katika ukuzaji, usanifishaji na uenezaji wa istilahi. In Mlacha, S. A. K. (Ed.), *Kiswahili na Vyombo vya Habari* (pp. 1–13). Dar es salaam, Tanzania: TUKI.

Mwanzi, H. A. (1977). *A history of the Kipsigis*. Nairobi, Kenya: East African Literature Bureau.

Mwarogo, P. (2007). *From behavior change communication to strategic behavioral communication on HIV in Kenya, 1999–2006: Impact, project experience, and lessons learned*. Nairobi, Kenya: USAID/FHI.

Mwesige, P. G. (2004). Cyber elites: A survey of Internet cafe users in Uganda. *Telematics and Informatics*, *21*(1), 83–101. doi:10.1016/S0736-5853(03)00024-8

Mwesige, P. G. (2009). The democratic functions and dysfunctions of political talk radio: The case of Uganda. *Journal of African Media Studies*, *1*(2), 221–245. doi:10.1386/jams.1.2.221_1

Mwita, M. (2010). *Breaking cultural barriers of HIV/AIDS intransigence: An examination of theatre-against-AIDS interventions in Kenya*. Unpublished PhD Thesis, University of Dar es Salaam.

Mwita, M. (2011). 'From a cathartic function to confronting stigma: Theorizing the conscietization efficacy of theater-against-AIDS interventions in Kenya.' in *East African literature: Essays in written and oral traditions*. Eds. J. K. S. Makokha, Egara Kabaji & Dominica Dipio. Berlin. Logos Verlag. pp. 319-344

My Uganda. (2010a). *Monarchies-Bunyoro kingdom.* Retrieved January 20, 2010, from http://www.myuganda.co.ug/about/peoples/ monarchies/bunyoro.php

My Uganda. (2010b). Monarchies-Tooro kingdom. Retrieved January 12, 2010, from http://www.myuganda.co.ug/about/peoples/ monarchies/toro.php

Myers-Scotton, C. (1989). Codeswitching with English: Types of switching, types of communities. *World Englishes*, *8*, 333–346. doi:10.1111/j.1467-971X.1989.tb00673.x

Myers-Scotton, C. (1990). Elite closure as a powerful language strategy: The African case. *International Journal of the Sociology of Language*, *103*, 149–163.

Myers-Scotton, C. (2002). *Contact linguistics: Bilingual encounters and grammatical outcomes*. Oxford, UK & New York, NY: Oxford University Press.

Myers-Scotton, C., & Bolonyai, A. (2001). Calculating speakers: Codeswitching in a rational choice model. *Language in Society*, *31*(1), 1–28. doi:10.1017/S0047404501001014

Myers-Scotton, C. (1992). Codeswitching in Africa: A model of the social functions of code relations. In Herbert, K. R. (Ed.), *Language and society in Africa*. Witwatersland, South Africa: Witwatersland University Press.

Myers-Scotton, C. (1993). Social motivations for codeswitching: Evidence from Africa. In Romaine, S., & Mulhausler, P. (Eds.), *Oxford studies in language contact*. Oxford, UK: Claredon Press.

Myhill, J. (2003). The native speaker, identity and the authenticity hierarchy. *Language Sciences*, *2*, 77–97. doi:10.1016/S0388-0001(01)00023-7

Naficy, H. (1993). *The making of exile cultures: Iranian television in Los Angeles*. Minneapolis, MN: University of Minneapolis Press.

Naficy, H. (Ed.). (1999). *Home, exile, homeland: Film media and the politics of place*. New York, NY: Routledge.

Nagra, N. (2003). Whiteness in Seattle: Anti-globalization activists examine racism within the movement. Alternatives Journal(Winter 2003), 27-28.

Nakamura, L. (2002). *Cybertypes: Race, ethnicity, and identity on the Internet*. New York, NY: Routledge.

Nalugo, M., & Mugerwa, Y. (October 24, 2010). Museveni orders re-opening of CBS radio station. Retrieved on March 13, 2011 from http://allafrica.com/stories/201011010005.html

Nam, S. (2008). The politics of compressed development in new media: A history of Korean cable television, 1992-2005. *Media Culture & Society*, *30*(5), 641–661.

Namutebi, J. (December 11, 2009). Regional government bill tabled. *The New Vision*. Retrieved June 10, 2010, from http://allafrica.com/stories/ 200912141161.html

National Bureau of Statistics Tanzania. (2009). *Census Population*. Retrieved October 25, 2009, from http://www.nbs.go.tz/population.htm

Ndigirigi, G. (2007). *Ngugi wa Thiong'o's drama and the Kamiriithu popular theater experiment*. Trenton, NJ: African World Press.

Ndlela, D. (2009, July 15). ZBC tv spruces up to win viewers, adverts. *Bizcommunity*. Retrieved from http://www.bizcommunity.com/article/238/66/37933.html

Nederveen Pieterse, J. (1995). Globalization as hybridization. In Featherstone, M. (Eds.), *Global modernities* (pp. 45–68). London, UK: Sage.

Nevin, T. (2002). Africa's new challenge: Crossing the digital divide. *African Business*, *277*, 24–26.

News, B. B. C. (2009, January 2). *Kenya 'media gag' law approved.* Retrieved February 14, 2010, from http://news.bbc.co.uk/go/pr/fr/-/2/hi/africa/7808815.stm

Ngesa, M. (2007, March 17). Setback for anti-AIDS fight as patients sell drugs to buy food. *Nation Newspapers*.

Ngũgĩ wa, T. (1981) *Writers in politics: Essays*. London, UK: Heinemann.

Ng'weno, H. B. (1969). The nature of the threat to press freedom in East Africa. *Africa Today*, *16*(3), 1–4.

Nielsen, J. (2006). *Participation inequality: Encouraging more users to contribute*. Retrieved February 21, 2010, from http://www.useit.com/ alertbox/ participation_inequality.html

Nissanke, M., & Thorbecke, E. (2006). Channels and policy debate in the globalization–inequality–poverty nexus. *World Development*, *34*, 1338–1360. doi:10.1016/j.worlddev.2005.10.008

Njogu, K. (Ed.). (2009). *Media and identity in Africa*. Edinburg University Press for the International African Institute.

Njogu, K. (2001). Discourse and serial drama in the promotion of health in Kenya. In Shitemi, N. L. (Eds.), *Kiswahili: A tool for development- a multidisciplinary approach* (pp. 181–187). Eldoret, Kenya: Moi University Press.

Nnyago, U. K. (February 15, 2008). The land amendment bill: Buganda's blessing in disguise. Retrieved February 15, 2008 from http://allafrica.com/stories/ 200802141011.html

Noelle-Neumann, E. (1993). *The spiral of silence: Public opinion – our social skin* (2nd ed.). Chicago, IL: University of Chicago Press.

Nonini, J., & Qchilla. (2007). Kumekucha. *Aluta Continua*. GMC Wasanii PromotersLtd.

Nordlinger, J. (April 2, 2007). China dreamin'. *National Review Online*. Retrieved March 14, 2010, from http:// nrd.nationalreview.com/ article/ ?q=NjdmMmI zOWZhM 2NkMjA5 Y2EyZDJm NDY 3MD AzN2 VkM2Y=

Norris, P. (2001). *Digital divide: Civic engagement, information poverty, and the Internet worldwide*. Cambridge, MA: Cambridge University Press.

Ntabo, M. M., & Ogechi, N. O. (2009). *Siege politics: The underlying problem in the Kenyan crisis*. Paper presented at the Moi University-Indiana University Purdue University Indianapolis conference at Moi University, June 2009.

Ntarangwi, M. (2007). Hip-hop, westernization and gender in East Africa. In Njogu, K., & Maupeu, H. (Eds.), *Songs and politics in Eastern Africa*. East Lansing, MI: Michigan State University Press.

Nwadiora, E. (1996). Therapy with African families. *The Western Journal of Black Studies*, *20*, 117–124.

Nwagwu, M. (2010, April 28). Where is Nigeria? *The Guardian*, p 67.

Nyabiaegea, J. (2009). *Mobile was a gift horse from the heavens*. Retrieved March 17, 2010, from http://www.nation.co.ke/news/-/1056/830852/-/vona75/-/index.html

Nyamnjoh, F. B. (2005). *Africa's media: Democracy and the politics of belonging*. London, UK: Zed Books.

O'Shaughnessy, M., & Stadler, J. (2008). *Media and society*. Oxford University Press.

Obanyi, G., & Pyne-Mercier, L. (2000). Realizing the HIV prevention-to-care continuum in Kenya. *Impact on HIV: Making a Difference*, *2*(2), 11–15.

Obbo, O. C. (February 24, 2010). Why CBS might be off air until 2016. Retrieved February 25, 2010 from http:// www.monitor.co.ug/OpEd/OpEdColumnists/ CharlesOnyangoObbo/-/878504/867528/-/xgnuus/-/index.html

Ochieng, P. (1992). *I accuse the press: An insider's view of the media and politics in Africa*. Initiative Publishers.

Ochieng, W. R. (1974). *An outline history of Nyanza up to 1914. Essays in memory of Professor G.S. Were*. Nairobi, Kenya: EAEP.

Ochieng, E. O. (1991). Economic adjustment programmes in Uganda, 1985-8. In Hansen, H. B., & Twaddle, M. (Eds.), *Changing Uganda: Between decay and development* (pp. 43–60). Oxford, UK: James Currey.

Ochieng, E. (2001). Tribalism and national unity: The Kenyan case. In Ojuka, A., & Ochieng, W. R. (Eds.), *Politics and leadership in Africa*. Nairobi, Kenya: East African Literature Bureau.

Odendaal, N. (2006). Towards the digital city in South Africa: Issues and constraints. *Journal of Urban Technology*, *13*(3), 29–48. doi:10.1080/10630730601145997

Odhiambo, C. J. (2004). Theatre for development in Kenya: In search of an effective procedure and methodology. Unpublished PhD Thesis, University of Stellenbosch.

Odoki Commission. (1993). *The report of the Uganda constitutional commission: Analysis and recommendations*. Entebbe, Uganda: Government Printer.

Ogechi, N. O., & Sang, J. K. (in press). Nonformal peace education through codeswitching and Sheng in FM radio for development in Kenya. *Africa Peace and Conflict Journal.*

Ogechi, N. (2005). On lexicalization in Sheng. *Nordic Journal of African Studies*, *14*(3), 334–355.

Ogechi, N. O. (2007). Building bridges through trichotomous youth identities in Kenya: Evidence from code-choice. In Njogu, K., & Oluoch-Olunya, G. (Eds.), *Cultural production and social change in Kenya: Building bridges* (pp. 129–150). Nairobi, Kenya: Twaweza Communications.

Ogechi, N. O. (2008). Sheng as a youth identity marker: Reality or misconception. In Njogu, K. (Ed.), *Culture, performance and identity: Paths of communication in Kenya* (pp. 75–92). Nairobi, Kenya: Twaweza Communications.

Ogechi, N. O. (in press). Linguistic human rights for development communication Kenya. *Frankfurter Afrikanistische Blaetter, 19.*

Ogola, G. (2009). *Why sands are shifting in the media.* (p. 4). Retrieved October 15, 2009, from www.multimediamarsgroupkenya.org/storyid265079

Ogot, B. A. (1967). *A history of the Luo*. Nairobi, Kenya: EAPH.

Ogundimu, F. (2009). African video, film cinema, and cultural repackaging in the diaspora. In Okpewho, I., & Nzegwu, N. (Eds.), *The new African Diaspora* (pp. 387–399). Bloomington, IN: Indiana University Press.

Ogunleye, F. (2003b). Female stereotypes in the Nigerian video film: A case for re-socialization. *Humanities Review Journal*, *3*(2), 1–14.

Ogunleye, F. (2003c). That we may serve him without fear: Nigerian Christian video film and battle against cultism. *International Journal of Humanistic Studies*, *2*, 16–27.

Ogunleye, F. (2003a). Christian video film in Nigeria: Dramatic sermons through the silver screen. In Ogunleye, F. (Ed.), *African video film today* (pp. 105–128). Manzini, Swaziland: Academic Publishers.

Ogunsola, L. A. (2005). Information and communication technologies and the effects of globalization: Twenty-first century "Digital Slavery" for developing countries-myth or reality? *Electronic Journal of Academic and Special Librarianship, 6,* 1-11. Retrieved June 4, 2008 from http://southernlibrarianship.icaap.org/ content/v06n01/ ogunsola_l01.htm

Ojebode, A. O. (2008). Low patronage of development radio programmes in rural Nigeria: How to get beyond the rhetoric of participation. *Journal of Multicultural Discourses*, *3*(2), 135–145. doi:10.1080/17447140802240019

Okello, D., & Owino, K. (2004). Socio-economic context of governance in Kenya. In Bujra, A. (Ed.), *Democratic transition in Kenya: The struggle from liberal to social democracy* (pp. 57–103). Nairobi & Addis Ababa, Kenya: African Centre for Economic Growth & Development Policy Management.

Okome, O. (2000a). Naming suffering and women in Nigerian video films: Notes on the interview with Emem Isong. *Hdunode: Calabar Journal of Humanities*, *3*(1), 43–56.

Okome, O. (2004). Women, religion and the video film in Nigeria. *Film International*, *7*(1), 4–13. doi:10.1386/fiin.2.1.4

Okome, O. (2005). *From the periphery of the visual space: Women and the video film in Nigeria. Film in Africa: A reader*. NEH Africa Cinema Institute Project.

Okome, O. (2000b). Onome: Ethnicity, class, gender. In Haynes, J. (Ed.), *Nigerian video films* (pp. 148–164). Athens, OH: Ohio University Press.

Okome, O. (2007a). The message is reaching a lot of people: Proselytizing and video films of Helen Ukpabio. *Postcolonial Text, 3*(2), 1-20. Retrieved on March 18, 2009, from http://journals,sfu.ca/ pocol/ indcx.php/ pct/ article/ view/ 750/419

Okome, O. (2007b). Nollywood: Spectatorship, audience and the sites of consumption. *Postcolonial Text, 3*(2), 1-21. Retrieved on March 18, 2009, from http://Joumals.sfu.ca/ pocol/ index.php/ pct/ article/ view/ 763/425

Okoro, O. I., & Chikuni, E. (2007). Power sector reforms in Nigeria: Opportunities and challenges. *Journal of Energy in Southern Africa*, *18*(3), 52–57.

Olupot, M. (July 23, 2008). Mengo, govt conflicts not new. Retrieved March 13, 2011 from http://newvision.co.ug/D/8/426/640785/Mengo

Omoera, O. S. (2009). Video film and African social reality: A consideration of Nigeria Ghana block of West Africa. *Journal of Human Ecology (Delhi, India)*, *25*(3), 193–199.

Ondimu, J. K. (2006). *Linguistic analysis of computer-mediated text messages.* (Unpublished MPhil thesis). Eldoret, Kenya: Moi University.

Oni, A. A., Oni, O., & Omidiji, O. G. (2009). *The World Wide Web.* Seminar Paper presented at the Department of Communication and Language Arts, University of Ibadan, Ibadan.

Onwujekwe, O., Uzochukwu, B., Ezumah, N., & Shu, E. (2005). Increasing coverage of insecticide-treated nets in rural Nigeria: Implications of consumer knowledge, preferences and expenditures for malaria prevention. *Malaria Journal*, *4*, 29. doi:10.1186/1475-2875-4-29

Onyango, J. O. (2008). *Ethnic conflict and ethnic discourse in Kenya: Advanced researchers fellowship report*. Dakar, Senegal: CODESRIA.

OpenNet Initiative. (2007). *South Korea*. Retrieved from February 27, 2010, from http://opennet.net/ sites/ opennet.net/ files/ south_korea.pdf

Opiyo, P., Mukabana, W. R., Kiche, I., Mathenge, E., Killeen, G. F., & Fillinger, U. (2007). An exploratory study of community factors relevant for participatory malaria control on Rusinga Island, Western Kenya. *Malaria Journal*, *6*, 48. doi:10.1186/1475-2875-6-48

Orao, J. (2009). The Kenyan indigenous languages and the mass media: Challenges and opportunities. *Stellenbosch Papers in Linguistics PLUS*, *38*, 77–86.

Osifo-Whiskey, O. (in press). *Witness to justice: An insider account of Nigeria's Truth Commission.*

Osofisan, F. (Ed.) and School of Post Graduate Studies UI. (2008). *Sixty iears of the muse at UI: A commemorative directory*. Lagos, Nigeria: Concept Publications.

Osotimehein, B. (2009, April 24). In Nigeria, others race against time in the fight against malaria. *The Guardian*, pp. 24-25.

Otieno, I. O. (2005). *Mobile telephony and democratic struggles: A case of 2002 elections in Kenya*. Retrieved on 17 March, 2010, from mokk.bme.hu/centre/conferences/reactivism/FP/fpOtieno.

Otiso, K. (2006). *Culture and customs of Uganda*. Westport, CT: Greenwood press.

Oyugi, E. (1997). *Ethnic politics: An ideological framework for antagonistic social mobility in Africa. Bujumbura workshop*. Bujumbura: CODESRIA.

Page, J. (June 24, 2008). Elder brother Hu Jintao makes Internet debut. *Times Online.* Retrieved January 31, 2010, from http://www.timesonline.co.uk /tol/ news/ world/ asia/ article41983 10.ece? print=yes & randnum= 1151003209000

Paracha, F. N. (2009, April 19). Smokers'corner: The creeping malaise. *Dawn News*.

Parayil, G. (2005). Digital divide and increasing returns: Contradictions of informational capitalism. *The Information Society*, *21*(1), 41–51. doi:10.1080/01972240590895900

Park, H. (1998). A Gramscian approach to interpreting international communication. *The Journal of Communication*, *48*(4), 79–99. doi:10.1111/j.1460-2466.1998.tb02771.x

Park, L., Crocker, J., & Mickelson, K. (2004). Attachment styles and contingencies of self-worth. *Personality and Social Psychology Bulletin*, *30*(10), 243–254. doi:10.1177/0146167204264000

Passage, D. (1998). *Africa: New realities and U.S. policy. Special Warfare, 11(3)*. Fort Bragg, NC: JFK Special Warfare Center.

Paterson, C. (2008). Introduction: Why ethnography? In C. Paterson & D. Domingo. *Making online news: An ethnography of new media production* (pp. 1-11). New York, NY: Peter Lang.

PATH. (2005). *Magnet Theatre: A facilitators training manual*. Nairobi, Kenya: PATH/UNAIDS.

Pavlik, J. (2000). The impact of technology on journalism. *Journalism Studies*, *1*(2), 229–237. doi:10.1080/14616700050028226

Pavlik, J. V. (2001). *Journalism and new media*. New York, NY: Columbia University Press.

Pavlik, J. V. (2005). Running the technological gauntlet: Journalism and new media. In De Burgh, H. (Ed.), *Making journalists* (pp. 245–263). London, UK: Routledge.

People's Daily Online. (June 20, 2008). Hu Jintao talks to netizens via People's Daily Online. Retrieved January 31, 2010, from http://english.peopledaily.com.cn/ 90001/ 90776/ 90785/ 6433952.html

People's Daily. (December 24, 2009). New highlights in China's democratic politics 2009. Retrieved January 31, 2010, from http://english.peopledaily.com.cn/ 90001/ 90780/ 91342/ 6851678.html

Perloff, R. M. (1998). *Political communication: Politics, press, and public in America*. Mahwah, NJ: Lawrence Earlbaum Associates, Publishers.

Perry, W. (2002). *The ethical use of military force*. International Studies: National and International Security. Maxwell Air Force Base, AL: Squadron Officer College (Air University).

Pfuwka, C. (2008). Black September et al: Chimurenga songs as historical narratives in the Zimbabwean liberation war. *Journal of Music Research in Africa, 5*(1), 30-61.

Poku, N. K. (2005). *Aids in Africa: How the poor are dying*. Cambridge, UK: Polity Press.

Poldervaart, S. (2001). Utopian aspects of social movements in postmodern times: Some examples of DIY politics in the Netherlands. *Utopian Studies, 12*(2), 143–164.

Polikanov, D., & Abramova, I. (2003). Africa and ICT: A chance for breakthrough? *Information Communication and Society, 6*(1), 42–56. doi:10.1080/1369118032000068778

Poplack, S., Sankoff, D., & Miller, C. (1988). The social correlates and linguistic processes of lexical borrowing and assimilation. *Linguistics, 26*, 47–104. doi:10.1515/ling.1988.26.1.47

Posteguillo, S. (2003). *Netlinguistics: Language, fiscourse and ideology in Internet*. Jaume, UK: Universitas Jaume.

Poster, M. (2001). *What's the matter with the Internet?*Minneapolis, MN: University of Minnesota Press.

Press Freedom Index. (n. d.). *Reporters without borders for press freedom*. Retrieved February 23, 2010, from http://www.rsf.org/ en-classement1003-2009.html

Professor Jay. (2003). *Mapinduzi Halisi*. GMC Wasanii Promoters Ltd.

Putnam, R. (2000). *Bowling alone: The collapse and revival of American community*. New York, NY: Simon & Schuster.

Qin, J., Liu, Q., & Ding, Q. (2010). Ministry refutes US claims China restricts Internet. *China Daily*. Retrieved January 26, 2010 from http://www.chinadaily.com.cn/ china/ 2010-01/23/ content_9365824.htm

Rais, R. (2007). Identity politics and minorities in Pakistan. *South Asia: Journal of South Asian Studies*, 111-125.

Rampton, B. (1985). *Crossing: Language and ethnicity among adolescents*. London, UK: Longman.

Ramstad, E. (2009, April 21). Leading the news: Korean judge finds blogger not guilty. *The Wall Street Journal Asia*, p. 3.

Ranger, T. (2004). Historiography, patriotic history and the history of nation. *Journal of Southern African Studies, 30*(2), 215–234. doi:10.1080/0305707042000215338

Ranger, T. (2005). The rise of patriotic journalism in Zimbabwe and its possible implications. *Westminister Papers in Communication and Culture*, 8-17.

Ratemo, J. (2010). Digital villages bring government to rural areas. Retrieved on June 16, 2010, from http://www.standardmedia.co.ke

Raudenbush, S., & Bryk, A. (2002). *Hierarchical linear models: Applications and data analysis methods* (2nd ed.). Thousand Oaks, CA: Sage Publications, Inc.

Reglitz, E. (2009). Households online. *Magazine-Deutschland: The New World of Media, 48.*

Ren, L. P., & He, Y. (2009). *Land for low-priced housing turned into land for villas while a reporter was questioned by an official, "Who are you speaking for?"* Retrieved January 26, 2010, from http://news.sina.com.cn/ c/ 2009-06-17/ 0751158 03313s.shtml

Rendon, L. (1996). Life on the border. *About Campus, 1*, 14–20.

Republic of Kenya. (2001). *AIDS in Kenya: Background, projections, impacts, interventions, policy* (6th ed.). Nairobi: Ministry of Health.

Reuters. (2008). *Up to 1,000 killed in Kenya crisis*. Retrieved January 18, 2010, from http://www.reuters.com/article/homepage crisis/iduslo743589-CH.2400

Reuters. (2008, January 13). *Kenya death toll hits 693: News for South Africa and the World*. Retrieved January 4, 2010 from http://www.101.co.za/index.phpset-id=nw2008011303/

Reuters. (Aug. 11, 2010). China can't both pull and push the yuan. Retrieved August 12, 2010, from http://blogs.reuters.com/ columns/ 2010/ 08/11/ china-cant-both-pull-and-push-the-yuan/

Rheingold, H. (1993). *The virtual community: Homesteading on the electronic frontier*. Reading, MA: Addison-Wesley.

Rheingold, H. (2002). *Smart mobs: The next social revolution*. New York, NY: Perseus.

Rhodes, T. (2010). *In African hot spots, journalists forced into exile*. Committee to Protect Journalists. Retrieved from: http://cpj.org/2010/02/in-african-hot-spots.php

Riccio, T. (2004). Kenya's community health awareness puppeteers. *A Journal of Performance and Art, 26*(1), 1-12.

Ricks, T. E. (2006). *Fiasco: The American military adventure in Iraq*. New York, NY: Penguin Books.

Rieder, R. (2007). *Counting the spoons in the American journalism review of February/ March 2007 issue*.

Riffe, D., Lacy, S., & Varouhakis, M. (2008, January). Media system dependency theory and using Internet for in-depth, specialized information. *Web Journal of Mass Communication Research*. Retrieved January 15, 2010 from http://www.scripps.ohiou.edu/wjmcr/vol11/11-b.html

Riggins, S. H. (1992). The promise and limits of ethnic minority media. In Riggins, S. H. (Ed.), *Ethnic minority media: An international perspective* (pp. 276–787). Newbury Park, CA: Sage.

Robbin, A., Courtright, C., & Davis, L. (2004). ICTs and political life. In B. Cronin (Ed.), *Annual Review of Information Science and Technology (ASIS&T), 38*, 411-482. Medford, NJ: Information Today Inc.

Roberts, A. (1962). The sub-imperialism of the Baganda. *Journal of African History, 8*, 435–450. doi:10.1017/S0021853700003340

Roberts, S. (2005, February 21). More Africans enter U.S. than in days of slavery. *The New York Times*, p. A1.

Rodriguez, N., & Villaverde, L. E. (2000). *Dismantling White privilege: Pedagogy, politics, and Whiteness*. New York, NY: Peter Lang.

RoK. (2009). *Economic survey*. Nairobi, Kenya: Government Printer.

Roll Back Malaria. (2008). *The global malaria action plan for a malaria free world.* Retrieved on November 19, 2009, from http://rollbackmalaria.org

Rothbart, D., & Bartlett, T. (2007). Rwandan radio broadcasts and Hutu/Tutsi positioning. In Moghaddam, F. M., Harre, R., & Lee, N. (Eds.), *Global conflict resolution through positioning analysis*. New York, NY: Springer.

Rukiramakuba, E. (1995). Mwanga na maendeleo ya Kiswahili, Rwanda. In Mlacha, S. A. K. (Ed.), *Kiswahili na Vyombo vya Habari* (pp. 83–92). Dar es salaam, Tanzania: TUKI.

Saco, D. (2002). *Cybering democracy: Public space and the Internet*. Minneapolis, MN: University of Minnesota Press.

Said, V. M. (2007). Matumizi ya Kiswahili nchini Kenya katika redio na runinga. In Njogu, K. (Ed.), *Kiswahili na Elimu Nchini Kenya* (pp. 20–27). Nairobi, Kenya: Twaweza Communications.

Saleh, N., & Kasisa, A. (1995). Usambazaji wa Kiswahili, Zaire. In Mlacha, S. A. K. (Ed.), *Kiswahili na Vyombo vya Habari* (pp. 63–71). Dar es salaam, Tanzania: TUKI.

Salmon, K. (2003, June 12). Drama-Africa: Persuading a man without hitting him on the head. *Global Information Network, 1*.

Salwen, M. B. (1987). Mass media issue dependency and agenda-setting. *Communication Research Reports, 4*(1), 26–31.

Sanchez, R. (2007, October 17). Iraq War a nightmare with no end in sight. *CNN*. Retrieved from www.cnn.com

Schiappa, E., Gregg, P. B., & Hewes, D. E. (2005). The parasocial contact hypothesis. *Communication Monographs*, *1*, 92–115. doi:10.1080/0363775052000342544

Schiller, D. (2000). *Digital capitalism: Networking the global market system*. Cambridge, MA: MIT Press.

Schudson, M. (2000). The sociology of news revisited. In Curran, J., & Gurevitch, M. (Eds.), *Mass media and society* (pp. 175–200). London, UK: Arnold.

Severin, W. J., & Tankard, J. W. Jr. (2001). *Communication theories: Origins, methods, and uses in the mass media*. New York, NY: Longman.

Shepperson, A., & Tomaselli, K. (2009). Media in Africa: Political, cultural and theoretical trajectories in the global environment. *International Communication Gazette*, *71*(6), 473–489. doi:10.1177/1748048509339790

Shim, D. (2002). South Korean media industry in the 1990s and the economic crisis. *Prometheus*, *20*, 337–350.

Shitemi, N. L. (Eds.). (2001). *Kiswahili: A tool for development- a multidisciplinary approach*. Eldoret, Kenya: Moi University Press.

Shohat, E. A. (1994). *Unthinking Eurocentrism: Muticulturalism and the media*. New York, NY: Routledge.

Shohat, E., & Stam, R. (1996). From the imperial family to the transnational imaginary: Media spectatorship in the age of globalization. In Wilson, R., & Dissanayake, W. (Eds.), *Global, local: Cultural production and the transnational imaginary* (pp. 145–171). Durham, NC: Duke University Press.

Sifry, M. L. (2009). A see-through society: How the Web is opening up our democracy. *Columbia Journalism Review*, *47*(5), 43–47.

Silberg, J. (Dir.). (1984). *Breakin'*. Cannon Films Inc.

Sinka, I. (2000). The search for cross-linguistic influences on the language of young Latvian-English bilinguals. In Dopke, S. (Ed.), *Cross-linguistic structures in simultaneous bilingualism* (pp. 149–174). Amsterdam, The Netherlands & Philadelphia, PA: John Benjamins.

Skjerdal, T. S. (2008). Self-censorship among news journalists in the Ethiopian state media. *African Communication Research*, *1*(2), 185–205.

Skumanich, S. A., & Kintsfather, D. P. (1998). Individual media dependency relations within television shopping programming: A causal model reviewed and revised. *Communication Research*, *25*(2), 200–219. doi:10.1177/009365098025002004

Slack, J., & Fejes, F. (1987). *The ideology of the Information Age*. New Jersey: Alex Publishing Corporation.

Slocum, R. (2006). Anti-racist practice and the work of community food organizations. *Antipode*, *38*(2), 327–349.

Smith, P., & Fagbule, T. (2010). Reform stutters amid calls for power shift. *Africa Report*, *22*, 45–48.

Smitherman, G. (2000). *Talkin that talk: Language, culture, and education in African America*. London, UK & New York, NY: Routledge.

Snow, D. M. (1993). *Peacekeeping, peacemaking and peace-enforcement: The U.S. role in the new international irder*. Carlisle Barracks, PA: Strategic Studies Institute (U.S. Army War College).

Snyder, W. P. (1995). Strategy: Defining it, understanding it and making it. Maxwell Air Force Base, AL: Air War College (Air University).

Spencer, R. (June 20, 2008). China: Internet debut for leader Hu Jintao. *Telegraph.* Retrieved January 31, 2010, from http://www.telegraph.co.uk/ news/ worldnews/ asia/ china/ 2164637/ China-Internet-debut-for-leader-Hu-Jintao.html

Spiegel, M. (1996). The dreaded comparison: Human and animal slavery (rev. and expanded ed.). New York, NY: Mirror Books.

Spitulnik, D. (1999). The language of the city: Town Bemba as urban hybridity. *Journal of Linguistic Anthropology*, *8*(1), 30–59. doi:10.1525/jlin.1998.8.1.30

Ssejjoba, E., & Lukwago, J. (April 13, 2008). Archbishop Lwanga warns oninsulting Kabaka. Retrieved April 13, 2008 from http://www.newvision.co.ug/ D/8/12/622091

Ssemujju, I. N. (January 3, 2008a). Why Kabaka refused to meet Museveni. *The Uganda Observer Online*. Retrieved January 3, 2008 from http://www.ugandaobserver.com/new/index.php

Ssemujju, I. N. (January 6, 2008c). Kabaka sets terms to meet Museveni. *The Uganda Observer Online*. Retrieved January 6, 2008, from http://www.ugandaobserver.com/new/index.php

Ssemujju, I. N. (March 7, 2008b). Swords out. *The Uganda Observer Online*. Retrieved

Ssemujju, I. N. (May 29, 2008d). Cabinet to delete "press freedom" from constitution. *The Uganda Observer Online*. Retrieved May 29, 2008, from http://www.observer.ug/index.php?option =com_content&view=article&id=103 :ssemujju- ibrahim-nganda&catid=34:news&Itemid=59

Ssengoba, N. (May 27, 2008). Press crackdown; NRM loses battle for hearts and minds. Retrieved May 27, 2008, from http://allafrica.com/stories/200805270747.html

Ssenkabirwa, A. M. (January 2, 2008). Mengo replies Museveni's letter to Kabaka. Retrieved January 2, 2008, from http://allafrica.com/stories/200801030314.html

Sserwanga, M. (May 27, 2008). Media crackdown? Bye-bye to democracy. *The Monitor Online*. Retrieved May 27, 2008, from http://allafrica.com/stories/200805270222.html

Stein, L. (2008). Speech without rights: The status of public space on the Internet. *Communication Review*, *11*, 1–23.

Stone, B., & Barboza, D. (January 15, 2010). Scaling the digital wall in China. *The New York Times*. Retrieved January 19, 2010, from http://www.nytimes.com/ 2010/01/16/technology/ internet/ 16evade.html

Subervi-Velez, F. A. (1994). Mass communication and Hispanics. In Padilla, F. (Ed.), *Handbook of Hispanic cultures in the United States: Sociology* (pp. 304–357). Houston, TX: Arte Publico.

Sunstein, C. (2001). *Republic.com*. Princeton, NJ: Princeton University Press.

Syed, J. (2008). The representation of cultural diversity in Urdu-language newspapers in. *South Asia: Journal of South Asian Studies*, 317-347.

Tai, Z., & Sun, T. (2007). Media dependencies in a changing media environment: The case of the 2003 SARS epidemic in China. *New Media & Society*, *9*(6), 987–1009. doi:10.1177/1461444807082691

Tajfel (Ed.). *Differentiation between social groups: Studies in the social psychology of intergroup relations*, (61-76). New York, NY: Academic Press.

Tajfel, H., & Turner, J. C. (1986). The social identity theory of intergroup behaviour. In Worchel, S., & Austin, G. (Eds.), *Psychology of intergroup relations*. Chicago, IL: Nelson-Hall.

Tajfel, H. (1978). Social categorization, social identity, and social comparison. In H.

Talley v. California, 362 U.S. 60 (1960).

Tamale, S. (1999). Towards legitimate governance in Africa: The case of affirmative action and parliamentary politics in Uganda. In Quashigah, E. K., & Okafor, O. C. (Eds.), *Legitimate governance in Africa: International and domestic legal perspective* (pp. 235–262). The Netherlands: Kluwer Law International.

Tan, A. S. (1984). *Mass communication theories and research*. Columbus.

Tangri, R., & Mwenda, A. (2010). President Museveni and the politics of presidential tenure in Uganda. *Journal of Contemporary African Studies*, *28*(1), 31–49.. doi:10.1080/02589000903542574

Taylor, P. (2003). *Applied theater: Creating transformative encounters in the community*. Portsmouth, NH: Heinemann.

Taylor, P. (2002). Afterthought: Evaluating applied theatre. *Applied Theatre Researcher, 6*(3). Retrieved on February 3, 2007, from http://www.gu.edu.au/centre/cpci/atr/journal/article6_number3.htm

Tedesco, J. C. (2004). Changing the channel: Use of the Internet for communicating about politics. In Kaid, L. L. (Ed.), *Handbook ofpolitical communication research* (pp. 507–532). Mahwah, NJ: Lawrence Erlbaum Associates, Publishers.

The Holy Bible, King James Version, Acts 16, 26-40.

The Observer. (October 4, 2009). *Untold story of Museveni Kabaka meet*. Retrieved on October 5, 2009, from http://www.observer.ug/index.php?option=com_content&view=article&id=5390:untold-story-of-museveni-kabaka-meet&catid=78:topstories&Itemid=59

The Sunday Vision Online. (February 9, 2008). *President Museveni to sue MP Beti Kamya*. Retrieved February 9, 2008, from http://allafrica.com/stories/2008021 Toro Kingdom. (2008). King Oyo Nyimba Kabamba Iguru Rukidi IV. Retrieved January 22, 2010, from http://www.torokingdom.org/Oyo.htm

The World Factbook. (2009). *Kenya*. Retrieved October 20, 2009, from https://www.cia.gov/library/publications/the-world-factbook/geos/ke.html

The World Factbook. (2009). *Tanzania*. Retrieved October 20, 2009 from https://www.cia.gov/library/publications/the-world-factbook/geos/tz.html

The World Factbook. (2009). *Uganda*. Retrieved October 20, 2009, from https://www.cia.gov/library/publications/the-world-factbook/geos/ug.html

Thram, D. (2006). Patriotic history and the politicization of memory: Manipulation of popular music to re-invent the liberation struggle in Zimbabwe. *Critical Arts: A South-North Journal of Cultural and Media Studies, 20*(2), 75-88.

Tomlinson, J. (1999). *Globalization and culture*. Cambridge, UK: Polity Press.

Tomlinson, J. (2003). *Globalization and cultural identity*. Retrieved on October 15, 2008, from http://www.polity.co.uk/ global/ pdf/ GTReader2eTomlinson.pdf

Top 20 countries with the highest number of Internet users. (June 2010). *Miniwatts Marketing Group*. Retrieved March 12, 2011 from http://www.internetworldstats.com/top20.htm

Travers, R. (1997). A new millennium and a strategic breathing space. *The Washington Quarterly, 20*(2), 97–114. doi:10.1080/01636609709550243

Tripp, A. M. (2000). *Women and politics in Uganda*. Madison, WI: University of Wisconsin Press.

Truman, J. S. (2008). *Political communication in American campaigns*. Los Angeles, CA: Sage Publications.

Tuhkanen, M. (2001). Of blackface and paranoid knowledge: Richard Wright, Jacques Lacan, and the ambivalence of black minstrelsy. *Diacritics, 31*(2), 9–34.

Turiera-Puigbo, T. (2009). How is Internet use changing the way in which politics is carried out and communicated? *Quaderns del CAC, 33*, 13–19.

Turner, J. C. (1987). *Rediscovering the social groups: A self-categorization theory*. New York, NY: Basil Blackwell Inc.

Turner, J. C., Hogg, M. A., Oakes, P. J., Reicher, S. D., & Wetherell, M. S. (1987). *Rediscovering the social group: A self-categorization theory*. Oxford, UK: Basil Blackwell.

Uganda Bureau of Statistics. (2003). *2002 population report*. Retrieved October 25, 2009, from http://www.ubos.org/index.php

UI. (2008). *60th anniversary brochure*. Ibadan, Nigeria: Publications Unit of the Registrar's office, University of Ibadan.

Ukadike, F. (1994). *Black African cinema*. Berkeley, CA: University of California Press.

Ukadike, F. (2000). Images of the reel thing: African video-films and the emergence of a new cultural art. *Social Identities, 6*(3), 237–240. doi:10.1080/13504630050137615

Ukadike, F. (2003). Video booms and the manifestations of first cinema in Anglophone Africa. In Guneratne, A. R., & Dissanayake, W. (Eds.), *Rethinking third cinema* (pp. 127–143). New York, NY: Routledge.

Ukadike, F. (forthcoming). Nollywood history criticism: Rethinking African film discourse. In Krings, M., & Okome, O. (Eds.), *Hollywood and beyond: Transnational dimensions of an African video film industry*.

Ukah, A. (2003). Advertising god: Nigerian Christian video-films and the power of consumer culture. *Journal of Religion in Africa. Religion en Afrique, 33*(2), 203–231. doi:10.1163/15700660360703141

Ukah, A. (2004). Pentecostalism, religious expansion, and the city: Lessons from the Nigerian bible belt. In Probst, P., & Spittler, G. (Eds.), *Between resistance and expansion* (pp. 415–442). Berlin, Germany: LIT Verlag.

Ultimate Media. (2006). Uganda business news: Bunyoro minister and kingdom press secretary resign over oil. Retrieved March 13, 2011 from http://www.ugpulse.com/articles/daily/ news.asp?about=Bunyoro+Minister+ and+ Kingdom+Press+Secretary+resign+over+oil&ID=2393

United Nation Population Division. (2000). *World population prospects: Population database*, 2002 revision. Retrieved on November 19, 2009, from http://esa.un.org/unpp

United Nations Information and Communication Technologies Task Force. (2002). *Definition of information communication and technologies (ICTs).* Retrieved June 12, 2008, from http://www.unicttaskforce.org/

Van der Veur, P. R. (2003). *Media and democracy in Africa, broadcasting and political reform*. New Brunswick, NJ: Transaction Publishers.

van Dijk, J. G. M. (2006). Digital divide research, achievements and shortcomings. *Poetics*, *34*(4-5), 221–235. doi:10.1016/j.poetic.2006.05.004

Volosinov, V. N. (1973). *Marxism and the philosophy of language*. New York, NY: Seminal Press.

von Clausewitz, C. (1962). *On war*. Chicago, IL: Regnery Press.

Vrij, A., Van Schie, E., & Cherryman, J. (1996). Reducing ethnic prejudice through public communication programs: A social-psychological perspective. *The Journal of Psychology*, 130.

wa Thiongó, N. (1986). *Decolonising the mind: The politics of language in African literature.* Nairobi, Kenya: Heinemann.

Wa'Njogu, K. (2004). Language and multiparty democracy in a multiethnic Kenya. *Africa Today*, *3*(50), 55–73.

Wachanga, D. N. (2011). 'Kenya's indigenous radio stations and their use of metaphors in the 2007 election violence'. *Journal of African Media Studies*, *3*(1), 109–125.

Wachanga, D. N. (in press). Global media and information ethics: Challenges re-examined. In Yin, D. Y. (Ed.), *Global media convergence and cultural transformation: Emerging social patterns and characteristics*. Hershey, PA: IGI Global Publishing.

Wachanga, D. N. (2007). *Sanctioned and controlled message propagation in a restrictive information environment: The small world of clandestine radio broadcasting.* Unpublished doctoral dissertation. University of North Texas.

Wachira, N. (2010, 14 August). He built Shs25bn empire from scratch. *Saturday Nation.*

Wagosi wa Kaya. (2005). *Nyeti*. GMC Wasanii Promoters Ltd.

Wainaina, E. (2001). *Nchi ya Kitu Kidogo. Sawa sawa.* Christian Kaufmann.

Waki, P. (2008). *Report by the Commission of Inquiry into Post-Election Violence* (CIPEV). Wa-Mungai, M. (2010). "Soft power," popular culture and the 2007 elections. In Kanyinga, K., & Okello, D. (Eds.), *Tensions and Reversals in Democratic Transitions: The Kenya 2007 General Elections*. Kenya: Nairobi. Institute for Development Studies, University of Nairobi.

Waki, P. (2008). *Commission report of the Kenyan 2007 post-election violence*. Retrieved December 2, 2009, from www.dialoguekenya.org/docs/pev%report.pdf

Walther, J. (2004). Language and communication technology: Introduction to the special issue. *Journal of Language and Social Psychology*, *23*(4), 384–396. doi:10.1177/0261927X04269584

Wambugu, S. (2010, 23 January). Why that SMS behind the wheels kills. *Daily Nation.*

Wang, S. (2009). Is convergence enough to make existing media a brighter future? *China Media Report Overseas, 5*(1).

Warschauer, M. (2003a). *Technology and social inclusion: Rethinking the digital divide*. Cambridge, MA: MIT Press.

Warschauer, M. (2003b). Dissecting the "digital divide": A case study in Egypt. [from http://olt-bta.hrdc-drhc.gc.ca/ download/ oltdualdivide.pdf]. *The Information Society*, *19*, 297–304. Retrieved June 14, 2008. doi:10.1080/01972240309490

Watts, J. (2003, February 24). World's first Internet president logs on: Web already shaping policy of new South Korean leader. *Guardian.* Retrieved February 27, 2010, from http://www.guardian.co.uk/ technology/ 2003/ feb/ 24/ newmedia.koreanews

Wa-Wamwere, K. (2008). *Kenya: Loving each other won't cure negative ethnicity*. Retrieved January 5, 2010, from www.allafrican.com/stories/2011002010256.html

Wei, C. Y., & Kolko, B. (2005). Resistance to language and Internet diffusion patterns in Uzbekistan. *New Review of Hypermedia and Multimedia, 11*(2), 205–220. doi:10.1080/13614560500402817

Wei, R. (2009). The influence of mobile phone advertising on dependency: A cross-cultural study of mobile phone use between American and Chinese youth. In Golan, G. J., Johnson, T. J., & Wanta, W. (Eds.), *International media communication in a global age* (pp. 401–421). New York, NY: Routledge.

Wei, C., & Kolko, B. E. (2005). Studying mobile phone use in context: Cultural, political, and economic dimensions of mobile phone use. In *Proceedings of the International Professional Communication Conference*.

Weicher, M. (2007). [Name withheld]: Anonymity and its implications. *Proceedings of the American Society for Information Science and Technology, 43*, 1–11.

Weinberg, J. (2009). *A language of its own: The Syrian mobile phone culture.*

Wendl, T. (2007). Wicked villagers and the mysteries of reproduction: An exploration of horror movies from Ghana and Nigeria. *Postcotonial Text, 3*(2), 1-21. Retrieved from http://Journals.sfu.ca/ pocol/ index php/ pct/ anicle/ vicw/529/420

Werbner, P., & Modood, T. (Eds.). (1997). *Debating cultural hybridity*. London, UK: Zed Books.

Were, G. S. (1967). [Nairobi, Kenya: E.A.P.H.]. *History of the Abaluyia of Western Kenya, C*, 1500–1930.

Whorf, B. L. (1956). *Language thought and reality: Selected writings* (Carrol, J. B., Ed.). Cambridge, MA: MIT Press.

William, A. (1999). Communication accommodation theory and miscommunication: Issues of awareness and communication dilemmas. *International Journal of Applied Linguistics, 9*(2), 151–165. doi:10.1111/j.1473-4192.1999.tb00169.x

Williams, R. (1990). *Television: Technology and cultural form*. New York, NY: Routledge.

Winston, B. (1998). *Media technology and society: A history: From the telegraph to the Internet*. New York, NY: Routledge.

Wodak, R. (2001). The discourse historical approach. In Wodak, R., & Meyer, M. (Eds.), *Methods of critical discourse analysis*. London, UK: Sage.

Woka, J., & Lion, R. (2002). *Nani Alaumiwe?* MJ Records.

Wolchok, S., Yao, R., & Halderman, J. A. (2009). *Analysis of the Green Dam censorware system.* Computer Science and Engineering Division, the University of Michigan. Retrieved March 8, 2010, from http://www.cse.umich.edu/ ~jhalderm/ pub/gd/

Woolard, K. (1985). Language variation and cultural hegemony: Toward an integration of sociolinguistic and social theory. *American Ethnologist, 12*(4), 738–748. doi:10.1525/ae.1985.12.4.02a00090

Woollacott, J. (1982). Messages and meanings. In Gurevitch, M. T. (Ed.), *Culture, society and the media.* London, UK: Methuen.

World Bank. (2008). *Facts and figures about Uganda.* Retrieved June 18, 2008, from http://www.worldbank.org/afr/ug

World Bank. (2009). *Information and communications for development (IC4D): Extending reach and increasing impact of ICTs.* Oxford, Uk: Oxford University Press. Retrieved from http://www.worldbank.org/ ic4d World History Archives. (March 2004). *The contemporary political history of the Kingdom of Bunyoro-Kitara*. Retrieved January 22, 2010, from http://www.hartford-hwp.com/archives/ 36/index-ijb.html

World Health Organization. (2005). *African Summit on Roll Back Malaria*, Abuja, Nigeria, April 25 2000. Geneva, Switzerland. (Document WHO/CDS/RBM/2000.17).

Worsley, K. J., & Taylor, J. E. (2007). Detecting sparse signals in random fields, with an application of brain mapping. *Journal of the American Statistical Association, 102*, 913–928. doi:10.1198/016214507000000815

Wright, B. (2008). Is triple play the answer? *African Business, 346*, 44.

Wright, M. (2005). Finding a place in cyberspace: Black women, technology, and identity. *Frontiers, 26*(1), 48–59.

Wu, C. (September 30, 2009). Han Han: finding happiness by being different. *The Daily Telegraph.* Retrieved January 31, 2010, from http://www.telegraph.co.uk/news/worldnews/ asia/china/6244112/ Han-Han-finding-happiness-by-being-different.html

Wu, J. (July 15, 2008). Public poll to help judge work of officials. *China Daily.* Retrieved January 28, 2010, from http://www.chinadaily.com.cn/ cndy/ 2008-07/15/ content_6846121.htm

Xinhua News Agency. (January 24, 2007). President Hu Jintao asks Chinese officials to better cope with Internet. Retrieved January 31, 2010, from http://news.xinhuanet.com/ english/ 2007-01/24/ content_5648674.htm

Xinhua News Agency. (January 13, 2010). CPC pushes Party leaders to report family information to stem corruption. Retrieved January 31, 2010, from http://news.xinhuanet.com/ english/ 2010-01/13/ content_12804650.htm

Xinhua News Agency. (January 22, 2010). 2009 Flashback: Public opinion and political views rocked the Internet in China. Retrieved January 28, 2010, from http://news.xinhuanet.com/ comments/ 2010-01/21/ content_12848338.htm

Xinhua News Agency. (July 26, 2009). Evaluation tells Chinese officials to listen to online opinions and act fast. Retrieved February 11, 2010 from http://english.people.com.cn/ 90001/90776/ 90785/ 6709971.html

Xinhua News Agency. (July 4, 2008). Guizhou officials admit residents harbored grievances. Retrieved January 31, 2010, from http://www.chinadaily.com.cn/ china/ 2008-07/04/ content_6817910.htm

Xinhua News Agency. (July 4, 2008). Police chief sacked after violent protest in Guizhou. Retrieved January 31, 2010, from http://www.chinadaily.com.cn/ china/2008-07/04/ content_6820613.htm

Xinhua News Agency. (September 25, 2005). Administrative rules for providing Internet information services. Retrieved March 14, 2010, from http://news.xinhuanet.com/ politics/ 2005-09/25/ content_3538899.htm

Yancy, G. (2004). *What white looks like: African-American philosophers on the whiteness question*. New York, NY: Routledge.

Yang, G. (2009). *The power of the Internet in China: Citizen activism online*. New York, NY: Columbia University Press.

Yansane, A. Y. (1990). Cultural, political, and economic universals in West Africa. In Asante, M. K., & Asante, K. W. (Eds.), *African culture: The rhythms of unity* (pp. 39–68). Trenton, NJ: Africa World Press.

Yi, L., & Chen, C. (September 30, 2009). How the Internet is shaping China's future. *The Daily Telegraph.* Retrieved January 31, 2010, from http://www.telegraph.co.uk/ news/worldnews/ asia/china/ 6244019/ How-the-internet-is-shaping-Chinas-future.html

Yieke, F. A. (2008). *The discursive construction of ethnicity: The case of the 2007 Kenyan general election*. Paper read at governing the African public sphere, Yaounde, Cameroon.

Yin, J. (2005). *Spiraling out of silence: The Internet is influencing news agenda in China.* Presentation at the 14th AMIC Annual Conference: Media and Society in Asia: Transformations and Transition, Beijing, China, July 18-21, 2005.

Yin, J., & Gao, J. (2004). *The one truly global media in China: The Internet is transforming the largest Communist country.* Presentation at the Conference on Communication & Globalization, Center for Global Media Studies, Washington State University, Seattle, Washington, July 16-17, 2004.

Young, C. (2003). Assimilation and social change dynamics in African and African American communities. *The Western Journal of Black Studies*, *27*, 164–175.

Young, R. J. (2001). *Postcolonialism: An historical introduction*. Oxford, UK: Blackwell Publishers.

Youtube. (2007a). *Si Lazima.* Retrieved on 19 September, 2008, from http://www.youtube.com/watch?v=UiTR2Lt_fMw

Youtube. (2007b). *Manzi wa Nairobi*. Retrieved on 29 September, 2008, from http://www.youtube.com/watch?v=-AExNsJLKho

Youtube. (2007c). *Keroro*. Retrieved on 29 September, 2008, from http://www.youtube.com/watch?v=KIQcPm_btm0

Youtube. (2007d). *We Kamu*. Retrieved on 30 September, 2008, from http://www.youtube.com/watch?v=Ne2DXjIzOxk

Youtube. (2007e). *Furahiday*. Retrieved on 1 October, 2008, from http://www.youtube.com/watch?v=HILOjXi9wDs

Zajc, M. (2009). Nigerian video film cultures. *Anthropological Notebooks*, *1*(15), 65–85.

Zaller, J. (1991). *The nature and origins of mass opinion*. Cambridge University Press.

Zavella, P. (1996). Feminist insider dilemmas: Constructing ethnic identity with Chicana informant. In Wolfe, D. (Ed.), *Feminist dilemmas in fieldwork*. Boulder, CO: Westview Press.

Zeleza, P. T. (2009). The media in social development in contemporary Africa. In Njogu, K., & Middleton, J. (Eds.), *Media and identity in Africa* (pp. 19–35). Edinburg University Press for the International African Institute.

Zembylas, M., & Vrasidas, C. (2005). Globalization, information and communication technologies and the prospect of a global village: Promises of inclusion or electronic colonization? *Journal of Curriculum Studies, 37*, 65–83. Retrieved June 16, 2008, from http://www.courses.unt.edu/efiga/ 6880_Globalization/zembylas_vrasidas_2005_global_village.pdf

Zheng, Y., & Walsham, G. (2008). Inequality of what? Social exclusion in the e-society as capability deprivation. *Information Technology & People*, *21*(3), 222–243. doi:10.1108/09593840810896000

Zheng, Y. (2008). *Technological empowerment: The Internet, stat, and society in China*. Stanford, CA: Stanford University Press.

Zhou, Y. (2006). *Historicizing online politics: Telegraphy, the Internet, and political participation in China*. Stanford, CA: Stanford University Press.

Zhou, K. L. (Oct. 16, 2009). *Investigation into fishing law enforcement: Fines and confiscations totaled more than 50 million yuan.* Retrieved February 21, 2010 from http://china.huanqiu.com/ roll/ 2009-10/ 604442.html

Zhu, H. X., Hu, J. C., & Sun, W. T. (2007). *2007 report on the analysis of Internet public opinion in China*. Published by the Chinese Academy of Social Sciences. Retrieved February 18, 2010, from http://media.people.com.cn/ GB/ 22114/ 52789/ 115416/6852601.html

Zhu, H. X., Shan, X. G., & Hu, J. C. (2008). *2008 report on the analysis of Internet public opinion in China*. Published by the Chinese Academy of Social Sciences. Retrieved February 18, 2010 from http://www.china.com.cn/ aboutchina/ zhuanti/ 09zgshxs/ content_17100922.htm

Zhu, H. X., Shan, X. G., & Hu, J. C. (Dec. 22, 2009). 2009 report on the analysis of Internet public opinion in China. *2010 Bluebook on China's Social Development and Forecast.* Published by the Chinese Academy of Social Sciences. Retrieved January 31, 2010, from http://yq.people.com.cn/ htmlArt/ Art392.htm

Zittel, T. (2009). Lost in technology? Political parties and the online campaigns of constituency candidates in Germany's mixed member electoral system. *Journal of Information Technology & Politics*, *6*, 298–311. doi:10.1080/19331680903048832

About the Contributors

D. Ndirangu Wachanga, PhD (University of North Texas, 2007) is assistant professor of Journalism and Information Science at the University of Wisconsin-Whitewater. He is currently directing and producing two major documentaries on Prof. Ali Mazrui and Prof. Ngugi Wa Thiong'o. These pioneering projects seek to preserve African memories through conversations with African sages and elders. They are being filmed in three continents: Africa, North America and Europe. Wachanga has taught at The Mayborn Graduate Institute of Journalism, and Department of Radio, TV, and Film at the University of North Texas. He has taught courses on media and society; media ethics; new communication technologies; culture, community and documentary film, and new Diasporas. Wachanga has published several essays in prestigious peer reviewed journals, has an essay in press at the Yale University Press as well as a chapter in *Global Media Convergence and Cultural Transformation: Emerging Social Patterns and Characteristics.* Wachanga has made several presentations at national and international conferences. He has written passionate journalistic pieces for various newspapers in Kenya and United States. He has been featured both by the BBC and VOA. Wachanga is writing a book on New African Diaspora as well as editing a monograph on media ethics in the Global South.

* * *

Adedayo "Dayo" Abah, PhD - Associate Professor in the Dept. of Journalism and Mass Communications at Washington and Lee University, Lexington, VA. She has a master's degree in Theological Studies from Southern Methodist University, Dallas, TX (1993), and another master's degree in Media Studies from Texas Christian University, Ft. Worth, TX (1996). She has a Ph.D. from University of South Carolina, Columbia, SC (2004). Her research interests include First Amendment and online speech issues as well as Nollywood and African pop culture. She is grateful to the Lenfest Fund at Washington and Lee University for making the research for this work possible.

Nwachukwu A. Egbunike - Graduate student in Communication and Language Arts, University of Ibadan, Ibadan. He was Commissioning Editor in Spectrum Books Ltd, until 2007. Egbunike has been the Managing Editor of *Feathers and Ink*, Ibadan since 2007. Egbunike is the Director of Irawo University Centre, Ibadan.

Amie Breeze Harper - PhD candidate at University of California-Davis. She researches critical food geographies with an emphasis on critical race and black feminist theory. She is the editor of *Sistah Vegan: Black Female Vegans Speak on Food, Identity, Health and Society*

Peter Githinji, PhD - Assistant professor of linguistics at Ohio Unviversity, Athens. Githinji has published on language attitudes, and on negotiation and construction of layered identities. He is a Swahili expert and has conducted research on sheng.

Frederick Kang'ethe Iraki, PhD - Associate Professor of French at the United States International University, Kenya. He holds a PhD in linguistics from the University of Geneva. He is the founder editor of the *Journal of Language, Technology and Entrepreneurship in Africa.* His areas of research and publication include language and culture, language & Technology, language & politics, and higher education issues of quality assurance.

Yusuf Kalyango Jr., PhD - Assistant professor of journalism in the E. W. Scripps School of Journalism at Ohio University, where he also serves as Director of the Institute for International Journalism. Dr. Kalyango published nine peer-reviewed journal articles and two scholarly book chapters between 2007 and 2010. His research focuses on comparative political communication, media and democratization, and international crises, which has won two AEJMC and two ICA awards.

Bernard J. Kikech - Graduate student in English and Linguistics, Kenyatta University, Kenya. His main area of interest in research is corpus linguistics and language proficiency.

Purity Kimani - Graduate student of Applied Linguistics, Kenyatta University, Kenya. She has taught English and Literature in Kenya for twelve years. Her main areas of interest in research are sociolinguistics and discourse analysis.

Rick Malleus, PhD - Assistant professor Seattle University. Malleus has developed and taught a special topics course in "Africa and Communication", teaches foundation courses in communication theory and media, and offers several social interaction classes. His current research interests include intercultural re-entry and cross-cultural comparison in technology/media use. Malleus has written several publications for the Zimbabwe Open University.

Hayes Mawindi Mabweazara, PhD - Edinburgh Napier University, Scotland, UK. He has written book chapters and journal articles on the theoretical and methodological aspects of the new media and journalism practice in Africa and the consumption of the tabloid press in Zimbabwe. His research interests also centre around the sociology of journalism, audiences and reception studies in Africa and the wider global South. *Address*: Edinburgh Napier University, Craighouse Campus, Edinburgh, EH10 5lG, UK. *Email:* h.mabweazara@napier.ac.uk

Sangai Mohochi, PhD - Senior Lecturer in the Department of Linguistics, Languages and Literature at Maseno University, Kenya. He came to Maseno from Stanford University, where he taught Swahili Language, Literature and Culture for three years (September 2006- September, 2009). He joined Stanford from St. Lawrence University in Canton, New York where he taught Swahili for 2 years (June 2004-June 2006). Sangai taught Swahili Language and Literature at his alma mater for 9 years before moving to the United States of America. He has been actively involved in the development of Swahili language as a teacher, researcher, and theatre practitioner. His areas of research interest include: language attitudes;

motivation and language learning; Language and the media; language choice for identity and development; Swahili literature. He has published and made several conference presentations in those areas.

George W. Musambira. PhD - Assistant professor in Nicholson School of Communication, University of Central Florida, Orlando, Florida. He taught in the communication department at Western Kentucky University. His research interests include bereavement communication, international and intercultural communication, and communication research productivity. He has presented over 30 refereed papers at national and international conferences and published in refereed. Musambira previously served on the editorial board of the journal World Communication and currently serves on the editorial board of the Journal of Intercultural Communication Research.

Samuel Muwanguzi - Doctoral candidate in the College of Information, University of North Texas, Denton, Texas. He was chief news editor for the government-owned Radio Uganda and Uganda Television until 2001. Before, he worked as a journalist for various newspapers and as a news editor for Radio Uganda. His research interests include social construction of New Communication and Information Technologies and health communication concerning HIV/AIDS. He has presented over ten refereed national and international conference papers and published in the refereed journal International Journal of Distance Education Technology.

Mahiri Mwita, PhD - Lecturer of Kiswahili and Comparative Literature at Princeton University. He is a theatre scholar and a long-time practitioner in community theatre, Theatre for Development and the Kenya National Drama Festival movement. His current research focuses on the use of theatre-based outreaches to mobilize people in low-income communities to confront and discourse on sensitive issues facing them such as HIV/AIDS, tribalism, civic education and bad governance. He teaches and writes on Kiswahili literature and theatre, poetry and critical theories that have impacted the development of creative writing and criticism of African literature and theatre.

Siho Nam,PhD - Assistant Professor of Communication at the University of North Florida. His main research program delves into the relationship between the media and social change/democracy, informed by critical and cultural studies perspectives. Within that context, his research has dealt with such issues as media diversity, international communication, media globalization, and critical media literacy.

Martin C. Njoroge, PhD- Senior Lecturer in the Department of English & Linguistics, Kenyatta University, Kenya. His main areas of interest include research on literacy of children in extremely difficult circumstances, communication, sociolinguistics, general linguistics and language-in-education issues and have published on the same.

Emily Bosire-Ogechi, PhD, - Lecturer, School of Information Sciences, Moi University, Kenya. She has research interests and publications in ICTs for development, publishing and language education, indigenous knowledge, and information and communication.

Nathan Oyori Ogechi, PhD - Associate Professor and chair of the Department of Kiswahili and Other African Languages, Moi University, Kenya. His research interests are in contact linguistics, linguistic human rights, language and politics, and language in education with a focus on Kenya. He recently co-authored a Linguistic Human Rights and the Language Policy in the Kenyan Educational System (OSSREA: Addis Ababa). He has also edited the Harmonization and Standardization of the orthographies of Kenya Languages (CASAS: Cape Town).

Uche Onyebadi, PhD - Assistant professor of journalism at the School of Journalism, Southern Illinois University Carbondale, U.S.A. He holds a Ph.D. in Journalism (Political Communication) from the School of Journalism, University of Missouri.

Jim Schnell, PhD - Professor in the Division of Social & Behavioral Sciences at Ohio Dominican University. He has 11 books in print, over 60 book chapters and journal articles and over 140 conference presentations. Schnell was a Fulbright Scholar to Cambodia in 2005-2006 and currently serves as an assessor for the Fulbright Program. He retired from the U.S. Air Force (Reserve), at the rank of Colonel, in 2007 after serving 30 years in the intelligence community. His final assignment was as Assistant Air Attache to the U.S. Embassy in Beijing, China.

Jiafei Yin,PhD - Assistant Professor of journalism, Central Michigan University, received her doctorate and master's degree in journalism from the University of Missouri. Her bachelor's degree was in English and American literature from Nanjing University in China. She worked for the *China Daily*, the country's leading national English-language daily newspaper, from 1982-93 as a wire editor, reporter, copy-editor and chief copy-editor. Yin has published in the *Journal of Advertising Research* and the *Asian Journal of Communication,* and contributed book chapters in the *Issues and Challenges in Asian Journalism, Emerging Issues in Contemporary Journalism, Terrorism, Globalization & Mass Communication, Handbook on Global Communications Ethics* (to be published by Blackwell), and *Global Journalism.*

Saman Talib, PhD - Assistant professor at the School of Media and Communication, Beaconhouse National University, Pakistan. She has worked at publishing houses, broadcast networks, radio stations, and digital news organizations. Her scholarship considers the conjunctions of organizations, media and politics. She is particularly interested in researching the effective use of new media to foster knowledge, literacy and participation in civil society.

Saadia Gardezi - Writer for The Friday Times in Lahore on issues relating to humanitarian aid, political economy and the media in politics. She teaches political theory and is involved in research in economics and industrial policy at the Lahore University of Management Sciences. Her research interests also include critical Marxist theory in international political economy and international security studies. She also writes fiction and has been published by the Oxford University press and various journals.

Index

G

H

I

K

L

M

N

CPSIA information can be obtained at www.ICGtesting.com
261698BV00003B/71-80/P